Methods of Introducing System Models into Agricultural Research

Methods of Introducing System Models into Agricultural Research

Laj R. Ahuja and Liwang Ma, Editors

Advances in Agricultural Systems Modeling 2
Transdisciplinary Research, Synthesis, and Applications

Laj Ahuja, Series Editor

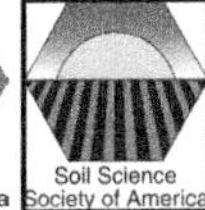

American Society of Agronomy, Inc.
Crop Science Society of America, Inc.
Soil Science Society of America, Inc.
5585 Guilford Road, Madison, WI 53711-5801 USA

doi:10.2134/advagricsystmodel2

ISBN: 978-0-89118-180-4

Library of Congress Control Number: 2011912699

Cover design: Patricia Scullion

Printed on recycled paper.
Printed in the United States of America.

Contents

Foreword

Agricultural systems are by nature complex ecosystems. Numerous interacting factors involving soil, plant, climate, and management components must be taken into account. These systems also need to consider production, environmental, and societal issues for the sustainability of agriculture. Because of the complex nature of agricultural systems, modeling is a key tool that aids in understanding the intricacies of the interactions and delivers a myriad of potential outcomes to users world-wide. This book is designed to guide scientists and other professionals in methods of parameter estimation, calibration, and validation of agricultural system models. As these models become more available to explore new management strategies and to extend information to larger scales, proper parameterization, calibration, and validation are critical to their use. This book is essential because as models are more widely used, they help advance research and guide producers and policy makers on agricultural systems that impact societal issues and needs. Because system modeling is vitally important to the understanding of agricultural systems, the American Society of Agronomy, the Crop Science Society of America and the Soil Science Society of America support the objectives of this book, which will help model users apply the proper techniques when using system models.

Charles W. Rice
2011 SSSA President

Maria Gallo
2011 CSSA President

Newell Kitchen
2011 ASA President

Preface

Potential Value of System Models in Agricultural Research and the Need for Helping Field Researchers in Using Models

To develop sustainable agricultural systems that address environmental challenges, the field of agricultural research needs to develop more quantitative guidance and site-specific decision tools to help producers. To this end, field research requires a quantitative whole-system approach to help optimize the complex interacting factors. The process level models of cropping systems, based on synthesis and quantification of disciplinary knowledge and important interactions among the system components, meet this challenge. These models integrated with cutting-edge field research will greatly enhance the value and efficiency of research for developing sustainable agriculture, enable a fast transfer of technologies to farmers, and inform policymakers and the general public on the major issues and tradeoffs of alternatives.

These system models require some input data about the system and need to be parameterized, calibrated, and validated correctly with good field experimental data at selected locations in the region. They can then be used to: (i) synthesize all other experiment data, (ii) extend current research results to multiple years of historical climate beyond the limited experimental years and to other soils in the area to evaluate the long-term sustainability of cropping and management systems in the area, (iii) derive the new optimum cropping and management strategies for future selective field testing, and (iv) derive simpler management guides or tools for producers. Furthermore, to prepare for projected climate-change effects on water availability and agricultural production with increased temperatures and carbon dioxide, there is an urgent need for validated models to project the effects of these changes on agricultural systems in the United States and the world.

A major difficulty that field scientists, students, and other users of agricultural system models face is the appropriate parameter estimation and correct model calibration. Many models are often misused and the results published without proper parameterization, calibration, and validation. Reviewers seldom look into parameterization, calibration, and validation of all important system components (e.g., water balance, N balance, phenology, yield components) other than the yield and some measured variables. Model users are not properly trained, often due to a lack of good documentation and guidance on systematic parameterization and calibration for a model. With development of Windows based user interfaces, it becomes easy for model users to run a model, but it does not guarantee correct use of a model. In most cases, a model user focuses more energy on the system components that he/she is most familiar with and leaves the rest untouched, using default values. There are science documentations available for some models, but how to interpret the science in terms of measurable or derivable model parameters in a given agricultural system is another, totally different process.

The knowledge and experience of model parameterization largely resides with the model developers. They are the ones who truly know the weaknesses and strengths of their models and the best ways to estimate and calibrate the model parameters. When the developers made the decisions on what components to include in their models and at what level of detail, they had an idea of the parameters needed and where to get them. This knowledge is usually not available to users in a systematic way and is scattered in many publications (usually, not in detail). Therefore, the concept of systematic parameterization and calibration, and even validation, of a system model is not fully comprehended by users. Furthermore, the processes and model parameters are well related in the user interfaces, and users have difficulty identifying the ones related to their particular studies. In addition, a lack of methodology in model parameterization makes model results not reproducible and least likely to generate new knowledge beyond what experimentalists already have gained in the field.

Objectives of this second volume of the Advances in Agricultural Systems Modeling series, therefore, were to create a "methodology volume" that contains the above much-needed information for model users. One objective of this book series is to promote the use of system models to enhance and extend field research. To accomplish this goal, it is vitally important to devote a volume to helping field scientists and other model users in proper methods of parameter estimation, calibration, validation, exploring new management options, and extension of experimental results to other weather conditions, soils, and climates. The proper methods are the key to realizing the great potential benefits of modeling an agricultural system. This volume contains information on the input data required and step-by-step procedures for parameter estimation, calibration, validation, and extended applications for the major models, illustrated with examples. To avoid duplication, we tried to describe these procedures in detail for one major model, and then special features of other models. This information will help users to correctly parameterize and use each model component without the need to fully understand the theory. This book also contains chapters that address further improvement of parameterization methods, field sampling and measurement of model inputs and parameters, effective properties for spatially variable soils, quality control of weather and other inputs, automatic parameter estimation software, and some new innovative methods to create simple management guidelines for farmers. We hope this book will serve as a linkage between models and field research. It will not only promote correct model application to complement and quantify field research, but also create new knowledge and identify knowledge gaps. We tried to have a user friendly and uniform format in presenting the methods as much as possible, to make it easier and attractive for the field scientists, currently not exposed to models, to start using them. Many of the models are provided on a companion CD.

The final synthesis chapter seeks to identify: (i) the best set of current procedures from across the models; (ii) knowledge gaps and research needed to further improve parameter estimation, calibration, and validation methods for users; (iii) required research to improve concepts in the models so as to make the parameters more biophysically based and less of empirical coefficients, and either directly measurable or related to some easily measured data; (iv) the means to balance the complexity of process and parameters; (v) determination of effective parameters for a spatially variable field and for temporally variable conditions; (vi) relationships of model results across different soils, climates, and scales; and (vii) simpler ways of documenting model components for users.

We wish to convey one important feature of the volumes in the new series Advances in Agricultural Systems Modeling. We have made an utmost effort to ensure the quality of each volume, with two to three (in some cases even more) anonymous peer reviews of each contribution. We have tried to ensure originality of each contribution in terms of new synthesis and quantification of knowledge in the topic area, which would be useful for further improving and advancing system models as well as advance science and further research. Each contribution is, therefore, treated as a review and synthesis paper, much like a review paper in a journal.

L.R. Ahuja and Liwang Ma, Editors
USDA-ARS, Agricultural Systems Research Unit
Fort Collins, Colorado

Contributors

Adiku, S.G.K.	Dep. of Civil and Environmental Engineering, Colorado State Univ., Fort Collins, CO 80523 (sadiku@mail.colostate.edu)
Ahuja, L.R.	USDA-ARS, Agricultural Systems Research Unit, Fort Collins, CO 80526 (Laj.Ahuja@ars.usda.gov)
Akkermans, Wies	Plant Research International, Bornsesteeg 47, 6700 AA Wageningen, The Netherlands
Andales, A.A.	Dep. of Soil and Crop Sciences, Colorado State Univ., Fort Collins, CO 80523 (allan. andales@ colostate.edu)
Bartling, P.N.S.	USDA-ARS, Agricultural Systems Research Unit, Fort Collins, CO 80526 (pat.bartling@ars.usda.gov)
Bergez, J.-E.	INRA, UMR 1248 AGIR, BP 52627, 31326 Castanet Tolosan Cedex, France
Boogaard, Hendrik	Alterra Research Institute, Droevendaalsesteeg 3, 6708 PB Wageningen, The Netherlands
Boote, Kenneth J.	Agronomy Dep., Univ. of Florida, Gainesville, FL 32611
Buis, S.	INRA, UMR 1114 EMMAH, Domaine Saint Paul- Site Agroparc, 84914 Avignon Cedex 9, France (samuel.buis@avignon.inra.fr)
de Wit, Allard	Alterra Research Institute, Droevendaalsesteeg 3, 6708 PB Wageningen, The Netherlands
Del Grosso, S.J.	USDA-ARS and Natural Resource Ecology Lab., Colorado State Univ., Fort Collins, CO 80523 (delgro@nrel.colostate.edu)
Derner, J.D.	USDA-ARS, High Plains Grasslands Research Station 8408 Hildreth Road Cheyenne, WY 82009-8899 (justin.derner@ars.usda.gov)
Dunn, G.H.	USDA-ARS, Agricultural Systems Research Unit, Fort Collins, CO 80526 (gale.dunn@ars.usda.gov)
Fienen, Michael N.	USGS, Wisconsin Water Science Center, 8505 Research Way, Middleton, WI 53562 (mnfienen@usgs.gov)
Flerchinger, G.N.	USDA-ARS, Northwest Watershed Research Center, Boise, ID 83712
Garcia, L.	Dep. of Civil and Environmental Engineering, Colorado State Univ., Fort Collins, CO 80523 (luis.garcia@colostate.edu)
Green, Christopher T.	USGS, Bldg 15, McKelvey Building, 345 Middlefield Road, Menlo Park, CA 94025-3561 (ctgreen@usgs.gov)
Green, T.R.	USDA-ARS, Agricultural Systems Research Unit, Fort Collins, CO 80526
Guérif, M.	INRA, UMR 1114 EMMAH, Domaine Saint Paul- Site Agroparc, 84914 Avignon Cedex 9, France
Guillaume, S.	INRA, UMR 1248 AGIR, BP 52627, 31326 Castanet Tolosan Cedex, France
Hatfield, Jerry L.	USDA-ARS, National Laboratory for Agriculture and the Environment, 2110 University Blvd., Ames, IA 50011 (jerry.hatfield@ars.usda.gov)
He, Jianqiang	INRA, UMR1095, Génétique, Diversité et Ecophysiologie des Céréales, F-63100 Clermont-Ferrand, France
Hessel, Rudi	Alterra Research Institute, Droevendaalsesteeg 3, 6708 PB Wageningen, The Netherlands
Hoogenboom, G.	AgWeatgerNet, Washington State Univ., Prosser, WA 99350
Hu, Z.	Dep. of Agricultural and Biological Engineering, Univ. of Florida, Gainesville, FL 32611
Hutson, John L.	School of the Environment, Flinders Univ., Adelaide 5001, Australia
Jabro, Ann D.	School of Communications and Information Systems, Robert Morris Univ., Moon Township, PA 15108.
Jabro, Jay D.	USDA-ARS, Northern Plains Agricultural Research Lab., 1500 N. Central Ave., Sidney, MT 59270 (jay.jabro@ars.usda.gov)
Jaynes, Dan B.	USDA-ARS, National Laboratory for Agriculture and the Environment, 2110 University Boulevard, Ames, IA 50011-0001 (dan.jaynes@ars.usda.gov)
Jones, James W.	Dep. of Agricultural and Biological Engineering, Univ. of Florida, Gainesville, FL 32611 (jimj@ufl.edu)
Justes, E.	INRA, UMR 1248 AGIR, BP 52627, 31326 Castanet Tolosan Cedex, France
Karlen, D.L.	USDA-ARS, National Laboratory for Agriculture and the Environment, 2110 University Boulevard, Ames, IA 50011-0001 (doug.karlen@ars.usda.gov)

Kemanian, Armen	Dep. of Crop and Soil Sciences, The Pennsylvania State Univ., 116 Agricultural Sciences & Industries Bldg., University Park, PA 16802
Keough, C.A.	Natural Resource Ecology Lab., Colorado State Univ., Fort Collins, CO 80523
Kersebaum, K.C.	Leibniz-Centre for Agricultural Landscape Research, Institute of Landscape Systems Analysis, Muencheberg, Germany (ckersebaum@zalf.de)
Launay, M.	INRA, Unité Agroclim, Domaine Saint Paul- Site Agroparc, 84914 Avignon Cedex 9, France
Lecharpentier, P.	INRA, UMR 1114 EMMAH, Domaine Saint Paul- Site Agroparc, 84914 Avignon Cedex 9, France
Ma, Liwang	USDA-ARS, Agriculture Systems Research, 2150 Centre Ave. Bldg. D, Fort Collins, CO 80526 (liwang.ma@ars.usda.gov)
Malone, Robert W.	USDA-ARS, National Laboratory for Agriculture and the Environment, 2110 University Boulevard, Ames, IA 50011-0001 (rob.malone@ars.usda.gov)
Meek, D.W.	USDA-ARS, National Laboratory for Agriculture and the Environment, 2110 University Boulevard, Ames, IA 50011-0001 (dave.meek@ars.usda.gov)
Murdock, L.	Univ. of Kentucky, Dep. of Plant and Soil Sciences, Research and Education Center at Princeton, Princeton, KY 42445
Nolan, Bernard T.	USGS, 413 National Center, Reston, VA 20192 (btnolan@usgs.gov)
Parton, W.J.	Natural Resource Ecology Lab., Colorado State Univ., Fort Collins, CO 80523
Porter, C.H.	Dep. of Agricultural and Biological Engineering, Univ. of Florida, Gainesville, FL 32611
Reyes-Fox,	USDA-ARS and Natural Resource Ecology Lab., Colorado State Univ., Fort Collins, CO 80523
Saseendran, S.A.	USDA-ARS, Agricultural Systems Research Unit, Fort Collins, CO 80526
Schwab, G.	Univ. of Kentucky, Dep. of Plant & Soil Sciences, Lexington, KY 40546
Shepherd, A.	Sustainable Soils and Grassland Systems Dep., Rothamsted Research, North Wyke, Okehampton, Devon EX20 2SB, UK
van Diepen, Kees	Alterra Research Institute, Droevendaalsesteeg 3, 6708 PB Wageningen, The Netherlands
Varella, H.	Météo-France, CNRM-GMAP, 42 av Gaspard Coriolis, 31057 Toulouse, France
Wallach, D.	INRA, UMR 1248 AGIR, BP 52627, 31326 Castanet Tolosan Cedex, France
Wang, Xiuying	Blackland Research and Extension Center, Texas AgriLIFE Research, 720 East Blackland Rd., Temple, TX 76502 (swang@brc.tamus.edu)
Wendroth, Ole	Univ. of Kentucky, Dep. of Plant and Soil Sciences, Lexington, KY 40546 (owendroth@uky.edu)
Wilkens, Paul	International Fertilizer Development Center, Muscle Shoals, AL 35662
Williams, Jimmy R.	Blackland Research and Extension Center, Texas AgriLIFE Research, 720 East Blackland Rd., Temple, TX 76502 (jwilliams@brc.tamus.edu)
Wolf, Joost	Wageningen Univ., group Plant Production systems, P.O. Box 430, 6700 AK Wageningen, The Netherlands (joost.wolf@wur.nl)
Wu, L.	Sustainable Soils and Grassland Systems Dep., Rothamsted Research, North Wyke, Okehampton, Devon EX20 2SB, UK (Lianhai.Wu@bbsrc.ac.uk)

Conversion Factors for SI and Non-SI Units

To convert Column 1 into Column 2 multiply by	Column 1 SI unit	Column 2 non-SI unit	To convert Column 2 into Column 1 multiply by
		Length	
0.621	kilometer, km (10^3 m)	mile, mi	1.609
1.094	meter, m	yard, yd	0.914
3.28	meter, m	foot, ft	0.304
1.0	micrometer, μm (10^{-6} m)	micron, μ	1.0
3.94×10^{-2}	millimeter, mm (10^{-3} m)	inch, in	25.4
10	nanometer, nm (10^{-9} m)	Angstrom, Å	0.1
		Area	
2.47	hectare, ha	acre	0.405
247	square kilometer, km^2 (10^3 m)^2	acre	4.05×10^{-3}
0.386	square kilometer, km^2 (10^3 m)^2	square mile, mi^2	2.590
2.47×10^{-4}	square meter, m^2	acre	4.05×10^3
10.76	square meter, m^2	square foot, ft^2	9.29×10^{-2}
1.55×10^{-3}	square millimeter, mm^2 (10^{-3} m)^2	square inch, in^2	645
		Volume	
9.73×10^{-3}	cubic meter, m^3	acre-inch	102.8
35.3	cubic meter, m^3	cubic foot, ft^3	2.83×10^{-2}
6.10×10^4	cubic meter, m^3	cubic inch, in^3	1.64×10^{-5}
2.84×10^{-2}	liter, L (10^{-3} m^3)	bushel, bu	35.24
1.057	liter, L (10^{-3} m^3)	quart (liquid), qt	0.946
3.53×10^{-2}	liter, L (10^{-3} m^3)	cubic foot, ft^3	28.3
0.265	liter, L (10^{-3} m^3)	gallon	3.78
33.78	liter, L (10^{-3} m^3)	ounce (fluid), oz	2.96×10^{-2}
2.11	liter, L (10^{-3} m^3)	pint (fluid), pt	0.473
		Mass	
2.20×10^{-3}	gram, g (10^{-3} kg)	pound, lb	454
3.52×10^{-2}	gram, g (10^{-3} kg)	ounce (avdp), oz	28.4
2.205	kilogram, kg	pound, lb	0.454
0.01	kilogram, kg	quintal (metric), q	100
1.10×10^{-3}	kilogram, kg	ton (2000 lb), ton	907
1.102	megagram, Mg (tonne)	ton (U.S.), ton	0.907
1.102	tonne, t	ton (U.S.), ton	0.907

Table cont.

To convert Column 1 into Column 2 multiply by	Column 1 SI unit	Column 2 non-SI unit	To convert Column 2 into Column 1 multiply by
	Yield and Rate		
0.893	kilogram per hectare, kg ha^{-1}	pound per acre, lb $acre^{-1}$	1.12
7.77×10^{-2}	kilogram per cubic meter, kg m^{-3}	pound per bushel, lb bu^{-1}	12.87
1.49×10^{-2}	kilogram per hectare, kg ha^{-1}	bushel per acre, 60 lb	67.19
1.59×10^{-2}	kilogram per hectare, kg ha^{-1}	bushel per acre, 56 lb	62.71
1.86×10^{-2}	kilogram per hectare, kg ha^{-1}	bushel per acre, 48 lb	53.75
0.107	liter per hectare, L ha^{-1}	gallon per acre	9.35
893	tonne per hectare, t ha^{-1}	pound per acre, lb $acre^{-1}$	1.12×10^{-3}
893	megagram per hectare, Mg ha^{-1}	pound per acre, lb $acre^{-1}$	1.12×10^{-3}
0.446	megagram per hectare, Mg ha^{-1}	ton (2000 lb) per acre, ton $acre^{-1}$	2.24
2.24	meter per second, m s^{-1}	mile per hour	0.447
	Specific Surface		
10	square meter per kilogram, m^2 kg^{-1}	square centimeter per gram, cm^2 g^{-1}	0.1
1000	square meter per kilogram, m^2 kg^{-1}	square millimeter per gram, mm^2 g^{-1}	0.001
	Density		
1.00	megagram per cubic meter, Mg m^{-3}	gram per cubic centimeter, g cm^{-3}	1.00
	Pressure		
9.90	megapascal, MPa (10^6 Pa)	atmosphere	0.101
10	megapascal, MPa (10^6 Pa)	bar	0.1
2.09×10^{-2}	pascal, Pa	pound per square foot, lb ft^{-2}	47.9
1.45×10^{-4}	pascal, Pa	pound per square inch, lb in^{-2}	6.90×10^3
	Temperature		
1.00 (K – 273)	kelvin, K	Celsius, °C	1.00 (°C + 273)
(9/5 °C) + 32	Celsius, °C	Fahrenheit, °F	5/9 (°F – 32)
	Energy, Work, Quantity of Heat		
9.52×10^{-4}	joule, J	British thermal unit, Btu	1.05×10^3
0.239	joule, J	calorie, cal	4.19
10^7	joule, J	erg	10^{-7}
0.735	joule, J	foot-pound	1.36
2.387×10^{-5}	joule per square meter, J m^{-2}	calorie per square centimeter (langley)	4.19×10^4
10^5	newton, N	dyne	10^{-5}
1.43×10^{-3}	watt per square meter, W m^{-2}	calorie per square centimeter minute (irradiance), cal cm^{-2} min^{-1}	698

Table cont.

To convert Column 1 into Column 2 multiply by	Column 1 SI unit	Column 2 non-SI unit	To convert Column 2 into Column 1 multiply by
	Transpiration and Photosynthesis		
3.60×10^{-2}	milligram per square meter second, mg m^{-2} s^{-1}	gram per square decimeter hour, g dm^{-2} h^{-1}	27.8
5.56×10^{-3}	milligram (H_2O) per square meter second, mg m^{-2} s^{-1}	micromole (H_2O) per square centimeter second, µmol cm^{-2} s^{-1}	180
10^{-4}	milligram per square meter second, mg m^{-2} s^{-1}	milligram per square centimeter second, mg cm^{-2} s^{-1}	10^4
35.97	milligram per square meter second, mg m^{-2} s^{-1}	milligram per square decimeter hour, mg dm^{-2} h^{-1}	2.78×10^{-2}
	Plane Angle		
57.3	radian, rad	degrees (angle), °	1.75×10^{-2}
	Electrical Conductivity, Electricity, and Magnetism		
10	siemen per meter, S m^{-1}	millimho per centimeter, mmho cm^{-1}	0.1
10^4	tesla, T	gauss, G	10^{-4}
	Water Measurement		
9.73×10^{-3}	cubic meter, m^3	acre-inch, acre-in	102.8
9.81×10^{-3}	cubic meter per hour, m^3 h^{-1}	cubic foot per second, ft^3 s^{-1}	101.9
4.40	cubic meter per hour, m^3 h-1	U.S. gallon per minute, gal min^{-1}	0.227
8.11	hectare meter, ha m	acre-foot, acre-ft	0.123
97.28	hectare meter, ha m	acre-inch, acre-in	1.03×10^{-2}
8.1×10^{-2}	hectare centimeter, ha cm	acre-foot, acre-ft	12.33
	Concentration		
1	centimole per kilogram, cmol kg^{-1}	milliequivalent per 100 grams, meq 100 g^{-1}	1
0.1	gram per kilogram, g kg^{-1}	percent, %	10
1	milligram per kilogram, mg kg^{-1}	parts per million, ppm	1
	Radioactivity		
2.7×10^{-11}	becquerel, Bq	curie, Ci	3.7×10^{10}
2.7×10^{-2}	becquerel per kilogram, Bq kg^{-1}	picocurie per gram, pCi g^{-1}	37
100	gray, Gy (absorbed dose)	rad, rd	0.01
100	sievert, Sv (equivalent dose)	rem (roentgen equivalent man)	0.01
	Plant Nutrient Conversion		
	Elemental	Oxide	
2.29	P	P_2O_5	0.437
1.20	K	K_2O	0.830
1.39	Ca	CaO	0.715
1.66	Mg	MgO	0.602

Chapter 1

A Protocol for Parameterization and Calibration of RZWQM2 in Field Research

L. Ma
L.R. Ahuja
S.A. Saseendran
R.W. Malone
T.R. Green
B.T. Nolan
P.N.S. Bartling
G.N. Flerchinger
K.J. Boote
G. Hoogenboom

Abstract

Use of agricultural system models in field research requires a full understanding of both the model and the system it simulates. This chapter provides a protocol on how to parameterize the USDA-ARS, Root Zone Water Quality Model (RZWQM2), which contains the DSSAT (Decision Support System for Agrotechnology Transfer) version 4.0 crop growth modules and the SHAW (Simultaneous Heat and Water) energy balance modules. It summarizes input data requirements of soil, weather, and plants, as well as the minimum data needed to calibrate the model. Calibration of each component or process is illustrated, along with possible adjustable parameters. An iterative procedure for model calibration is emphasized. Difficulties in model application are presented, along with possible solutions. However, the steps for model calibration are still in the trial-and-error stage and can vary from application to application. Model users should benefit from the materials in this chapter by combining them with those in the RZWQM2 user manual released with the model.

L. Ma (Liwang.ma@ars.usda.gov), L.R. Ahuja, S.A. Saseendran, T.R. Green, and P.N.S. Bartling, USDA-ARS, Agricultural Systems Research Unit, Fort Collins, CO 80526; R.W. Malone, USDA-ARS, Natl. Lab. for Agric. and the Environment, Ames, IA 50011; B.T. Nolan, USGS, 413 National Center, Reston, VA 20192; G.N. Flerchinger, USDA-ARS, Northwest Watershed Research Center, Boise, ID 83712; K.J. Boote, Agronomy Dep., Univ. of Florida, Gainesville, FL 32611; G. Hoogenboom, AgWeatherNet, Washington State Univ., Prosser, WA 99350.

doi:10.2134/advagricsystmodel2.c1

Methods of introducing system models into agricultural research. L.R. Ahuja and L. Ma (ed.)
Advances in Agricultural Systems Modeling Series 2.

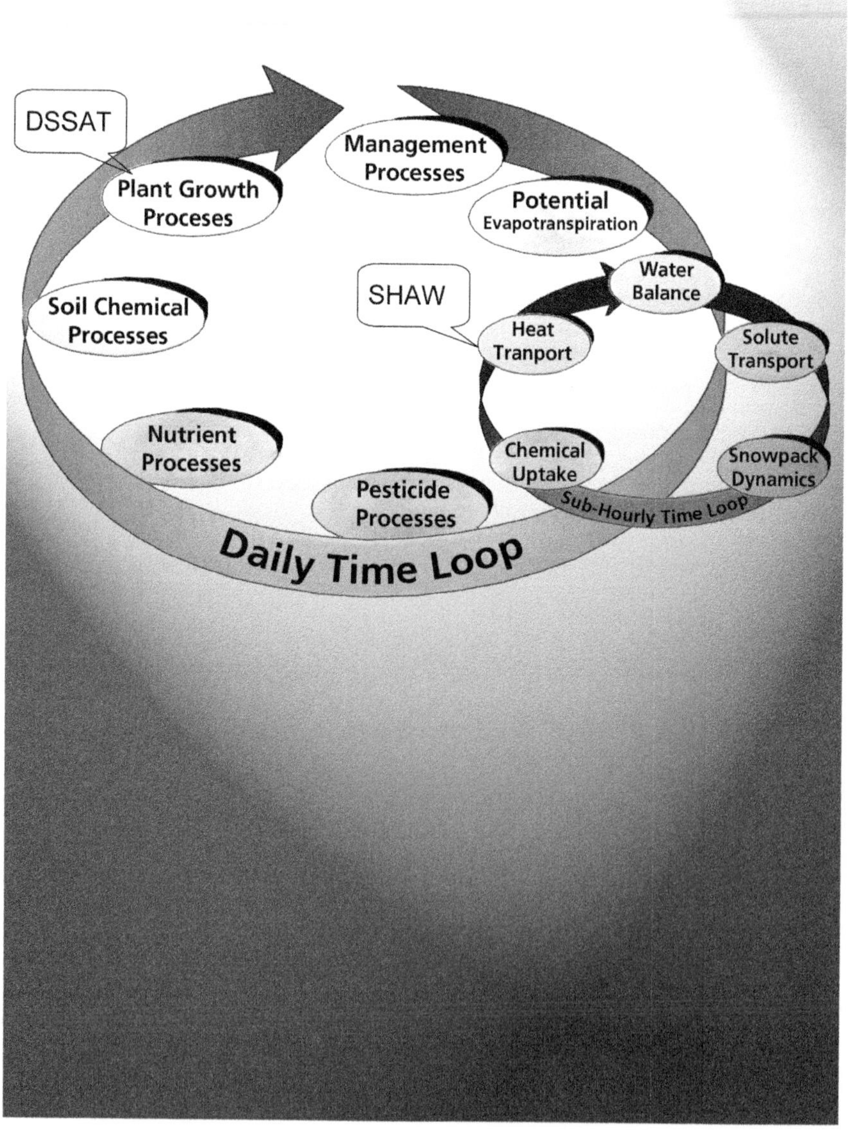
DSSAT
Management Processes
Plant Growth Proceses
Potential Evapotranspiration
Water Balance
SHAW
Soil Chemical Processes
Heat Tranport
Solute Transport
Nutrient Processes
Chemical Uptake
Snowpack Dynamics
Pesticide Processes
Sub-Hourly Time Loop
Daily Time Loop

Since the 1960s, agricultural system models have increased tremendously in their complexity due to greater understanding of the processes simulated, their application to real world problems, and acceleration of modern computing power. As a result, the parameterization of a system model is now more complex, requiring more input variables and providing more outputs to verify. Some of the physical or biological parameters are measurable, but difficult and time-consuming to obtain. Furthermore, these parameters when measured vary spatially and temporally due to soil and microclimate heterogeneity and management effects. Many model parameters in agricultural system models do not need calibration and are assumed correct. However, in most cases, one has to calibrate a few parameters (usually by trial-and-error) to obtain the desired results under different conditions. Because of interactions among system components, it is not a simple or sequential step-by-step procedure, and the calibration has to be iterative (Cameira et al., 2005; Hanson et al., 1999). The accuracy of such a calibration procedure depends primarily on the type and quality of measured data, but also on the interactions of the system components in the model and the experience of the user. An issue with automated calibration of agricultural system models is the lack of criteria and schemes for an objective optimization of all different parameters. A system model simulates a wide range of biotic and abiotic processes that vary greatly in magnitude and are highly interdependent, which makes optimization nearly impossible.

Some of the best guidance documents for calibration were provided for the CROPGRO model by Boote (1999) and Boote et al. (1998), who detailed how to parameterize the growth of legumes using CROPGRO. Hunt et al. (1993) described an optimization program for the DSSAT crop growth models called GENCALC. It can optimize the cultivar parameters given measured crop phenology, biomass, and yield data. However, their suggested procedure was only for crop parameters. Ahuja and Ma (2002) provided a status of model parameterization and guidelines on how to parameterize system components, but they did not detail step-by-step instructions on calibrating a particular model. Although the majority of agricultural system models are calibrated manually (Saseendran et al., 2010), such a manual calibration procedure may not be transferable to another model or model user (Boyle et al., 2000). Automated parameter optimization software such as LHS (Latin Hypercubic Sampling) and PEST (Parameter Estimation Software;

Doherty, 2010) are available with limited applications (Nolan et al., 2010). However, the automatic methods may not be transferable to other users either because of personal preferences in selecting an objective function to optimize, and the optimized parameters may have no physical or biological meanings (Boyle et al., 2000). Several automatic methods for parameterizing crop models are discussed in the book by Wallach et al. (2006).

On the other hand, data quantity and quality are important for a rigorous model calibration (Gan et al., 1997). Because of the cost of collecting data, crop scientists tend to focus efforts on plant related data collection and pay less attention to soil mineralization and soil water related data, whereas soil scientists may do just the opposite. Unless there is a group of scientists with broad expertise, it is difficult to collect data for all the system components with the same resolution. As a practical matter, collecting data on every aspect of the system is not possible either. Therefore, when data are not collected for a system component, there is a need to provide a typical range for the simulated process under a particular experimental condition for the users. The suggested minimum measured data requirements to calibrate a complete system model are soil moisture, soil nutrient, soil chemicals of interest, and plant growth (e.g., phenology, leaf area index, biomass, yield) (Ahuja and Ma, 2002).

In this chapter, we outline an iterative step-by-step procedure on how to manually calibrate a system model, using RZWQM2 as an example. Automated optimization methods of system models are discussed in other chapters in this volume (Nolan et al., 2011, Chapter 12; Jones et al., 2011, Chapter 13; Buis et al., 2011, Chapter 14). We first give a summary of the model, followed by agricultural system characterization and guidelines and procedures for calibrating key parameters.

Overview of RZWQM

RZWQM (Root Zone Water Quality Model) was initially developed in the early 1990s, followed by improvement over two decades (Ahuja et al., 2000b; Ma et al., 2000). The most recent version was released in September 2010 as Version 2.0 at http://ars.usda.gov/Main/docs.htm?docid=17740 or http://arsagsoftware.ars.usda.gov/agsoftware/ (verified 10 Mar. 2011). It is a research-level model containing physical, chemical, and biological processes for simulating agricultural management effects on soil processes, crop production, and water quality. The first DOS version of the model was released in 1992 and evaluated in collaboration with the Management Systems Evaluation Areas (MSEA) project in the U.S. Midwest (Watts et al., 1999). A Windows user interface was developed in 1998, and a book documenting the science in RZWQM was published in 2000 (Ahuja et al.,

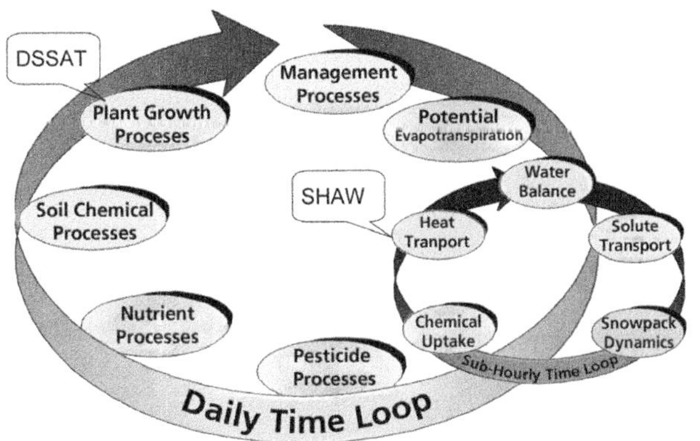

Fig. 1–1. A diagram to show the processes and execution time steps in RZWQM2 (adapted from Ahuja et al., 2000a).

2000b). The recently released RZWQM2 model (Ma et al., 2009) now includes the DSSAT V4.0 crop growth models (Jones et al., 2003; Hoogenboom et al., 1994, 2004) and the SHAW (Simultaneous Heat and Water) energy balance model (Flerchinger et al., 200; Kozak et al., 2007) with soil depth extended from 3 to 30 m (Nolan et al., 2010). RZWQM2 consists of seven main components: water balance, heat and chemical transport, nutrient processes (carbon and nitrogen), plant growth processes, soil chemical processes, evapotranspiration processes, pesticide processes, and management (Fig. 1–1).

The soil water module calculates the water balance using the Green–Ampt equation during infiltration and the Richards equation during redistribution (between rainfall or irrigation events) (Ahuja et al., 2000a). Plant water uptake is treated as a sink term in the Richards equation. If rainfall or irrigation intensity is greater than soil infiltration capacity, surface water ponding is generated, which is first diverted to macropore flow (if macropores exist) to meet the macropore flow rate estimated by Poiseuille's equation. The remainder is off-site runoff. A portion of water flowing through macropores is absorbed laterally by the soil. Surface retention storage of water is not currently considered in RZWQM2. The model also uses the steady-state Hoogenhout equation (Ahuja et al., 2000a) for subsurface tile drainage. A modified Shuttleworth–Wallace equation (Farahani and DeCoursey, 2000) is used to estimate daily or hourly potential evapotranspiration.

The soil temperature and surface energy balance modules are interrelated. Soil temperature is updated during infiltration according to heat movement with water infiltration and is calculated by solving the heat equation during redistribution. RZWQM2 can estimate a surface heat flux and surface temperature using either SHAW (Kozak et al., 2007; Yu et al., 2007) or the PENFLUX module (Aiken et al., 1997). Otherwise, soil surface temperature is set equal to air temperature.

The soil nutrient module is characterized by its flexibility in simulating soil carbon and nitrogen dynamics in conjunction with soil microbial growth (Shaffer et al., 2000). There are two surface residue pools, three soil humus pools, and three soil microbial pools. Harvested crop residue and applied manure are partitioned between the two surface residue pools based on the C/N ratio. First-order decomposition rates for each organic C pool are assumed, with rate coefficients as functions of soil temperature, soil oxygen concentration, soil C substrate amount, soil pH, and soil moisture. Urea hydrolysis, nitrification, denitrification, and ammonia volatilization are also simulated. RZWQM2 assumes no movement of ammonium unless it is nitrified to nitrate. There is no simulation of mobile organic carbon in the model either. Plants take up both ammonium and nitrate in proportion to their concentrations present in each soil layer.

RZWQM2 has a generic plant growth module that can be calibrated for any annual crop, although it has been used mostly for corn (*Zea mays* L.), soybean (*Glycine max* L.), and wheat (*Triticum aestivum* L.). The generic plant growth model has also been calibrated for cotton (*Gossypium hirsutum* L.) (Abrahamson et al., 2005, 2006). The model also contains the DSSAT4.0 crop growth modules that were calibrated for 22 crops. When the DSSAT 4.0 is used, RZWQM2 feeds daily soil moisture, soil N, and daily potential evapotranspiration (ET) to the plant modules and then retrieves daily plant N uptake and plant transpiration for its water and N balance calculation (Ma et al., 2005, 2006, 2009). RZWQM2 also has simple algorithms that simulate crop effects on water quality, but not biomass and yield production (i.e., Quick Plant, Quick Tree, and Quick Turf).

The pesticide module simulates pesticide movement in the soil, as well as pesticide uptake by the plant. The model accounts for both equilibrium and kinetic adsorption as specified by users. Daily degradation half-life of pesticides can also be specified by users and can be modified based on soil pH and soil moisture with depth. At a given soil pH, a pesticide may have a positive, negative, or neutral charge. Plant uptake of pesticide is based on its octanol–water coefficient. Volatilization of pesticide is also simulated (Wauchope et al., 2000, 2004a).

The management practices module modifies the soil status and controls the application of irrigation, fertilization, and pesticides; planting and harvest of plants; and crop rotation. Tillage affects soil porosity and bulk density, which in turn changes soil hydraulic properties. It also destroys macropore continuity. A tilled soil will restore to its original bulk density by reconsolidation with subsequent rainstorms and macropore continuity is regenerated. Plant management, such as planting, harvest, irrigation and fertilization, modifies the soil water and nutrient status (Rojas et al., 2000). It should be noted that RZWQM2 cannot do intercropping, only one crop at a given time. The soil chemistry module, rarely used, calculates long-term effects of management on chemical equilibrium of ions and soil pH.

Table 1–1. Minimum data required to run RZWQM2.

Date type	Minimum data required
Rainfall	Amount and intensity
Daily or hourly weather data	Daily meteorology data (minimum and maximum air temperature, wind run, solar radiation, and relative humidity)
Site description	Latitude, elevation, longitude, slope
Soil properties	Soil horizon delineation, soil texture, and bulk density. Optional soil hydraulic properties: 330- or 100 cm-suction water content and saturated hydraulic conductivity.
Pesticide properties	General pesticide data such as common name, half-life , K_d, dissipation pathway
Plant	Specifying a crop cultivar from supplied database
Management practices	Estimate of dry mass and age of residue on the surface, tillage, irrigation, planting/harvest, fertilization, irrigation, pesticide application
Initial soil conditions	Initial soil moisture contents/water table, temperatures, pH, cation exchange capacity values; initial nutrient status, including soil residue, humus, microbial populations, mineral NO_3–N and NH_4–N

Characterizing the System: Input Data Requirements

RZWQM2 needs quantitative information on each part of the system and the interactions among its components. Depending on scientific approaches used in the model, a model like RZWQM2 generally requires (i) weather information, including air temperature, solar radiation, wind speed, relative humidity, rainfall and rainfall intensity; (ii) site description, including slope, elevation, latitude, and longitude; (iii) soil information, including soil bulk density, texture, horizons, hydraulic properties, thermal properties, and chemical properties; (iv) plant information, including plant cultivar characteristics, sensitivity to daylength, phenology, especially thermal units to anthesis and physiological maturity; and (v) management information, including tillage, plant management, irrigation, fertilization, and pesticide applications. Table 1–1 lists the minimum basic information needed to define a system in RZWQM2. Although RZWQM2, like other models, has built-in default databases for soil, plant, and pesticide parameters, it is rare to use the database alone to simulate a specific experiment or condition with confidence.

Weather Information

RZWQM2 has options for either daily or hourly weather inputs. For the daily option, the model needs daily minimum and maximum air temperature (°C), solar radiation (MJ m^{-2} d^{-1}), daily wind run (km d^{-1}), average daily relative humidity (%, if available), daily total rainfall (mm), and rainfall duration (h). If the weather station is too far away from the experimental site, measured weather may not reflect local conditions, especially rainfall amount and intensity. If the experimental site is on a sloping surface, aspect of the slope should be provided to calculate solar energy on the surface. For the hourly option, the model needs hourly air temperature (°C), solar radiation (W m^{-2}), wind run (m s^{-1}), relative humidity (%, if available), and rainfall (mm h^{-1}). Rainfall intensity is calculated if hourly or subhourly rainfall is provided or by assuming daily storms of a uniform duration specified by the user. The RZWQM2 user interface also provides a tool to convert daily or hourly rainfall into breakpoint format. The model can also estimate solar radiation for a location if not given. The weather generator, CLIGEN, is also provided in RZWQM2 for the United States.

Performing quality control (QC) procedures for weather input should help improve agricultural system model simulations. Weather QC is often reported for ET research but not for agricultural system model research. Table 1–2 shows changes in yield and biomass in a wheat–corn–millet rotation system with 10% variation in rainfall, radiation, and humidity in a semiarid Colorado conditions. Similar simulation results were obtained with RZWQM2 by Malone et al. (2011) under Iowa conditions. When there is doubt on weather data, it is recommended to check with a nearby weather station or a weather generator (e.g., CLIGEN).

Soil Information

Characterizing several physical and chemical properties of the soil component of the system is essential because it is where many of the processes are taking place. Spatial and temporal variability is the main obstacle in characterizing the soil for field-scale modeling. When the field is nonuniform, it should be subdivided into smaller plots and the model run for each plot, instead of using soil parameters measured at one location. However, effective average soil properties may be used (Ahuja et al., 2010). Soil parameters include soil hydraulic properties, soil heat properties, and soil chemical properties. A USDA-NRCS Pedon Soil Properties Database is available for users who do not have any soil information for the study sites. Besides the parameters described below, initial values of soil water content, soil temperature, soil carbon pools, and concentrations of soil chemicals are needed to start the simulation process.

Soil Matrix Hydraulic Properties

In RZWQM2, suction head is used to track soil water movement during both infiltration (Green–Ampt equation) and redistribution (Richards' equation). Thus, there is a need to specify the relationship of soil moisture and soil hydraulic conductivity with suction head. In RZWQM2, slightly modified Brooks–Corey

Table 1–2. Change in RZWQM2-CERES simulated grain yield and biomass in a wheat–corn–millet rotation with 10% changes in weather data in Akron, CO.†

% Change in weather variables			Yield			Biomass		
Rain	Solar radiation	Relative humidity	Wheat	Corn	Millet	Wheat	Corn	Millet
−10	−10	−10	−38.8	−11.3	−30.0	−37.9	−8.6	−28.0
−10	−10	0	−29.6	−5.6	−23.6	−28.5	−4.1	−22.8
−10	−10	10	−20.5	1.5	−16.4	−19.4	1.2	−17.2
−10	0	−10	−32.9	−11.0	−22.9	−32.3	−11.5	−18.5
−10	0	0	−24.0	−6.5	−15.1	−23.4	−7.2	−12.2
−10	0	10	−13.7	−-2.9	−6.4	−13.1	−3.3	−5.5
−10	10	−10	−31.8	−12.7	−21.6	−31.2	−14.5	−13.4
−10	10	0	−23.2	−9.1	−12.9	−22.7	−11.4	−5.9
−10	10	10	−13.6	−5.8	−3.8	−13.0	−7.9	2.2
0	−10	−10	−18.2	−3.4	−19.7	−17.4	−1.5	−19.2
0	−10	0	−5.7	4.4	−12.0	−5.1	4.0	−13.2
0	−10	10	4.6	10.7	−4.5	5.6	8.9	−7.4
0	0	−10	−11.5	−4.9	−9.8	−11.3	−4.5	−7.5
0	0	0	0.0	0.0	0.0	0.0	0.0	0.0
0	0	10	11.6	5.0	9.7	11.8	5.0	7.5
0	10	−10	−10.8	−7.1	−5.8	−10.6	−8.0	1.4
0	10	0	−0.3	−3.3	4.1	−0.1	−4.3	9.8
0	10	10	11.0	1.0	14.4	11.3	−0.5	18.0
10	−10	−10	3.7	5.7	−8.9	3.9	5.8	−10.2
10	−10	0	16.0	11.9	−1.2	16.7	10.5	−4.7
10	−10	10	33.8	9.1	32.7	34.2	7.3	32.2
10	−10	10	31.7	20.6	7.1	32.4	16.7	1.3
10	0	−10	10.0	2.0	4.7	9.8	2.9	4.3
10	0	0	21.8	6.8	13.6	22.0	7.1	11.1
10	0	10	35.8	13.3	24.0	36.2	11.8	18.8
10	10	−10	11.5	−1.1	10.0	11.7	−1.8	15.1
10	10	0	22.3	3.3	20.1	22.7	2.5	22.9

† Simulation was based on Saseendran et al. (2010).

equations are used to relate volumetric soil water content (θ) and soil suction head (h) according to Ahuja et al. (2000a)

$$\theta = \theta_s - \lambda_1 |h| \qquad \text{when } |h| < |h_b|$$
$$\theta - \theta_r = B|h|^{-\lambda_2} \qquad \text{when } |h| \geq |h_b| \qquad [1]$$

where θ_s and θ_r are saturated and residual soil water contents (cm^3 cm^{-3}), λ_1 is a constant, h_b is the air-entry water suction for the θ–h curve (cm), and λ_2 is the slope of the log(θ)–log(h) curve (dimensionless). By imposing continuity at h_b, B

$= (\theta_s - \theta_r - \lambda_1 h_b)h_b^{\lambda 2}$. The unsaturated hydraulic conductivity versus suction head $[K(h)]$ is related as

$$K(h) = K_{sat}|h|^{-N_1} \quad \text{when } |h| < |h_{bk}|$$
$$K(h) = C_2|h|^{-N_2} \quad \text{when } |h| \geq |h_{bk}| \qquad [2]$$

where K_{sat} is the saturated hydraulic conductivity (h = 0) (cm h^{-1}) and h_{bk} is the air-entry water suction for the K–h curve (cm). N_1 and N_2 are the slopes of the two segments of the log(K)–log(h) curve divided at h_{bk}. C_2 is a constant which can obtained by imposing continuity at h_{bk}, C_2 becomes

$$C_2 = K_{sat} h_{bk}^{N_2 - N_1} \qquad [3]$$

N_2 in RZWQM2 is obtained as

$$N_2 = 2 + 3\lambda_2 \qquad [4]$$

The parameters h_b and h_{bk} are usually assumed to be equal. Based on soil texture, the RZWQM2 user interface provides a set of default parameters for eleven soil texture classes based on Rawls et al. (1982) (Table 1–3). However, if 1/3 bar soil water content ($\theta_{1/3}$) of a specific soil is known, the default (reference) soil water retention curve will be scaled to the given $\theta_{1/3}$ value (Ahuja et al., 2000a). Users may also choose the option in the interface to use the one-parameter model to derive the Brooks–Corey parameters from the $\theta_{1/3}$ value (Ahuja and Williams, 1991; Kozak and Ahuja, 2005). Whenever possible, independent measured processes should be used to derive these model parameters, such as water infiltration and redistribution data for soil hydraulic properties (Cameira et al., 2005).

The RZWQM2 User Interface can also calculate the Brooks–Corey parameters from soil water contents at 1/3 bar ($\theta_{1/3}$) and 15 bar suction (θ_{15}). Default saturated soil hydraulic conductivity (K_{sat}) is invoked when a soil texture is chosen (Rawls et al., 1982), but the users may enter their own estimated K_{sat}. Users may have the interface calculate a K_{sat} from effective porosity using either of the following equations (Ahuja et al., 1989, 2010):

$$K_{sat} = 764.5(\theta_s - \theta_{1/3})^{3.29} \qquad [5]$$

$$K_{sat} = 509.4(\theta_s - \theta_{1/3})^{3.633} \qquad [6]$$

where there is subsurface flow (tile flow), lateral K_{sat} needs to be defined. Users may set it equal to K_{sat} as initial values.

Table 1–3. Soil hydraulic parameter default values based on soil texture (Rawls et al., 1982).

Soil texture class	θ_s	θ_r	h_b	λ_2	$\theta_{1/3}$	θ_{15}	K_{sat}
	—— $cm^3\ cm^{-3}$ ——		cm		—— $cm^3\ cm^{-3}$ ——		$cm\ h^{-1}$
Sand	0.437	0.020	7.26	0.592	0.0632	0.0245	21.00
Loamy sand	0.437	0.035	8.69	0.474	0.1063	0.0467	6.11
Sandy loam	0.453	0.041	14.66	0.322	0.1916	0.0852	2.59
Loam	0.463	0.027	11.15	0.220	0.2334	0.1163	1.32
Silty loam	0.501	0.015	20.76	0.211	0.2855	0.1361	0.68
Silt	0.501	0.015	20.76	0.211	0.2855	0.1361	0.68
Sandy clay loam	0.398	0.068	28.08	0.250	0.2457	0.1366	0.43
Clay loam	0.464	0.075	25.89	0.194	0.3119	0.1882	0.23
Silty clay loam	0.471	0.040	32.56	0.151	0.3433	0.2107	0.15
Sandy clay	0.43	0.109	29.17	0.168	0.3221	0.2214	0.12
Silty clay	0.479	0.056	34.19	0.127	0.3727	0.2513	0.09
Clay	0.475	0.090	37.30	0.131	0.3789	0.2655	0.06

Soil Macropore Properties

Macropores are large noncapillary soil pores, such as root channels, earthworm holes, and structural cracks. Macroporosity may be measured with a tension infiltrometer at 3 cm water tension or other water tension (Cameira et al., 2000; Malone et al., 2001b) or counting the number of macropores and their diameters directly (Malone et al., 2003) or other methods listed by Ma and Selim (1997). However, there is a possibility that not all the macropores are active during water movement (Malone et al., 2003). The top soil horizons are generally assumed to have cylindrical macropores and the subsoil lateral cracks, although just one type can be accommodated. RZWQM2 requires the following macropore properties for each soil horizon:

- total macroporosity as volume fraction of soil ($cm^3\ cm^{-3}$);
- dead-end macroporosity as fraction of the total macroporosityaverage radius of cylindrical macropores (cm) and average width of cracks (cm);
- a sorptivity-flow correction factor for reducing lateral absorption of water from macropores to surrounding soil caused by compaction of macropore walls, SFCT (0–1);
- the number of macropores per unit area and average radius of soil around a macropore are calculated from the above information inside the model.

It has been observed that not all apparently continuous macropores produce percolate, and some macropores produce only a very small amount (e.g., Quisenberry et al., 1994). Malone et al. (2001b) concluded that percolate-producing continuous macroporosity was an important RZWQM2 parameter for long-term no-till corn and recommended using one-half of the apparently continuous macroporosity to the 30-cm soil depth. The number of macropores per unit area affects both the maximum flow rate into the macropores at the surface (Poiseuille's law), lateral water absorption into the soil, and chemical sorption to macropore walls. The radius mostly affects maximum flow rate through macropores and to a lesser extent chemical sorption to macropore walls.

Water and chemicals vertically moving through macropores mix and react with a user-defined radial width of the macropore wall. Ahuja et al. (2000a) found this to be about 1 mm for Br exchange. Malone et al. (2001b) called this the effective soil radius (ESR) and found that 0.6 cm produced good simulations for pesticide movement in a silt loam soil in a long-term no-till maize field. The ESR of 0.6 cm was attributable to (i) greater partitioning of pesticide from macropores to macropore walls, partially because of organic lining inside the macropores; (ii) blockage and tortuosity of natural macropores; and (iii) lateral water movement through soil into macropores, rather than just ponded surface water moving into macropores as simulated by RZWQM2 (Malone et al., 2001b; Stehouwer et al., 1993, 1994).

Chemical Partitioning, Degradation, and Chemical Transfer to Overland Flow

Chemical sorption to soil may be instantaneous (linear adsorption) or kinetic (instantaneous plus a time-dependent adsorption). The model also simulates pesticide degradation and irreversible adsorption (Wauchope et al., 2004a). For the instantaneous adsorption, the amount adsorbed to soil (S, μg g^{-1}) is expressed as $S = K_d C_s$, where K_d (mL g^{-1}) is the partition coefficient, and C_s (μg mL^{-1}) is the chemical concentration in solution. The partition coefficient for various chemicals and soils may be obtained from the literature (e.g., Wauchope et al., 1992) and are supplied with default values in the model. Another way is to directly measure K_d using batch-type procedures (e.g., Roy et al., 1992). Batch procedures combine water, chemical, and soil; then the chemical in water and chemical adsorbed to soil after 24 h of continuous mixing is measured. The K_d is often expressed relative to organic carbon ($K_{oc} = K_d$/oc), where oc is the mass of organic carbon per mass of soil (g g^{-1}). The kinetic and irreversible adsorptions and pesticide degradation are first order with respect to pesticide concentration in the soil solution.

Chemical transferred from soil surface to overland flow and macropore flow by rainfall impact is assumed to occur within the top 2 cm of soil, and contribution decreases exponentially within this depth increment (0–2 cm). This can be expressed as the nonuniform mixing model (Ahuja, 1986)

$$M_{avg} = e^{-bz} \quad [7]$$

where M_{avg} is the average degree of mixing between rainfall and soil solution at the 1-cm depth increment, b is a nonuniform mixing factor (defaulted to 4.4 cm^{-1}), and z is the center of depth increment (0.5 or 1.5 cm). The chemical is transferred to runoff water at each time increment and may be determined using a mass balance approach (Heathman et al., 1986).

Potential Evapotranspiration and Soil Heat Properties

The Suttleworth–Wallace approach is used to calculate potential evapotranspiration (PET) in RZWQM; this is an extension of the Penman–Monteith to partial canopy (Farahani and DeCoursey, 2000). The parameters needed to calculate the PET from the weather data described above are the daily leaf area index (LAI), canopy height, minimum stomatal resistance, and the albedo (reflectance of total solar energy) values of soil, crop residue, plant canopy, and snow. A guideline for soil albedo is provided by Farahani and DeCoursey (2000). Soil albedo varies from 0.05 to 0.45, depending on soil type, surface roughness, organic matter content, soil moisture content, and color (Farahani and DeCoursey, 2000). Usually soil albedo is between 0.1 and 0.3 (Monteith and Unsworth, 2008). High organic matter content and soil moisture decrease albedo. Surface roughness also reduces albedo. In RZWQM2, users may enter albedo at dry (15 bar) and wet (1/3 bar) soil conditions, and the model linearly interpolates albedo between these two extremes. Ranges of albedo for dry soils suggested by Farahani and DeCoursey (2000) are: light sand (0.25–0.45), clay or gray soil (0.25–0.30), dark clay (0.14–0.2), and general dark soil (0.05–0.15). Albedo for crop residue depends on residue type, water content, geometry, and age, ranging from 0.6 for a fresh bright and flat residue to 0.2 as residue ages (Farahani and DeCoursey, 2000). Albedo for plant canopy is around 0.3, ranging from 0.15 to 0.34 and decreases as plant height increases (Monteith and Unsworth, 2008). Range of Albedo for wheat, barley (*Hordeum vulgare* L.), oats (*Avena sativa* L.), and corn was listed from 0.20 to 0.26 by Farahani and DeCoursey (2000). Albedo of snow is not a user-defined parameter in RZWQM2 and is calculated based on the density and age of the snow as in the PRMS model (Precipitation-Runoff Modeling System), and it ranges between 0.4 and 0.95 (Gray and Male, 1981).

Soil heat properties from de Vries (1963) were used as default values in RZWQM2 with volumetric heat capacity of 0.002 J mm^{-3} K^{-1} for sand, silt, and clay, and 0.0025 J mm^{-3} K^{-1} for organic matter; and thermal conductivity of 31.7 J mm^{-1} h^{-1} K^{-1} for sand, 10.4 J mm^{-1} h^{-1} K^{-1} for silt and clay, and 0.9 J mm^{-1} h^{-1} K^{-1} for organic matter. For agricultural soils, the contribution of soil organic matter is negligible. Users generally do not need to calibrate these parameters.

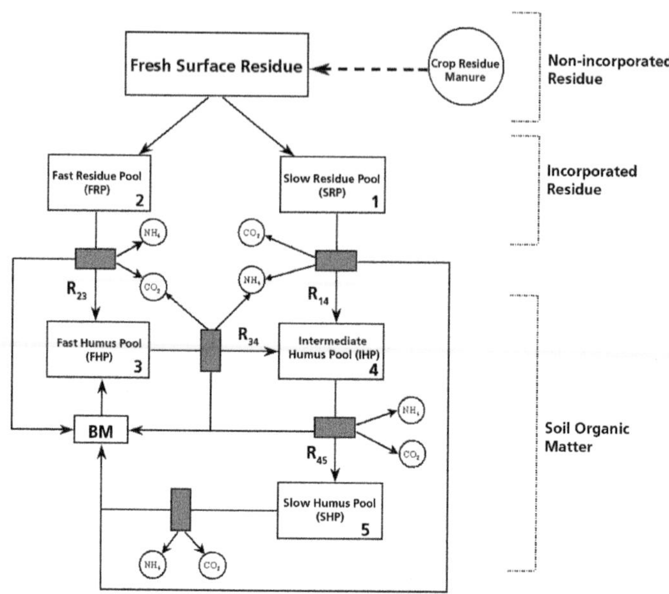

Fig. 1–2. A diagram to show the soil C pools in RZWQM2. R_{14}, R_{23}, R_{34}, and R_{45} are inter-pool transfer coefficients. BM is biomass (from Ma et al., 1998a).

Soil Chemical Properties

If salinity and soil pH simulation are not a focus of the study, the equilibrium chemistry module should be disabled. Thus, the only soil chemical properties needed are soil pH, initial soil nitrate and ammonium, and initial soil pesticide concentration. Otherwise, users need to input initial soil anion and cation concentrations, addition of these ions from irrigation and rain water, and soil cation-exchange capacity.

Soil Nutrient Properties

Soil N is the only nutrient currently considered in RZWQM2. Since soil organic N (or C) is heterogeneous in nature, RZWQM2 partitions soil organic N into eight pools based on state and C/N ratio of the components: two surface residue pools, three soil humus pools, and three soil microbial pools (Fig. 1–2). Generally, we only know the total soil organic C (or N) content or organic matter (OM), which is the sum of the three humus pools. There is a need to partition the known soil C into the three pools. The user interface provides guidance to do this. In general, a rough partitioning of the total soil organic C among fast, intermediate, and slow humus pools is assumed based on soil type and soil management history. The soil microbial populations are generally initialized by running the model for multiple years (10–12 yr) (Ma et al., 1998a). Where there are less than 10 yr of weather and management information, it is suggested to run the model multiple times with initial conditions reset from the previous run.

An option is provided in RZWQM to run the model with constant microbial populations (Ma et al., 2009). When this option is chosen, users should check for annual mineralization rate according to local conditions. If it is too high, more C should be partitioned to the slow pools. If it is too low, more C should be partitioned to the fast or intermediate pools. The RZWQM2 user interface provides a wizard to partition these pools.

It is generally suggested not to modify the C/N ratios of the carbon pools. However, if extremely low C/N ratio of manure is used, the interface will automatically lower the C/N ratio of the slow residue pool to that value so that manure will be partitioned correctly between the fast and slow residue pools. Other soil N dynamics parameters should be kept as they are unless there are strong reasons to change them, except for the interpool transfer coefficients that determine the fraction of degraded C transferred to the next less degradable pool. Parameters to delay nitrification due to nitrification inhibitor and anhydrous application may be adjusted for local conditions.

Crop Information

Users may choose three types of crop modules: (i) a simple known growth curve approach that may be used to extract water and nutrients from the soil with time, without simulating plant growth (i.e., Quick Plant, Quick Turf, and Quick Tree Modules); (ii) a generic plant growth module that can be used to simulate any annual crop; and (iii) the DSSAT crop growth modules (i.e., CROPGRO and CERES) that simulate specific crops. Users are encouraged to select the cultivar characteristics that are most similar to the one they are using and then calibrate the individual cultivar coefficients using observed data. When the crop parameters are calibrated, it is important to use the least stressed treatment, so that the cultivar parameters only reflect the genetic variability, not genotypic by environment (G × E) interactions (Boote, 1999).

The CSM-CROPGRO model simulates soybean, drybean (*Phaseolus vulgaris* L.), fababean (*Vicia faba* L.), velvetbean (*Mucuna pruriens* L.), cabbage (*Brassica oleracea* L.), chickpea (*Cicer arietinum* L.), cowpea [*Vigna unguiculata* (L.) Walp.], peanut (*Arachis hypogaea* L.), pepper (*Capsicum annuum* L.), tomato (*Lycopersicon esculentum* Mill.), bahiagrass (*Paspalum notatum* Flügge), brachiaria spp., cotton, and canola (*Brassica napus* L.). CSM-CERES simulates triticale (× *Triticosecale* Wittm.), wheat, corn, sorghum [*Sorghum bicolor* (L.) Moench], barley, proso millet (*Panicum miliaceum* L.), pearl millet (*Pennisetum glaucum* L.), and foxtail millet [*Setaria italica* (L.) Beauv.]. Simulations of these crops are achieved by specifying the species, ecotype, and cultivar-specific developmental and growth traits. Coefficients of base and optimum temperatures for developmental processes (rate of emergence, rate of leaf appearance, and rate of progress toward flowering and maturity) and growth processes (e.g., photosynthesis, tissue composition, growth and maintenance respiration, nodule growth, N fixation, leaf expansion, pod addition, seed growth, N mobilization) are provided through the species file. The species parameters are not recommended to be changed when developing the coefficients for a new cultivar (Boote, 1999).

Within a given species, there may be a group of cultivars that share similar characteristics. Parameters related to each group of cultivars are included in the ecotype files of each crop. The most critical parameters that define the unique characteristics of a cultivar are included in the cultivar file and can be modified through calibration. These parameters determine the vegetative and reproductive stages, yield components, and leaf appearance rate (Jones et al., 2003). The DSSAT model provides a database of cultivar parameters for the crops and cultivars determined in previous modeling studies. The user should select parameters of a cultivar closest to his/her crop cultivar as the starting value of each parameter.

Depending on the crop module selected in RZWQM2, observed plant data required for calibration might be different. If the simple plant growth curve approach is used, the user should provide maximum LAI, maximum rooting depth, amount of crop residue and root returned to the soil and their C/N ratios, and seasonal N uptake. When the generic plant growth model is chosen, final crop yield and biomass, anthesis and maturity dates, LAI, and plant height are needed. For parameterizing cultivars for the DSSAT model, more detailed plant information, such as flowering date, maturity date, number of leaves, yield, and yield components, is needed (Table 1–4). Of course, additional measurements of biomass and LAI during the growing seasons are helpful.

Management Information

Agricultural management practices are the interface between humans and nature. Users need to specify exactly what has been done to the field, in terms of tillage, irrigation, fertilization, planting, harvest, and pesticide application. Rule-based management practices include auto-irrigation based on soil moisture depletion, ET requirement, or fixed dates; autofertilization based on leaf chlorophyll content and soil N test; and autoharvest based on maturity date. The RZWQM2 user interface facilitates inputs of these managements.

In RZWQM2, implemented management practices are recorded in a file (MANAGE.OUT) and users should check this file to make sure all management practices are implemented as intended. Some exceptions to management inputs

Table 1–4. Data required to calculate crop genotype coefficients (from Hunt and Pararajasingham, 1994).

Crop	Required data
Grain legumes (e.g, soybean, peanut)	flowering date
	maturity date
	seed yield
	pod yield
	biomass at maturity
	seed number/m^2
	number of seeds per pod
	seed size
Corn	silking date
	maturity date
	grain yield
	biomass at maturity
	grain number/m^2
	grain number per ear
	grain/kernel size
Wheat	anthesis date
	maturity date
	grain yield (dry)
	biomass at maturity
	grain number/m^2
	grain number per spike
	grain dry weight

are default tillage and irrigation during fertilization. For example, when NH_3 anhydrous injection is prescribed, the model will automatically add a tillage event for NH_3 application on that day. When fertigation is prescribed but without irrigation on that day, the model will add a 1-cm irrigation event. When fertilizer is incorporated without a tillage event, the model will assume that it is field cultivator. When auto-irrigation is used, the model also records when and how much water is applied each day, to meet the prescribed rules of irrigation according to time interval, soil water depletion, or reference ET. Scheduled management practices may be delayed when the soil surface moisture is above the management soil moisture threshold.

Plant management includes planting density (seeds ha^{-1}), row spacing (cm), planting depth (cm), method of planting (seeds, transplant, and pregerminated), time of harvest, harvest efficiency, type of harvest (seeds, biomass, or root). Irrigation management includes type of irrigation (sprinkler, flood, and drip), application rate (cm h^{-1}), and amount (cm). Fertilizer management includes methods of application (surface broadcast, surface incorporated, injection, fertigation, and best management practices), and NO_3–N, NH_4–N, and urea-N amounts. Information on organic waste and bedding materials are needed in terms of amount and C/N ratios for manure management. Tillage is defined by commonly used names, but users may modify the tillage depth and intensity. Pesticides can be applied based on crop stage as preplanting, pre-emergence, post-emergence, after harvest (lay-by), and on a specific date, and by the way of surface broadcast, surface incorporated, foliar application, with irrigation water, and in slow-release forms.

There is a pesticide database available in RZWQM2 (Wauchope et al., 1992). Users may modify or calibrate these parameters, including washoff fraction from plant canopy or crop residue, adsorption constant (K_{oc}), half-lives for kinetic desorption, degradation, and irreversible bound adsorption. Fraction of kinetic adsorption is another parameter needing to be specified.

In preparing experimental data for model calibration, users may organize their data for convenient automatic comparison with simulation results. This can be done by entering experimental data in the A and T files (e.g., *.MZA and *.MZT) if DSSAT crop growth modules are used or the EXPDATA.DAT file in RZWQM2. Users are also encouraged to develop an easy way to post-process the simulation results for quick comparison using Microsoft Excel or an equivalent for graphing and statistical analysis. Consult the user's manual released with RZWQM2 or DSSAT V4.0 on how to use these files.

Model Calibration

General Considerations

After a system is characterized with the best available information, the next step is to run the model and check if the results are in reasonable agreements with the observed data for different parts of the system. Field-measured data are needed to verify model outputs at this stage. Basic field measurements are soil moisture, evapotranspiration, leaf area index (LAI), phenology at main growth stages, and soil inorganic N, crop biomass, yield at harvest, and total N uptake at harvest. Surface runoff is desirable when runoff may affect soil water balance. When there is subsurface tile flow, measurements should be available along with water table depth. Yield components are important for calibrating the DSSAT4.0 crop models (Table 1–4). Where there is damage due to pests, flood, or hail, users should not use these data for model calibration because RZWQM2 does not account for these effects.

Users should first select a dataset for model calibration and the rest for model evaluation. A calibration dataset could be a treatment in one or multiple years (Hu et al., 2006; Saseendran et al., 2010; Fang et al., 2010b) or one or multiple treatments in single year (Ma et al., 2003; Saseendran et al., 2004, 2005; Sophocleous et al., 2009). When multiple years of data are used for calibration, the data should include wet, average, and dry years (Moriasi et al., 2007). Calibration should also start with a sensitivity analysis followed by manual or automated adjustment of parameters (Moriasi et al., 2007; Sophocleous et al., 2009). Our experience with RZWQM2 has shown that the calibration should be in the order: soil water, soil nutrients, plant growth, and pesticide, and then iterated a couple of times (Hanson et al., 1999), although Cameira et al. (2007) iterated parameterization in the sequence: soil water, plant growth, evapotranspiration, soil nutrient, and pesticide. Before adjusting any parameters, a thorough examination of the inputs and outputs is essential, including mass balance of water, N, and other chemicals (e.g., MASSBAL.OUT, MBLNIT.OUT, MBLWAT.OUT, MBLP*.OUT), soil water and N processes, and plant growth. The RZWQM2 user interface can also facilitate summarizing model output at various temporal resolutions. When there are multiple treatments, it is a good practice to run all the treatments and examine the results before parameter calibration. This will provide users possible clues to which parameters need adjustment. Questions to be asked at this stage are:

- Are total rainfall and irrigation amounts correct in the water mass balance?
- Are fertilizer/manure inputs correct in the N balance file?
- Are pesticide amount correct in the pesticide balance file?
- Are all the management practices implemented correctly?
- Are initial soil water, soil nutrients, and soil chemical status correct?
- Are all soil water components reasonable even if there are no measurements?
- Are all soil N components reasonable even if there are no measurements?
- Are crop phenology, yield, and biomass reasonable even if there are no measurements?
- Are soil chemicals (e.g., pesticides) reasonable?

Table 1–5 lists the processes or variables that users should look at before calibration so that the appropriate process or parameters can be identified and calibrated. It is desirable to complete the table for all simulation runs. It is common for users to focus on one or a few system components and ignore the rest, which can contribute to inappropriate parameterization and result in "good results" for "the wrong reasons." Model users should be more sensitive to any irregularity in simulation results in all components, not only those with field measurements. Once the modeler is satisfied with the general simulation of all

Table 1–5. Processes and variables that should be checked in system model outputs.

Annual or total soil N balance
Initial soil N (organic and inorganic)
End of simulation soil N (organic and inorganic)
N loss to runoff/erosion
N loss to leaching
N loss to denitrification
N Inputs (crop residue, fertilizer, and manure)
Annual N mineralization
Annual or total plant N balance
Total N uptake
N in root
N in grain
N in biomass
N returned to soil at harvest
N fixation
Annual or total water balance
Initial Soil water
End Soil water
Runoff
Seepage
Evapotranspiration
Subsurface drainage (tile flow)
Water Inputs (rain, irrigation, water table)
Annual or total soil chemical balance
Initial amount
End amount
Chemical loss to runoff
Chemical loss to leaching
Chemical loss to air
Chemical loss to tile flow
Chemical Input

system components, he or she can then focus on the processes or variables that have measured values available for comparison. Table 1–6 lists the choice of parameters and variables for calibration of the major processes and outcomes in an agricultural system.

Usually, the weather data are inputs and by default should be correct, although they can contain errors as well (Malone et al., 2011). Generally, no adjustment to the weather files is needed unless there are questions, such as rainfall intensity and solar radiation. When rainfall duration needs to be adjusted to match runoff, it should be noted that extended rainfall duration may affect plant growth because water uptake is not simulated during rainfall events in RZWQM2. Management practices also should be fixed inputs and should be known before running the model. If crop management information is not available or incomplete, such as the dates and amount of irrigation, one should not

Table 1–6. Choice of parameters and variables to be considered for calibration for various processes or outcomes.

Processes or outcomes to calibrate	Related parameters or variables
Soil water dynamics	Brooks–Corey parameters, especially $\theta_{1/3}$ and θ_{15}, pore size distribution (λ_2), bubbling pressure, N_2, and K_{sat}, bulk density or porosity, water inputs to the soil and losses from the soil including plant water uptake
Runoff	K_{sat} at surface layer, rainfall intensity, presence of macropore flow, and surface crusting
Tile flow	K_{sat} and lateral K_{sat}, tile spacing and depth, lateral flow below tile controlled by a lateral hydraulic gradient, drainable porosity (porosity $\theta_{1/3}$) and water table leakage rate
Water table fluctuation	K_{sat} at lower soil layers, tile flow amount, lateral flow below the tile lines, and leakage rate
Evapotranspiration	Albedos, residue cover, LAI simulation, stomatal resistance, rooting depth
Water uptake	PET, rooting depth, soil moisture, $\theta_{1/3}$ and θ_{15}, K_{sat} by layers, soil root growth factor (SRGF)
N uptake	N supply from soil, N demand from daily plant growth as defined by N concentrations in each tissue, passive uptake through transpiration and active uptake parameters
Annual N mineralization	Soil C pool partitioning, interpool transfer coefficients, crop residue returned to the soil, decomposition rates of each pool if needed, and tillage to mix crop residues
Soil inorganic N (NO_3 and NH_4) dynamics	Plant N uptake, N leaching, denitrification, volatilization, nitrification, and hydrolysis rate constants, and N applications and methods
Plant development	Thermal or minimum days between growth stages
Plant biomass accumulation	Daily photosynthesis, length of vegetative growth, plant water/N stresses, and rooting depth
Plant yield formation	Parameters related to yield formation, such as maximum kernel number, kernel weight, daily partitioning of biomass to yield (generic growth model), grain filling duration and rate, length of reproductive growth, plant water/N stresses.
Rooting distribution	Maximum rooting depth, relative SRGF in each layer, and partitioning of photosynthate to root (generic growth model)
Pesticide processes	Adsorption constant, kinetic adsorption, macropore flow, volatilization, runoff loss, leaching loss, and plant uptake

proceed with model calibration unless the dates are obtained with reasonable confidence. Practical issues such as spatial and temporal variability in the field (e.g., soil heterogeneity, uneven irrigation and fertilization, planting density) need to be dealt with separately. Model calibration usually starts with unknown or less certain parameters for the experiment. For example, soil hydraulic conductivity is usually not measured in field research. Thus, it is reasonable to calibrate K_{sat} within a certain range. It is better not to change the non-site-specific default

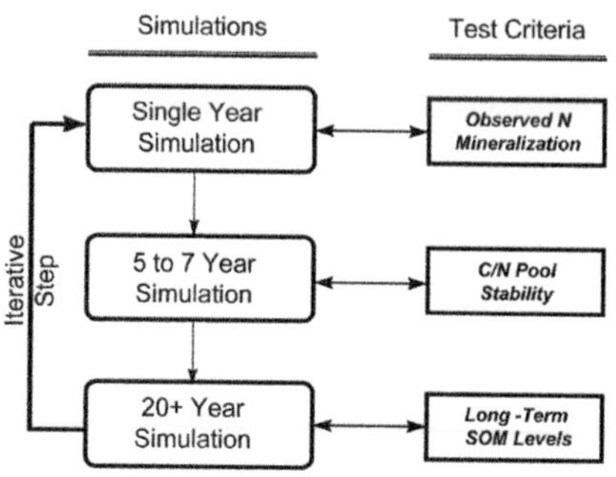

Fig. 1–3. Procedures for calibrating the N submodel of RZWQM (Hanson et al., 1999).

parameters suggested by model developers unless there is a strong reason to do so. When there are multiple measurements of an input parameter, users may need to develop an average value or run the model multiple times and then average the model outputs. If the users do not know the implications of changing a parameter, a sensitivity analysis of the parameter may help (Boote et al., 2008; Ma et al., 1998a). Sensitivity analysis works best when there are multiple treatments. Longer term simulations may also be needed to check system stability for some outcomes such as changes in soil organic carbon and nitrogen content (Fig. 1–3).

Soil water should be calibrated first, followed by soil nutrient and plant growth. Since pesticide is only affected by water and does not affect soil nutrient and plant growth in RZWQM2, it should be calibrated after the soil water–soil nutrients–plant growth calibration. However, because of the interaction among the model components, an iterative procedure should be used. Both the automated calibration procedure of PEST (Parameter Estimation Software; Doherty, 2010) and use of Latin Hypercubic Sampling (LHS) to identify realistic parameter sets for RZWQM2 have shown promise, but more research in this area is needed (Nolan et al., 2010, 2011; Fang et al., 2010a). These procedures face challenges from the difficulty in defining an objective function that optimizes interdependent processes. High intercorrelation can cause "nonuniqueness" issues, where different combinations of parameters result in the same objective function minimum. However, inverse modeling with PEST is an efficient and effective means of identifying highly correlated and insensitive parameters, which can be handled appropriately during the calibration process. For example, singular value decomposition in PEST can mitigate effects from highly correlated and insensitive parameters (Nolan et al., 2010, 2011). In addition, optimized parameters should be constrained to have physical and biological meanings, and a combined manual and automatic method may be more effective (Boyle et al., 2000).

One important, but less discussed, issue in model calibration is how to quantify the goodness of a calibration. In most cases, a model is considered well calibrated if it responds to management practices with reasonable accuracy in terms of root mean squared error (RMSE), relative biases, model efficiency, coefficient of determination (r^2), and the D index (Ahuja and Ma, 2002). For a data set with N measured points, RMSE is defined as

$$\mathrm{RMSE}=\sqrt{\frac{\sum_{i=1}^{N} w_i(P_i-O_i)^2}{N}} \qquad [8]$$

where w_i is the weight factor often set equal to 1.0, and P_i and O_i are the model predicted and experimental measured or observed points, respectively. The N observed data points may be from one treatment or multiple treatments (Ma and Selim, 1997). The RMSE reflects a magnitude of the mean difference between experimental and simulation results. A normalized objective function (NOF) or relative RMSE (RRMSE) may be calculated from RMSE as (Ma et al., 1998b)

$$\mathrm{NOF}=\frac{\mathrm{RMSE}}{O_{\mathrm{avg}}} \qquad [9]$$

where O_{avg} is the averaged observed value. NOF = 0 indicates a perfect match between experimental and modeling results. NOF < 1 may be interpreted as simulation error of less than one standard deviation around the experimental mean. A similar measure is the mean absolute error (MAE):

$$\mathrm{MAE}=\frac{\sum_{i=1}^{N}\left|P_i-O_i\right|}{N} \qquad [10]$$

Another index is the mean bias error (ME) (Shen et al., 1998):

$$\mathrm{ME}=\frac{\sum_{i=1}^{N}\left(P_i-O_i\right)}{N} \qquad [11]$$

The value of ME indicates whether there is a systematic bias in the prediction. A positive value means an overprediction, and a negative value indicates an overall underprediction. Another commonly used approach is to conduct regression analysis between measured and predicted outputs. A coefficient of determination (r^2) is then calculated as

$$r^2=\frac{\left[\sum_{i=1}^{N}\left(O_i-O_{\mathrm{avg}}\right)\left(P_i-P_{\mathrm{avg}}\right)\right]^2}{\sum_{i=1}^{N}\left(O_i-O_{\mathrm{avg}}\right)^2\sum_{i=1}^{N}\left(P_i-P_{\mathrm{avg}}\right)^2} \qquad [12]$$

The r^2 value ranges from 0 to 1. $r^2 = 1$ indicates a perfect correlation between experimental and simulation results, and $r^2 = 0$ means no correlation between the two results. The r^2 approach alone can be misleading, as it does not account for a systematic bias. Some model users simply use the percentage difference between simulated and measured results as a criterion for goodness of model parameterization (Hanson et al., 1999). Wu et al. (1996, 1999) used three other statistical indices to compare RZWQM simulation results with measured ones. They are the maximum difference (MD), the Nash–Sutcliffe modeling efficiency (EF) (Nash and Sutcliffe, 1970), and the *D* index (Willmott, 1981), which are defined by:

$$\mathrm{MD} = \max \left| P_i - O_i \right|_{i=1}^{N} \qquad [13]$$

$$\mathrm{EF} = 1.0 - \frac{\sum_{i=1}^{N} \left(P_i - O_i \right)^2}{\sum_{i=1}^{N} \left(O_i - O_{\mathrm{avg}} \right)^2} \qquad [14]$$

$$D = 1.0 - \frac{\sum_{i=1}^{N} \left(P_i - O_i \right)^2}{\sum_{i=1}^{N} \left(\left| P_i - O_{\mathrm{avg}} \right| + \left| O_i - O_{\mathrm{avg}} \right| \right)^2} \qquad [15]$$

EF is a measure of the deviation between model predictions and measurements in relationship to the scattering of the observed data. EF = 1 indicates a perfect match between simulation and observed results. The *D* index is similar to EF but more sensitive to systematic model bias. It also has values ranging from 0 to 1, where $D = 1$ means perfect simulation. A slightly different version of MD is to add all the absolute differences between the simulated and observed results (Johnsen et al., 1995). Loague and Green (1991) used a coefficient of residual mass (CRM):

$$\mathrm{CRM} = \frac{\left(\sum_{i=1}^{N} O_i - \sum_{i=1}^{N} P_i \right)}{\sum_{i=1}^{N} O_i} \qquad [16]$$

The coefficient of residual mass (CRM) tests more integrated values. It may be used for chemical load to groundwater. Malone et al. (2010) used two other statistics in evaluating RZWQM. One is the percent bias (PBIAS) and the other is the ratio of RMSE to the standard deviation of measured data (RSR):

$$\text{PBIAS} = \frac{\sum_{i=1}^{N} \frac{1}{N}(P_i - O_i)100}{\sum_{i=1}^{N} O_i} \qquad [17]$$

$$\text{RSR} = \frac{\sqrt{\sum_{i=1}^{N} \frac{1}{N}(P_i - O_i)^2}}{\sqrt{\sum_{i=1}^{N} \frac{1}{N}(O_i - O_{\text{avg}})^2}} \qquad [18]$$

Multiple statistics should be used in model calibration to avoid biases (Moriasi et al., 2007). Another issue with agricultural system models is balancing the goodness of simulation among various components of the system, such as soil water, soil N, and plant growth. Efforts to improve one component may worsen the simulation of others, which is mainly due to the interdependence of calibrated model parameters. Given that all field experiments have uncertainty in the data collected, a model should not be overcalibrated for one component; rather, model users should strive for balanced simulation of all the components. The goodness of model parameterization is quantified by some statistical indices (e.g., root mean squared deviation) and further judged by model prediction for other experimental conditions, including different soils, weather, and treatments (Ahuja and Ma, 2002). When a composite statistical index is used for optimization (e.g., PEST), users should pay more attention to the weight factor assigned to the statistics of each variable (Nolan et al., 2010, 2011).

What is considered "acceptable" for a model calibration may vary with the objective of the study and experimental errors in the field. It is certainly acceptable when simulation error is within the experimental error of the replicates (Cameira et al., 2007). When treatment effects are the main objective, a model may be considered acceptable when it responds correctly to the differences among treatments (Cameira et al., 2007; Ma et al., 2007b). Moriasi et al. (2007) rated model performance as acceptable when $r^2 > 0.5$, EF > 0.5, –25 < PBIAS < 25, and RSR < 0.7 for watershed models. However, the rating may vary with output variables and time and space resolution, and it should vary according to the uncertainty in measured data (Moriasi et al., 2007). For a point model like RZWQM2, it is preferred to use a more stringent rating, such as $r^2 > 0.8$, EF > 0.7, –15 < PBIAS < 15, and RSR < 0.5. Based on simulation data from Saseendran et al. (2010), the D index should be greater than 0.7 for an "acceptable" calibration. Qi et al. (unpublished data, 2011) recommended both r^2 and D of 0.65 as "satisfactory," 0.8 as "good," and

0.9 as "very good." For pesticide simulation, a PBIAS of 100 may be considered "acceptable" due to large experimental errors (Malone et al., 2004b). Graphical presentation of the simulated and observed results should also be used for visual comparison and is essential for model calibration (Loague and Green, 1991; Legates and McCabe, 1999; Moriasi et al., 2007). Among the above-mentioned statistics, RMSE (or MAE) is the most unbiased model comparison, whereas r^2 and the D index tend to make mediocre models look good because of their high values (Legates and McCabe, 1999). If RMSE or MAE is within 10% of the measured mean values for all measurements, it is a "very good" calibration. A value of 15% is considered "good" and 20% is "satisfactory" for agricultural models. Hanson et al. (1999) recommended 15% error for biomass, yield, and leaf area index (Cameira et al., 2005). The following two modified, but more conservative statistics proposed by Legates and McCabe (1999) have not been used much in the literature, however.

$$\mathrm{EF}_1 = 1.0 - \frac{\sum_{i=1}^{N} |P_i - O_i|}{\sum_{i=1}^{N} |O_i - O_{\mathrm{avg}}|} \qquad [19]$$

and

$$D_1 = 1.0 - \frac{\sum_{i=1}^{N} |P_i - O_i|}{\sum_{i=1}^{N} \left(|P_i - O_{\mathrm{avg}}| + |O_i - O_{\mathrm{avg}}| \right)} \qquad [20]$$

Calibrating Soil Water Parameters

Accurate simulation of the soil water balance is critical for any agricultural system model. Obtaining average effective soil hydraulic properties for a site is often more difficult due to spatial and temporal variability and uncertainty in soil horizons (Ahuja et al., 2010). In addition, it is rare for an experimentalist to measure representative soil hydraulic properties for a given field with confidence. Due to spatial variability, laboratory measured soil hydraulic properties from a few samples may not be representative of the entire field, not to mention the hysteresis behavior in the field. While we suggest the use of locally field measured soil hydraulic properties if available, the soil hydraulic properties based on soil texture class (Table 1–3) may be sufficient in many general applications (e.g., Saseendran et al., 2009). However, some calibration may be needed for specific situations.

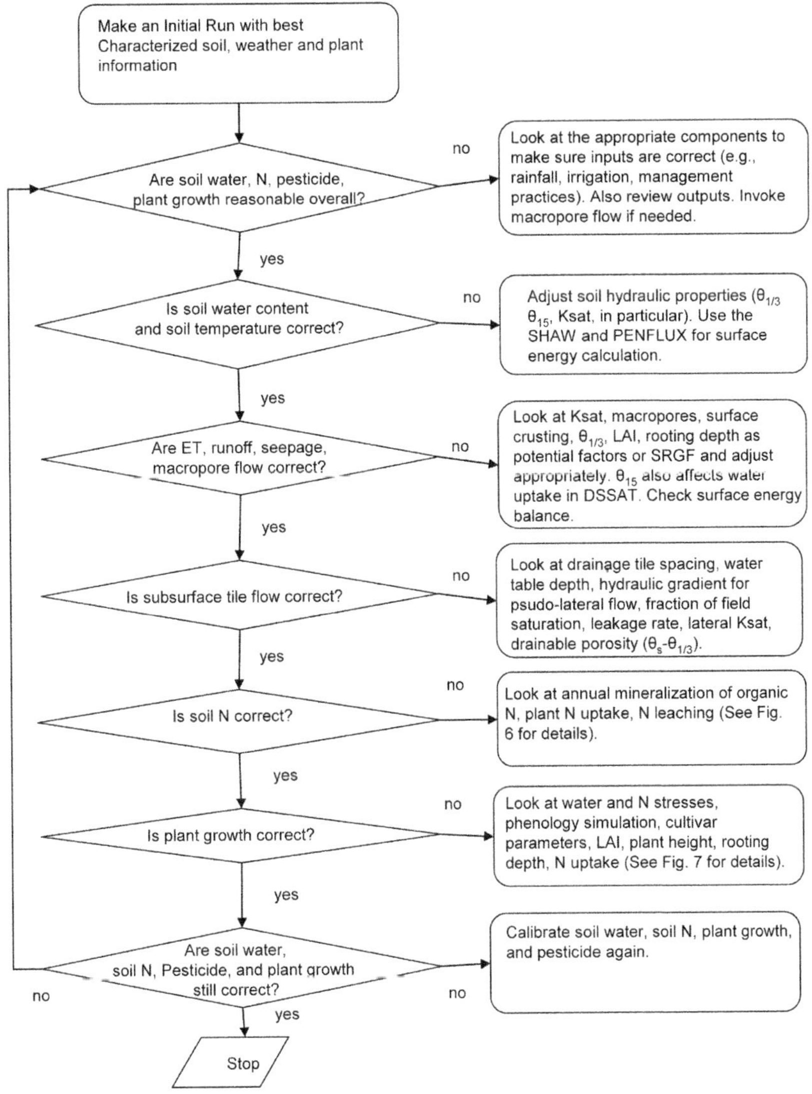

Fig. 1–4. Recommended calibration steps for soil water in RZWQM2.

The variables to calibrate for soil water balance are soil moisture content, ET, runoff, deep seepage, water table if present within the simulated soil profile, and subsurface flow (tile flow) if measured (Fig. 1–4). Even though there are no measured data for some of the water components, it is still a good practice to review them to make sure that they are reasonable for the soil and climate. Depending on the purpose and data availability, users may aggregate temporal simulated results to daily, monthly, or yearly values, and soil depth results to soil horizon or whole soil profile.

Soil Moisture Distribution

If there are soil moisture measurements in the experiment, they should be compared first. If not well simulated, users should check water inputs (rainfall and irrigation) first to make sure the inputs are correct. Then, review water outputs from the system (ET, runoff, drainage).

1. If measured soil hydraulic property is not available, calibrate the soil water contents at 1/3 bar and 15 bar suction (also known as the drained upper limit and lower limit of plant available water). The model can then calculate the Brooks–Corey parameters from these values. These two soil moisture contents also affect plant water uptake when DSSAT crop growth modules are selected.
2. For experienced users, adjusting λ_2 and K_{sat} of appropriate layers should also help water distribution. λ_2 can be adjusted through $\theta_{1/3}$ based on the one-parameter model (Ahuja and Williams, 1991). However, if λ_2 is adjusted, make sure the continuity of the Brooks–Corey equation is maintained at the bubbling pressure head (h_b) in Eq. [1] (RZWQM2 user interface can help check the continuity). In RZWQM2, the slopes for both soil water retention curve and hydraulic conductivity curve are related by Eq. [4], which is commonly assumed, but may not be the case for some soils (Malone et al., 2004b) (Fig. 1–5). The RZWQM2 user interface provides four different methods of calibrating soil hydraulic properties.
3. Surface soil moisture can be affected by ET and surface runoff, so check ET and runoff simulations. If needed, adjust soil albedo or add surface crusting for better simulation. Leaf area index contributes to ET simulation as well. When runoff is too high or too low, adjust rainfall intensity if not measured or K_{sat} of surface layer.
4. Soil moisture at lower horizons is affected by root distribution and depth. However, soil hydraulic conductivity and soil water retention curves (the Brooks–Corey parameters) are definitely the most important parameters controlling the soil water distribution.
5. If there is macropore flow, macropore flow should be activated and macropore parameters need to be looked at.
6. Fraction of field saturation may be used when full saturation in the field is not obtained.
7. Heterogeneity of field soil moisture measurement should also be taken into account (including row vs. interrow). It is reasonable to vary the soil properties within the experimental error or spatial variability (e.g., soil bulk density). Temporal variability can also contribute to the accuracy of measured hydraulic properties, depending on when the samples are taken (e.g., before or after tillage, before or after rainfall).

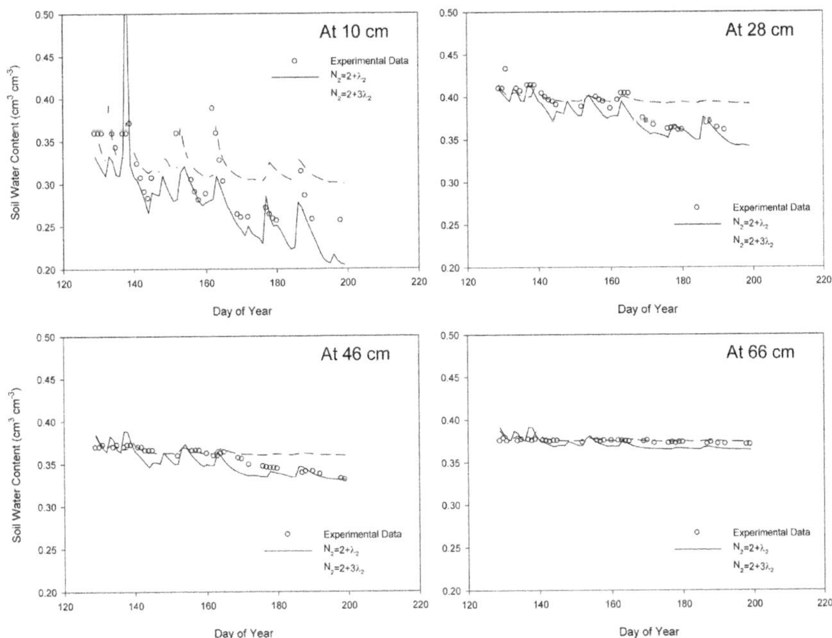

Fig. 1–5. Comparing soil water content simulations using Eq. [4], $N_2 = 2 + 3\lambda_2$ and $N_2 = 2 + \lambda_2$ (Malone et al., 2004b).

8. Another important aspect is soil horizonation. A poor simulation may reflect an error in the definition of soil horizons in terms of soil bulk density and soil texture. If simulated soil moisture in a particular layer is too high, more roots should be grown in that layer. This can be accomplished by adjusting the root distribution parameters in the model (maximum rooting depth in the generic model or SRGF in the DSSAT crop models).

Evapotranspiration

In RZWQM2, potential evapotranspiration (PET) is estimated using the Shuttleworth–Wallace method, and actual root water uptake is either from the Nimah–Hanks equation when the generic plant growth module is invoked or from an empirical equation when the DSSAT crop growth modules are used.

1. When actual ET is not simulated correctly, PET should be checked first to make sure that weather inputs (radiation, wind speed, relative humidity) and soil albedo and the minimum stomatal resistance are correct. Table 1–7 shows reference PET for different weather regions, which should be the upper boundary for the PET calculated in RZWQM2.
2. Examine the water balance output files to see each component of soil water balance, including runoff and soil water storage in the soil profile. If one of them is not reasonable, correct that component.

Table 1–7. Average reference evapotranspiration for different agroclimate regions (Allen et al., 1998).

Region		Mean daily temperature (°C)		
		Cool 10°C	Moderate 20°C	Warm 30°C
		mm d^{-1}		
Tropic and subtropics	humid and subhumid	2–3	3–5	5–7
	arid and semiarid	2–4	4–6	6–8
Temperate region	humid and subhumid	1–2	2–4	4–7
	arid and semiarid	1–3	4–7	6–9

3. Since LAI and canopy height enter into the calculation of PET, plant growth needs to be checked as well, including rooting depth. Soil moisture content and root length density distribution are important to check when actual simulated ET is not adequate.
4. Hydraulic conductivity is used in the Nimah–Hanks water uptake equation, and the lower limit of plant available water (15 bar soil moisture content) is used for water uptake in DSSAT. These two parameters should be reviewed along with their roles in other simulations in the system (e.g., soil water distribution and plant growth).
5. Surface crop residue cover should be included if it exists.

Surface Water Runoff

In RZWQM2, runoff occurs when rainfall or irrigation intensity exceeds the infiltration capacity. Therefore, rainfall or irrigation intensity should be accurate if runoff is an objective of model simulation. Surface water detention (surface roughness) is not simulated.

1. Increasing K_{sat} of the surface horizon will reduce runoff. Adding surface crusting and reducing K_{sat} in the surface soil layer will increase runoff.
2. Tillage temporarily changes soil hydraulic properties and can affect runoff for a short time until the surface layer reconsolidates.
3. Macropores, if present, can greatly increase infiltration and decrease runoff. Macroporosity and the macropore dimension determine the flow capacity by Poiseuille's law, so users need to pay attention to these when macropore flow is invoked and runoff is simulated poorly.
4. Currently, the soil slope does not affect surface runoff in RZWQM2.

Deep Seepage

Deep seepage is obviously affected by K_{sat} and soil water retention (especially the λ_2 and $\theta_{1/3}$). Plant root water uptake will reduce deep seepage.

1. Reducing K_{sat}, a smaller λ_2, and higher $\theta_{1/3}$ all reduce deep seepage.

2. Rainfall or irrigation intensity and runoff should also be estimated reasonably when their actual value is unknown for better simulation of deep seepage.
3. Reduction of field saturation during infiltration caused by air entrapment in the soil also affects deep seepage, as it reduces K_{sat} (Ahuja et al., 2000a).
4. Macropore flow helps direct more water into the soil and reduce runoff loss.
5. Increased plant water use and evaporation reduces deep seepage.

Subsurface Drainage

Subsurface drainage (i.e., tile flow) is affected by water table simulation and LKsat (lateral K_{sat}). Drainage spacing and depth are important parameters for estimating subsurface drainage in the Hooghoudt's steady state equation in RZWQM (Ahuja et al., 2000a).

1. First, water table simulation should be evaluated.
2. Field capacity (or $\theta_{1/3}$) and the λ_2 (slope of the log-log soil water retention curve) are also important in determining tile flow amount (Bakhsh et al., 1999) because they control how fast water moves through the restricting layer to refill the saturated soil zone.
3. The leakage rate of water table and lateral hydraulic gradient have a major effect on maintaining the water table and partitioning water between tile flow and groundwater lateral flow.
4. Calibrating lateral K_{sat} for each soil layer may be needed for correct tile flow simulation.
5. Check on ET and runoff to verify the correct amount of water entering the water table.
6. Reducing deep seepage and lateral flow below the tiles helps maintain a water table and consequently increasing tile flow if needed.
7. Calibrating K_{sat} may help refill water table, but should be done in context with other water components.

Water Table

Hydraulic conductivities at the bottom of the soil profile and water leakage rate through the restricting layer, if present, are important for building a water table. Subsurface groundwater flow also affects water table fluctuation. RZWQM2 simulates quasi-steady-state lateral flow to account for groundwater flow beneath the tile lines.

1. Make sure a constant flux boundary condition is used at the bottom of the soil profile, and field saturation fraction is set to 1.0. Otherwise, the model may not simulate soil water saturation correctly.

2. Increasing the lateral flow component by increasing lateral hydraulic gradient reduces both the water table level and tile flow amount. However, the lateral flow component only directs excess water out of the system, not water from adjacent plots flowing into the system.
3. Increasing K_{sat} directs more water into the soil profile and increases the water table.
4. Make sure the water table is within the defined soil profile at all times during simulation.
5. Reducing the leakage rate helps build up a water table too.

Calibrating Water and Chemical Movement through Macropores if Present

In heavy-textured soils, there may be a need to activate and parameterize the macropore flow (Ahuja et al., 1993, 1995). The first step is to use parameters that are reasonably estimated from the literature. These include the reduction factor (SFCT = 0.1), effective soil radius (ESR = 0.6), and nonuniform mixing factor (B = 4.4 for tilled and 6.0 for no-till) (Malone et al., 2003). The number of percolate-producing continuous macropores may increase with increasing initial soil water, but the number of macropores between 0.01 and 0.04 cm^{-2} has been used for several soils without a clear difference among tillage methods (Malone et al., 2003).

1. Since rainfall intensity determines the amount of water available for macropore flow, check rainfall intensity and K_{sat}.
2. Macroporosity, fraction of dead macropores, dimension of macropores, and lateral sorptivity reduction factor should be calibrated for the soil first in the order for obtaining correct water flow through macropores.
3. The ESR value of 0.6 by Malone et al. (2001b) was reasonable for several subsequent macropore simulations (Malone et al., 2003, 2004b), but can be calibrated if needed. The ESR affects only the chemical transport through the macropores.
4. Individually adjusting macropore radius (r_p), K_{sat}, and surface crust hydraulic conductivity by about 50% resulted in metribuzin transport in percolate (macropore and matrix flow) to vary by 100% or more (Malone et al., 2004b).
5. Pesticide transport through macropores is very sensitive to soil parameter changes, such as surface crust hydraulic conductivity, K_{sat}, and λ_2.
6. The nonuniform mixing factor (4.4) may be adjusted when chemical extraction from the soil is thought to be a contributing factor. However, chemical concentration in the top soil layer should be checked before the mixing factor is adjusted.
7. Absorption of chemicals to the macropore walls may be reduced and chemical in tile flow increased by allowing a fraction of the chemical entering to tiles directly without any exchange with the maropore walls.

Calibrating Soil Temperature and Surface Energy Balance

Surface temperature is determined by water movement, solar energy, partitioning of the surface energy, soil heat capacity, and soil thermal conductivity. RZWQM2 calculates soil heat capacity and thermal conductivity based on soil texture and soil moisture content. The effect of soil organic matter on soil heat transfer is not considered in the model. Therefore, it is important to know the soil texture in each soil horizon and have correct soil moisture simulations.

1. Since soil thermal property is determined by soil texture, make sure soil texture for each layer is correct.
2. Check on soil moisture simulation and soil water balance before adjusting soil thermal properties. Low surface soil moisture simulation tends to result in high surface temperature.
3. When transpiring plants are present, soil water content within the root zone becomes important, with inaccurate water contents having similar effects as in the bare soil situation.
4. Albedo of the soil surface, residue, and canopy cover should be checked to make sure that they are within reasonable ranges, especially when there is residue and canopy covering the soil surface.
5. If needed, the PENFLUX or SHAW option may be used to calculate the top boundary condition for surface temperature via soil surface heat flux calculation, rather than assuming it equals air temperature.
6. When hourly air temperature is available as input, it should be used in soil temperature and surface energy balance calculations. The model can also convert daily weather inputs to hourly if desired.
7. The amplitude of the diurnal soil temperature variation will be influenced not only by soil water content, but by residue and plant cover. Insufficient plant and residue cover will cause simulated midday soil temperatures to be lower than when there is plant and residue cover to shade and insulate the soil surface in the winter; the reverse trend will be simulated in the summer.
8. When there is a problem in partitioning net shortwave and longwave radiation among latent heat (associated with ET), sensible heat, and soil heat flux, then the ET calculation and albedo should be re-evaluated.

When the SHAW option for energy balance is invoked, energy components should be checked and calibrated. The latent heat is determined by ET and is affected by LAI, soil moisture, plant height, albedo, rooting depth, and net radiation (R_n). Latent heat and sensible heat should be calibrated together. Net radiation may be calibrated by adjusting albedo and LAI. Surface soil heat flux (G_0) is affected by surface residue, canopy cover, and soil albedo.

Calibrating Soil Nutrient Parameters

Nitrogen Distribution in the Soil Profile

Soil N contents and distribution are affected not only by soil water movement, but also by the soil C/N dynamics and plant N uptake. Plant N uptake from each layer is an important factor. Therefore, it is very difficult to simulate N distribution in the soil profile correctly.

1. Soil water balance should be checked first when N distribution in the soil profile is not adequate.
2. Overall soil N balance in the soil should be evaluated at this point to identify any unreasonable components, such as NH_3 volatilization, N leaching, and denitrification. If volatilization is too high, the volatilization constant may be reduced. Changing fertilization application method from broadcast to injection or surface incorporation also reduces volatilization, so the appropriate method should be specified. Fertigation (fertilizer application with irrigation water) will leave NH_4 on the soil surface because NH_4 is assumed to be immobile in the soil.
3. Nitrogen supply from fertilizer and soil organic N mineralization should be checked, followed by plant N uptake (Fig. 1–6). If all three are simulated reasonably well, but N distribution in the soil profile is still not good enough, the user should try to adjust the nitrification rate or activation energy to reduce or increase nitrification. When urea is applied, urea hydrolysis rate constant may need to be modified for the condition. These parameters are in the detailed soil C/N parameters screen in RZWQM2 user interface. It is also essential to run the model for 10 yr to initialize the soil organic pools for a reasonable mineralization rate for the system of study.
4. Denitrification may be calibrated by modifying the denitrification constant or the activation energy for temperature response.
5. Adjusting rooting depth or root length distribution may not improve N distribution in the soil profile if soil water is simulated well because roots are also responsible for plant water uptake. In RZWQM2, soil N is assumed to be in equilibrium between soil micropores (immobile) and mesopores (mobile). However, soil N in the micropores does not exchange with the mobile water during infiltration. Therefore, adjusting the microporosity in each soil layer may improve the N distribution in the soil profile to a certain extent.
6. For a short simulation period, the correct partitioning of soil organic pools and checking all fates of soil N are essential for a reasonable N distribution in the soil profile. Use constant soil microbial populations if needed.

Soil Organic C/N

Soil organic C/N dynamics are controlled by C/N ratios of each pool, microbial populations, rate constants, soil temperature, soil moisture, and activation energy. Increasing or decreasing activation energy controls sensitivity of the process to

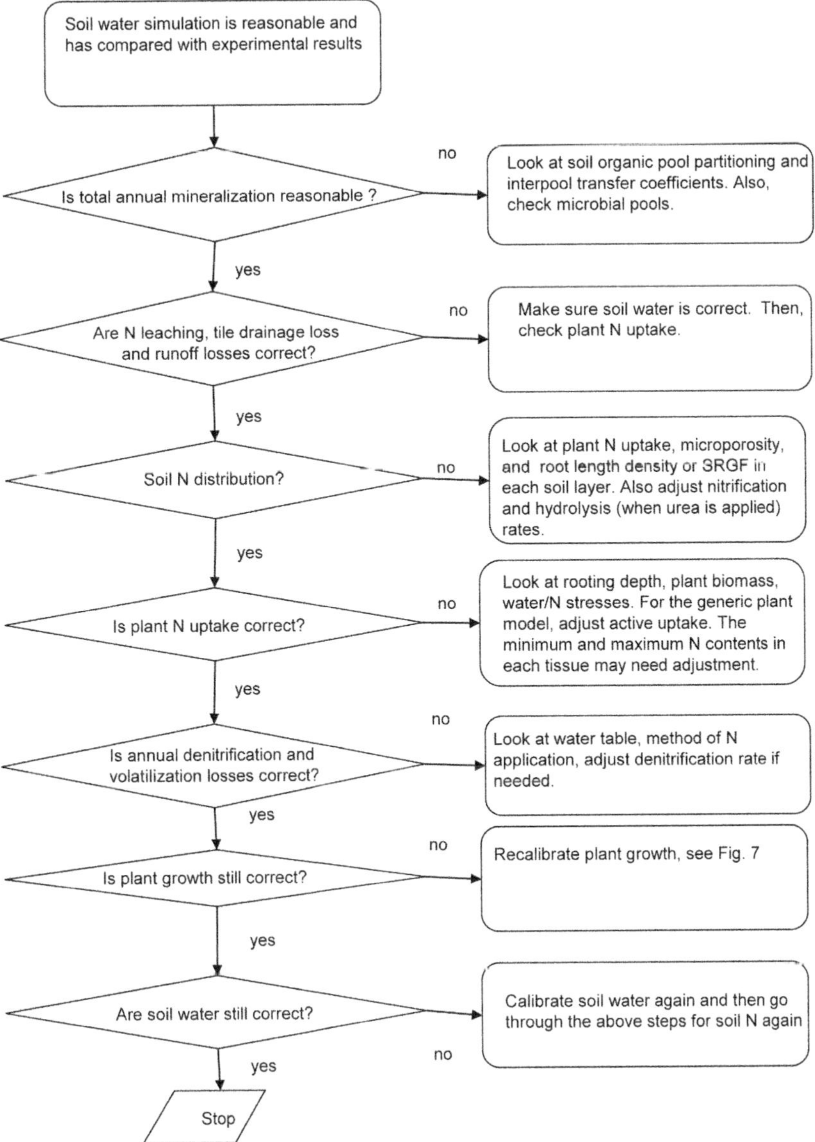

Fig. 1–6. Recommended calibration steps for soil N in RZWQM2.

soil temperature. It is generally recommended not to modify the rate constants or activation energies. Instead, the C/N pool sizes and fraction of mass transfer from one soil organic pool to another (interpool transfer coefficients) may be calibrated. Partitioning of the soil C/N pools, including microbial pools, significantly affects mineralization rate. Since the soil pool sizes and microbial populations are generally unknown, the user interface provides a wizard to partition total organic C into various pools. The users can refine the pools as follows:

1. When microbial population is simulated, it is recommended to pre-run the model for 10 to 12 yr using previous or current management, soil type, and weather data. Then initialize the pool structure for your simulation from the results of the pre-run (Ma et al., 1998a).
2. When constant microbial populations are used, the user should manually partition the pools using the wizard and then compare annual mineralization rate with field-estimated N supply under the soil and weather conditions of interest. Partition more of the total soil C to the fast pool or intermediate pool if simulated annual mineralization rate is too low and to the slower pool if the rate is too high (Ma et al., 2008). A rule of thumb is that for every 1% OM in the top 0.3 m of soil, annual mineralization is ~20 kg ha^{-1} (Schepers and Mosier, 1991). Where there is manure application, it is expected that 50% of the organic N in manure will be released in the first year (Karlen et al., 2004).

Steps to calibrate soil organic C/N are as follows:

1. Set the initial soil organic C pools based on measured soil organic matter content at each soil depth using the wizard provided in RZWQM2. Initial amount, type, height, age, and C/N ratio of crop residue may be needed as well on the surface and in the soil (root biomass, if any).
2. Initialize the C pools as shown above with or without soil microbial growth simulation. Failure to initialize the pools will result in unrealistic N mineralization and N supply to the soil.
3. Return of crop residue to the soil surface and tillage will also affect soil C/N dynamics. Check soil C and N balance files to make sure crop residue return is correct, including surface residue and root biomass.
4. Calibrate soil C pool dynamics by primarily adjusting the interpool transfer coefficients. Experienced users may also want to modify the rate constants and activation energy for each process.
5. When soil microbial growth is simulated, make initialization runs for 10 to 12 yr before real time simulation for the experiment and then load the initialized values to the user interface to replace the original guesses.
6. The C/N ratio of each organic C pool should not be modified.

Nitrification, Denitrification, and Ammonia Volatilization

Nitrification and denitrification rate constants may be adjusted if NH_4 and NO_3 concentrations in the soil differ from the measured values when soil moisture contents, N leaching, and plant N uptake appear to be predicted adequately. Usually there are no measurements for these processes, especially nitrification.

Table 1–8. Approximate N denitrification loss for various soils (Meisinger and Randall, 1991).

Soil organic matter content	Soil drainage classification				
	Excessively well-drained	Well-drained	Moderately well-drained	Somewhat poorly drained	Poorly drained
%	% inorganic N denitrified				
<2%	2–4	3–9	4–14	6–20	10–30
2–5%	3–9	4–16	6–20	10–25	15–45
>5%	4–12	6–20	10–25	15–35	25–55

Denitrification occurs only under saturated water conditions. Guidelines on denitrification loss are shown in Table 1–8.

When there is surface application of manure or NH_3, NH_3 volatilization loss is expected. Injection of liquid manure or NH_3 will considerably reduce volatilization. When urea is applied on the surface, volatilization is higher than when it is incorporated because of high concentration of NH_4 from urea hydrolysis. Table 1–9 provides some guidelines on ammonia losses in soils.

Nitrogen Leaching and Nitrogen in Runoff

Nitrogen leaching is driven by soil water movement. Timing and amount of leaching are highly dependent on water input to the system via rainfall and irrigation, in addition to soil horizon texture affecting soil hydraulic properties.

1. Check deep seepage and runoff amount before working on soil N parameters.
2. Nitrogen-producing processes should be scrutinized. Annual soil organic N mineralization and fertilizer and/or manure application should be checked along with plant N uptake. Nitrification affects the conversion from NH_4 (immobile) to NO_3 (mobile).
3. Nitrogen-consumption processes, such as N uptake, volatilization, and denitrification should be checked. When soil is saturated, denitrification can be high enough to affect the leaching result.
4. Nitrogen in runoff is controlled by the soil N concentration in the top 2 cm and runoff water. Therefore, the ability of top 2 cm to hold NO_3 and the degree of mixing are important in determining runoff N. Factors affecting NO_3 in the top soil layer are extent of leaching to deeper soil horizons and volatilization if NH_3 is present.
5. Macropore flow may reduce runoff and subsequently N loss to runoff.

Nitrogen Loss in Tile Flow and in Subsurface Flow

1. Check on the amount of tile flow or subsurface flow below the tile first.

Table 1–9. Approximate ammonia losses from fertilizer N for various fertilizers, application methods, soils, and climate scenarios (Meisinger and Randall, 1991).†

Fertilizer N source	Ferilizer application method	Likely precipitation after application		
		Humid climate, 13 mm or more rain within 2 d	Subhumid, 0–6 mm of rain within 7 d	Dry climate, little or no rain likely within 7 d
		% of fertilizer N lost		
Soil pH > 7				
Urea or UAN	Surface broadcast	0–20	2–30	2–40
	Surface dribble	0–15	2–20	2–30
	Incorporated	0–10	0–10	0–10
Ammonium sulfate	Surface broadcast	0–40	2–50	5–60
	Incorporated	0–10	0–20	0–30
Ammonium nitrate	Surface broadcast	0–20	2–25	5–30
	Incorporated	0–10	0–15	0–20
Anhydrous ammonia	Injected	0–2	0–3	0–5
Soil pH < 7				
Urea	Surface broadcast	0–5	5–30	5–40
	Surface dribble	0–5	2–20	2–30
	Incorporated	0	0–2	0–2
UAN	Surface broadcast	0–5	2–15	2–20
	Surface dribble	0–5	2–10	2–15
	Incorporated	0	0–2	0–2
All other N sources	Any method	0	0–2	0–2

† Adjust for soil texture and surface residue as follows: for low CEC soils (<10 meq/100 g) use upper end of range, for high CEC soils (> 25 meq/100 g) use lower end of range; for > 50% surface residue cover (no-tillage, etc.) use upper end of range; for paddy conditions use values in no precipitation column for surface broadcast application.

2. Check on N supply from fertilizer and soil organic N (mineralization).
3. Soil N concentrations in the soil profile should be calibrated if necessary, especially around the tile location.
4. Plant N uptake and rooting depth also affect the amount and where soil N is taken out of the soil and is not available for leaching.

Calibrating Crop Parameters

It is recommended to calibrate plant parameters under the least stressed conditions so that soil water deficit and N availability do not affect the calibration (Boote, 1999). It is hoped that calibrated plant parameters under these conditions are only determined by the cultivar species and can be carried over from location to location. Phenology should be calibrated first because it affects not only crop production (yield and biomass), but also plant responses to water, N, and temperature stresses as plant's sensitivity to stresses varies with growth stage (phenology) (Fig. 1–7).

The Quick Plant Parameters

QUICK PLANT. Quick Plant is a module that mimics plant growth through extraction of water and nutrients from the soil. It does not simulate actual plant growth per se. Users need to define maximum plant height, maximum rooting depth, maximum leaf area index, and total plant N uptake (both aboveground and belowground) during a growing season. The module will grow a plant based on a user-defined growing period. It also provides a simple way for vernalization such that the plant will enter dormancy when the 5-d average temperature is below zero and out of dormancy when the 10-d average temperature is above zero. The module also allows user-specified stover and dead roots to be returned to the soil at harvest. If users do not specify root biomass, roots are assumed to be 25% of aboveground biomass. The user must specify:

- Length of growing season (days)
- Total N uptake during the growing season (kg N ha^{-1})
- Maximum plant height (cm)
- Maximum leaf area index
- An estimate of the amount and C/N ratio of leftover crop residue at harvest (kg ha^{-1})
- An estimate of when winter dormancy ends (day of the year)
- Maximum rooting depth (cm)
- Root biomass and C/N ratio at harvest (kg ha^{-1})

QUICK TURF. The Quick Turf module is designed for perennial grass and has a constant rooting depth that the user provides. The LAI and height at each cut should also be defined. The module assumes a uniform cut to a given LAI and height. LAI and height will "grow" after cutting until the next scheduled cutting. However, the cutting intervals can vary. Cutting dates are specified as day of the year and are applicable to every year of the simulation. The interval between the last cutting of the previous year and the first cutting of the following year is

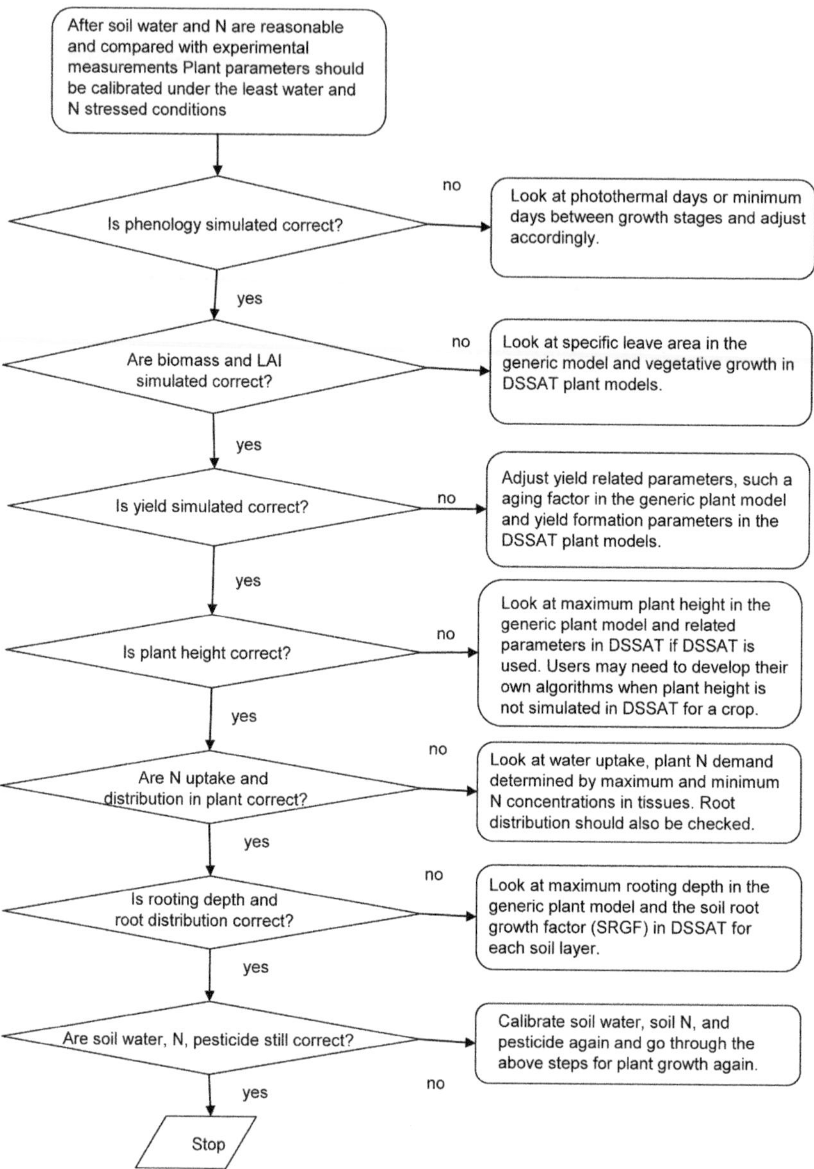

Fig. 1–7. Recommended calibration steps for plant growth in RZWQM2.

defined as the growing period for the first cutting of the following year. Grass can experience a period of dormancy at beginning of the year only. Users may define when grass has completed its dormancy, otherwise, the module will bring the plants out of dormancy when the average 10-d temperature is above 0°C. The module also allows user-specified stover and dead roots to be returned to the soil at harvest. Dead roots are assumed to be zero if not specified. User-defined plant N uptake is total uptake between cuttings. If user-defined initial plant height is greater than specified height at cutting, the module will automatically cut it to the height after cutting. The user must specify:

- Grass height (cm) and LAI after each cutting
- Initial height at simulation (cm)
- An estimate of when winter dormancy ends (day of the year)
- Total N uptake between cuttings (kg N ha^{-1})
- Grass height (cm) and LAI when the grass needs to be cut
- Amount and C/N ratio of the grass cut as stover
- Maximum rooting depth (cm)
- Root biomass and C/N ratio at cutting (kg ha^{-1})

QUICK TREE. The Quick Tree module provides a way to simulate shrubs and trees where LAI may be the only variable during a simulation year. The user needs to provide a constant plant height and a constant rooting depth. The LAI will change during the growing season. Users still need to "plant" and "harvest" a tree or shrub every year to start and end leaf growth. The module allows users to specify dead leaves (stover) and dead roots returned to the soil at harvest. Dead roots will be zero if not specified. Plant N uptake has to be provided and is defined as the total amount during the growing season. The user must specify:

- Length of growing season (days)
- Total N uptake during the growing season (kg N ha^{-1})
- Tree height (cm)
- Maximum leaf area index
- An estimate of the amount and C/N ratio of dead leaves at the end of growing season (kg ha^{-1})
- An estimate of when winter dormancy ends (day of the year)
- Maximum rooting depth (cm)
- Dead root biomass and C/N ratio at end of growing season (kg ha^{-1})

The Generic Plant Growth Parameters

The generic plant growth module can be calibrated for any annual crop. However, it only simulates the growth of roots, stems, leaves, and grains. There is no simulation of leaf number or yield components. Crop parameters are in a text file called PLGEN.DAT with a few variables allowed to be calibrated by users in the Window interface. Table 1–10 shows the 10 parameters recommended for

Table 1–10. Crop growth-related model parameters in RZWQM.DAT for users to calibrate RZWQM2.

Parameter name	Definition
CNUP1	Maximum daily N uptake (g plant^{-1} d^{-1}) used in the Michaelis–Menten equation. Increasing this parameter causes an increase in active uptake, which will result in an increase in yield.
ALPHA	Proportion of photosynthate used for maintenance respiration. Increasing this parameter results in a decrease in biomass.
CONVLA	Conversion factor from biomass to leaf area index (g LAI^{-1}). Increasing this parameter causes a decrease in total plant production.
CLBASE	Plant density on which CONVLA is based (no. plants ha^{-1})
SLA3	Factor to reduce photosynthetic rate at end of reproductive stage. This parameters along with SLA3 will adjust harvest index.
SLA4	Factor to reduce photosynthetic rate at seed production stage (0–1). Same as SL3 used to adjust harvest index.
RDX	Maximum rooting depth under optimal conditions (cm). Increasing rooting depth will promote early penetration of roots and generally increase total plant production.
RST	Minimum leaf stomatal resistance (s m^{-1}). Increasing this parameter decreases potential evapotranspiration.
SUFNDX	Nitrogen sufficiency index threshold below which automatic fertilization is triggered (0–1).
EFFLUX	Plant luxurious nitrogen uptake efficiency factor (0–1).

calibrating at a site. However, calibrating only these 10 parameters does not always result an adequate calibration unless some additional parameters in PLGEN.DAT are also modified using a text editor provided in the user interface. The CO_2 effect on photosynthesis is not explicitly simulated in the generic plant model (Hanson, 2000).

PHENOLOGY. The generic plant growth module defines five growth stages: germination, emergence, 4-leaf stage, end of vegetative growth, and end of reproductive growth. Phenology is determined by the minimum days or thermal time (growing degree days) between the five stages in PLGEN.DAT. Calibrating these parameters will also affect yield and crop growth because many processes are phenology dependent. Phenology is also affected by water and N stresses. Germination parameters should also be provided in the file.

LEAF AREA INDEX. Maximum leaf area index (LAI) is specified in PLGEN.DAT. Experienced users can also adjust root/shoot and leaf/shoot ratios to calibrate LAI for a crop. Of course, duration of vegetative growth also affects LAI. The most sensitive parameter is the specific leaf weight in Table 1–10. Increasing this parameter will decrease LAI.

BIOMASS. After phenology is correct, users can calibrate biomass by adjusting some of the parameters in Table 1–10. More maintenance respiration will reduce biomass production, so does a shallow rooting depth under water or N stress conditions. Less growing days in the vegetative stage will also reduce total biomass. Parameters related to photosynthesis (light extinction coefficient, maximum photosynthetic rate, and light use efficiency) can greatly affect biomass production. It is always a good idea to check for N and water stress factors before adjusting the growth parameters (PLANT2.PLT). When adjusting growth parameters is inevitable, calibrate only those in doubt and use values from literature as much as possible.

GRAIN YIELD. he generic model partitions photosynthate to grain using an aging factor (Table 1–10). Another two parameters in PLGEN.DAT related to yield formation are "proportion of photosynthate to propagule" and "proportion of propagule to seed" in PLGEN.DAT. When there is inaccuracy in predicting yield, biomass should be checked first along with water and N stresses.

PLANT HEIGHT. An exponential function is used to scale plant height from shoot biomass (leaf and stem) (Eq. [21]). Therefore, correct shoot biomass simulation is the key for accurate simulation of plant height.

$$\text{Height} = H_{\max}\left(1 - e^{-\frac{\alpha \text{SH}_{\text{bio}}}{2H_{\max}}}\right) \qquad [21]$$

where SH_{bio} is shoot biomass simulated in the model, Height is canopy height (cm), and $H_{\max}$ is the maximum plant height at maturity, and α is a constant that may be estimated from the shoot biomass at one-half $H_{\max}$, which is listed in PGLEN.DAT.

WATER UPTAKE (TRANSPIRATION). Water uptake is determined by several factors. Users should be careful in identifying the real reasons for poor simulation of transpiration. The suggested steps would be to check (i) PET calculation, (ii) soil water balance, and (iii) rooting depth. The Generic plant model has a parameter for maximum rooting depth (Table 1–10). When transpiration is low, possible reasons are low LAI, low soil moisture supply, low K_{sat}, and too shallow rooting depth. The model assumes the lower limit of plant available water is at 15 bar (θ_{15}). Plant water uptake can only take place during redistribution period after infiltration. Soils having high water holding capacity above θ_{15} should have higher available water for plant extraction. Surface energy balance may also need to be improved using the SHAW or PENFLUX modules.

NITROGEN UPTAKE. RZWQM2 has two N uptake mechanisms: a primary passive uptake based on N concentration in transpiration water and, if not sufficient, a secondary active uptake based on soil N concentration via the Michaelis–Menton equation (Hanson, 2000). Water uptake plays a major role in N uptake. When there is a problem with N uptake, one should check (i) actual transpiration, (ii) soil N availability in the root zone, (iii) plant growth, (iv) rooting depth, and (v) maximum daily N uptake rate (Table 1–10). If users have no measured plant N uptake, they should find an approximate N uptake for the produced biomass to compare with modeling results. The ratios of N in grain to total aboveground N uptake for several crops are listed in Table 1–11, and N fixation is shown in Table 1–12.

Plant N demand is critical for N uptake as well (Ma et al., 2008). The generic plant growth module has parameters to quantify minimum and maximum N concentrations in each plant tissue, root/shoot ratio, and leaf/shoot ratio. Calibrating these ratios in the PLGEN.DAT file affects N uptake as well as plant growth. The DSSAT model also has plant N concentrations in the species file.

Calibrating DSSAT-CSM Models in RZWQM2

The DSSAT-CSM incorporated into RZWQM2 has enhanced the application of RZWQM2 over the last few years. The most common method for calibrating cultivar parameters of the CSM-DSSAT crop models is through manual adjustment (trial and error) of the parameters to match the simulated outputs closely with the field-measured data (Boote et al., 1997, 1998). A more objective and systematic procedure for parameter estimation is to use optimization techniques (Mavromatis et al., 2002; Hunt et al., 1993; Anothai et al., 2008; Jones et al., 2011), which is not available in RZWQM2. Notwithstanding the methods used for parameter estimation, a minimum dataset is needed to obtain cultivar-specific parameters that are stable (Boote et al., 1998). Records of phenological observations and yield and yield components from variety trials are well suited for determination of cultivar coefficients (Mavromatis et al., 2002). However, growth analysis data from crop trials or other experiments conducted under well-watered conditions (with minimum constraints due to water deficit) are ideal for calibration of the cultivar parameters (Table 1–4).

The timings of anthesis and maturity are essential minimum data needed to calibrate the phenological component of the model, and the final yield and yield components at harvest are essential for calibrating the biomass growth

Table 1–11. Ratio of grain N to total aboveground N uptake (Schepers and Mosier, 1991).

Crop	Mean	Range
Winter wheat	0.80	0.74–0.86
Grain sorghum	0.58	0.52–0.64
Corn	0.71	0.64–0.78
Soybean	0.70	0.55–0.85

Table 1–12. Rate of biological N_2 fixation for various legumes (Schepers and Mosier, 1991).

Legume crop	N_2 fixed
	kg^{-1} ha^{-1} yr^{-1}
Soybean	
Midwestern conditions	57–94
Southeastern conditions	73–218
Cowpea	84
Clover (*Trifolium*)	104–160
Alfalfa (*Medicago sativa* L.)	128–336+
Lupin (*Lupinus angustifolius* L.)	150–169

and yield components. The CERES and CROPGRO crop models supported in RZWQM2 have different numbers of cultivar coefficients. These parameters broadly fall under two categories: phenological and growth parameters. The first step in calibration of all CSM-DSSAT crop models in RZWQM2 is to select a cultivar in the "cultivar file" that in general is similar to the cultivar of your experiment or predicts anthesis and maturity close to the observed data, as well as yield and yield components (Boote, 1999). When calibration is needed, it is suggested to vary the value within the range given in the cultivar database that is distributed with DSSAT for each crop. Correct simulation of LAI is critical for light interception, ET, and photosynthesis. The soil fertility factor (SLPF) and soil root growth distribution (SRGF) may also need to be adjusted to further improve simulations (Malone et al., 2010). After calibrating for unstressed conditions, a sensitivity analysis of the calibrated model to management practices is recommended, especially to water and N treatments. Users may also want to check upper drain limit (DUL) and lower limit of plant available water (LL), which are assumed to be soil water content at 1/3 and 15 bar suction. The amount of available soil water (DUL-LL) is often underestimated by pedotransfer functions proposed in the literature and also used in DSSAT, possibly because there is considerable "usable" water in soil above 1/3 bar (Boote, 2011, personal observation).

Table 1–13. Genetic coefficients used in the CSM-CROPGRO model for the simulations of soybean, drybean, fababean, velvetbean, cabbage, chickpea, cowpea, peanut, pepper, tomato, canola, bahiagrass, brachiaria, and cotton. (Example values are representative of soybean.)

Acronyms used and definitions of traits		Units	Range
CSDL	Critical short day length below which reproductive development progresses with no daylength effect (for short day plants)	h	11–15
PPSEN	Slope of the relative response of development to photoperiod with time (positive for short day plants)	h^{-1}	0.129–0.349
EM-FL	Time between plant emergence and flower appearance (R1)	ptd†	15–29
FL-SH	Time between first flower and first pod (R3)	ptd	5-10
FL-SD	Time between first flower and first seed (R5)	ptd	11–16
SD-PM	Time between first seed (R5) and physiological maturity (R7)	ptd	27–38
FL-LF	Time between first flower (R1) and end of leaf expansion	ptd	15–35
LFMAX	Maximum leaf photosynthesis rate at 30°C, 350 ppm CO_2, and high light	mg CO_2 m^{-2} s^{-1}	0.9–1.4
SLAVR	Specific leaf area of cultivar under standard growth conditions	cm^2 g^{-1}	300–400
SIZELF	Maximum size of full leaf (three leaflets)	cm^2	140–230
XFRT	Maximum fraction of daily growth that is partitioned to seed shell		1.0
WTPSD	Maximum weight per seed	g	0.153–0.195
SFDUR	Seed filling duration for pod cohort at standard growth conditions	ptd	17–26
SDPDV	Average seed per pod under standard growing conditions	no. seed pod^{-1}	1.9–2.4
PODUR	Time required for cultivar to reach peak pod number under optimal conditions	ptd	8–12

† ptd, photothermal days.

Calibration of Cultivar Parameters for CROPGRO Models

The CROPGRO model was originally developed to simulate legume crops, but it has also been parameterized for nonlegume crops. In general, seven cultivar parameters (CSDL, PPSEN, EM-FL, FL-SH, FL-SD, SD-PM, and FL-LF) define the crop life cycles, and eight parameters (LFMAX, SLAVR, SIZELF, XFRT, WTPSD, SFDUR, SDPDV, and PODUR) define the growth traits (biomass assimilation and partitioning) of the crops (Table 1–13). Calibration of CROPGRO is conducted in the order of phenology, LAI and biomass growth, and yield and yield components (Boote, 1999).

Phenology

1. Select a maturity group based on the cultivar in the experiment, and compare simulated and measured flowering and maturity dates.

2. Adjust the EM-FL to match the date of flowering with that measured. If the value of the parameter fails to stay within the range of values for all the cultivars in the database, you may also have to adjust the CSDL and PPSEN parameters, depending on the photoperiod sensitivity of the crop and/or cultivar. An increase in EM-FL delays flowering.
3. Adjust the SD-PM to match simulated and measured date of maturity. If the value of the parameter fails to stay within the range of values for all the cultivars in the database, one may also adjust the CSDL and PPSEN parameters for daylength-sensitive crops. However, note that this will also affect the time from EM-FL. An increase in SDM-PM delays maturity.
4. If CSDL and PPSEN were adjusted, recheck the simulation of the anthesis date and make adjustments in EM-FL if needed.

Growth—Leaf Expansion and Biomass

1. If the slope of the simulated biomass progression with time is too rapid or slow, as in Fig. 1–8, adjust SLPF in the specific DSSAT soil profile if you think that poor soil fertility has contributed to the observed mismatch in simulated and observed biomass. Please note that SLPF is defined as a soil profile input parameter and is site specific.
2. If there are two cultivars that show different rates of dry matter accumulation when grown on the same soil, and the slope of the simulated biomass progression with time is too rapid or slow, then you can adjust LFMAX, the cultivar-specific light-saturated leaf photosynthesis rate. If you have only one cultivar, adjust SLPF.
3. To match the simulated and measured LAI, increase the SLAVR parameter if the simulated SLA is too high or decrease SLAVR if the simulated SLA is too low. If the peak in the simulated LAI growth with time does not match the measured LAI, increase or decrease FL-LF to delay or accelerate the time of maximum LAI, respectively. The SIZELF parameter can be used to increase or decrease leaf area expansion during early vegetative stage of development of the plant when leaf area expansion is "sink-limited." If the cultivar is a dwarf type, the RWDTH (relative width) and RHGHT (relative height) parameters in the "ecotype" might need adjustment.
4. At this stage, go back and readjust SLPF (or LFMAX) if necessary for better simulation of biomass.

Yield

Correct information for WTPSD and SDPDV is needed for matching final seed size and seeds per pod. If measured information for these parameters is not available, try to obtain approximate data from the literature or from the seed companies and refine to match simulation with measured data for final seed size and seeds per pod. (The model is intentionally not very sensitive to WTPSD or

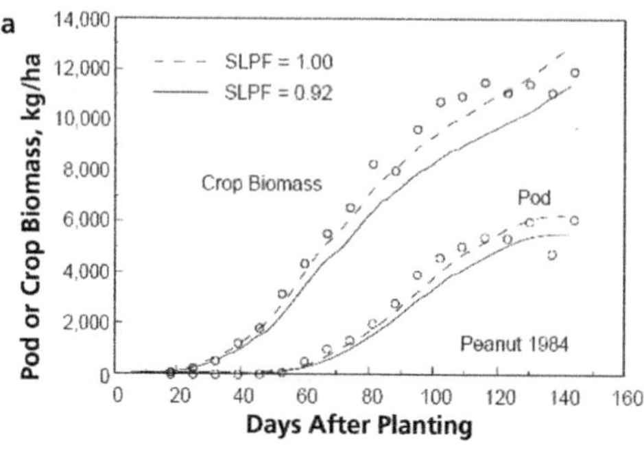

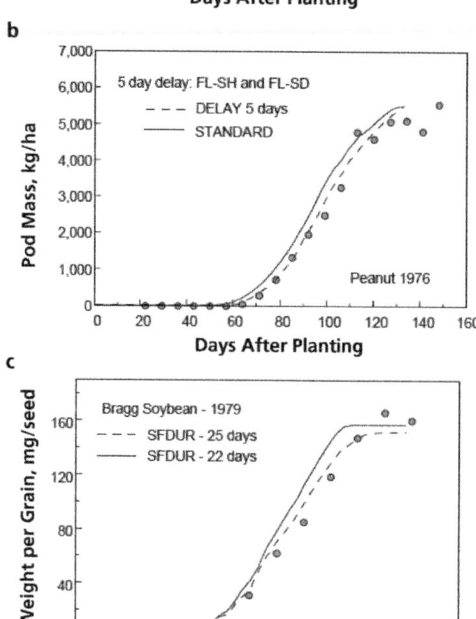

Fig. 1–8. Comparison between simulated and measured peanut biomass and pod yield progression with time (a) at two values of SLPF, and for a delay of 5 d in (b) both FL-SH and FL-SD (middle), and (c) soybean seed size for two different values of SFDUR (Boote, 1999).

SDPDV.) Also, provide a first approximation for the SFDUR, which should be about 70 to 80% of the SD-PM.

1. To match the pod and seed dry weights and their growth pattern with the measured data, adjust PODUR, FL-SH, and FL-SD. Decrease FL-SH and FL-SD to start pod and seed growth sooner, and increase to delay (Fig. 1–8). In general, both parameters need to be changed in the same direction because the duration FL-SH to FL-SD sets the lag from start of first pod to start of seeds for all cohorts of pods, and as such, decreasing FL-SH for early pod and increasing FL-SD to delay the start of seed for one cohort would slow down the process for all of them. Reducing PODUR will cause the number of pods and seeds to be set more rapidly, as in a more determinate manner.

2. Re-adjust SD-PM if necessary to match simulated and measured physiological maturity periods.
3. At this stage, go back and re-adjust parameters for seed size and shelling percentage. In the CROPGRO model, seed growth becomes limited as cohorts reach their maximum shelling percentage (THRESH in "eco" file). Make sure that the simulated shelling percentage does not exceed the value for THRESH. If simulated shelling percentage is too high, increase SFDUR until the shelling percentage comes down. The WTSPD may be adjusted to obtain the correct final seed size. If simulated shelling percentage and seed size through time are too high and reach a maximum long before maturity, increase SFDUR, which will decrease the slope of dry matter increase per seed and, in fact, sets single seed growth rate (Fig. 1–8).
4. If the simulation for phenology, seed size, shelling percentage, biomass, LAI, and grain yield match the observed data, the calibration of the cultivar parameters is complete. If there are still inaccuracies, make adjustments based on the procedures described previously.

Nitrogen uptake is the minimum of soil N supply based on soil NH_4 and NO_3 concentrations and N demand based on daily crop growth. Soil N supply from each soil layer is a function of soil water contents (DUL, LL, and θ_s), RLV, and concentrations of NH_4 and NO_3. Crop N demand is based on given N or protein concentrations in each plant tissue, which is specified in the species file. When there is a discrepancy between measured and simulated N uptake, users should check for soil N supply first before working on any of the parameters in the species file. Nitrogen fixation parameters are in the species file, and users should be reluctant to modify these parameters, as the model is very robust at fixing N needed to make up the deficit of soil mineral N not taken up by roots. Table 1–12 lists some estimated N fixation values.

Experienced users may calibrate the parameters in the species and ecotype files when there is strong evidence to do so. These parameters are used to simulate photosynthesis, respiration, partitioning of assimilates among plant tissues and yield components, leaf growth and senescence, seed and shell growth, N fixation, and root growth.

Calibration of Cultivar Parameters for CERES Models

CERES has slightly different cultivar parameters for different crops (Ritchie et al., 1998). Growth and development of crops in the CERES-Maize (corn) model is specified with six cultivar parameters, i.e., P1, P2, P5, G2, G3, and PHINT, as defined in Table 1–14.

Table 1–14. Cultivar coefficients used in the CSM-CERES-Maize model for simulations of corn.

Acronyms used and definitions of traits		Units	Range
P1	Degree days (base temperature of 8°C) from seedling emergence to end of juvenile phase.	ptd†	100–450
P2	Daylength sensitivity coefficient [the extent (days) that development is delayed for each hour increase in photoperiod above the longest photoperiod (12.5 h) at which development proceeds at maximum rate].		0–2
P5	Degree days (base temperature of 8°C) from silking to physiological maturity	ptd	600–1000
G2	Potential kernel number	kernel no. plant^{-1}	440–1000
G3	Potential kernel growth rate	mg/(kernel d)	5–16
PHINT	Degree days required for a leaf tip to emerge (phyllochron interval)	ptd	38–55

† ptd, photothermal days.

The sorghum model includes proso millet and foxtail millet as well as sorghum, while there is a separate model for pearl millet. Growth and development for these crops are specified with seven cultivar parameters, including P1, P2O, P2R, P5, G1, G2/G4, and PHINT, as defined in Table 1–15. Five parameters, including P1, P2O, P2R, P5, and PHINT, affect phenological development, while the remaining parameters, including G1 and G2/G4 define the growth traits, including biomass assimilation and partitioning.

CERES-Wheat also includes barley and triticale. Growth and development of these crops in the CSM-CERES models are specified with seven cultivar parameters P1D, P1V, P5, G1, G2, G3, and PHINT. The parameters P1D, P1V, P5, and PHINT define the crop life cycles, while the parameters G1, G2, and G3 define the growth traits (biomass assimilation and partitioning) (Table 1–16).

When calibrating the CERES models, like CROPGRO, phenology should be calibrated first followed by biomass, LAI, and yield. Calibration of these models is generalized as follows:

Phenology

1. Select a cultivar closest to the variety used, and compare simulated versus measured flowering and maturity dates.
2. Get a reasonable match between simulated versus measured anthesis and physiological maturity dates by adjusting P1, P2, and P5 within the range available in the cultivar parameter database (cultivar file) or elsewhere in the literature. An increase in P1 and P5 in general will delay flowering and maturity.

Table 1–15. Cultivar coefficients used in the CSM-CERES model for simulations of pearl millet, proso millet, foxtail millet and sorghum.

Acronyms used and definitions of traits		Units	Range
P1	Degree days from seedling emergence to the end of the juvenile phase (base temperature of 10 ºC) during which the plant is not responsive to changes in photoperiod.	ptd†	100–500
P2O	Critical photoperiod or the longest day length (in hours) at which development occurs at a maximum rate. At values greater than P2O, the rate of development is reduced.		12.5–16.0
P2R	Extent to which phasic development leading to panicle initiation (expressed in degree days) is delayed for each hour increase in photoperiod above P2O.		1–300
P5	Degree days (base temperature of 10 ºC) from beginning of grain filling (3-4 days after flowering) to physiological maturity.	ptd	300–650
G1	Scaler for relative leaf size.		0–20
G2	Scaler for partitioning of assimilates to the panicle (head) (for proso millet, foxtail millet and sorghum).		1–8
G4	Scaler for partitioning of assimilates to the panicle (head) (for pearl millet)		0–1.0
PHINT	Degree days required for a leaf tip to emerge (phyllochron interval)	ptd	38–55

† ptd, photothermal days.

Table 1–16. Cultivar coefficients used in the CSM-CERES model for simulations of wheat, barley and triticale.

Acronyms used and definitions of traits.		Units	Range
P1D	Percentage reduction in development rate when photoperiod is shorter than the threshold of 20 h (% reduction/h near threshold).		1–100
P1V	Days at optimum vernalizing temperature required to complete vernalization.	d	0–60
P5	Grain filling (excluding lag) phase duration	ptd†	170–600
G1	Kernel number per unit stem + spike weight at anthesis	no. g^{-1}	15–30
G2	Standard kernel size under optimum conditions	mg	20–60
G3	Standard, nonstressed dry weight (total, including grain) of a single tiller at maturity	g	1.0–2.5
PHINT	Thermal time between the appearance of leaf tips (8°C d).	ptd	60–100

† ptd, photothermal days.

3. For maize, adjust the daylength sensitivity for those cultivars where P2 > 0.
4. For sorghum and millet models, adjust the photoperiod coefficients P2R and P2O.
5. For wheat, adjust the coefficients P1D for daylength sensitivity and P1V for vernalization.
6. Recheck the predicted flowering and maturity dates, and adjust the cultivar parameters if needed.

Growth—Leaf Expansion and Biomass

1. Plant biomass can now be calibrated in the CERES models by adjusting the SLPF value in the soil file as described above for the CROPGRO crops. The RUE (radiation use efficiency) value is now available in the ecotype file, but is a stable parameter that is not recommended for calibration except in the case of reducing RUE for inbreeds or open-pollinated maize. Be sure that LAI is simulated correctly, as this influences light capture and dry matter gain. PHINT has some impact on LAI, but also impacts time of anthesis.
2. Adjust RUE in the ecotype file for maize or PARUV and PARVR in the ecotype file for wheat, but only where you know hybrids or cultivars differ in dry matter accumulation when grown on the same soil.

Yield

1. Define the potential kernel growth rate (mg d^{-1}) by setting G3 to match the final observed kernel size, as it is critical for the correct simulation of yield and yield components. An increase in G3 will increase grain yield and vice versa.
2. For maize, the last step in calibrating grain yield is to adjust maximum potential kernel number per plant G2. An increase in G2 will increase grain yield, and a decrease will reduce yield. Note that adjusting SLPF will also influence grain yield, mostly through kernel number.
3. For sorghum and millet, grain yield can be calibrated by adjusting the "scaler for partitioning assimilates to panicle" (G2 for pearl millet, and G4 for proso millet, foxtail millet, and sorghum).
4. For wheat G1 and G2 should be calibrated with further improvement by adjusting G3.

Plant height is not simulated in CERES-Maize and CERES-Sorghum models. In RZWQM2, an empirical equation was developed based on biomass from the studies in the Central Great Plains by Dr. David Nielsen (personal communication) from Eq. [21]. H_{max} is 246 cm for maize and 93 cm for sorghum and millet. The α values are 0.0636 for maize; 0.222 for sorghum, foxtail millet, and proso millet; and 0.042 for pearl millet. In CERES-Wheat, daily plant height increment is calculated from a user-provided standard height (in the ecotype file) and growth stage.

However, it should be noted that calibrated plant parameters may not transferrable to other locations because (i) there are many parameters sets that can provide acceptable simulation for a calibration dataset, (ii) the calibrated parameters may reflect the interaction between N and water stresses, and (iii) inputs of other plant and soil parameters may affect the calibrated parameters. Unless the plant parameters are well evaluated under other experimental conditions (water and N treatments) and in multiple years, one should not expect to transfer the calibrated plant parameters to other locations. In addition, methods of estimating

Table 1–17. Pesticide washoff parameters (Wauchope et al., 2000).

Pesticide solubility range	Washoff power	Examples
<10	0.002	pyrethroid, organochlorine
10–100	0.005	organophosphorus
100–1000	0.013	–
1000–10^4	0.033	carbamate,organophosphorus
10^4–10^5	0.085	organophosphorus
>10^5	0.2	organophosphorus

K_{sat}, PET, and other soil parameters should be the same among locations for the plant parameters to be transferrable.

Calibrating Pesticide Parameters

Pesticide parameters are provided in RZWQM2 from a database compiled by Wauchope et al. (1992), which defines washoff from leaves and crop residue, soil degradation, soil adsorption, and irreversible binding. Washoff is defined by the following equation:

$$P = P_0 F e^{-w*R} \quad [22]$$

where P_0 is the total amount (μg/g) and F is washable fraction of pesticide on leaves or crop residue (100% as default), w is a washoff power factor (see Table 1–17), and R is rainfall (mm) for each storm event. Due to washoff and spatial variability, initial pesticide amount applied onto the soil surface may be unknown. Therefore, measuring or calibrating initial pesticide amount at time = 0 should be provided. Degradation may occur at plant canopy, surface residue, and in soil layers. Degradation rate in each soil layer is generally proportional to the amount present (first order) and determined by the half-life of the pesticide. Pesticide degradation rate in soil can be adjusted for soil moisture and soil depth. Degradation pathways (e.g., anaerobic, aerobic) in each component (e.g., soil, canopy, and residue) are lumped together to give an overall degradation constant. When the pesticide molecules are electrically charged, users may supply an adsorption constant for the anion or cation. The model calculates an overall equilibrium adsorption constant for the pesticide. The model also has an option for using kinetics of adsorption–desorption, in addition to the equilibrium assumption. Users may also provide a diffusion coefficient to control mass flux between immobile micropores and mobile water. For those who are interested in pesticide uptake, an octanol–water partition coefficient (K_{ow}) is needed. Calibration of the pesticide component should be done after soil water, soil N, and plant growth were calibrated satisfactorily (Fig. 1–9).

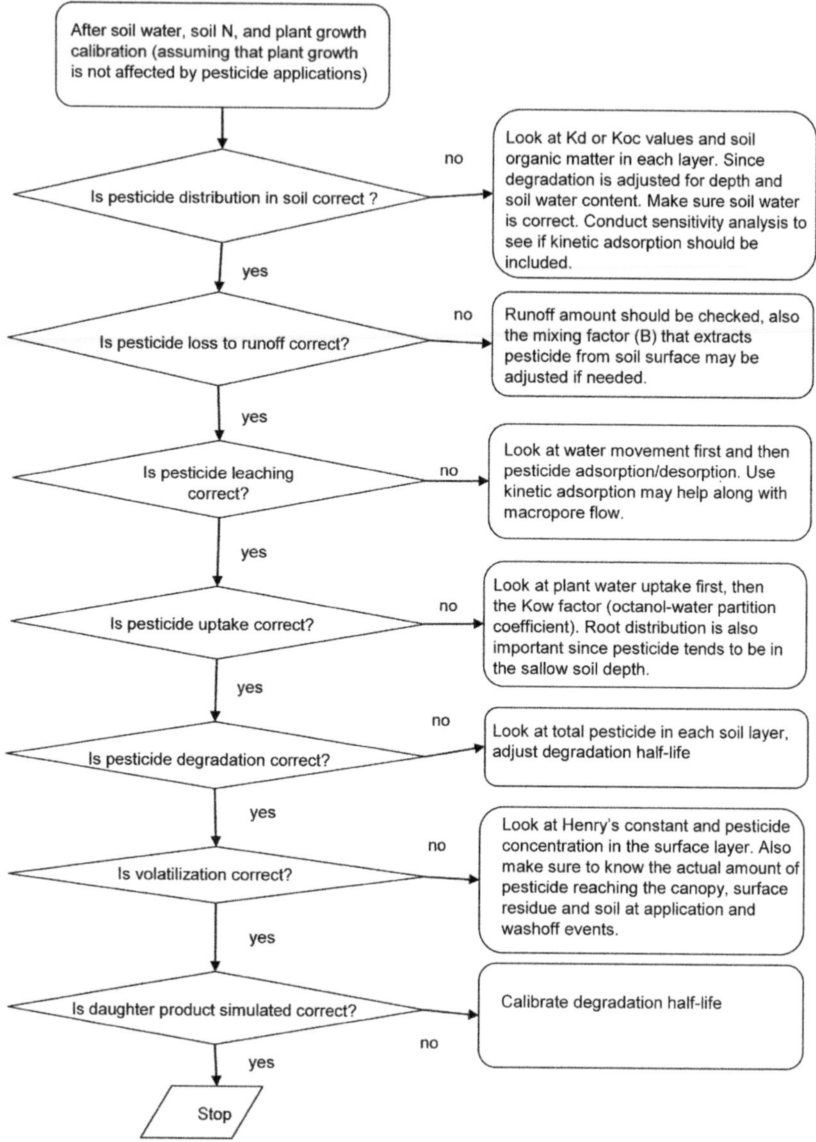

Fig. 1–9. Recommended calibration steps for pesticide in RZWQM2.

1. Select a pesticide name from the pesticide database or define your own name. A default set of pesticide parameters is then loaded into the model.
2. Define parent or daughter or granddaughter products if necessary. When daughter or granddaughter is defined, a pathway from parent to daughter

or granddaughter should be defined along with a percentage of degradation products that is the daughter or granddaughter.

3. Define half-lives for degradation processes for each component (foliar, surface residue, soil surface, and soil subsurface).
4. Define the conditions (soil pH, soil temperature, soil water content) where the above degradation half-lives are measured. Also, provide the Walker constant to adjust the degradation half-lives with soil moisture.
5. Define the octanol carbon partition constant (K_{oc}) values for each species of the pesticide (neutral, cationic, or anionic), half-life of absorbed pesticide for irreversible pesticide degradation, kinetic absorption–desorption constants, diffusion rate between micropores and mesopores, and octanol–water partitioning coefficient for plant pesticide uptake.
6. When a slow release form of a pesticide is applied, a proportion of daily pesticide release should be provided.
7. When a pesticide is applied to a plant canopy or on surface residue, wash-off coefficient should be defined.
8. Henry's law constant is needed for volatile pesticide.

Soil Pesticide Distribution

It is generally difficult to know exactly how much pesticide is applied to a soil due to canopy interception, drifting, spatial variability, and quick volatilization from the air and soil surface. It is good to put filter papers on the soil surface to measure the amount applied to the soil. Otherwise, comparing the patterns of measured and simulated relative amounts and pesticide distributions in the soil are more appropriate than matching the exact amount in each soil layer. Parameters that affect pesticide movement in soils are adsorption–desorption constants (K_d values) and degradation half-lives. The soil surface has other pesticide dissipation mechanisms such as volatilization (from Henry's law constant or user defined) and photodegradation. Soil water movement and plant pesticide uptake are also important mechanisms in defining a pesticide distribution in the soil profile. Including kinetics in pesticide adsorption may improve simulation results as well (Fig. 1–10).

1. Check pesticide input to make sure that drifting is not an issue during application.
2. Check soil water movement before adjusting pesticide parameters.
3. Calibrate K_{oc} value first, and then add kinetics if necessary.
4. Depending on recovery of pesticide from the soil, degradation and irreversible adsorption may be needed to provide correct pesticide soil simulation.

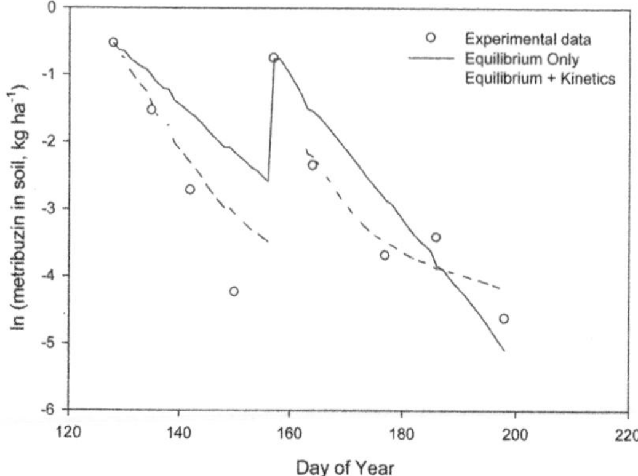

Fig. 1–10. Effects of kinetic adsorption on metribuzin disappearance in soils (Malone et al. 2004b).

5. The Walker constant may need to be modified to adjust for soil moisture difference.
6. Check if it is necessary to modify degradation constants with soil depth.
7. If plant uptake is simulated, check the octanol–water partitioning coefficient (K_{ow}) and root distribution (by adjusting SRGF).
8. It is important to match the first measurement of soil pesticide concentration, preferably immediately after application so that drifting or application efficiency can be calculated.

Pesticide Leaching and Runoff

The total amount of pesticide leaching from the soil profile is affected primarily by soil water movement, degradation, and adsorption. The irreversible binding of pesticides is also a major sink term in pesticide balance. Generally there are many parameters that can be calibrated for a given experiment. However, users should refer more to literature values to make the calibration credible. Absorption kinetics should also be considered if pesticide leaching is too fast. The pesticide runoff amount is determined by runoff water and pesticide concentration in the top 2 cm of the soil profile, and adjusting the nonuniform mixing factor (*B*) may help simulate pesticide concentrations in runoff simulation. Degradation, volatilization, adsorption, and leaching of pesticide affect pesticide concentration in the top soil layer.

Model Responses to Water, Nitrogen, and Temperature Variations

Table 1–18 provides examples of model parameters that have been calibrated for various model outputs to be compared with experimental measurements. Malone et al. (2001a, 2004a) also reviewed parameterization of RZWQM for various processes and associated parameters. However, these calibrations are highly

Table 1–18. Examples of calibration in RZWQM applications.

Parameters calibrated	Measurements to match	Authors
K_{sat}, θ_s, macroporosity, fraction of dead macropores	Water redistribution and infiltration	Cameira et al. (2000, 2005) Kumar et al. (1998)
Stomatal and soil resistance, maximum rooting depth	Evapotranspiration	Cameira et al. (2005) Ma et al. (2003)
Soil C pool size, nitrification and urea hydrolysis rate coefficients	Total soil C, Soil nitrate distribution and plant N uptake	Cameira et al. (2007)
Interpool transfer coefficients among soil C pools (R_{14}, R_{23}, R_{34}, R_{45})	N loss in tile flow, soil N distribution	Malone et al., (2011 Ma et al. (1998a,b,c)
Lateral hydraulic gradients and lateral K_{sat}, N_2, λ_2	Tile flow	Malone et al. (2010) Ma et al. (2007b)
K_{sat}, λ_2, bulk density	Percolate from soil block, Soil moisture distribution	Malone et al. (2003) Sophocleous et al. (2009)
K_{sat} for crusted soil surface	Surface runoff	Ma et al. (1995, 2004)
$\theta_{1/3}$ or field capacity θ_{15} or wilting point	Soil moisture	Ma et al. (2003) Hu et al. (2006) Fang et al. (2010b)
Porosity and field capacity	Subsurface drainage	Kumar et al. (1998) Bakhsh et al. (2004)
K_{oc} and pesticide half live	Pesticide loss in tile drainage	Bakhsh et al. (2004)
Macroporosity	Till and no-till effects on pesticide leaching	Malone et al. (2004a)
Soil root growth factor (SRGF)	Plant growth, soil moisture distribution	Malone et al. (2010) Fang et al. (2010b)
Lateral sorptivity factor	Pesticide leaching into tile flow	Kumar et al. (1998) Malone et al. (2001a)
Nonuniform mixing factor (*B*)	Chemical in runoff	Ghidey et al. (1999) Malone et al. (2001a)
Soil horizon depth	Soil moisture and N	Martin and Watts (1999)
Albedo	Soil moisture	Nokes et al. (1996)
Initial soil moisture if not measured	Pesticide soil distribution Tile flow	Azevedo et al. (1997) Singh et al. (1996)
Washoff factor and macropore radius	Pesticide leaching	Malone et al. (2001a)

dependent on experience of the users and data availability. This calibration process is not complete either until the model is evaluated for its performance against some additional experimental data and for its responses to water, N, and temperature stresses. This can be done by simulating different water, N, and temperature treatments in the study or by a sensitivity analysis. Such a sensitivity analysis should reduce the degree of freedom in model parameterization because there could be many sets of parameters that produce similar accuracy in simulating one treatment. A long-term simulation with generated or recorded weather is also needed to test the stability of the model and simulation results (e.g., soil organic C). Since there are many sets of model parameters that can provide acceptable calibration for a dataset, evaluating the calibrated model for other treatments or under other conditions is essential.

When the model is not responsive to additional water and N management, users should check if the phenology is simulated correctly under different treatments and then look at simulated water and N stresses in output files to make sure the stresses are simulated correctly for the treatments. Variables to look for are potential and actual transpiration. The generic plant growth model also has a water stress index (WSI) that can be used to adjust plant responses to water stresses (Kozak et al., 2006).

RZWQM2 has the capability to modify management practices to conduct sensitivity analysis of the model to fertilization and irrigation and to modify weather driving variables such as air temperature, solar radiation, rainfall, relative humidity, and atmosphere CO_2. These features facilitate sensitivity analysis but are not automated.

No Convergence and Mass Balance Issues

As the Richards equation is used during water redistribution, no convergence can happen when unusual soil hydraulic properties are used in the model or when there is too much difference between soil hydraulic properties of adjacent soil layers. This problem may be solved by inserting an intermediate soil layer between the two distinct layers. Another known cause of mass balance is due to too high or too low rainfall intensity. When there is a very small amount of rain in a few minutes, the rainfall intensity is unreasonably high. On the contrary, when there is a small rainfall event over many hours, the rainfall intensity can be unreasonably low. Both cases can cause mass balance problems. Users may adjust the rainfall duration to avoid mass balance problems for those days.

It is also known that these problems occur more often when there are water table and tile flow simulations. If that is the case, users may want to pay attention to water table leakage rate and lateral hydraulic gradient for simulating pseudo-lateral flow below the tile. Also, field saturation should be set to 1.0.

Conclusions

It is a constant challenge for experimental scientists to use and incorporate system models in their research. This step-by-step procedure was developed for RZWQM2 to assist and encourage more users of agricultural system models. The effort to apply a system model to a given field site can be great, as are the potential benefits. Keeping these principles in mind will make model application less frustrating:

- Modeling is a scientific discipline by itself and requires education and training.
- A successful modeling study requires accurate characterization of the agricultural system.
- Calibration is an iterative process, and all simulated system components should be critically reviewed.
- There will be uncertainty in model inputs and model parameters due to experimental error, lack of understanding in our current knowledge, spatial–temporal variability, and unknown history of crop and soil management prior to the experiment.

There will be errors in simulation results, but hopefully within the experimental error tolerance. It should be kept in mind that models are simplifications of the real complex agricultural system and may never be perfect. There may always be factors that cannot be represented in the model, and the complex interactions may never be fully understood.

Despite these shortcomings, a model that is calibrated and evaluated for an experimental site with good field data can greatly enhance and extend agricultural research, education, and extension (Ma et al., 2007a). The benefits of using a modeling tool are (i) to quantify the interactions among system components, (ii) to help design better and well-balanced field experiments and data collection, (iii) to extend limited field results beyond the experimental period and site and to project system behavior under future weather and management conditions, (iv) to conduct virtual experiments for new hypotheses before conducting costly experiments (Boote et al., 2001), (v) to identify knowledge gaps in integrating all system components, and (vi) to suggest possible soil–crop processes that may have occurred parallel to some observed processes.

References

Abrahamson, D.A., D.E. Radcliffe, J.L. Steiner, M.L. Cabrera, D.M. Endale, and G. Hoogenboom. 2006. Evaluation of the RZWQM for simulating tile drainage and leached nitrate in the Georgia Piedmont. Agron. J. 98:644–654.

Abrahamson, D.A., D.E. Radcliffe, J.L. Steiner, M.L. Cabrera, J.D. Hanson, K.W. Rojas, H.H. Schomberg, D.S. Fisher, L. Schwartz, and G. Hoogenboom. 2005. Calibration of the root zone water quality model for simulating tile drainage and leached nitrate in the Georgia Piedmont. Agron. J. 97:1584–1602.

Ahuja, L.R. 1986. Characterization and modeling of chemical transfer to runoff. Adv. Soil Sci. 4:149–188.

Ahuja, L.R., D.K. Cassel, R.R. Bruce, and B.B. Barnes. 1989. Evaluation of spatial distribution of hydraulic conductivity using effective porosity data. Soil Sci. 148:404–411.

Ahuja, L.R., D.G. DeCoursey, B.B. Barnes, and K.W. Rojas. 1993. Characteristics of macropore transport studied with the ARS root zone water quality model. Trans. ASAE 36:369–380.

Ahuja, L.R., K.E. Johnsen, and G.C. Heathman. 1995. Macropore transport of a surface-applied bromide tracer: Model evaluation and refinement. Soil Sci. Soc. Am. J. 59:1234–1241.

Ahuja, L.R., K.E. Johnsen, and K.W. Rojas. 2000a. Water and chemical transport in soil matrix and macropores. p. 13–50. *In* L.R. Ahuja et al. (ed.) The Root Zone Water Quality Model. Water Resources Publ., Highlands Ranch, CO.

Ahuja, L.R., and L. Ma. 2002. Parameterization of agricultural system models: Current Issues and techniques. p. 273–316. *In* L.R. Ahuja et al. (ed.) Agricultural system models in field research and technology transfer. CRC Press, Boca Raton, FL.

Ahuja, L.R., L. Ma, and T.R. Green. 2010. Effective soil properties of heterogeneous areas for modeling infiltration and redistribution. Soil Sci. Soc. Am. J. 74:1469–1482.

Ahuja, L.R., K.W. Rojas, J.D. Hanson, M.J. Shaffer, and L. Ma (ed.) 2000b. The Root Zone Water Quality Model. Water Resources Publ., Highlands Ranch, CO.

Ahuja, L.R., and R.D. Williams. 1991. Scaling water characteristic and hydraulic conductivity based on Gregson–Hector–McGowan approach. Soil Sci. Soc. Am. J. 55:308–319.

Aiken, R.M., G.N. Flerchinger, H.J. Farahani, and K.E. Johnsen. 1997. Energy balance simulation for subsurface and residue temperatures with incomplete cover. Agron. J. 89:404–415.

Allen, R.G., L.S. Pereira, D. Raes, and M. Smith. 1998. Crop evapotranspiration: Guidelines for computing crop water requirements. United Nations FAO, Irrigation and Drainage Paper 56. FAO, Rome, Italy.

Anothai, J., A. Patanothai, S. Jogloy, K. Pannangpetch, K.J. Boote, and G. Hoogenboom. 2008. A sequential approach for determining the cultivar coefficients of peanut lines using end-of-season data of crop performance trials. Field Crops Res. 108:169–178.

Azevedo, A.S., R.S. Kanwar, P. Singh, L.R. Ahuja, and L.S. Pereira. 1997. Simulating atrazine using root zone water quality model for Iowa soil profiles. J. Environ. Qual. 26:153–164.

Bakhsh, A., R.S. Kanwar, and L.R. Ahuja. 1999. Simulating the effect of swine manure application on NO_3–N transport to subsurface drainage water. Trans. ASAE 42:657–664.

Bakhsh, A., L. Ma, L.R. Ahuja, J.L. Hatfield, and R.S. Kanwar. 2004. Using RZWQM to predict herbicide leaching losses in subsurface drainage water. Trans. ASAE 47:1415–1426.

Boote, K.J. 1999. Concepts for calibrating crop growth models. p. 179–200. *In* G.Y. Tsuji et al. (ed.) DSSAT Version 3. Vol. 4–6. International benchmark sites network for agrotechnology transfer. Univ. of Hawaii, Honolulu, HI.

Boote, K.J., G. Hoogenboom, J.W. Jones, and K.T. Ingram. 2008. Modeling nitrogen fixation and its relationship to nitrogen uptake in the CROPGRO model. p. 13–46. *In* L. Ma et al. (ed.) Quantifying and understanding plant nitrogen uptake for systems modeling. CRC Press, Boca Raton, FL.

Boote, K.J., J.W. Jones, G. Hoogenboom, and G.G. Wilkerson. 1997. Evaluation of the CROPGRO-Soybean model over a wide range of experiments. p. 113–133. *In* Kropff et al. (ed.) Systems approaches for sustainable agricultural development: Applications of systems approaches at the field level. Kluwer Academic Publishers, Boston.

Boote, K.J., J.W. Jones, G. Hoogenboom, and N.B. Pickering. 1998. The CROPGRO model for grain legumes. p. 99–128. *In* G.Y. Tsuji, G. Hoogenboom, and P.K. Thornton (ed.)Understanding options for agricultural production. Kluwer Academic Publishers, Boston.

Boote, K.J., M.J. Kropff, and P.S. Bindraban. 2001. Physiology and modelling of traits in crop plants: Implications for genetic improvement. Agric. Syst. 70:395–420.

Boyle, D.P., H.V. Gupta, and S. Sorooshian. 2000. Toward improved calibration of hydrologic models: Combining the strengths of manual and automatic methods. Water Resour. Res. 36:3663–3674.

Buis, S., D. Wallach, S. Guillaume, H. Varella, P. Lecharpentier, M. Launay, M. Guérif, J.-E. Bergez, and E. Justes. 2011. The STICS crop model and associated software for analysis, parameterization, and evaluation. p. 395–426. *In* L.R. Ahuja and L. Ma (ed.) Methods of introducing system models into agricultural research. Advances in Agricultural Systems Modeling 2. ASA, CSSA, and SSSA, Madison, WI.

Cameira, M.R., L.R. Ahuja, R.M. Fernando, and L.S. Pereira. 2000. Evalauting field measured soil hydraulic properties in water transport simulations using RZWQM. J. Hydrol. 236:78–90.

Cameira, M.R., R.M. Fernando, L.R. Ahuja, and L. Ma. 2007. Using RZWQM to simulate the fate of nitrogen in field soil–crop environment in the Mediterranean region. Agric. Water Manage. 90:121–136.

Cameira, M.R., R.M. Fernando, L.R. Ahuja, and L.S. Pereira. 2005. Simulating the fate of water in field soil–crop environment. J. Hydrol. 315:1–24.

De Vries, D.A. 1963. Thermal properties of soils. p. 210–235. *In* W.R. van Wijk (ed.) Physics of plant environment. North-Holland Publishing Company, Amsterdam.

Doherty, J. 2010. PEST: Model independent parameter estimation. Watermark Numerical Computing, Brisbane, Australia.

Fang, Q.X., T.R. Green, L. Ma, R.W. Malone, R.H. Erskine, and L.R. Ahuja. 2010a. Optimizing soil hydraulic parameters in RZWQM2 using automated calibration methods. Soil Sci. Soc. Am. J. 74:1897–1913.

Fang, Q.X., L. Ma, Q. Yu, L.R. Ahuja, R.W. Malone, and G. Hoogenboom. 2010b. Irrigation strategies to improve the water use efficiency of wheat–maize double cropping systems in North China Plain. Agric. Water Manage. 97:1165–1174.

Farahani, H.J., and D.G. DeCoursey. 2000. Evaporation and transpiration processes in the soil-residue-canopy system. p. 51–80. *In* L.R. Ahuja et al. (ed.) The Root Zone Water Quality Model. Water Resources Publ., Highlands Ranch, CO.

Flerchinger, G.N., R.M. Aiken, K.W. Rojas, and L.R. Ahuja. 2000. Development of the root zone water quality model (RZWQM) for over-winter conditions. Trans. ASAE 43:59–68.

Gan, T.Y., E.M. Dlamini, and G.F. Biftu. 1997. Effects of model complexity and structure, data quality, and objective functions on hydrologic modeling. J. Hydrol. 192:81–103.

Ghidey, F., E.E. Alberts, and N.R. Kitchen. 1999. Evaluation of RZWQM using field measured data from the Missouri MSEA. Agron. J. 91:183–192.

Gray, D.M., and D.H. Male. 1981. Handbook of snow: Principles, processes, management and use. Pergamon Press, New York.

Hanson, J.D. 2000. Generic crop production. p. 81–118. *In* L.R. Ahuja et al. (ed.) The Root Zone Water Quality Model. Water Resources Publ., Highlands Ranch, CO.

Hanson, J.D., K.W. Rojas, and M.J. Shaffer. 1999. Calibration and evaluation of the root zone water quality model. Agron. J. 91:171–177.

Heathman, G.C., L.R. Ahuja, and J.L. Baker. 1986. Test of a non-uniform mixing model for transfer of herbicides to surface runoff. Trans. ASAE 29:450–455.

Hoogenboom, G., J.W. Jones, P.W. Wilkens, W.D. Batchelor, W.T. Bowen, L.A. Hunt, N.B. Pickering, U. Singh, D.C. Goldwin, B. Baker, K.J. Boote, J.T. Ritchie, and J.W. White. 1994. Crop models. p. 95–129. *In* G.Y. Tsuji et al. (ed.) DSSAT version 3, Vol. 2. Univ. of Hawaii, Honolulu.

Hoogenboom, G., J.W. Jones, P.W. Wilkens, C.H. Porter, W.D. Batchelor, L.A. Hunt, K.J. Boote, U. Singh, O. Uryasev, W.T. Bowen, A.J. Gijsman, A. du Toit, J.W. White, and G.Y. Tsuji. 2004. Decision Support System for Agrotechnology Transfer Version 4.0 [CD-ROM]. Univ. of Hawaii, Honolulu.

Hu, C., S.A. Saseendran, T.R. Green, L. Ma, X. Li, and L.R. Ahuja. 2006. Evaluating nitrogen and water management in a double-cropping system using RZWQM. Vadose Zone J. 5:493–505.

Hunt, L.A., S. Pararajasingham, J.W. Jones, G. Hoogenboom, D.T. Imamura, and R.M. Ogoshi. 1993. GENCALC: Software to facilitate the use of crop models for analyzing field experiments. Agron. J. 85:1090–1094.

Johnsen, K.E., H.H. Liu, J.H. Dane, L.R. Ahuja, and S.R. Workman. 1995. Simulating fluctuating water tables and tile drainage with a modified Root Zone Water Quality Model and a new model WAFLOWN. Trans. ASAE 38:75–83.

Jones, J.W., G. Hoogenboom, C.H. Porter, K.J. Boote, W.D. Batchelor, L.A. Hunt, P.W. Wilkens, U. Singh, A.J. Gijsman, and J.T. Ritchie. 2003. The DSSAT cropping system model. Eur. J. Agron. 18:235–265.

Jones, J.W., J. He, K.J. Boote, P. Wilkens, C.H. Porter, and Z. Hu. 2011. Estimating DSSAT cropping system cultivar-specific parameters using Bayesian techniques. p. 365–394. *In* L.R. Ahuja and L. Ma (ed.) Methods of introducing system models into agricultural research. Advances in Agricultural Systems Modeling 2. ASA, CSSA, and SSSA, Madison, WI.

Karlen, D.L., C.A. Cambardella, and R.S. Kanwar. 2004. Challenges of managing liquid swine manure. Appl. Eng. Agric. 20:693–699.

Kozak, J.A., and L.R. Ahuja. 2005. Scaling of infiltration and redistribution across soil textural classes. Soil Sci. Soc. Am. J. 69:816–827.

Kozak, J.A., L. Ma, L.R. Ahuja, G.N. Flerchinger, and D.C. Nielsen. 2006. Evaluating various water stress calculations in RZWQM and RZ-SHAW for corn and soybean production. Agron. J. 98:1146–1155.

Kozak, J.A., R. Aiken, G.N. Flerchinger, D.C. Nielsen, L. Ma, and L.R. Ahuja. 2007. Quantifying residue architecture effects on soil temperature and water. Soil Tillage Res. 95:84–96.

Kumar, A., R.S. Kanwar, and L.R. Ahuja. 1998. Evaluation of preferential flow component of RZWQM in simulating water and atrazine transport to subsurface drains. Trans. ASAE 41:627–637.

Legates, D.R., and G.J. McCabe. 1999. Evaluating the use of "goodness-of-fit" measures in hydrologic and hydroclimatic model validation. Water Resour. Res. 35:233–241.

Loague, K., and R.E. Green. 1991. Statistical and graphical methods for evaluating solute transport models: Overview and application. J. Contam. Hydrol. 7:51–71.

Ma, L., L.R. Ahuja, J.C. Ascough II, M.J. Shaffer, K.W. Rojas, R.W. Malone, and M.R. Cameira. 2000. Integrating system modeling with field research in agriculture: Applications of Root Zone Water Quality Model (RZWQM). Adv. Agron. 71:233–292.

Ma, L., L.R. Ahuja, and R.W. Malone. 2007a. Systems modeling for soil and water research and management: Current status and needs in the 21st century. Trans. ASABE 50:1705–1713.

Ma, L., G. Hoogenboom, L.R. Ahuja, J.C. Ascough II, and S.S. Anapalli. 2006. Evaluation of RZWQM-CERES-maize hybrid for corn production. Agric. Syst. 87:274–295.

Ma, L., G. Hoogenboom, L.R. Ahuja, D.C. Nielsen, and J.C. Ascough II. 2005. Development and evaluation of RZWQM-CROPGRO hybrid for soybean production. Agron. J. 97:1172–1182.

Ma, L., G. Hoogenboom, S.A. Saseendran, P.N.S. Bartling, L.R. Ahuja, and T.R. Green. 2009. Estimates of soil hydraulic property and root growth factor on soil water balance and crop production. Agron. J. 101:572–583.

Ma, L., R.W. Malone, P. Heilman, D.L. Karlen, R.S. Kanwar, C.A. Cambardella, S.A. Saseendran, and L.R. Ahuja. 2007b. RZWQM simulation of long-term crop production, water and nitrogen balances in Northeast Iowa. Geoderma 140:247–259.

Ma, L., R.W. Malone, D.B. Jaynes, K.R. Thorp, and L.R. Ahuja. 2008. Simulated effects of nitrogen management and soil microbes on nitrate-N loss in tile drainage and crop production. Soil Sci. Soc. Am. J. 72:1594–1603.

Ma, L., D.C. Nielsen, L.R. Ahuja, R.W. Malone, S.A. Saseendran, K.W. Rojas, J.D. Hanson, and J.G. Benjamin. 2003. Evaluation of RZWQM under varying irrigation levels in eastern Colorado. Trans. ASAE 46:39–49.

Ma, L., H.D. Scott, M.J. Shaffer, and L.R. Ahuja. 1998b. RZWQM simulations of water and nitrate movement in a manured tall fescue field. Soil Sci. 163:259–270.

Ma, L., and H.M. Selim. 1997. Physical nonequilibrium modeling approaches to solute transport in soils. Adv. Agron. 58:95–150.

Ma, L., M.J. Shaffer, J.K. Boyd, R. Waskom, L.R. Ahuja, K.W. Rojas, and C. Xu. 1998a. Manure management in an irrigated silage corn field: Experiment and modeling. Soil Sci. Soc. Am. J. 62:1006–1017.

Ma, Q.L., L.R. Ahuja, K.W. Rojas, V.F. Ferreira, and D.G. DeCoursey. 1995. Measured and RZWQM predicted atrazine dissipation and movement in a field soil. Trans. ASAE 38:471–479.

Ma, Q.L., R.D. Wauchope, J.E. Hook, A.W. Johnson, C.C. Truman, C.C. Dowler, G.J. Gascho, J.G. Davis, H.R. Summer, and L.D. Chandler. 1998c. Influence of tractor wheel tracks and crusts/seals on runoff: Observation and simulations with the RZWQM. Agric. Syst. 57:77–100.

Ma, Q.L., R.D. Wauchope, L. Ma, K.W. Rojas, R.W. Malone, and L.R. Ahuja. 2004. Test of the Root Zone Water Quality Model (RZWQM) for predicting runoff of atrazine, Alachlor, and fenamiphos species from conventional-tillage corn mesoplots. Pest Manage. Sci. 60:267–276.

Malone, R.W., L.R. Ahuja, L. Ma, R.D. Wauchope, Q.L. Ma, and K.W. Rojas. 2004a. Application of the Root Zone Water Quality Model (RZWQM) to pesticide fate and transport: An overview. Pest Manage. Sci. 60:205–221.

Malone, R.W., D. Jaynes, L. Ma, B.T. Nolan, D. Meek, and D. Karlen. 2010. Soil-test N recommendations augmented with PEST-optimized RZWQM simulations. J. Environ. Qual. 39:1711–1723.

Malone, R.W., S. Logsdon, M.J. Shipitalo, J. Weatherington Rice, L. Ahuja, and L. Ma. 2003. Tillage effect on macroporosity and herbicide transport in percolate. Geoderma 116:191–215.

Malone, R.W., L. Ma, L.R. Ahuja, and K.W. Rojas. 2001a. Evaluation of the Root Zone Water Quality Model (RZWQM): A review. *In* J.E. Parsons et al. (ed.) Agricultural non-point source water quality models: Their use and application. Southern Coop. Ser. Bull. 398. Available at http://www.bae.ncsu.edu/www3/acad/Regional-Bulletins/Modeling-Bulletin/RZWQM2-word.html (verified 14 Mar. 2011).

Malone, R.W., L. Ma, R.D. Wauchope, L. Ahuja, K. Rojas, Q. Ma, R. Warner, and M. Byers. 2004b. Modeling hydrology, metribuzin degradation, and metribuzin transport in macroporous tilled and no till silt loam soil using RZWQM. Pest Manag. Sci. 60:253–266.

Malone, R.W., D. Meek, L. Ma, D. Jaynes, B.T. Nolan, and D. Karlen. 2011. Quality assurance of weather data for agricultural system model input. p. xxx–xxx. *In* L.R. Ahuja and L. Ma (ed.) Methods of introducing system models into agricultural research. Adv. Agric. Systems Modeling 2. ASA, CSSA, SSSA, Madison, WI.

Malone, R.W., M.J. Shipitalo, L. Ma, L.R. Ahuja, and K.W. Rojas. 2001b. Macropore component assessment of the Root Zone Water Quality Model (RZWQM) using no-till soil blocks. Trans. ASAE 44:843–852.

Martin, D.L., and D.G. Watts. 1999. Application of the root zone water quality model in central Nebraska. Agron. J. 91:201–211.

Mavromatis, T., K.J. Boote, J.W. Jones, G.G. Wilkerson, and G. Hoogenboom. 2002. Repeatability of model genetic coefficients derived from soybean performance trials across different states. Crop Sci. 42:76–89.

Meisinger, J.J., and G.W. Randall. 1991. Estimating nitrogen budget in soil–crop systems. p. 85–124. *In* R.F. Follett et al. (ed.) Managing nitrogen for groundwater quality and farm profitability. SSSA, Madison, WI.

Monteith, J.L., and M.H. Unsworth. 2008. Principles of environmental physics. 3rd ed. Academic Press, New York.

Moriasi, D.N., J.G. Arnold, M.W. Van Liew, R.L. Gingner, R.D. Harmel, and T.L. Veith. 2007. Model evaluation guidelines for systematic quantification of accuracy in watershed simulation. Trans. ASABE 50:885–900.

Nash, J.E., and J.V. Sutcliffe. 1970. River flow forecasting through conceptual models. Part 1. A discussion of principles. J. Hydrol. 10:282–290.

Nokes, S.E., F.M. Landa, and J.D. Hanson. 1996. Evaluation of the crop component of the root zone water quality model for corn in Ohio. Trans. ASAE 39:1177–1184.

Nolan, B.T., R.W. Malone, L. Ma, C.T. Green, M.N. Fienen, and D.B. Jaynes. 2011 Inverse modeling with RZWQM2 to predict water quality. p. 327–364. *In* L.R. Ahuja and L. Ma (ed.) Methods of introducing system models into agricultural research. Advances in Agricultural Systems Modeling 2. ASA, CSSA, and SSSA, Madison, WI.

Nolan, B.T., L.J. Puckett, L. Ma, C.T. Green, E.R. Bayless, and R.W. Malone. 2010. Predicting unsaturated zone nitrogen mass balances in diverse agricultural settings of the United States. J. Environ. Qual. 39:1051–1065.

Quisenberry, V.L., R.E. Phillips, and J.M. Zeleznik. 1994. Spatial distribution of water and chloride macropore flow in a well–structured soil. Soil Sci. Soc. Am. J. 58:1294–1300.

Rawls, W.J., D.L. Brakensiek, and K.E. Saxton. 1982. Estimation of soil water properties. Trans. ASAE 25:1316–1320, 1328.

Ritchie, J.T., U. Singh, D.C. Godwin, and W.T. Bowen. 1998. Cereal growth, development and yield. p. 19–88. *In* G.Y. Tsuji et al. (ed.) Understanding options for agricultural production. Kluwer Academic Publ., Dordrecht, The Netherlands.

Rojas, K.W., L. Ma, J.D. Hanson, and L.R. Ahuja. 2000. RZWQM98 user guide. *In* L.R. Ahuja et al. (ed.) The Root Zone Water Quality Model. Water Resources Publ., Highlands Ranch, CO.

Roy, W.R., I.G. Krapac, S.F.J. Chou, and R.A. Griffin. 1992. Batch-type procedures for estimating soil adsorption of chemicals. EPA/530-SW-87-006-F. USEPA, Cincinnati, OH.

Saseendran, S.A., L. Ma, D.C. Nielsen, M.F. Vigil, and L.R. Ahuja. 2005. Simulating planting date effects on crop production using RZWQM and CERES-maize models. Agron. J. 97:58–71.

Saseendran, S.A., D.C. Nielsen, D. Lyon, L. Ma, D.G. Felter, D.D. Baltensperger, G. Hoogenboom, and L.R. Ahuja. 2009. Modeling responses of dryland spring triticale, proso millet, and foxtail millet to initial soil water in the High Plains. Field Crops Res. 113:48–63.

Saseendran, S.A., D.C. Nielsen, L. Ma, L.R. Ahuja, and A.D. Halvorson. 2004. Modeling nitrogen management effects on a winter wheat cropping system using RZWQM and CERES-wheat. Agron. J. 96:615–630.

Saseendran, S.A., D.C. Nielsen, L. Ma, L.R. Ahuja, and M.F. Vigil. 2010. Simulating alternative dryland rotational cropping systems in the Central Great Plains with RZWQM2. Agron. J. 102:1521–1534.

Schepers, J.S., and A.R. Mosier. 1991. Accounting for nitrogen in nonequilibrium soil-crop systems. p. 125–138. *In* R.F. Follett et al. (ed.) Managing nitrogen for groundwater quality and farm profitability. SSSA, Madison, WI.

Shaffer, M.J., K.W. Rojas, D.G. DeCoursey, and C.S. Hebsen. 2000. Nutrient chemistry processes–OMNI. p. 119–144. *In* L.R. Ahuja et al. (ed.) The Root Zone Water Quality Model. Water Resources Publ., Highlands Ranch, CO.

Shen, J., W.D. Batchelor, J.W. Jones, J.T. Ritchie, R.S. Kanwar, and C.W. Mize. 1998. Incorporation of a subsurface tile drainage component into a soybean growth model. Trans. ASAE 41:1305–1313.

Singh, P., R.S. Kanwar, K.E. Johnsen, and L.R. Ahuja. 1996. Calibration and evaluation of subsurface drainage component of RZWQM v2.5. J. Environ. Qual. 25:56–63.

Sophocleous, M., M.A. Townsend, F. Vocasek, L. Ma, and K.C. Ashok. 2009. Soil nitrogen balance under wastewater management: Field measurements and simulation results. J. Environ. Qual. 38:1286–1301.

Stehouwer, R.C., W.A. Dick, and S.J. Traina. 1993. Characteristics of earthworm burrow lining affecting atrazine sorption. J. Environ. Qual. 22:181–185.

Stehouwer, R.C., W.A. Dick, and S.J. Traina. 1994. Sorption and retention of herbicides in vertically oriented earthworm and artificial burrows. J. Environ. Qual. 23:286–292.

Wallach, D., D. Makowski, and J.W. Jones. 2006. Working with dynamic crop models: Evaluation, analysis, parameterization, and applications. Elsevier, Amsterdam. The Netherlands.

Wauchope, R.D., T.M. Buttler, A.G. Hornsby, P.W.M. Augustijn-Beckers, and J.P. Burt. 1992. The SCS/ARS/CES pesticide properties database for environmental decision-making. Rev. Environ. Contam. Toxicol. 123:1–164.

Wauchope, R.D., R.G. Nash, L.L. McDowell, K.W. Rojas, L.R. Ahuja, G.H. Willis, T.B. Moorman, and Q. Ma. 2000. Pesticide processes. p. 163–244. *In* L.R. Ahuja et al. (ed.) The Root Zone Water Quality Model. Water Resources Publ., Highlands Ranch, CO.

Wauchope, R.D., K.W. Rojas, L.R. Ahuja, Q.L. Ma, R.W. Malone, and L. Ma. 2004a. Documenting the pesticide processes module of the ARS RZWQM agroecosystem model. Pest Manag. Sci. 60:222–239.

Watts, D.G., N.R. Fausey, and D.A. Bucks. 1999. Background of the MSEA-RZWQM modeling project. Agron. J. 91:169–170.

Willmott, C.J. 1981. On the validation of models. Phys. Geogr. 2:184–194.

Wu, L., R.R. Allmaras, J.A. Lamb, and K.E. Johnsen. 1996. Model sensitivity to measured and estimated hydraulic properties of a Zimmerman fine sand. Soil Sci. Soc. Am. J. 60:1283–1290.

Wu, L., W. Chen, J.M. Baker, and J.A. Lamb. 1999. Evaluation of RZWQM field measured data from a sandy soil. Agron. J. 91:177–182.

Yu, Q., G.N. Flerchinger, S. Xu, J. Kozak, L. Ma, and L.R. Ahuja. 2007. Simulation of energy balance and water and heat transfer by SHAW and RZ-SHAW. Trans. ASABE 50:1507–1516.

Chapter 2

Special Features of the HERMES Model and Additional Procedures for Parameterization, Calibration, Validation, and Applications

K.C. Kersebaum

Abstract

The HERMES model is an agroecosystem model designed to operate under data-limited conditions of agricultural practice. The fundamentals of the model components are given, and the data requirements, data structure, and user interface are described. A general guideline for how the model can be calibrated is presented, and examples demonstrate how the model can be calibrated and validated with field experimental data. The use of the model to derive fertilizer recommendations for winter wheat (*Triticum aestivum* L.) is described, and examples using spatially variable data demonstrate the sensitivity of the model and the uncertainty of predictions due to annual weather variability.

K.C. Kersebaum, Leibniz-Centre for Agricultural Landscape Research, Institute of Landscape Systems Analysis, Muencheberg, Germany (ckersebaum@zalf.de).

doi: 10.2134/advagricsystmodel2.c2

Methods of introducing system models into agricultural research. L.R. Ahuja and L. Ma (ed.)
Advances in Agricultural Systems Modeling Series 2.

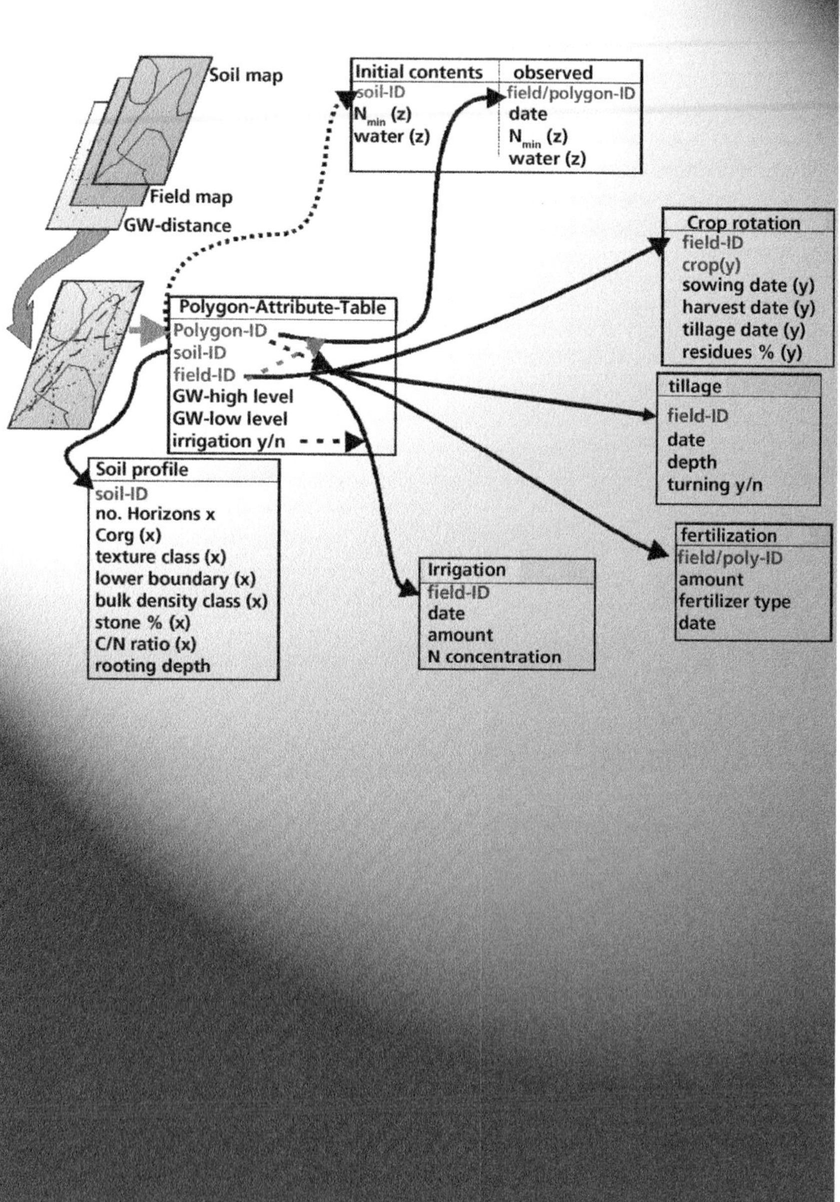

Soil map
Field map
GW-distance
Initial contents
soil-ID
N_{min} (z)
water (z)
observed
field/polygon-ID
date
N_{min} (z)
water (z)
Crop rotation
field-ID
crop(y)
sowing date (y)
harvest date (y)
tillage date (y)
residues % (y)
Polygon-Attribute-Table
Polygon-ID
soil-ID
field-ID
GW-high level
GW-low level
irrigation y/n
tillage
field-ID
date
depth
turning y/n
Soil profile
soil-ID
no. Horizons x
Corg (x)
texture class (x)
lower boundary (x)
bulk density class (x)
stone % (x)
C/N ratio (x)
rooting depth
Irrigation
field-ID
date
amount
N concentration
fertilization
field/poly-ID
amount
fertilizer type
date

There are many models to simulate processes in agroecosystems, and they differ in their complexity depending on their purpose, their scale of application, and their potential end users. Models for testing scientific hypotheses on processes are more complex and mechanistic than those designed to be used operationally for practical applications in agriculture, like irrigation or fertilization, or those that give answers for large areas with limited or generalized data availability. Regarding their potential errors, both more complex and simpler models have their advantages and disadvantages. While the numerous parameters of complex models and the interaction of processes involved increase the risk of error propagation, which can be summarized as parameter error, the neglect of important processes in simplified models can lead to an increasing structural error (Grunwald, 1997).

The HERMES model has been developed primarily for its application in agricultural practice where detailed information on management is usually available but knowledge about soil properties at the field scale is often limited to a few basic properties like texture and soil organic matter content. The objective of this chapter is to describe the main characteristics of the present version of HERMES, its data structure and requirements, and to give examples of how to calibrate the model for new crops and environmental conditions and how it can be validated. This includes applications for practical use in agricultural practice.

Characteristic of HERMES

The HERMES model was initially developed in the mid 1980s from a simple model to estimate soil mineral nitrogen on arable soils in early spring (Nuske 1983) as a foundation for fertilizer recommendations. Based on a simple model approach from Richter et al. (1978) describing nitrate transport assuming quasi-stationary flux conditions during winter and experimental work on nitrogen mineralization of Richter et al. (1982), a preliminary model was developed by Nuske (1983) to simulate nitrogen dynamics in loess soils over the winter period. The aim was to create a tool for the calculation of soil mineral nitrogen content in early spring, which is usually determined by soil sampling, as a basis for fertilizer recommendation (Wehrmann and Scharpf, 1986). Combining this concept with a complete water balance submodel for transient water flux and a module for the dynamic simulation of crop growth for winter wheat, a new model (HERMES) was built by Kersebaum (1989) which was able to simulate nitrogen dynamics in the soil–crop system for the whole growing season. The simulated deficiency between nitrogen supply from soil and the demand of the growing crop was used to calculate recommendations for nitrogen fertilizer applications for the whole growing season. In the following years the model was modified for crops other than winter wheat, and a module for denitrification was added (Kersebaum, 1995). In its present version the model is able to simulate crop rotations using a generic approach for crop growth that can be extended by the user for other crops by creating their individual crop parameter files. The model operates on a daily time step and requires daily weather data (Table 2–1).

Thus, HERMES (Kersebaum, 1995, 2007) is a process-oriented functional model designed to simulate crop growth, water, and nitrogen dynamics for arable land. The model has been used under central European conditions to calculate N fertilization recommendations at the field and subfield scale (Kersebaum and Beblik, 2001; Kersebaum et al., 2003a) and for regional calculations of N pollution under different management options (Kersebaum, 2000; Kersebaum et al., 2003b). Recently, it was adapted for applications in other regions, including Canada (Kersebaum et al., 2008), India, and China. A special version NVINO exists for simulating vine growth and nitrogen dynamics in vineyard soils (Nendel and Kersebaum, 2004). The model has been evaluated in several international model comparison studies providing datasets to different modelers and showed its capability to simulate water and nitrogen dynamics in arable fields (de Willigen, 1991; Diekkrüger et al., 1995; Kersebaum et al., 2007a; Palosuo et al., 2011).

The model is comprised of submodules for the water balance, N transport and transformations, and crop development and growth, including N uptake. The main algorithms were described previously in detail by Kersebaum and Richter (1991), Kersebaum (1995), and Kersebaum and Beblik (2001), and therefore only a brief description is presented here.

Water and Nitrogen Balance

Soil water is simulated by a simple capacity-based approach. The capacity parameters required by the model are attached to the model by external data files which are consistent with the German soil texture classification and their capacity parameters (Ad-hoc-AG Boden, 2005). An example for some soils is given in Table 2–2. The basic values are modified by soil organic matter content and bulk density. In the case of water logging, water holding capacity of single horizons can be raised by a hydromorphic factor. For shallow groundwater sites, capillary rise is calculated dependent on soil texture and the distance to the groundwater using tabulated values of Ad-hoc-AG Boden (2005). These fluxes are defined for a water

Table 2–1. Weather data requirements.

Variable	Characteristic	Unit	Comment	Recommended distance
Air temperature (T_m)	Avg. 2-m height	°C	Essential	Max. 20 km
Air temperature min. (T_{min})	Daily minimum	°C	Essential	Max. 20 km
Air temperature max. (T_{max})	Daily maximum	°C	essential	Max. 20 km
Precipitation (P)	Daily sum	mm	Essential	Max. 2 km
Global radiation (GR)	Daily sum	J cm^{-2}	Essential, can be replaced by SD	Max. 50 km
Sunshine duration (SD)	Daily sum	h	Optional, replaces GR	Max. 50 km
Vapor pressure deficit (VPD)	At 2 pm	mm Hg	Only for HAUDE PET†	Max. 2 km
Rel. humidity (rH)	Daily average	%	Only Penman–Monteith ET_0‡	Max. 2 km
Wind speed	Daily average	m s^{-1}	Only Penman–Monteith ET_0‡	Max. 2 km

† HAUDE method for PET according to Haude (1955).

‡ ET_0 Penman–Monteith according to Allen et al. (1998).

content of 70% of crop available water and are used by the model as a steady-state flux up to the deepest soil layer with less than 70% of available water capacity. An example for different soils is given in Table 2–2. Because soil information is often very limited under practical conditions, the chosen capacity approach seemed to be appropriate for field use, as well as on a regional scale, because parameters are more robust and can easily be derived from rough basic soil information (Addiscott and Wagenet, 1985). Functional approaches of this kind have been shown to perform as well as mechanistic water models based on the potential concept under field conditions (de Willigen, 1991; Diekkrüger et al., 1995; Kersebaum et al., 2007a).

Regarding the driving weather input variables, there is an option to correct the daily precipitation data for the systematic error of Hellmann measurements at the 1-m height compared to soil surface precipitation. For Germany the model uses an external table according to Richter (1995) with monthly correction factors that have to be modified when the model is used in other regions. However, the correction can be switched off to use the original data. A crop specific potential evapotranspiration (PET) is calculated from daily weather data. Presently, four options for evapotranspiration formulas are included in the model:

- Calculation of PET with the empirical HAUDE formula (Haude, 1955) using crop specific monthly coefficients of Heger (1978) and the daily vapor pressure deficit at 1400 h.

Table 2–2. Selected texture classes (10 of 34) from Ad-hoc-AG Boden (2005) with their particle-size distribution, capacity parameters depending on bulk density classes, and daily capillary rise.

Texture class:	SG†	SM†	SF†	SL2	LS2	LT2	UT3	UT4	TU3	TU2
	% (w/w)									
Sand	>85	>85	>85	67–85	25–43	15–45	<23	<18	<20	<25
Silt	<10	<10	<10	10–25	40–50	30–50	65–88	65–83	50–65	30–55
Clay	<5	<5	<5	5–8	17–25	25–35	12–17	17–25	30–45	45–65

Bulk density class‡:	Capacity parameters								
	Field capacity			Wilting point			Total pore space		
	1–2	3	4–5	1–2	3	4–5	1–2	3	4–5
	% (v/v)								
Texture class									
SG† (coarse sand)	12	8	6	4	3	2	46	41	35
SM† (middle sand)	14	10	8	5	4	3	50	42	34
SF† (fine sand)	16	14	12	6	5	4	50	45	35
SL2 (loamy sand)	28	25	23	8	7	6	51	43	36
LS2 (sandy loam)	40	34	31	19	18	17	53	43	37
LT2 (clay loam)	42	36	32	24	22	21	53	43	37
UT3 (loamy silt)	39	37	35	13	12	12	50	43	38
UT4 (clay silt)	39	37	35	16	16	16	51	44	38
TU3 (silty clay)	45	38	35	28	25	25	53	44	38
TU2 (slight silty clay)	47	42	36	31	30	26	52	46	39

	Daily rates of capillary rise									
Groundwater distance (cm) line 1	10	20	30	40	50	60	70	80	90	100
line 2	110	120	130	140	150	160	170	180	190	200
	mm d^{-1}									
Texture class										
SG† 0.1–1 m	5.5	5.0	1.5	0.9	0.4	0.2	0	0	0	0
1.1–2 m	0	0	0	0	0	0	0	0	0	0
SM†	5.5	5.3	5.1	3.0	1.0	0.5	0.3	0.2	0.1	0
	0	0	0	0	0	0	0	0	0	0
SF†	5.5	5.3	5.2	5.1	3.5	1.9	0.8	0.4	0.3	0.2
	0.1	0	0	0	0	0	0	0	0	0
SL2	5.5	5.3	5.1	3.1	1.7	1.0	0.6	0.4	0.3	0.2
	0.1	0.1	0	0	0	0	0	0	0	0
LS2	5.5	5.3	5.1	2.6	1.6	1.3	1.0	0.7	0.4	0.2
	0.1	0	0	0	0	0	0	0	0	0
LT2	5.2	5.0	2.3	1.2	0.7	0.4	0.3	0.2	0.1	0.1
	0	0	0	0	0	0	0	0	0	0
UT3	5.5	5.5	5.5	5.5	5.4	5.3	5.2	5.1	5.0	4.2
	3.4	2.9	2.4	2.0	1.5	1.2	1.2	1.1	1.0	0.8
UT4	5.5	5.4	5.3	5.2	5.1	5.0	4.2	3.3	2.6	2.1
	1.7	1.4	1.2	1.0	0.8	0.7	0.6	0.5	0.45	0.4
TU3	5.2	5.0	3.1	2.4	1.8	0.7	0.4	0.3	0.2	0.2
	0.1	0	0	0	0	0	0	0	0	0
TU2	5.0	3.2	1.3	0.7	0.5	0.3	0.2	0.2	0.1	0.1
	0	0	0	0	0	0	0	0	0	0

† Name modified for model use.

‡ Bulk density classes: 1–2 = <1.4 g cm^{-3}, 3 = 1.4–1.6 g cm^{-3}, 4–5 = >1.6 g cm^{-3}.

- Calculation of a reference evapotranspiration for grass with the TURC-WENDLING formula (Wendling et al., 1991). The formula requires the diurnal average temperature and global radiation sum.
- Calculation of grass reference evapotranspiration using the Penman–Monteith approach (Allen et al., 1998), which requires additionally the daily average wind speed and average relative humidity.
- Calculation of grass reference evapotranspiration according to Priestley–Taylor (Priestley and Taylor, 1972), which requires daily minimum and maximum temperature and global radiation only.

For the latter three formulas the model uses crop coefficients (k_c) describing the relation between crop and reference grass evapotranspirations to calculate potential evapotranspiration specific to the crop growth stage and soil coverage. The relative factors can be defined for bare soil and within the crop parameter file for the end points of development stages. Within the end points of development stages there is a linear interpolation for k_c values. Sunshine duration can be used instead of global radiation using the Ångström formula (Ångström, 1924). As an additional option a grass reference evapotranspiration from any other externally calculated formula can be given in the weather data file. Recent versions of HERMES (Kersebaum et al., 2009) allow consideration of the effect of changing CO_2 concentrations of the atmosphere on crop transpiration using an approach of Yu et al. (2001) to modify the stomatal conductance within the Penman–Monteith calculation procedure.

Partitioning between evaporation and transpiration is a function of leaf area index (LAI), estimated by the crop growth model. The calculation of actual evaporation and transpiration considers the soil water status and vertical root distribution of specific crops, using an empirical function (Gerwitz and Page, 1974) to distribute the dry matter allocated to the roots over depth. Maximum rooting depth can be defined for soil profiles as a default for wheat and modified if crop specific rooting depth is different from cereals.

Nitrogen Dynamics

The N mineralization submodule follows the concept of net mineralization, simulating mineral N release from two pools of potentially decomposable organic matter according to first-order kinetic reactions. Pools are conceptually derived from incubation experiments, and mathematically separating N release into two pools with different reaction coefficients (Richter et al., 1982). The basic size of the slow mineralizable pool is derived from soil organic matter N using a standard percentage of 13% according to Nuske (1983). However, this percentage has to be adjusted depending on site conditions and history. Another possibility is to run the model for a longer period of about 20 yr in advance with the typical weather and management to let the pools equilibrate to the conditions.

Resistant compounds from crops are added to this slow decomposing pool at harvest time. A smaller, much faster mineralizing pool is fed by easily decomposable compounds of different crop residues and manure. Nitrogen in crop residues recycled to the soil is calculated automatically using the simulated N uptake and a crop specific relation of the N export with the yield allowing simulation of complete rotations. Daily mineralization coefficients are calculated based on the mean air temperature using two Arrhenius functions from Nuske (1983) and Nordmeyer and Richter (1985). Soil moisture effects on N mineralization are implemented according to Myers et al. (1982). Daily denitrification N_{den} (kg N ha^{-1}) is simulated for the top soil (0–30 cm depth) using Michaelis–Menten kinetics, modified by reduction functions dependent on water-filled pore space (Θ_r) and temperature (T) (Richter and Söndgerath, unpublished data, cited in Schneider, 1991):

$$N_{\mathrm{den}} = \frac{V_{\max}(\mathrm{NO_3})^2}{(\mathrm{NO_3})^2 + K_{\mathrm{NO_3}}} f(\Theta_r) f(T) \qquad [1]$$

with a maximum denitrification rate V_{max} (1.274 kg N ha^{-1} d^{-1}), the soil nitrate content NO_3 (kg NO_3–N ha^{-1}) in 0 to 30 cm and the Michaelis–Menten coefficient for nitrate K_{NO3} (74 kg NO_3–N ha^{-1}), and functions for water content and temperature:

$$f(\Theta_r) = 1 - e^{-\left(\frac{\Theta_r}{\Theta_{\mathrm{crit}}}\right)^6} \qquad [2]$$

$$f(T) = 1 - e^{-\left(\frac{T}{T_{\mathrm{crit}}}\right)^{4.6}} \qquad [3]$$

Critical values T_{crit} for air temperature and Θ_{crit} for water-filled pore space are set to 15.5°C and 0.766 cm^3 cm^{-3} pore space, respectively.

Plant Growth

The submodel for crop growth represents a generic approach, which is able to simulate different crops using external crop parameter files. The submodel was built on the basis of the SUCROS model (Van Keulen et al., 1982). Driven by global radiation and temperature, the daily net dry matter production by photosynthesis and respiration is simulated. Dry matter production is partitioned depending on crop development stage, which is calculated from a thermal sum or degree

Table 2–3. Basic crop parameters (head of crop parameter file) of HERMES (name of crop compartments can be given by the user).

Crop: winter wheat	
Crop no./abbreviation	1 WW
AMAX: Max. CO_2 assimilation rate (kg CO_2/ha leave/h)	52
Type of temperature dependency (C3 = 1/C4 = 2)	1
Minimum temperature crop growth (degree Celsius)	4
Maximum effective rooting depth (dm)	12
Root distribution function no. (only 1 available)	1
Crop N-content function no. (critical and max. N-contents)	1
Aboveground organs (numbers of compartiments increasing order)	234
Start concentration N in ab. gr. biomass (% i. d.m.)	2.0
Start concentration N in roots (% i. d.m.)	2.0
Number of crop compartiments	4
Compartments	**1**root **2**leave **3**stem **4**ears **5** –
Initial weight kg d.m./ha	00053.. 00053.. 00000.. 00000.. 00000
Maintenance rates of organs	0.010.. 0.030.. 0.015.. 0.010.. 0.000
Initial k_c factor for evapotranspiration	0.5
Number of development phases	6

days (°C days), modified for each stage by daylength and vernalization if applicable for a specific crop. Up to five different crop organs and 10 development stages can be defined in the parameter file for partitioning (Tables 2–3 and 2–4). Root dry matter is exponentially distributed over depth according to Gerwitz and Page (1974), with the rooting depth increasing with the abovementioned modified thermal sum until a crop and soil specific maximum is reached. Grain yield is estimated at harvest from the weight of the storage organ.

Nitrogen recycling through crop residues is calculated automatically from the simulated crop N uptake minus the N exported at harvest with yield and removed byproducts (e.g., straw, leaves). The distribution of N in the residues to the pools is set in an external parameter file for residues.

Crop growth is limited by water and N stress. Drought stress is indicated by the ratio of actual/potential transpiration. Sensitivity of the crop to this ratio can be adjusted for each development stage in the parameter file (Table 2–4). Temporary limitation of soil air by water logging is considered through reducing transpiration and photosynthesis according to Supit et al. (1994). For this, a critical percentage of air-filled pores has to be defined in the crop parameter file. Water and N uptake are calculated from potential transpiration and crop N status, depending on the simulated root distribution and water and N availability in different soil layers (Kersebaum, 1995). The concept of critical N concentration in plants as a function

Table 2–4. Specific crop parameters required for each development stage (number of stages are defined in basic parameters, see Table 2–3). Example for winter wheat.

Development phase 1: sowing to emergence	
Development phase 1 temperature sum	140
Base temperature in phase 1	1
Vernalization requirements (days)	0
Daylength requirements (hours)	0
Base day length in phase 1(hours)	0
Drought stress below AET/PET-quotient of	1
Critical air content in topsoil (cm^3 cm^{-3})	0.08
Specific leave area (area per d.m. mass) (m^2 kg^{-1})	0.002
N-content root end of 1. phase	0.02
Compartments……… **1**root **2**leave **3**stem **4**ears	**5** –
Partitioning at emergence……… 0.500.. 0.500.. 0.000.. 0.000..	0.000
Death rate at end of phase 1……… 0.000.. 0.000.. 0.000.. 0.000..	0.000
k_c factor for evapotranspiration at end of phase 1	0.65
Development phase 2: emergence to double ridge	
Development phase 2 temperature sum	284
Base temperature 2	1
Verbalization requirements (days)	50
Daylength requirements (hours)	20
Base day length in phase 2 (hours)	0
Drought stress below AET/PET-quotient of	0.8
Critical air content in topsoil (cm^3 cm^{-3})	0.08
Specific leave area (area per d.m. mass) (m^2 kg^{-1})	0.002
N-content root end of 2. phase	0.02
Compartments……… **1**root **2**leave **3**stem **4**ears	**5** –
Partitioning at end of phase 2……… 0.200.. 0.600.. 0.200.. 0.000..	0.000
Death rate at end of phase 2……… 0.000.. 0.000.. 0.000.. 0.000..	0.000
k_c factor for evapotranspiration at end of phase 2	1.1
Development phase 3: double ridge to ear emergence	
.	
.	
Development phase 4: ear emergence to flowering	
.	
.	
Development phase 5: grain filling	
.	
.	
Development phase 6: senescence	
development phase 6 temperature sum	25
Base temperature 6	9
Verbalization requirements (days)	0
day length requirements (hours)	0
Base day length in phase 6(hours)	0
drought stress below AET/PET-quotient of	0.6
critical air content in topsoil(cm^3 cm^{-3})	0.08
specific leave area (area per d.m. mass) (m^2 kg^{-1})	0.002
N-content root end of 6. phase	0.010
Compartments……… **1**root **2**leave **3**stem **4**ears	**5** –
Partitioning at end of phase 6……… 0.000.. 0.000.. 0.000.. 0.000..	0.000
death rate at end of phase 6……… 0.000.. 0.050.. 0.000.. 0.000..	0.000
k_c factor for evapotranspiration at end of phase 6	0.25

of crop developmental stage (Kersebaum and Beblik, 2001) or as a function of crop biomass (Greenwood et al., 1990) is applied to assess the impact of N shortage. First attempts have been made using the latter type of relation in HERMES to consider N shortage for potatoes (*Solanum tuberosum* L.) and sugarbeets (*Beta vulgaris* L. subsp. *vulgaris*) from literature (Kabat et al., 1995; Greenwood and Draycott, 1995; Institut International de Recherches Betteravières, 2003). Duval et al. (2003) found a strong similarity between the critical N dilution curve of sugarbeets and potatoes. Leviel et al. (2003) also used the generalized curve of Greenwood et al. (1990) to consider N stress in a CERES approach for sugarbeets.

Model Limitations

The model is focused on the most relevant processes in the soil–crop system under central European conditions. Therefore, several processes which might be important under different climatic and soil conditions are not sufficiently covered or have not yet been sufficiently tested. Therefore, some information about the main limitations and current constructions of the model should be given here. For example, a soil temperature module has recently been included and is being tested presently. At some specific sites the model requires additional testing, for example, the simulation of nitrogen dynamics in organic soils. A clear gap is that the model presently does not cover surface runoff and erosion processes. Actually the capacity approach does not allow for water logging at surface or in the profile. A modified approach using retardation factors is currently under construction which will also contribute to close the surface runoff gap. Although the model includes a simplified procedure to consider capillary rise from shallow groundwater, the consideration of fluctuating groundwater level is still problematic. Finally, N dynamics are not linked to C dynamics with the consequence that short-term immobilization is not considered.

Data Requirements and Structure

Data Requirements

Data required by the model can be separated into three parts: weather data, soil information, and management data. The model operates on a daily weather data basis. Input data required are precipitation, average and minimum and maximum air temperature, global radiation or alternatively sunshine duration, and depending on the evapotranspiration formula used for the simulation either the vapor pressure deficit at 1400 h (for Haude formula only) or the average relative humidity and wind speed (Penman–Monteith formula). A list of weather data required by the model is given in Table 2–1.

Table 2–5. Examples of fertilizer parameterization.

Code	Total N	Mineral N N_{tot}^{-1}	N_{fast} N_{org}^{-1}	N_{slow} N_{org}^{-1}	NH_4 N_{min}^{-1}	Loss NH_4^{-1}	Application unit	Full name
	kg N unit^{-1}							
CAN	1.0	1.0	0	0	0.5	0.1	kg N ha^{-1}	Calcium ammonium nitrate
CM	0.60	0.15	0.2	0.8	1.0	0.4	t ha^{-1}	Cattle manure
CS	4.70	0.43	0.0	1.0	1.0	0.4	m^3 ha^{-1}	Cattle slurry
PDM	2.30	0.15	0.20	0.80	1.0	0.4	t ha^{-1}	Poultry dry manure
BCO	1.63	0.0	0.05	0.70	0.0	0.0	t ha^{-1}	Bio-waste compost

Soil information is required at a depth resolution of 10 cm for the profile. For the plow layer, the organic matter content and its C/N ratio should be given. The soil texture class of Ad-hoc-AG Boden (2005) is the most important soil information required. Groundwater depth, which is defined as totally water-filled pore space, and soil texture are needed to calculate capillary rise if applicable. Bulk density classes, organic matter content, groundwater table, and percentage of stones are used to modify capacity parameters automatically within the model. Alternatively, the user can modify the default capacity values in the Windows-based user interface. Table 2–2 shows an example how soil hydraulic parameters are linked to the texture class for some important texture classes.

Mandatory data for management are crop species; dates of sowing, harvest, and soil tillage; nitrogen fertilizer; and water application (kind of fertilizer, quantity and date of application/incorporation). Fertilizers can be specified regarding their mineral/organic composition and fractionation (Table 2–5). Measured vertical distribution of mineral nitrogen content and soil moisture are used as initial values for the simulation.

Data Structure

The input data are almost entirely organized in projects, which means they are located in the respective project folders. A project is defined by a common location (same latitude and weather station). The simulation period as well as some options (output interval, evapotranspiration method, and precipitation correction) can be defined in the driver file for each project. Within a project folder separate folders are available for annual weather data (\WEATHER\) and output files and measured values (\RESULT\). These folders are automatically generated if a new project is created with the user interface.

The structure of the files allows linking the model to a Geographic Information System (GIS) via polygons, that is, fields with the same management but

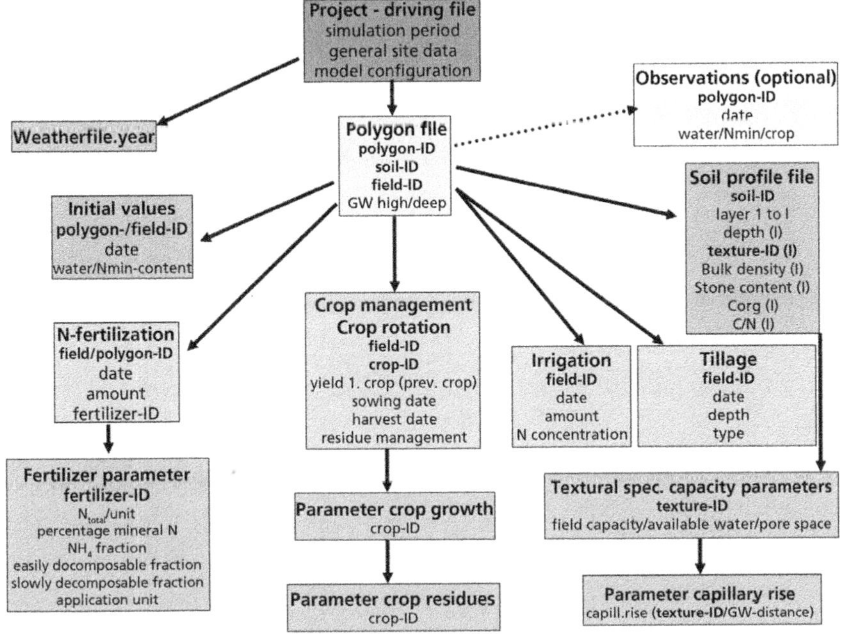

Fig. 2–1. Data structure of the HERMES model.

different soil properties. Figure 2–1 shows the data structure of the model, Fig. 2–2 shows the relations of GIS oriented data.

All data are available in ASCII format and can thus be edited with a simple text editor. Since doing it this way may lead to accidental changes in the file structure followed by read-in errors of the model, the user is advised to use the Windows-based user interface for any editing.

Graphical User Interface

A graphical user interface (GUI) can be used to generate the ASCII input files for the model. While weather files for the simulation can be easily generated by an Excel sheet, the main inputs can be generated by the GUI. Figure 2–3 shows the initial sheet of the HERMES user interface, which starts with the selection of the project. After that the user has to define the general settings for a project in the driver file. The driver file contains information about the location of the site, the general settings of atmospheric N deposition, length of the simulation period, depth for seepage and N leaching calculation, correction of precipitation, and output definitions, Furthermore, the user can select between different methods for evapotranspiration calculation and CO_2 effect on photosynthesis. Initial conditions for the simulation for soil water and mineral N distribution within the soil

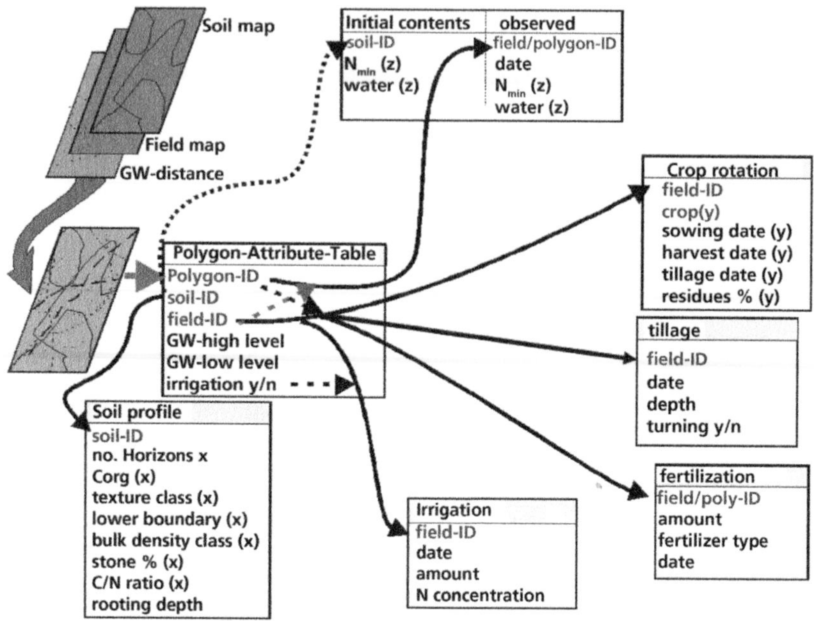

Fig. 2–2. GIS oriented relations between input and parameter files.

Fig. 2–3. Initial window of the HERMES graphical user interface (selection of project).

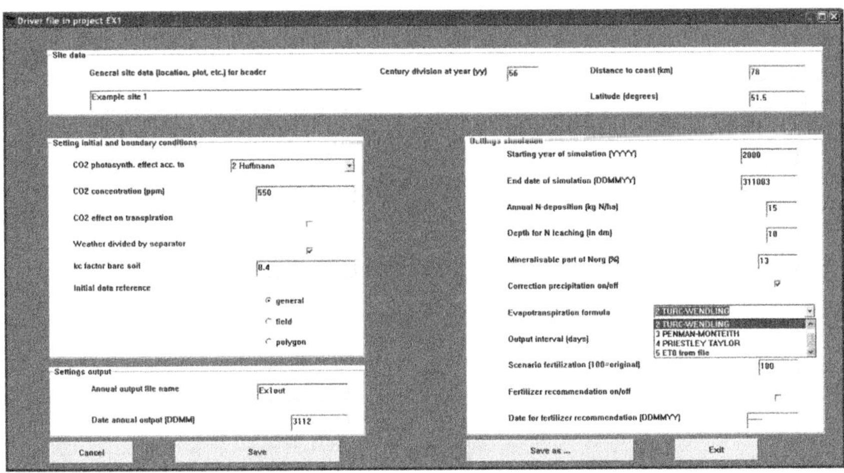

Fig. 2–4. Input window of the HERMES for model settings in the project driver file.

profile can be defined either for a single polygon or field or can be set as general information for all polygons if detailed information is missing, as in the case of a regional application. This setting alters the input sheet for initial values under the site data menu. The input sheet for the driver file is shown in Fig. 2–4, which gives an example of the choice of the evapotranspiration formula.

Figure 2–5 gives an example of the definition of polygons that can be used in conjunction with a GIS. This offers different options for the simulations. For a classical field trial, different managements (e.g., fertilization levels) can be defined through the field ID and used with a uniform soil. On the other hand, a heterogeneous field can be separated into different zones with different soil profiles which can be uniformly (i.e., only one field ID) or site-specifically managed (i.e., different field IDs).

Figure 2–6 shows the input sheet for the soil profile data. Usually the user can select the soil texture class from the selection list without giving details for soil particle-size distribution. However, one can add these details, which leads to an automatic selection of the texture class. If no specific values for the capacity parameters are given by the user, the model uses the default values from the German soil taxonomy (Ad-hoc-AG Boden, 2005). However, the user can give his/her own parameters in the input fields of the sheet, which are then used by the model instead of the default values.

Within the user interface, the user can check if all required files are available before starting the simulation. After running the model for selected or all polygons he can use the menu for graphical output to have a fist view on the model results. Graphical outputs are available for the time course of the main output

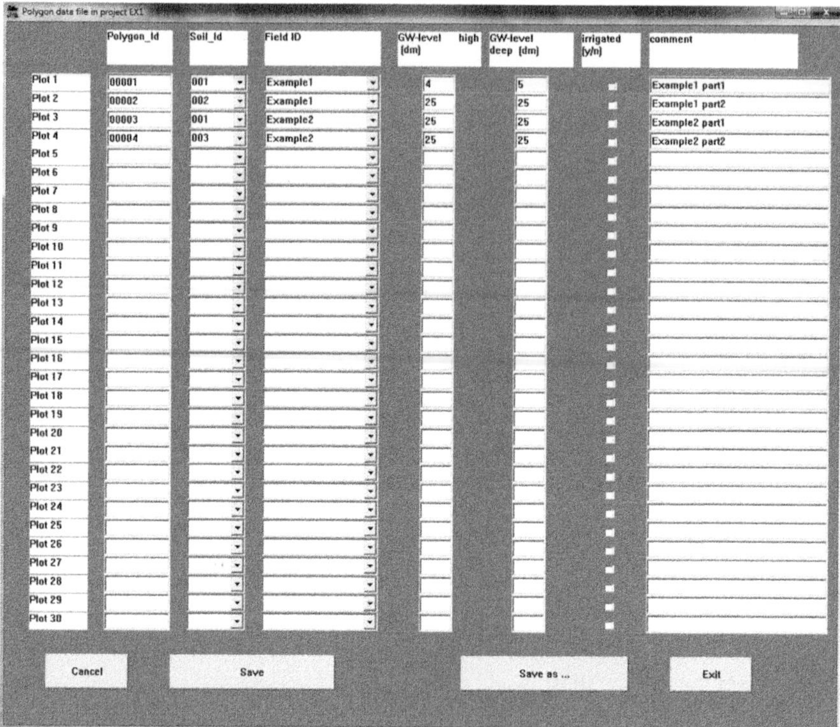

Fig. 2–5. Input window for the definition of polygons for HERMES (soil and fields have to be defined before this step).

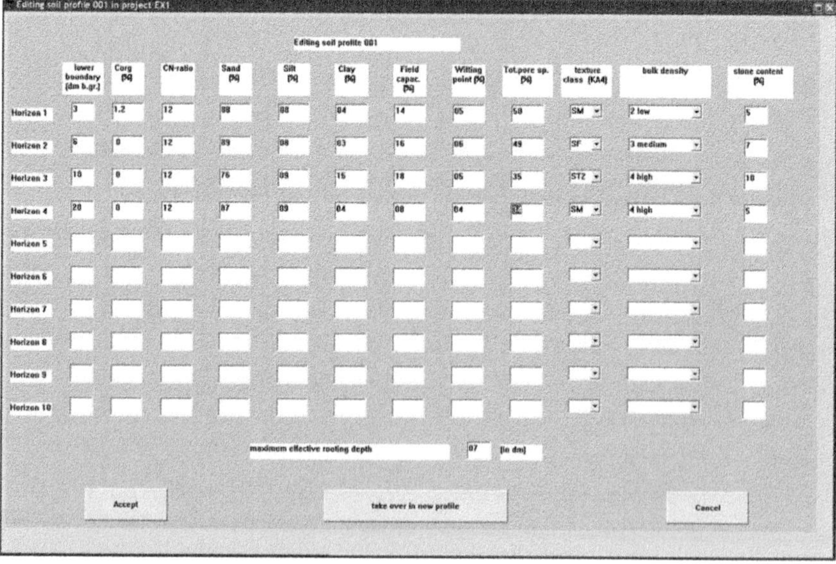

Fig. 2–6. Input window of HERMES for soil profile definition (soil texture data and capacity parameter are optional, to modify default parameters from texture class).

variables (water and nitrogen contents in different depth increments, crop organs, total biomass, N uptake, cumulative leaching, mineralization and denitrification), as depth profiles for soil water and nitrogen or as measured versus simulated graph with a linear trendline and R^2 estimation. Measurements are taken from a separate file, which has to be an external, prepared Excel sheet.

Potential users may contact the developer by email to ckersebaum@zalf.de to obtain the latest version of the HERMES model.

Model Calibration

Because the model is designed to be used by nonscientific users, the parameters that can be adjusted in the external parameter files are limited. Mainly the most important parameters for soil hydrology and the mineralization of soil organic matter, residues and fertilizers, and for crop growth can be adjusted. Therefore, this section addresses how to adjust these parameters using field observations.

In general, the calibration of a model integrating different processes has to be done using observed data of more than one output state variable. For example, it makes no sense to adjust some model parameters to fit only data of annual yields if soil water dynamics are not realistically simulated at the same time. Therefore, calibration requires balanced datasets that contain as much of the different observed state and flux variables as possible. An example of a typical dataset that was used for a model comparison (Kersebaum et al., 2007a) is shown in Fig. 2–7. The importance of using such datasets of different variables was also documented by the variation of single processes simulated by the participating models (Kersebaum et al., 2007a).

Measurements should be performed to reflect the dynamics of the observed system in time and space. Time resolution of observations should consider the relevant changes over time. However, temporal and spatial scales have to be comparable. Many processes show high spatial and temporal variability. For example, denitrification requires a coincidence of higher temperatures, anoxic conditions, and the presence of nitrate at a certain location. On a very small scale these conditions may vary within very small distances, leading to a high spatial variability of measurements. To integrate this variability to a larger scale, such as field scale, measurements should cover a representative elementary volume or area to avoid too much noise in the observations due to microscopic effects (Wagenet, 1998). Also the vertical resolution should be adequate for the scale, the problem to be addressed, and the model approach used. If a simplified functional model approach for water dynamics, like the capacity approach in HERMES, is used, a depth increment of 30 cm for the observations is better than 5- or 10-cm resolution,

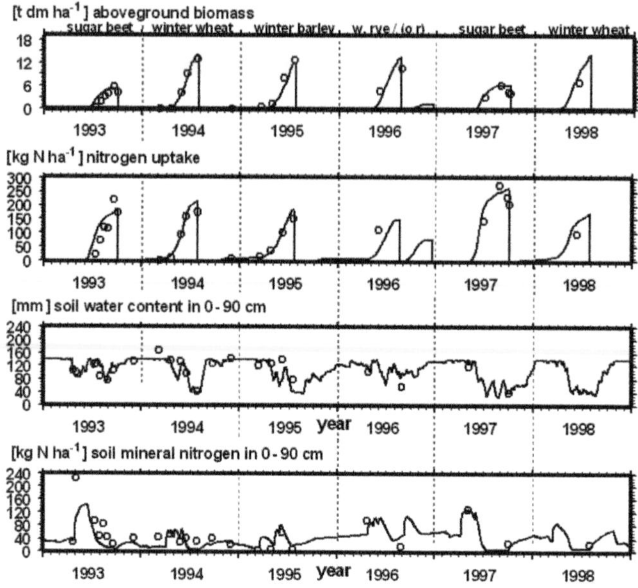

Fig. 2–7. Example of a dataset for model calibration and validation, using a sandy soil at Muencheberg, Germany (data: Mirschel et al., 2007; simulation: Kersebaum, 2007).

which might be suitable for a model using the Richards equation. In agricultural practice increments of 20 or 30 cm are widely used for measurements of soil nutrients, which is usually sufficient for practical purposes on a field scale.

A general scheme for model calibration is shown in Fig. 2–8. Calibration of parameters should start with the adjustment of soil water dynamics and crop phenology because k_c factors for evapotranspiration are linked to the crop development stage. In a next step crop growth parameters can be adjusted using total biomass and crop organ (leaves, stem, storage organ) weight data. This adjustment of crop parameters should be done preferably with data from field trials under optimum growth conditions, which means that at least the N supply should be sufficient. Finally soil mineral N and crop N data can be used to calibrate parameters for N mineralization and transport.

Soil Hydraulic Parameters

Usually, basic soil data such as texture and organic matter content are used in HERMES to derive model parameters automatically as described above. For German conditions the standard parameters for field capacity, wilting point, and total pore space given for three different bulk density classes in German soil taxonomy (Ad-hoc-AG Boden, 2005) can be applied. However, the relationship of these param-

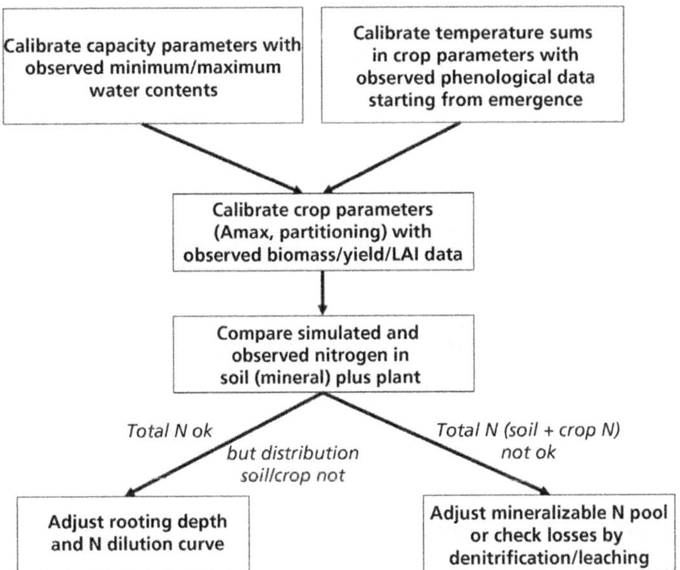

Fig. 2–8. Scheme of model calibration.

eters to soil texture can be different, especially under different climatic conditions. Therefore, the values stored in the external parameter file can be adjusted to specific field conditions in the specific soil profile description file as described above. One advantage of the capacity approach used in HERMES is that the hydraulic parameters can be easily obtained from simple soil water content measurements. Values for field capacity can be determined gravimetrically during the drainage season, usually in winter a few days after the last rainfall, while water content at wilting point can be obtained after a longer dry period in summer, such as after harvest of cereals. In the case of a shallow water table, parameters for capillary rise have to be calibrated if the default values are not suitable.

Crop Phenology

As shown in Table 2–2 the length of the development stages are mainly dependent on accumulated daily average air temperatures. At some growth stages daylength requirements and vernalization requirements might influence the summation of daily temperature and crop development. These crop-specific requirements are normally derived from literature. Therefore, the calibration can be focused on the adjustment of the required thermal sum for each development stage. In the output file the modified temperature sum and the "number of development stage" is given for each day of a simulation. By comparing the temperature sum of the simulated date of a development stage with the observed date development stages can be adjusted one by one starting from emergence.

Soil Water Dynamics

In the next step the k_c values to transfer reference grass evapotranspiration into crop evapotranspiration can be adjusted starting from suggested values given in the literature. Micrometeorological measurements or lysimeter data can be used for this adjustment. However, under field conditions measured water content data for the root zone are a good alternative. The most reliable data come from gravimetric values taken in reasonable time intervals. Although continuous measurements from sensors like time domain reflectometry or frequency domain reflectometry provide data with a high temporal resolution, they should be calibrated with gravimetric values to avoid systematic errors in the measurements (Wegehenkel, 2005). Soil water dynamics in the lower part of the profile can be used to find out the effective depth of water extraction by crop roots. The setting of the effective rooting depth is very important because it may influence not only soil water dynamics but also nutrient uptake and crop growth significantly.

Crop Growth and Partitioning

One of the most sensitive parameters concerning biomass formation is the maximum carbon dioxide assimilation rate at light saturation per leaf area (A_{max} in kg CO_2 h^{-1} ha^{-1} leaf). Data of the total aboveground biomass can be used for a first adjustment. However, this requires more or less optimal growth conditions during the observation period. Measurements of dry matter weights per area of the different crop organs are required to calibrate the parameters for biomass partitioning to the defined crop organs. Figure 2–9 shows an example of measured and HERMES simulated corn leaf, stem, and cob for a maize (*Zea mays* L.) FACE experiment performed in Braunschweig, Germany (Manderscheid and Weigel, 2010, unpublished data). Because the LAI of the crop is the main interface to absorb radiation for photosynthesis, the leaf weight has to be transformed into a leaf area using the specific leaf area (cm leaf kg^{-1} leaf weight), which should be defined for each development stage. This parameter can be estimated by measuring the weight and area of leaves that are formed during a specific development stage. This parameter is often taken from the literature.

Root distribution can affect water and nutrient uptake and consequently crop growth. However, root distribution measurements are laborious and rarely performed in field research. In addition to the problem of determining the fibrous

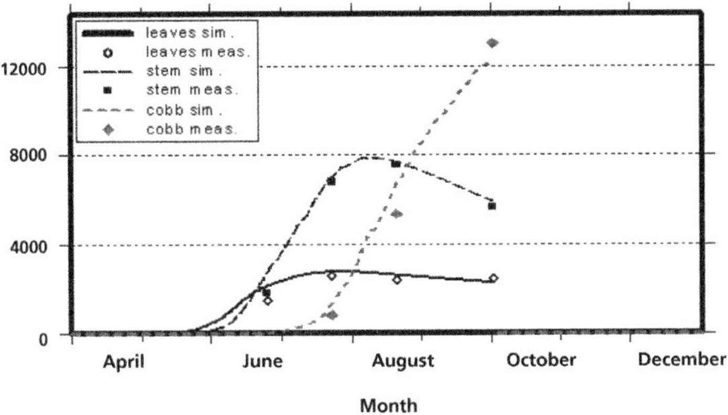

Fig. 2–9. Measured dry matter biomass of different crop organs of maize to calibrate biomass partitioning in crop parameter file of HERMES (data: Manderscheid and Weigel, 2010, unpublished data).

roots, the measurement always shows a snapshot of a present situation, while there is continuous root growth and decay. In practice, it is more important to estimate maximum rooting depth for a specific site as described above.

Nitrogen Dynamics

Calibration of nitrogen related processes is more difficult because so many different processes are involved. First, summarized annual model outputs of each process should be compared with literature values under similar conditions. Second, simulated N in the soil–crop system at different times should be compared with experimental measurements. This would require time series measurements of N in the crop (mainly the aboveground parts) and soil mineral N content in the root zone at the same time. If the combined soil mineral N and N in crops shows differences between observations and simulations, the total N balance has to be calibrated. This can be done by changing the percentage of the mineralization potential of the total soil N (slowly decomposable pool) or the fractionation into the different N pools of the crop residue parameters or fertilizers (see fertilizer parameters in Table 2–5). Also one has to check whether simulated losses by leaching as well as by denitrification are in a reasonable range. However, even when the total sum is estimated fairly accurate, the distribution between soil and crop may not be reflected well. This can be a problem of crop growth or, if this has been checked before, of a wrong root distribution.

For long-term simulations long and dense time series might be too laborious. Another possibility for examining model results for a longer period is to extend the depth of sampling using the deep profile as a temporal summary of the N dynamics over a longer period. This requires at least two sampling dates,

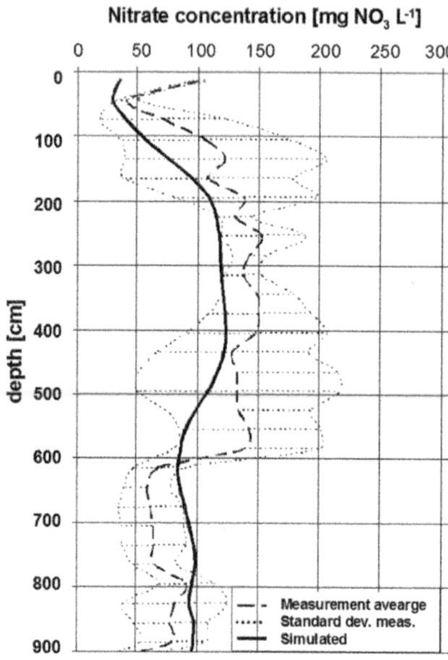

Fig. 2–10. Measured and simulated nitrate concentrations in soil solution of a 9-m-deep profile at Getelo-Itterbeck (sandy Podzol soil with loamy underground, Grimm 2008, modified).

one to estimate the initial distribution and a second one at the end of the simulation period. However, sampling has to consider the spatial variability even within a soil unit. Figure 2–10 shows an example of a deep profile of 9 m covering a crop rotation of maize, spring barley (*Hordeum vulgare* L.), grass, winter rye (*Secale cereale* L.), mustard (*Brassica* L.), winter rye on a Podzol soil (Grimm, 2008). It demonstrates the relatively high variability of measured nitrate concentrations in different layers. However, using the HERMES model with uniform soil data gives only one line, which is mostly within the standard deviation of the observations. Although, the consideration of variable soils is possible, as will be shown in the next section, large differences can occur even within very small distances, making calibration difficult.

Model Applications and Validation

One possibility to assess the validity of the model and its sensitivity to different site conditions is to use spatially variable input data, which can typically be obtained for precision farming within a heterogeneous field. Figure 2–11 shows results of soil moisture and soil mineral N measured at harvest at 60 locations on a 50 by 50 m grid on a 20-ha field. Additionally spatial distribution of crop yields were determined using a yield monitor on a combine harvester. From these data, the average yield and standard deviation of all measurements within a radius of 10 m around the soil sampling grid points were determined. Simulations with

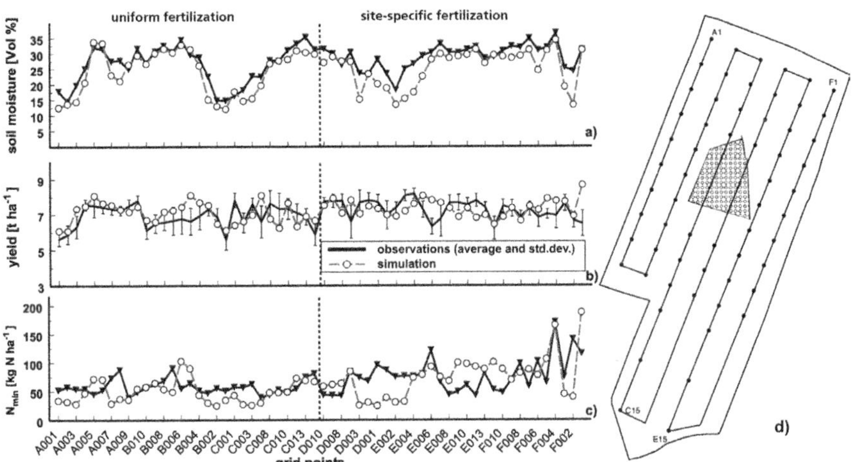

Fig. 2–11. Comparison of observed and simulated (a) soil moisture, (b) grain yields of winter wheat, and (c) soil mineral N along a transect of 60 grid points (d) on a field in North Rhine-Westphalia, Germany in the summer of 2000.

HERMES were performed using the texture and organic matter contents measured around the grid points as described above using the standard parameters of German soil taxonomy. Variability of texture is reflected by the differences in soil moisture after harvest. Yields were mostly in the range of one standard deviation of observations within the surrounding area. Low yields occurred at the southern part of the field (C–E15) due to a shallow marl layer.

The original purpose of the HERMES model was to derive N fertilizer requirements, which requires the model to assess plant N demand at a specific date for a future period, for example, until the next phenological stage when N is usually applied. The procedure to derive fertilizer recommendations is described in the following paragraphs.

The basic concept for calculating fertilizer recommendations was described in detail by Kersebaum and Beblik (2001). The simulation of N dynamics in the soil–crop system usually starts at known initial mineral N contents in the soil profile, for example, after harvest of the previous crop, and using actual weather data from a neighboring weather station. To make a N recommendation, the model predicts N uptake and soil mineral N changes operating with typical site-specific weather scenarios until the next predicted development stage for fertilization (e.g., stem elongation for winter wheat fertilization). The model accumulates the daily deficiency between N uptake required for N-unlimited growth and available N in soil for the prediction period. The total deficiency is recommended to be applied before the model indicates a critical deficiency period (Fig.

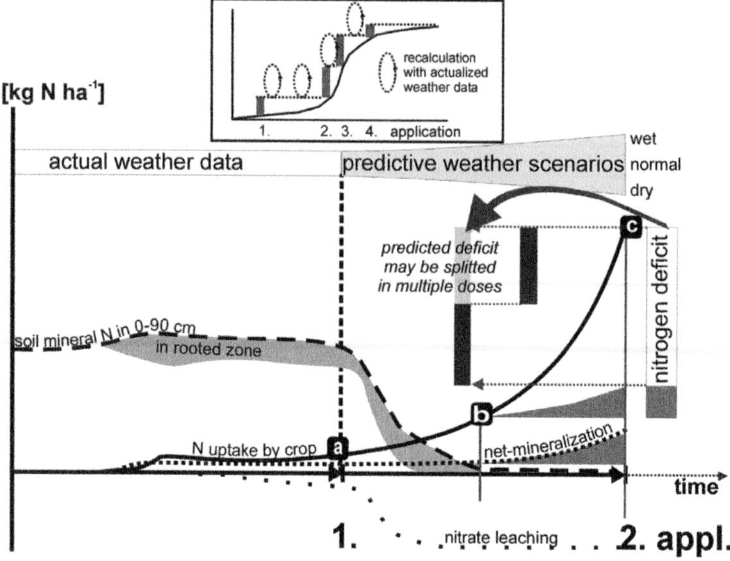

Fig. 2–12. Scheme of calculating N fertilizer recommendations with HERMES (modified from Kersebaum et al., 2007b).

2–12). High recommended amounts might be divided into two doses. Between the relevant stages the model should be run after shorter intervals (e.g., 10–15 d) with actualized weather data. If the actual weather deviates greatly from the assumed typical weather scenario (e.g., N is leached due to high precipitation), the next fertilization has to be given earlier. If soil N supply is higher than predicted crop demand, the next fertilizer recommendation is reduced automatically.

Usually, a typical "representative" weather dataset is selected from long-term data. However, the weather varies substantially across the years. Alternatively, the user can select a typical dry or wet year from the weather database to assess the uncertainty in N recommendation due to the weather variability. This uncertainty was assessed by running the model with a 30-yr weather dataset (Kersebaum et al., 2007b) for selected grid points of the precision farming field investigation mentioned in Fig. 2–11. In Fig. 2–13a the variability of model derived fertilizer recommendations from 1973 to 2001 is shown for the wet, dry, and normal weather scenarios at grid location A1, which is a loamy sand with a high yield potential. The same is shown for grid location E15 (Fig. 2–13b), which is a silty clay loam with a low yield potential due to a shallow root zone. The different yield potentials are reflected by the difference in the average "normal" recommendation, which is 135 kg N ha^{-1} for A1 compared to 53 kg N ha^{-1} for the E15 site. Recommendations for A1 varied from 82 to 180 kg N ha^{-1}, with a maximum difference among the scenarios of 40 kg N ha^{-1}, which reflects the uncertainty due to weather variability. For the E15

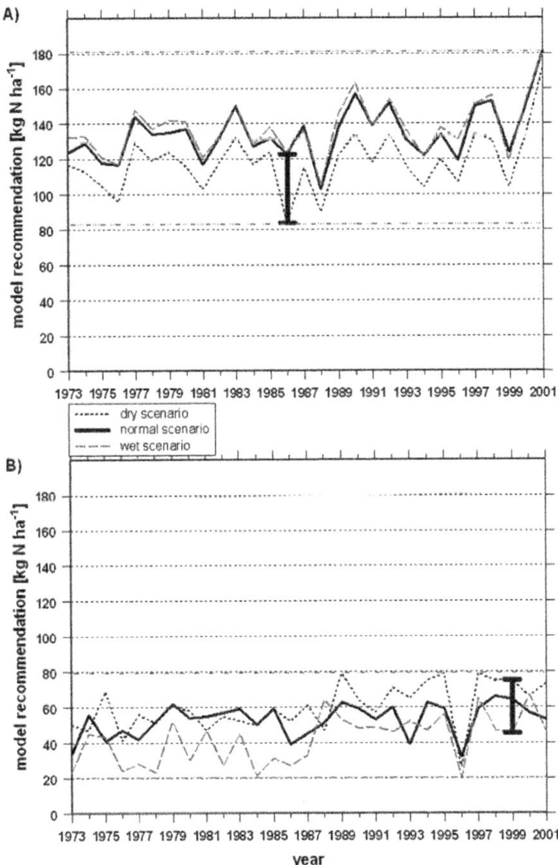

Fig. 2–13. Simulated N fertilizer recommendations for two grid locations on field Autobahn (see Fig. 2–6) using local weather data from 1971 to 2001: (A) loamy sand site A1 and (B) clay loam with shallow root zone at E15 location. Error bars show largest deviation between scenarios used for each prediction (Kersebaum et al., 2007b, modified).

location the maximum deviation among scenarios was only 31 kg N ha^{-1}. However, the average "normal" recommendation for the E15 site was only 53 kg N ha^{-1}, and the range of all recommendations was between 20 and 80 kg N ha^{-1}, which means that the potential error in N recommendation was about 50% of the total variation.

The procedure described above was also applied for real-time recommendations of a field trial in Saxony on spatial variable N fertilization using the same crop parameters for winter wheat in 2 yr (2000 and 2002). As inputs we used soil organic matter (0–30 cm) and soil texture from three depth increments (0–30, 30–60, and 60–90 cm) at 64 locations of a grid auger sampling across the field. Because of the hummocky relief of the field we used the topographic wetness index as a hydromorphic correction and a generalized spatial correction of the measured radiation to consider shading effects at footslope positions. For the field trial we separated 64 grid cells around the grid points of our auger sampling for site-specific management. And run the model for each grid cell to derive fertilizer recommendations as described above. In this field trial the model was

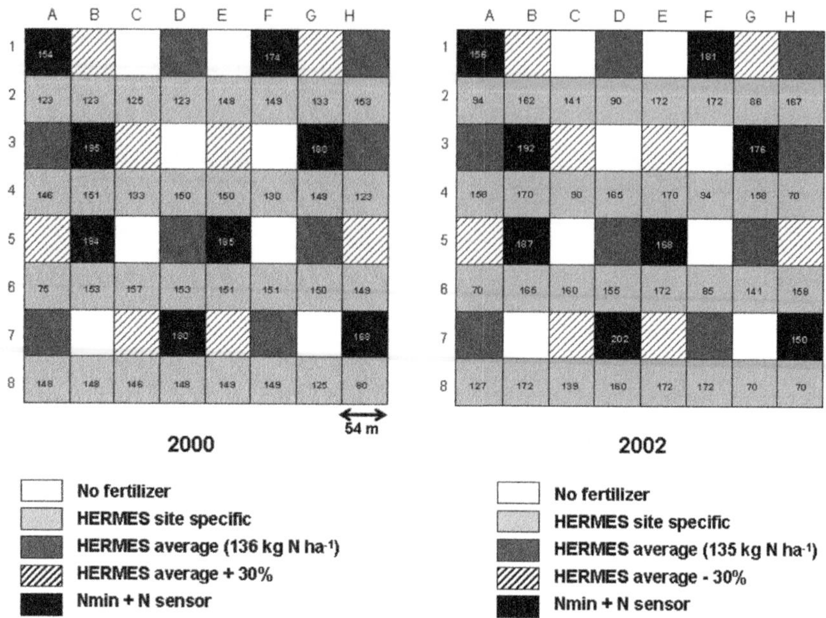

Fig. 2–14. Experimental design of a winter wheat field trial in 2000 and 2002 at Luettewitz, Saxony, Germany with real-time N fertilizer recommendations of the HERMES model and recommendations based on soil mineral N measurements and N sensor measurements (N_{min} + N sensor). Number show the site specific N fertilization for the whole season.

compared with a recommendation based on combined soil analysis and crop N status measurements using the Hydro N-sensor (e.g., Link et al., 2002). The spatial experimental design and the given total seasonal amounts of nitrogen fertilizer are given in Fig. 2–14. In both years model recommendations achieved the same yield on average, with 40 kg N ha^{-1} less than the other method (Kersebaum et al., 2005). Especially areas of reduced yield in the wet footslope positions require less fertilizer than high yield areas, which was considered by the model recommending lower fertilizer recommendations as for the rest of the field. In contrast, the N sensor increased the recommendation at these locations. Nevertheless, yield performance at these locations was poor despite the high fertilization while the lower recommendation (not applied) of the model was in line with the lower yield level. More details of these study were given by Kersebaum et al. (2005).

Conclusions

Agroecosystem models are increasingly applied to support decision-making at different scales. Applications range from fertilizer recommendations for farmers on a field scale and water suppliers on a catchment scale to a landscape or regional scale for strategic policy decision support. There is an increasing demand from model users and decision-makers that the validity of models used for decision support has been proven by comparing uncalibrated modeling results with field observations outside the range of model development, which should enhance the reliability of model calculations for decision-making. However, the question of model validation is contradictorily discussed in literature. Oreskes et al. (1994) generally disavowed the possibility of validating models for natural systems; they argued that natural systems are generally open systems, always involving uncertain inputs. Rykiel (1996) concluded that validation is neither a method to test model theory nor a certificate of truth. He mentioned that validation is just an indication that a model can give acceptable results for a specific purpose. He also stated that this might not be scientifically necessary, but leads to a higher confidence and acceptance of model results. Holling (1978) proposed investigating the nonvalidity of models by applying them without calibration to a large number of sites and conditions and accepting the risk of not reflecting the observations very well. This is thus suitable to demonstrate the degree of confidence and the limitations of a model approach.

Putting a model into a new environment or applying it for a new crop requires the adaptation and calibration of model parameters. Calibration and validation of agroecosystem models require consistent datasets of multiple state variables and fluxes. For field research the assessment of "effective" parameters for the field scale is usually more suitable than laborious laboratory measurements of single point locations because of the spatial and temporal variation of parameters. Using a functional approach may alleviate parameter estimation at the field scale because it is based mainly on easily measurable variables. Although spatial variability complicates model calibration, it also provides an opportunity to evaluate how sensitive a model reacts to variable inputs on a field, which often contain most of the variability of a whole landscape. Comparing model results with measured values must include a consideration of the uncertainties of inputs as well as the small-scale variability of observations.

References

Addiscott, T.M., and R.J. Wagenet. 1985. Concepts of solute leaching in soils: A review of modelling approaches. J. Soil Sci. 36:411–424.

Ad-hoc-AG Boden. 2005. Bodenkundliche Kartieranleitung. 5th ed. Schweizerbart, Stuttgart. 438p.

Allen, R.G., L.S. Pereira, D. Raes, and M. Smith. 1998. Crop evapotranspiration. Guidelines for computing crop water requirements. Irrigation and Drainage Paper 56. FAO, Rome.

Ångström, A. 1924. Solar and terrestrial radiation. Q. J. R. Meteor. Soc. 50:121–126.

de Willigen, P. 1991. Nitrogen turnover in the soil–crop system; Comparison of fourteen simulation models. Fert. Res. 27:141–149.

Diekkrüger, B., D. Söndgerath, K.C. Kersebaum, and C.W. McVoy. 1995. Validity of agroecosystem models—A comparison of results of different models applied to the same data set. Ecol. Model. 81:3–29.

Duval, R., J.M. Machet, and F. Maupas. 2003. Critical curve for the dilution of nitrogen under sugar beet. p. 47–61. *In* Advances in sugar beet research. Vol. 5. IIRB, Brussels, Belgium.

Gerwitz, A., and E.R. Page. 1974. An empirical mathematical model to describe plant root systems. J. Appl. Ecol. 11:773–781.

Greenwood, D.J., and A. Draycott. 1995. Modelling uptake of nitrogen, phosphate and potassium in relation to crop growth. p. 155–175. *In* P. Kabat et al. (ed.) Modelling and parameterization of the soil plant system—A comparison of potato growth models. Wageningen Press, Wageningen, The Netherlands.

Greenwood, D.J., G. Lemaire, G. Gosse, P. Cruz, A. Draycott, and J.J. Neeteson. 1990. Decline of percentage N of C3 and C4 crops with increasing plant mass. Ann. Bot. (Lond.) 66:425–436.

Grimm, M. 2008. Modellierung landwirtschaftlicher Maßnahmen zur Minderung der Nitratauswaschung im Wasserschutzgebiet Getelo-Itterbeck in Niedersachsen. (In German.) Ph.D. diss. Univ. Giessen, Germany.

Grunwald, S. 1997. GIS-gestützte Modellierung des Landschaftswasser- und Stoffhaushaltes mit dem Modell AGNPSm. (In German.) Boden und Landschaft 14.

Haude, W. 1955. Zur Bestimmung der Verdunstung auf möglichst einfache Weise, Mitteilgn. Dtsch. Wetterdienst 11.

Heger, K. 1978. Bestimmung der potentiellen Evapotranspiration über unterschiedlichen landwirtschaftlichen Kulturen. Mitteilgn. Dtsch. Bodenkundl. Ges. 26:21–40.

Holling, C.S. 1978. Adaptive environmental assessment and management. John Wiley and Sons, New York.

Institut International de Recherches Betteravières. 2003. Sugar beet growth and growth modelling. *In* Advances in sugar beet research Vol. 5. IIRB, Brussels, Belgium.

Kabat, P., B. Marshall, B.J. van den Broek, J. Vos, and H. van Keulen (ed.) 1995. Modelling and parameterization of the soil plant system—A comparison of potato growth models. Wageningen Press, Wageningen, The Netherlands.

Kersebaum, K.C. 1989. Die Simulation der Stickstoffdynamik in Ackerböden. Ph.D. diss. University Hannover, Germany.

Kersebaum, K.C. 1995. Application of a simple management model to simulate water and nitrogen dynamics. Ecol. Model. 81:145–156.

Kersebaum, K.C. 2000. Model-based evaluation of land use strategies in a nitrate polluted drinking water catchment in North Germany. p. 223–238. *In* R. Lal (ed.) Integrated watershed management in the global ecosystem. CRC Press, Boca Raton, FL.

Kersebaum, K.C. 2007. Modelling nitrogen dynamics in soil–crop systems with HERMES. Nutr. Cycling Agroecosyst. 77:39–52.

Kersebaum, K.C., and A.J. Beblik. 2001. Performance of a nitrogen dynamics model applied to evaluate agricultural management practices. p. 551–571. *In* M. Shaffer et al. (ed.) Modeling carbon and nitrogen dynamics for soil management. CRC Press, Boca Raton, FL.

Kersebaum, K.C., J.-M. Hecker, W. Mirschel, and M. Wegehenkel. 2007a. Modelling water and nutrient dynamics in soil-crop systems: A comparison of simulation models applied on common data sets. p. 1–17. *In* K.C. Kersebaum et al. (ed.) Modelling water and nutrient dynamics in soil crop systems. Springer, Dordrecht, The Netherlands.

Kersebaum, K.C., K. Lorenz, H.I. Reuter, J. Schwarz, M. Wegehenkel, and O. Wendroth. 2005. Operational use of agro-meteorological data and GIS to derive site specific nitrogen fertilizer recommendations based on the simulation of soil and crop growth processes. Phys. Chem. Earth 30:59–67.

Kersebaum, K.C., K. Lorenz, H.I. Reuter, O. Wendroth, A. Giebel, and J. Schwarz. 2003a. Site specific nitrogen fertilisation recommendations based on simulation. p. 309–314. *In* J. Stafford and A. Werner (ed.) Precision agriculture. Proc. 4th European Conference on Precision Agriculture, Berlin. Wageningen Academic Publishers, Wageningen, The Netherlands.

Kersebaum, K.C., C. Nendel, W. Mirschel, R. Manderscheid, H.-J. Weigel, and K.O. Wenkel. 2009. Testing different CO_2 response algorithms against a FACE crop rotation experiment and application for climate change impact assessment at different sites in Germany. Idöjárás 113:79–88.

Kersebaum, K.C., H.I. Reuter, K. Lorenz, and O. Wendroth. 2007b. Model based nitrogen fertilization considering agro-meteorological data. p. 1–9. *In* T. Bruulsema (ed.) Managing crop nitrogen for weather. Proc. Soil Sci. Soc. Am. Symposium Integrating weather variability into nitrogen recommendations. IPNI, Norcross, GA.

Kersebaum, K.C., and J. Richter. 1991. Modelling nitrogen dynamics in a plant–soil system with a simple model for advisory purposes. Fert. Res. 27:273–281.

Kersebaum, K.C., J. Steidl, O. Bauer, and H.-P. Piorr. 2003b. Modelling scenarios to assess the effects of different agricultural management and land use options to reduce diffuse nitrogen pollution into the river Elbe. Phys. Chem. Earth 28:537–545.

Kersebaum, K.C., A. Wurbs, R. De Jong, C.A. Campbell, J. Yang, and R.P. Zentner. 2008. Long term simulation of soil–crop interactions in semiarid southwestern Saskatchewan, Canada. Eur. J. Agron. 21:1–12.

Leviel, B., C. Crivineanu, and B. Gabrielle. 2003. CERES-BEET, a prediction model for sugar beet yield and environmental impact. p. 143–152. *In* Advances in sugar beet research Vol.5. IIRB, Brussels, Belgium.

Link, A., M. Panitzki, and S. Reusch. 2002. Hydro-N-Sensor: Tractor mounted remote sensing for variable nitrogen fertilization. p. 932–937. *In* P.C. Roberts (ed.) Proc. 6th Int. Conf. Precision Agriculture, Minneapolis, MN. ASA, CSSA, and SSSA, Madison, WI.

Mirschel, W., K.-O. Wenkel, M. Wegehenkel, K.C. Kersebaum, U. Schindler, and J.-M. Hecker. 2007. Müncheberg field trial data set for agro-ecosystem model validation. p. 219–243. *In* K.C. Kersebaum et al. (ed.) Modelling water and nutrient dynamics in soil crop systems. Springer, Dordrecht, The Netherlands.

Myers, R.J.K., C.A. Campbell, and K.L. Weier. 1982. Quantitative relationship between net nitrogen mineralization and moisture content of soils. Can. J. Soil Sci. 62:111–124.

Nendel, C., and K.C. Kersebaum. 2004. A simple model approach to simulate nitrogen dynamics in vineyard soils. Ecol. Model. 177:1–15.

Nordmeyer, H., and J. Richter. 1985. Incubation experiments on nitrogen mineralization in loess and sandy soils. Plant Soil 83:433–445.

Nuske, A. 1983 Ein Modell für die Stickstoff-Dynamik von Acker-Lössböden im Winterhalbjahr—Messungen und Simulationen. Ph.D. diss. Univ. Hannover, Germany.

Oreskes, N., K. Shrader-Fechette, and K. Belitz. 1994. Verification, validation and confirmation of numerical models in earth sciences. Science 263:641–646.

Palosuo, T., K.C. Kersebaum, C. Angulo, P. Hlavinka, M. Moriondo, J.E. Olesen, R.H. Patil, F. Ruget, C. Rumbaur, J. Takáč, M. Trnka, M. Bindi, B. Caldag, F. Ewert, R. Ferrise, W. Mirschel, L. Saylan, B. Šiška, and R. Rötter. 2011. Simulation of winter wheat yield and its variability in different climates of Europe. A comparison of eight crop growth models. Eur. J. Agron. (in press.)

Priestley, C.H.B., and R.J. Taylor. 1972. On the assessment of surface heat flux and evaporation using large-scale parameters. Mon. Weather Rev. 100:81–92.

Richter, D. 1995. Ergebnisse methodischer Untersuchungen zur Korrektur des systematischen Messfehlers des Hellmann-Niederschlagsmessers. Ber. Dtsch. Wetterd. 159:93.

Richter, J., A. Nuske, W. Habenicht, and J. Bauer. 1982. Optimized N-mineralization parameters of loess soils from incubation experiments. Plant Soil 68:379–388.

Richter, J., H.C. Scharpf, and J. Wehrmann. 1978. Simulation der winterlichen Nitratverlagerung in Böden. Plant Soil 49:381–393.

Rykiel, E.J., Jr. 1996. Testing ecological models: The meaning of validation. Ecol. Model. 90:229–244.

Schneider, U. 1991. Messungen von Denitrifikations- und Nitratauswaschungsverlusten in einem landwirtschaftlich genutzten Wassereinzugsgebiet, Ph. D. diss. Univ. Bonn, Germany.

Supit, I., A.A. Hooijer, and C.A. van Diepen (ed.) 1994. System description of the WOFOST 6.0 crop simulation model implemented in CGMS. Vol. 1. Theory and algorithms. EC Publication EUR 15956, Luxemburg.

Van Keulen, H., F.W.T. Penning de Vries, and E.M. Drees. 1982. A summary model for crop growth. p. 87–97. *In* F.W.T. Penning de Vriew and H.H. van Laar (ed.) Simulation of plant growth and crop production. PUDOC, Wageningen.

Wagenet, J. 1998. Scale issues in agroecological research chains. Nutr. Cycl. Agroecosyst. 50:23–34.

Wegehenkel, M. 2005. Validation of a soil water balance model using soil water content and pressure head data. Hydrol. Processes 19:1139–1164.

Wehrmann, J., and H.C. Scharpf. 1986. The Nmin-method—An aid to integrating various objectives of nitrogen fertilization. Z. Pflanzenernaehr. Bodenkd. 149:428–440.

Wendling, U., H.-G. Schellin, and M. Thomä. 1991. Bereitstellung von täglichen Informationen zum Wasserhaushalt des Bodens für die Zwecke der agrarmeteorologischen Beratung. Z. Meteorol. 41:468–475.

Yu, Q., J. Goudriaan, and T.D. Wang. 2001. Modelling diurnal courses of photosynthesis and transpiration of leaves on the basis of stomatal and non-stomatal responses, including photoinhibition. Photosynthetica 39:43–51.

Chapter 3

Parameterizing LEACHM Model for Simulating Water Drainage Fluxes and Nitrate Leaching Losses

Jay D. Jabro
John L. Hutson
Ann D. Jabro

Abstract

Model calibration, validation, and sensitivity analysis are essential processes in the application and evaluation of computer simulation models. The utility of a model requires consideration of parameter selection, sensitivity analysis, calibration, and validation, but the extent to which each of these aspects is pursued depends on the reasons for model use. This chapter discusses calibration, validation, and sensitivity analysis for the LEACHM model and an example of its performance using field measured annual water drainage fluxes and NO_3–N leaching losses under a continuous corn (*Zea mays* L.) cropping system. LEACHM calibration was performed to optimize the match between simulated and measured field data collected from control, fertilized, and manure treatments. The model was calibrated to the field site conditions using 1989–1990 data from a long-term leaching experiment conducted on a Hagerstown silt loam soil. The initial stage of calibration focused on small changes to soil water flow parameters in the model. The second phase, calibration, focused on input parameters controlling soil N transformation processes and rate constants in the model. After the calibration process was completed, the model was validated using 1988–1989, 1990–1991, 1991–1992, and 1992–1993 data by comparing model simulations with field measured data. Sensitivity analyses were performed on the calibrated LEACHM model by changing the values of input parameters within acceptable ranges. The LEACHM model was slightly sensitive to most input parameters; however, it was very sensitive to denitrification rates, initial soil water potential/water content, and the *b* parameter in the Campbell equation.

J.D. Jabro (jay.jabro@ars.usda.gov), USDA-ARS, Northern Plains Agricultural Research Lab., 1500 N. Central Ave., Sidney, MT 59270; J.L. Hutson, School of the Environment, Flinders Univ., Adelaide 5001, Australia; A.D. Jabro, School of Communications and Information Systems, Robert Morris Univ., Moon Township, PA 15108.

doi:10.2134/advagricsystmodel2.c3

Methods of introducing system models into agricultural research. L.R. Ahuja and L. Ma (ed.)
Advances in Agricultural Systems Modeling Series 2.

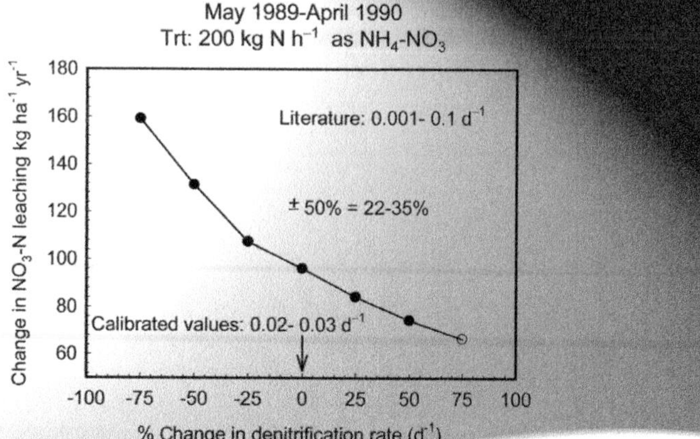
May 1989-April 1990
Trt: 200 kg N h⁻¹ as NH4-NO3
Change in NO3-N leaching kg ha⁻¹ yr⁻¹
Literature: 0.001- 0.1 d⁻¹
± 50% = 22-35%
Calibrated values: 0.02- 0.03 d⁻¹
% Change in denitrification rate (d⁻¹)

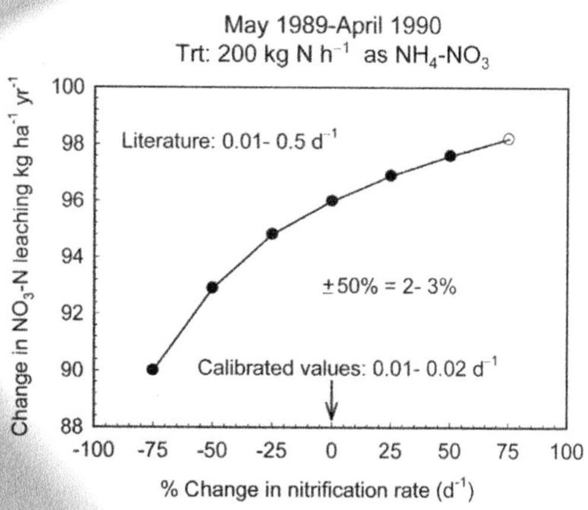
May 1989-April 1990
Trt: 200 kg N h⁻¹ as NH4-NO3
Change in NO3-N leaching kg ha⁻¹ yr⁻¹
Literature: 0.01- 0.5 d⁻¹
±50% = 2- 3%
Calibrated values: 0.01- 0.02 d⁻¹
% Change in nitrification rate (d⁻¹)

With the development and implementation of environmental regulations stemming from unintended consequences of agricultural and manufacturing operations, environmental awareness has heightened, prompting enhanced attention to environmental quality. Measures to conserve water resources have increased the need for a better understanding of human impacts on water quality. In water conservation efforts, one area of concern is the impact of crop production practices on NO_3–N contamination of surface and groundwaters.

Innovations in technology and software applications provided the impetus for the cluster technology of modeling. In the last two decades, processing speed to run sophisticated models has increased significantly, making practical applications of complex models affordable and accessible. Computer models serve as powerful tools to simulate complex systems for the purpose of understanding the linkages and interrelations of the components of an agricultural system, such as water usage, fertilizer amount, and temperature. Computer models can be adapted and adjusted to understand, analyze, and optimize systems where traditional experimental tools fail. Today, with the influx of model choices, proper use, calibration, field validation, and sensitivity analysis of such models are needed to ensure better quality output and promote more efficient and effective decision-making (Pennell et al., 1990; de Willigen, 1991; Khakural and Robert, 1993; Lengnick and Fox, 1994; Jabro et al., 1993, 1994, 1995, 1998, 2006; Jemison et al., 1994a,b). In turn, reliable computer model output serves to promote more efficient and effective decision-making (Pennell et al., 1990; de Willigen, 1991; Khakural and Robert, 1993; Lengnick and Fox, 1994; Jabro et al., 1993, 1994, 1995, 1998, 2006; Jemison et al., 1994a,b). The benefits of computer simulations are many. Models can be used to simulate site-specific management practices to optimize yields and minimize environmental impacts (e.g., Adapt-N, Melkonian), to assess relative differences between alternative management practices, or to assess temporal and spatial differences in water and nutrient balances across regional or watershed scales. The level of accuracy varies according to the application and the questions that are to be addressed.

The number of water quality models used to predict nitrate leaching through the root zone and into underlying unsaturated soil zone has grown rapidly over the last two decades. The creation, calibration, and validation of water quality computer models relating to agricultural practices lead to greater awareness and potential control of environmental impacts. These include LEACHM (Leaching Estimation And CHemistry Model) by Hutson and Wagenet (1992) and Hutson (2003), PRZM (Pesticide Root Zone Model) by Carsel et al. (1985), GLEAMS (Groundwater Loading Effects of Agricultural Management System) by Leonard et al. (1987); SLIM by Addiscott (1977) and Addiscott and Whitmore (1987), NLEAP (Nitrate Leaching and Economic Analysis Package) by Shaffer et al. (1991), SOIL-SOILN by Eckersten et al. (1996); GRASIM (GRAzing SImulation Model) by Mohtar et al. (1997a,b), Root Zone Water Quality Model (RZWQM) by Ahuja et al. (2000), and NCSWAP (Nitrogen and Carbon cycling in Soil Water And Plant) by Molina and Richards (1984). Evaluation of these models has also received increasing attention during the last two decades (Khakural and Robert, 1993; Jemison et al., 1994a,b; Ahuja et al., 1995; Jabro et al., 1998, 2001, 2006; Sogbedji et al., 2001; Mahmood et al., 2002; Larocque et al., 2002).

Overview of LEACHM Model

The focus of this discussion is the LEACHM model, which will be explained and linked to a research study to demonstrate the usability and effectiveness of the model. LEACHM is a general acronym that refers to several subversions of a simulation model that describes the water regime and the chemistry and transport of solutes in unsaturated and partially saturated soils (Hutson and Wagenet, 1992; Hutson, 2003). LEACHN describes N and P transport and transformation; LEACHP simulates pesticide displacement, degradation and movement; LEACHC describes the movement and leaching of major inorganic ions; and LEACHB describes microbial population dynamics in the presence of a single growth-supporting substrate (Hutson, 2003).

LEACHM is a one-dimensional deterministic model describing the storage, transport, and distribution of water and solute in a soil profile (Hutson and Wagenet, 1992; Hutson, 2003). The model either solves Richards' equation for water flow and the convection–dispersion equation (CDE) for chemical transport or utilizes a mobile–immobile capacity model (Addiscott, 1977; Hutson and Wagenet, 1991; Jabro et al., 1995). LEACHM uses either Campbell's equation (Campbell, 1973) or the van Genuchten water retention and hydraulic conductivity function (van Genuchten, 1980). In the nutrient model (LEACHN), nitrogen transformations are described in terms of fluxes between soil organic N pools (manure, litter and humus) and inorganic pools (urea, NH_4–N and NO_3–N). A complete description of the LEACHM model, equations, and estimation of the parameters, organic N pools and initial and boundary conditions are given in the LEACHM manual (Hutson and Wagenet, 1992; Hutson, 2003).

A plethora of reasons support the selection of the LEACHM model for this study:

- It is a field-scale research-type model.
- It is well documented.
- It is user-friendly.
- It requires manageable computer time.
- It is able to simulate water and nitrate leaching, N transformations (i.e., nitrification, denitrification, mineralization), soil nitrate distribution, and plant total N uptake.
- It uses different approaches and equations for water flow and N transport.
- It uses a manageable number of input parameters that are relatively measured, estimated or obtained from standard data.
- It has been evaluated and widely used in diverse geographic settings under various conditions.

Field Leaching Experiment and Data Collection

Soil and Site Description

The experimental site where data was collected and used for model evaluation is located in the R.E. Larson Agricultural Research Center of Pennsylvania State University (40°42′55″ N, 77°56′15″ W) at Rock Springs in central Pennsylvania about 16 km west of downtown State College, Pennsylvania. The field site is nearly level, and the soil series is mapped as a Hagerstown silt loam (fine, mixed, mesic, Typic Hapludalf) developed from limestone residuum parent material. The Ap horizon is about 20 to 30 cm deep and has a weak, fine, granular structure. The B horizon is silty clay to clay textured with well-developed blocky structural peds.

Leaching Experiment and Nitrogen Treatments

Water drainage and NO_3–N leaching data were collected from a long-term water quality study established in April 1988 to monitor nitrate leaching losses from corn as a function of N fertilizer rate and manure use. The experimental design was a split-plot with three replicates, with the whole plot treatments being manure or no manure and the split plot treatments being five application rates of NH_4NO_3 fertilizer. The water quality experiment, extending from 1988 to 1993, examined nitrate leaching from nonmanured, fertilized corn receiving 0 to 200 kg fertilizer N ha^{-1} annually in 50-kg increments and from manured, fertilized corn receiving 0 to 100 kg fertilizer N ha^{-1}annually in 25-kg increments. In addition to the fertilizer N, manured corn received dairy manure slurry in amounts supplying 264, 132, and 158 kg total N ha^{-1} in 1988, 1989, and 1990, respectively (Jemison et al., 1994a,b).

Soil water samples were collected using zero-tension pan lysimeters placed 1.2 m below the soil surface following each precipitation incident with an amount adequate to cause leaching. Water samples were then measured and analyzed for NO_3–N by an automated Cd reduction method (USEPA, 1979). The water drainage and NO_3–N were used to calibrate, validate, and assess the performance and accuracy of the LEACHM model.

Three sets of results within the N leaching treatments were selected for model evaluation. The data set included (i) the control treatment consisting of zero N addition for all years, (ii) the fertilizer treatment with N supplied as NH_4NO_3 at a rate of 200 kg N ha^{-1} for all years, and (iii) the manure treatment with N supplied as dairy manure slurry for the years 1988–1989, 1989–1990, and 1990–1991. In addition, two treatments from one array of nine lysimeters planted to corn were selected to evaluate the accuracy of the LEACHM model. This data set included treatments of 0 and 200 kg N ha^{-1} for the years 1991–1992 and 1992–1993 (Jemison et al., 1994a,b; Jabro et al., 2006).

Input Data Requirements

LEACHN requires input data describing soil properties for each layer, crop growth and phenological data, soil management information, N transformation and their rate constants, and boundary conditions (fertilizer applications, the lower boundary, and temperature and weather data). Some considerations for each of these data categories are discussed.

Soil Data

Depth-related data can be derived either from soil survey information or site-specific, measured data. At a minimum, soil texture can be translated into particle-size data for each depth, which can in turn be used, along with bulk density and organic carbon (OC), in conjunction with various pedotransfer functions to estimate water retention. These data can be evaluated either by comparing simulated water contents and fluxes with measured soil water contents and potentials or, more often, by performing some simple simulations that should lead to water content and matric potential distributions known to be representative of that soil type. Plots of water content distribution (e.g., at saturation and −5, −10, and −1500 kPa) allow approximate plant-available water to be evaluated through the profile and compared with guidelines for soils of that texture. Unsaturated hydraulic conductivity is rarely measured directly. A good starting point is to assume that, at field capacity, drainage from the lower boundary is a few millimeters per day, which implies that the hydraulic conductivity, under unit gradient, will also be a few millimeters per day. A useful simulation is that of drainage from saturation. Assuming unit gradient drainage at the lower boundary we expect that most soils will reach a field capacity of about −5 to −10 kPa after one or two days of drainage. Such a simulation enables saturation drainage to be checked, provides an indication of the effective field capacity (drained upper limit), and provides a means to estimate matric potentials at field capacity for a profile if the capacity flow option is to be used.

Initial Chemical Conditions

Initial chemical conditions are usually more difficult to estimate because they not only have to be realistic for the soil but need to be in equilibrium with the boundary conditions. Organic nitrogen and carbon pools, for example, must not be subject to abnormally high or low initial mineralization rates, as these give rise to a flush of mineralized N, which can be confused with the effects of an application of inorganic fertilizer or a change in land use. Our usual practice is to simulate N, C, and P dynamics by using repetitions of weather data until a dynamic equilibrium is attained. The N, C, and P concentrations obtained are then used as starting conditions for simulations to study the effects of changes in boundary conditions, land use, or management practices. Depending on the initial level of disequilibrium, this can require several hundred years of simulation. Also, when plant residues are recycled into the root zone, assumptions need to be made as to their depth distribution. In LEACHM, vegetation is defined and not explicitly simulated, so there is no feedback between soil conditions and plant growth except in the case of N uptake, which is reduced during periods of N deficiency and subsequently reduces plant C as well. Plant parameters such as C/N ratios in plant residues harvested and recycled fractions and how plant residues are allocated to the soil surface and through the profile all play a role, along with mineralization and transformation rate constants, in determining the "equilibrium" distribution of organic nutrient pools.

Weather Data

Weather data are crucially important. While rain is usually measured, potential evapotranspiration (ET) is usually based on estimates derived from evaporation pans, temperature, radiation, and humidity. While the FAO56 procedures (see http://www.kimberly.uidaho.edu/ref-et/fao56.pdf, verified 28 Mar. 2011), among others, are commonly used, other input data in LEACHM determine the split between potential transpiration and potential evaporation. Preliminary simulations should be conducted to ensure that the overall water balance corresponds to that expected for a specific environment and cropping pattern. This is particularly true for irrigated systems, where the long-term average leaching fraction could logically expect to lie within certain limits. Thus a mismatch between applied irrigation water and ET could lead to abnormally low or high drainage values or transpiration to evaporation ratios. LEACHM does have an automated triggered irrigation option, and it is often useful to compare these amounts and frequencies with a defined schedule. Soil temperature is used in LEACHM solely to adjust rate constants. The surface boundary temperature is air temperature, but the lower boundary is a constant, assumed to be the long-term average at a hypothetical soil depth of 2 m. This temperature is best obtained from a long-term average air temperature.

Crop Growth and Phenological Data

The crop growth and phenological data include crop type, crop cover fraction, planting, germination, emergence and harvest dates, root depth and distribution, and N crop uptake. As mentioned above, the temporal variation in crop cover fraction partly determines transpiration. Because the drainage depends on rain–evapotranspiration (R–ET), the pattern of both evapotranspiration and temperature through the growing season needs consideration. One issue sometimes ignored is the development of effective mulch as the crop approaches and passes senescence and harvest, although there is a mulching factor that reduces potential evaporation.

Management and Nitrogen Transformation Rate Constants

The management information for the simulation site includes type and timing of tillage and amounts, dates and rate of fertilization.

Nitrogen transformation rate constants include data for nitrification; denitrification; mineralization for residue, manure, and humus; urea hydrolysis; ammonia volatilization from the soil surface; the humification fraction; the synthesis efficiency factor; and additional rates and constants used to calculate N transformation, such as denitrification, half-saturation constant, and NO_3/NH_4 ratio in solution for nitrification.

The input parameters used by the model were measured, estimated, obtained from literature sources, or suggested by the model developers (Hutson and Wagenet, 1992; Hutson, 2003).

In summary, the procedure for estimating data for the compilation of an initial data set is:

1. Define the soil profile, and ensure that the water retention data patterns with depth are similar to those expected.

2. Using best estimates of an unsaturated conductivity–matching factor, simulate drainage from saturation in the absence of vegetation and evaporation. Ensure that saturated conductivity values are sufficient to allow the infiltration rates envisaged. Evaluate the predicted field capacity values through soil profile to ensure that drainage is neither too fast nor too slow. These field capacity values may also be used if the capacity model is to be used in place of the Richards equation or CDE.
3. Create the rain and ET data files. There is a utility in LEACHM that simplifies this procedure. By default, rain is assumed to start at 0.3 d and continue for 0.2 d, so change this if necessary. Rainfall amounts and frequency should be examined to ensure that trace amounts of rain are handled appropriately. As there is no ET during rain periods, an excessive number of very small rain events can reduce total ET merely by reducing the time available for evapotranspiration. Thus, specifying rain at night instead of during the day can lead to differences in the overall water balance. These issues are not a question of whether or not the model is realistic—they are a function of how the data files are set up.
4. Define the vegetation. The main decision here is whether to define constant or growing crop cover and roots. The period of active transpiration, the crop cover fraction and root depth are of prime importance. For LEACHN the period during which the plants will take up N and P is important. This can change with growth or be constant during the growth period, so careful thought is required. For deciduous or perennial crops it may be appropriate to define a constant root distribution and a series of short growth periods to represent cutting periods. Because the potential N uptake may be uniform over a period it may be appropriate to define a very short growth to maturity period (e.g., 1 d) and a long period of constant cover to harvest. Adjust starting C and N pools. Starting conditions are important to prevent an unnatural flush of water and nutrients from the profile owing to the fact that they are not in equilibrium with the boundary conditions. Decisions need to be made as whether to bring the pools into equilibrium with a specific land use (e.g., long-term corn) or whether changes are to be simulated, such as by bringing it into equilibrium with a representation of natural vegetation and then imposing a land use change, such as clearing for annual crops. This is important, as the effect of a fertilizer or manure management practice is superimposed on the organic pool cycling and mineralization. Nitrate leaching from applied fertilizer cannot easily be distinguished from that derived from mineralization of organic matter, so it is crucial to have the relative amounts correct. Maximum errors will occur at the start of a simulation in which pool levels (and perhaps water) are not in equilibrium with past practice, as mineralization rate constants vary with water content and temperature.
5. In this case study, the LEACHM model was run with the simulation period beginning in May of one year and continuing through April of the follow-

ing year. The simulation accuracy and performance of the LEACHM model to predict water drainage and NO_3–N leaching losses were evaluated and compared to field data from a 5-yr (1988–1992) experiment. The amount of water drainage and NO_3–N losses past the 1.2-m depth from N-fertilized and manured corn were measured with zero-tension pan lysimeters for 5 yr (Jabro et al., 2006).

Model Evaluation

The model evaluation process includes both calibration and validation. Parameterization of the model is one of the weakest points of modeling (Van Nes and Scheffer, 2005; Sourisseau et al., 2008), and identification of reliable estimates of different input parameters is a key issue and thus a challenging task in the calibration process (Nielsen, 1994; Sourisseau et al., 2008).

Model Calibration

The model calibration process allows users to describe the system under study. Model calibration is the first stage of adjusting the input parameters to a set of field data within expected values to minimize the difference between simulated and measured values. The first phase of calibration focuses on small changes to the soil water flow parameters in the models. Calibration then focuses on the input parameters controlling soil N transformation processes and rate constants in the model. The calibration process is complete when finite adjustments to the specific input parameter no longer reduce the mean differences or increase correlation coefficients and/or root mean square error.

The input parameters dealing with water movement through the soil profile, soil N transformation processes and their constants used to calibrate the model are presented in Table 3–1.

The model was calibrated to the field site conditions for each treatment using the 1989–1990 data. The 1989–1990 data set was selected for calibration because the environmental and weather conditions in 1989–1990 were normal and less extreme than in the dry 1988 season (Jabro et al., 1995, 2006).

Model Validation

The model validation process is necessary for accurate model application and evaluation. The model validation process is defined as an independent test of the calibrated model using additional set(s) of data under various environmental conditions to further examine the validity of the calibrated model.

The input parameters given in Table 3–1 determined through the calibration process (1989–90 data) were applied to the model in validation years. After the

Table 3–1. Input parameter values used in the model during calibration process.

Input parameter	Values	Literature values
Nitrification rate, d^{-1}	0.01–0.02	0.01–0.5
Denitrification rate, d^{-1}	0.02–0.03	0.01–0.75
Mineralization rate (humus) × 10^{-4}, d^{-1}	0.02–0.03	7–1.7
Mineralization rate (plant residue), d^{-1}	0.02–0.03	0.01–0.2
Mineralization rate (manure), d^{-1}	0.02	0.01–0.03
Ammonia volatilization rate, d^{-1}	0–0.04	0–0.4
Q10	2	2- 3
Initial water potential, kPa	35–10	vary
Air-entry value (Campbell equation)	0.3–3	vary
b parameter (Campbell equation)	7.8–18.7	vary
K_s, mm d^{-1}	2184–4 0	vary
Partition coefficient (NH_4–N), L kg^{-1}	3.0	3.0
Partition coefficient (NO_3–N), L kg^{-1}	0.0	0.0
C/N ratio for biomass and humus	10	10
Molecular diffusion coefficient, mm^2 d^{-1}	120	120

Table 3–2. Measured and LEACHM simulated values of cumulative (annual) drainage losses below the 1.2-m soil depth.

Year	Cumulative water drainage	
	Measured (Mean ± 95% C.I.)	Simulations
	mm	
1988–1989	472 ± 112	356
1989–1990†	429 ± 98	380
1990–1991	634 ± 108	582
1991–1992	510 ± 106	415
1992–1993	244 ± 47	245

† Calibration year.

model was made operational and calibrated for an individual year, the model was then validated for its simulative capability and accuracy using 1988–1989, 1990–1991, 1991–1992, and 1992–1993 NO_3–N data by comparing model-simulated drainage and NO_3–N leaching results against the mean of three replicated field measurements (Tables 3–2 and 3–3).

The calibration and validation results for zero N addition (0-N), the fertilizer treatment with N supplied as NH_4NO_3 at a rate of 200 kg N ha^{-1} (200-N), and manure treatments for the years 1988–89 and 1989–90 are presented in Fig. 3–1.

Sensitivity Analysis

Sensitivity analysis is a process used to identify the most sensitive or influential parameters that need to be determined with the greatest accuracy (McCuen,

Table 3–3. Measured and simulated mass values of cumulative nitrate leached below the 1.2-m soil depth.

Year	Treatment†	Cumulative mass of NO_3–N leached	
		Measured (Mean ± 95% C.I.)	LEACHM simulations
		kg ha^{-1}	
1988–89	Control	49.3 ± 10.7	42.2
	Fertilizer	107.7 ± 44.2	87.2
	Manure	70.6 ± 26	56.1
1989–90‡	Control	37.7 ± 20.1	42.5
	Fertilizer	89 ± 38.1	95
1990–91	Control	24 ± 6.8	47§
	Fertilizer	111.8 ± 47.4	120.3
	Manure	41.8 ± 22.4	51.2
1991–92	Control	11 ± 2.9	14.1§
	Fertilizer	81.9 ± 13.8	77.8
1992–93	Control	4.3 ± 2.8	4.9
	Fertilizer	25.8 ± 8.2	31.1

† Control, 0 kg N ha^{-1}; Fertilizer, 200 kg N ha^{-1} as NH_4NO_3; Manure, N supplied as dairy manure slurry containing 264, 132, and 158 kg total N ha^{-1} in 1988, 1989, and 1990, respectively.

‡ Calibration year.

§ Values indicate that simulated values are not within the 95% C.I of the measured values.

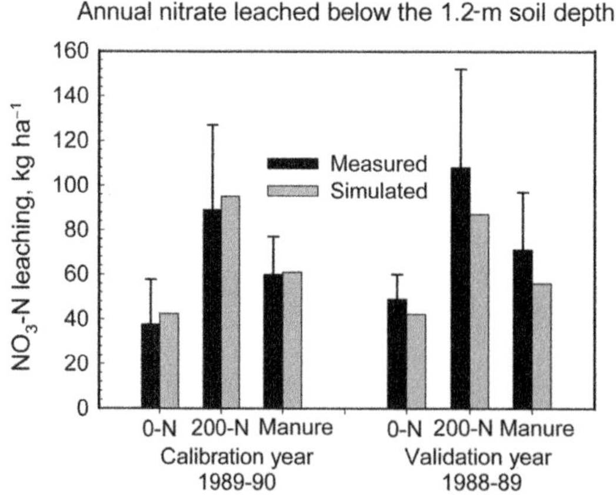

Fig. 3–1. The calibration and validation results for the 0-N, 200-N, and manure treatments. The data set includes the control treatment consisting of zero N addition (0-N), the fertilizer treatment with N supplied as NH_4NO_3 at a rate of 200 kg N ha^{-1} (200-N), and the manure treatment as 132 kg N ha^{-1} as dairy manure slurry for the years 1988–1989 and 1989–1990.

1973; Larocque et al., 2002) and has been performed using different approaches by numerous researchers.

Sensitivity analysis was performed for four input parameters: Campbell's *b* parameter, the initial soil water potential, and nitrification and denitrification rates. Sensitivity analyses of the *b* parameter of Campbell's equation and the initial soil water potential indicated that within acceptable limits (±50% change) the calibrated values resulted in changes of 6 to 14% and 3 to 6% in the amounts of water

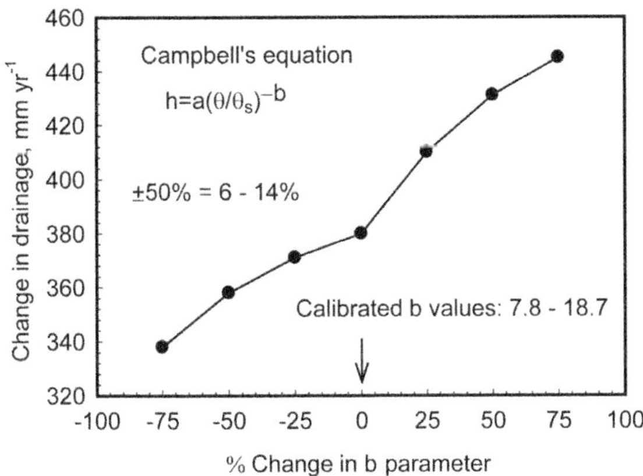

Fig. 3–2. Effect of *b* parameter in Campbell's equation on simulated water drainage.

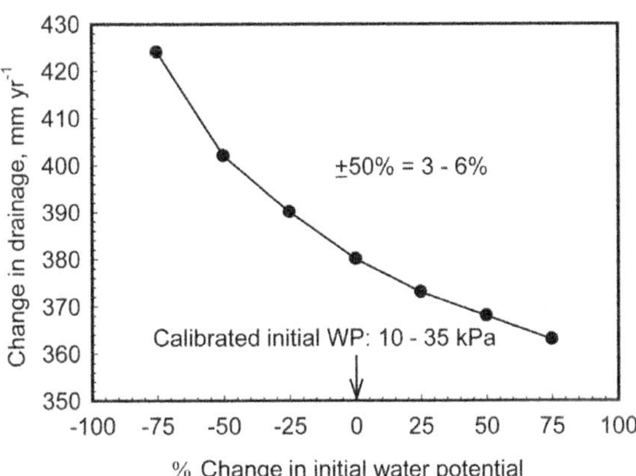

Fig. 3–3. Effect of soil water potential on simulated water drainage.

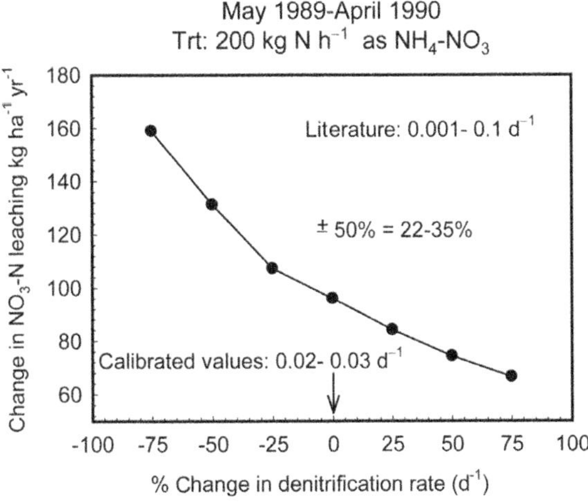

Fig. 3–4. Effect of denitrification rate on simulated NO_3–N leaching.

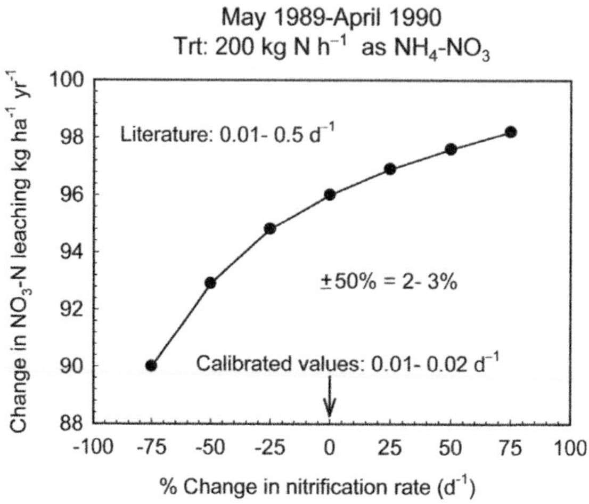

Fig. 3–5. Effect of nitrification rate on simulated NO_3–N leaching.

drainage, respectively (Fig. 3–2 and 3–3). A change of ±20% in denitrification values resulted in a 22 to 35% change in the amount of NO_3–N leaching losses (Fig. 3–4). On the other hand, a change of ±20% in the calibrated values of the nitrification rate resulted in a 2 to 3% variation in the amount of NO_3–N leaching losses (Fig. 3–5).

Sensitivity analysis of soil saturated hydraulic conductivity (K_s) indicated that within measured calibrated values of K_s (±20%), changes in these values have a small influence on the amounts of drainage and NO_3–N leaching losses (data not shown).

Further, sensitivity analyses of the molecular diffusion (D_0), dispersivity coefficients, air-entry value of Campbell's equation, and Q10 indicated that changes in these coefficients did not influence the simulated model output (data not shown). Ng et al. (2000) tested the sensitivity of the model to the molecular diffusion coefficient using the D_0 values of 60, 120, and 150 mm^2 d^{-1}. They found no significant differences in the simulated leaching outputs. Ng et al. (2000) also found that the model was not sensitive to dispersivity values between 10 and 80 mm. These results indicated that the LEACHM response and output of NO_3–N was slightly affected by changes in nitrification and mineralization rate constants (Fig. 3–5, 3–6, and 3–7). Conversely, sensitivity analyses indicated that most of other remaining input parameters used in the model had relatively little or no influence on drainage and NO_3–N leaching results (data not shown).

Model Simulation Accuracy and Statistical Assessment Tools

Although there is no standard protocol for statistically evaluating simulation accuracy of the model, a combination of statistical criteria have been used to assess the model's accuracy, performance, and simulation capability. Goodness-of-fit statistical measures were used to quantify how closely the overall model simulations

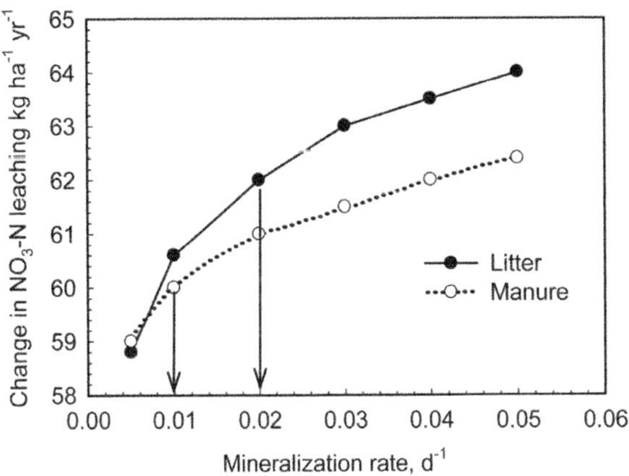

Fig. 3–6. Effect of litter and manure mineralization rates on simulated NO_3–N leaching.

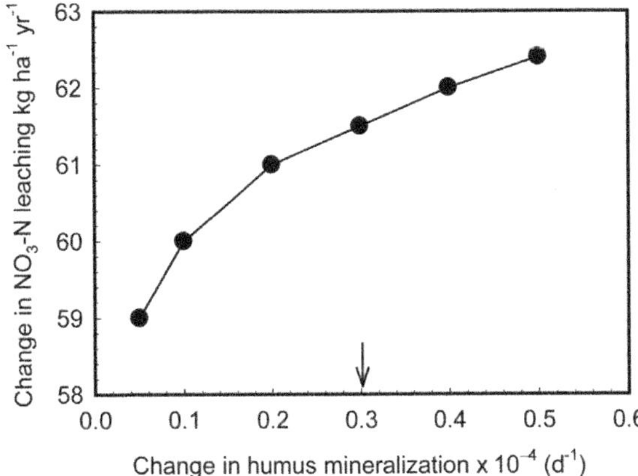

Fig. 3–7. Effect of humus mineralization rate on simulated NO_3–N leaching.

matched the measured drainage and NO_3–N leaching results (Loague and Green, 1991; Smith et al., 1996; Jabro et al., 1998; 2006).

Linear regression equations are generated for the simulated and measured values. The correlation coefficient (r) and comparison of the estimated intercept and slope with zero and one, respectively, were used as measures of the degree of association and coincidence between simulated and measured values. The null hypothesis of an intercept of zero and a slope of one was evaluated using a t test at the 0.05 probability level (Smith et al., 1996). The relative root mean square error (RMSE, Eq. [1]) provides a percentage for the total difference between simulated and measured values proportioned against the mean observed values (Smith et al., 1996). The lower limit for RMSE is zero, which denotes no difference between measured and simulated values. A smaller RMSE indicates a more accurate simulation. The modeling efficiency (EF, [Eq. 2]) is a measure for assessing the accuracy of simulations. The maximum value for EF is one, which occurs when the simulated values perfectly match the measured values (Smith et al., 1996). The mean difference (M_d, [Eq. 3]) is a measure of the average difference between the simulated and the measured values for each year (Addiscott and Whitmore, 1987). A small, nonsignificant M_d (H_0: M_d = 0, $p > 0.05$) statistically verifies the accuracy of the model simulation. The M_d can be positive or negative, the signs indicating whether the model tends to overestimate or underestimate the measured values. A t test was used to check the null hypothesis that $M_d = 0$ (Addiscott and Whitmore, 1987; Smith et al., 1996). These statistical measures utilized are defined as

$$\mathrm{RMSE}=\left[\frac{1}{n}\sum_{i=1}^{n}\left(M_i-S_i\right)^2\right]^{0.5}\left(\frac{100}{\overline{M}}\right) \qquad [1]$$

$$\mathrm{EF}=\frac{\sum_{i=1}^{n}\left(M_i-\overline{M}\right)^2-\sum_{i=1}^{n}\left(M_i-S_i\right)^2}{\sum_{i=1}^{n}\left(M_i-\overline{M}\right)^2} \qquad [2]$$

$$M_\mathrm{d}=\frac{\sum_{i=1}^{n}(S_i-M_i)}{n} \qquad [3]$$

where S is the model simulated value, M is the corresponding measured value, n is the number of measurements, M_d is a mean difference, and $\overline{M}$ is the mean of the measured data defined as:

$$\overline{M}=\frac{1}{n}\sum_{i=1}^{n}M_i \qquad [4]$$

In addition to the above statistical measures, model simulations were also assumed to be accurate if the predicted NO_3–N leached values fell within the 95% confidence intervals of the measured data (approximately ± 2 standard error). Smith et al. (1996) suggested that these statistical analyses be used for assessing model performance and accuracy, and for comparison among models. Addiscott and Whitmore (1987), Donigian and Rao (1988), Loague and Green (1991), Smith et al. (1996), and Jabro et al. (1998) found these statistical methods useful and efficient for evaluating model performance and accuracy.

LEACHM Performance and Accuracy Assessment

Water Drainage Fluxes

The LEACHM model drainage simulations were compared with the mean measured drainage water values of 18 lysimeters in 1988–1989, 1989–1990, and 1990–1991, and nine lysimeters in 1991–1992 and 1992–1993. Use of the 95% confidence interval criterion indicated that the LEACHM model accurately simulated annual drainage losses below the 1.2-m depth under the corn cropping system in both calibration (1989–1990) and validation years (1988–1989, 1990–1991, 1991–1992, and 1992–1993), as given in Table 3–2. The validation simulations of drainage fluxes were approximately as accurate as calibration simulations. These results suggest that the LEACHM model can be used to simulate cumulative annual drainage fluxes from the soil profile in validation years under corn without the need for yearly calibration.

Nitrate-Nitrogen Leaching Losses

The LEACHM simulations of NO_3–N leaching losses were compared with the mean of the lysimeter field-measured values for each year. Each measured mean value of NO_3–N was calculated from three replications for each treatment, and modeled annual NO_3–N values were compared to the 95% confidence interval of measured mean values.

In the validation years (1988–1989, 1990–1991, 1991–1992, and 1992–1993), LEACHM accurately simulated NO_3–N leached below a 1.2-m depth for 8 of the 10 cases (Table 3–3). The LEACHM simulated annual NO_3–N leaching losses fell within the 95% confidence interval of the measured values for these treatments (Table 3–3). However, the model inaccurately simulated the annual NO_3–N leached for the two control treatments in 1990–1991 and 1991–1992, indicating the LEACHM model was inadequate for simulating these two treatments (Table 3–3).

The LEACHM model overestimated the mass of NO_3–N leached for three control treatments in the validation years (Table 3–3). The simulation error in these treatments in 1990–1991 and 1991–1992 appeared to be related to the N mineralization process in the N submodels of the LEACHM model (Jabro et al., 2006). The N mineralization rate parameter in the LEACHM model appears to affect the production of N and the amount of nitrate leached in treatments where amendments (i.e., feces, manure, compost) are added (Jabro et al., 1995, 1997, 2006).

LEACHM Accuracy and Performance

Linear equations (Fig. 3–8 and Table 3–4) were generated from the regression analysis of overall simulated NO_3–N leach obtained from the LEACHM model and field-measured values (Littell et al., 1996). A high degree of association and coinci-

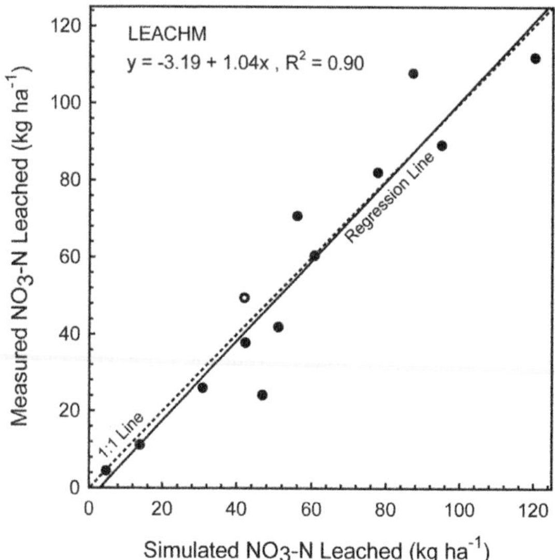

Fig. 3–8. Relationship between the overall measured and the model's simulated NO_3–N masses leached below 1.2 m for all 5 yr and 13 treatments of LEACHM.

Table 3–4. Values of the goodness-of-fit criteria computed for overall performance for LEACHM simulated values of water drainage and NO_3–N losses.

Statistical parameter†	NO_3–N	Water drainage	Perfect simulation‡
	kg ha^{-1}	mm	
r	0.95	0.95	1
a	-3.19	28.5	0
b	1.04	0.88	1
RMSE	10.7	9.2	0
EF	0.90	0.89	1
M_d	-1.14 $t = -0.37$ $p = 0.72$	-0.24 $t = -0.05$ $p = 0.96$	0

† *r*, correlation coefficient; *a*, intercept; *b*, slope; RMSE, root mean square error; EF, modeling efficiency; M_d, mean difference.

‡ Theoretical values when simulated results are the same as the measured results.

dence between annual model-simulated and measured NO_3–N leaching values is indicated by a high correlation coefficient, an intercept not significantly different from zero, and a slope not significantly different from one at $p > 0.05$ using a *t* test (Table 3–4). The correlation coefficient between the overall model's simulated and measured values of NO_3–N leached was high and significant (Table 3–4).

Conclusions

1. Calibration and sensitivity analyses are necessary for any model evaluation including LEACHM.
2. A universal model is a misnomer; no model can work under all conditions without calibration.
3. The modelers should always use values within the known or acceptable range to calibrate the model.
4. Calibration cannot be accomplished without the availability of experimental data for comparison.
5. Calibration and sensitivity analysis are time-consuming to perform, in some instances requiring more than a thousand runs and endless hours.
6. The LEACHM model was slightly sensitive to most input parameters; however, it was very sensitive to the denitrification rate, the initial soil water potential and water content and the *b* parameter in the Campbell equation.
7. The statistical analyses indicated that, overall, the LEACHM model performed well and accurately predicted NO_3–N leaching below the 1.2-m depth under corn in most treatments in the validation years.
8. The LEACHM model was able to simulate predictions of annual water drainage and NO_3–N leaching losses below the 1.2-m depth under corn in the validation years without the need for year-to-year model calibration.

References

Addiscott, T.M. 1977. A simple computer model for leaching in structured soils. J. Soil Sci. 28:554–563.

Addiscott, T.M., and A.P. Whitmore. 1987. Computer simulation of changes in soil mineral nitrogen and crop nitrogen during autumn, winter and spring. J. Agric. Sci. Camb. 109:141–157.

Ahuja, L.R., K.E. Johnson, and G.C. Heathman. 1995. Macropore transport of surface applied bromide tracer: Model evaluation and refinement. Soil Sci. Soc. Am. J. 59:1234–1241.

Ahuja, L.R., K.W. Rojas, J.D. Hanson, M.J. Shaffer, and L. Ma. 2000. RZWQM98 user guide. The Root Zone Water Quality Model. Water Resources Publ. LLC, Highland Ranch, CO.

Campbell, G. 1973. A simple method for determining unsaturated conductivity from moisture retention data. Soil Sci. 117:311–314.

Carsel, R.F., L.A. Mulkay, M.N. Lorber, and L.B. Baskin. 1985. The pesticide root zone model (PRZM): A procedure for evaluating pesticide leaching threats to groundwater. Ecol. Model. 30:49–69.

de Willigen, P. 1991. Nitrogen turnover in the soil–crop system: Comparison of fourteen simulation models. Fert. Res. 27:141–149.

Donigian, A.S., and P.S.C. Rao. 1988. Selection, application, and validation of environmental models. *In* Proceedings of the international symposium on water quality modeling of agricultural non-point sources. Part 2. 19–23 June 1988. USDA-ARS 81. Utah State Univ., Logan, UT.

Eckersten, H., P.E. Jansson, and H. Johnsson. 1996. SOILN model, user's guide (Version 9.1). Dep. of Soil Sciences, Swedish Univ. of Agricultural Sciences, Uppsala.

Hutson, J.L. 2003. LEACHM. Leaching Estimation And CHemistry Model: Model description user guide. School of Chemistry, Physics and Earth Sciences, The Flinders Univ. of South Australia, Adelaide.

Hutson, J.L., and R.J. Wagenet. 1991. Simulating nitrogen dynamics in soils using a deterministic model. Soil Use Manage. 7:74–78.

Hutson, J.L., and R.J. Wagenet. 1992. LEACHM. Leaching Estimation And CHemistry Model: A process based model of water and solute movement, transformations, plant uptake and chemical reactions in unsaturated zone. Version 3. Dep. of Agronomy, Cornell Univ., Ithaca, NY.

Jabro, J.D., A.D. Jabro, and R.H. Fox. 2006. Accuracy and performance of three water quality models for simulating NO_3–N losses under corn. J. Environ. Qual. 35:1227–1236.

Jabro, J.D., J.M. Jemison, Jr., R.H. Fox, and D.D. Fritton. 1994. Predicting bromide leaching under field conditions using SLIM and MACRO. Soil Sci. 157:215–223.

Jabro, J.D., J.M. Jemison, Jr., L.L. Lengnick, R.H. Fox, and D.D. Fritton. 1993. Field validation and comparison of LEACHM and NCSWAP models for predicting nitrate leaching. Trans. ASAE 36:1651–1657.

Jabro, J.D., W.L. Stout, S.L. Fales, and R.H. Fox. 1997. Nitrate leaching from soil core lysimeters treated with urine or feces under orchardgrass: Measurement and simulation. J. Environ. Qual. 26:89–94.

Jabro, J.D., W.L. Stout, S.L. Fales, and R.H. Fox. 2001. SOIL-SOILN simulations of water drainage and nitrate-nirogen transport from soil core lysimeters. J. Environ. Qual. 30:584–589.

Jabro, J.D., J.D. Toth, Z. Dou, R.H. Fox, and D.D. Fritton. 1995. Evaluation of nitrogen version of LEACHM for predicting nitrate leaching. Soil Sci. 160:209–217.

Jabro, J.D., J.D. Toth, and R.H. Fox. 1998. Evaluation and comparison of five simulation models for estimating water drainage fluxes under corn. J. Environ. Qual. 27:1376–1381.

Jemison, J.M., Jr., J.D. Jabro, and R.H. Fox. 1994a. Evaluation of LEACHM: I. Simulation of drainage, bromide leaching, and corn uptake. Agron. J. 86:843–851.

Jemison, J.M., Jr., J.D. Jabro, and R.H. Fox. 1994b. Evaluation of LEACHM: II. Simulation of nitrate leaching from nitrogen fertilized and manured corn. Agron. J. 86:852–859.

Khakural, B.P., and P.C. Robert. 1993. Soil nitrate leaching potential indices: Using a simulation model as a screening system. J. Environ. Qual. 22:839–845.

Larocque, M., O. Banton, J. Gagnon, and C. Camire. 2002. Using models to manage soil inorganic nitrogen in forest tree nurseries. Soil Sci. Soc. Am. J. 66:602–612.

Lengnick, L.L., and R.H. Fox. 1994. Simulation by NCSWAP of seasonal nitrogen dynamics in a corn crop: I. Soil nitrate. Agron. J. 86:167–175.

Leonard, R.A., W.G. Knisel, and A.D. Still. 1987. GLEAMS: Groundwater loading effects of agricultural management systems. Trans. ASAE 30:1403–1417.

Loague, K., and R.E. Green. 1991. Statistical and graphical methods for evaluating solute transport models: Overview and application. J. Contam. Hydrol. 7:51–73.

Littell, R.C., G.A. Milliken, W.W. Stroup, and R.D. Wolfinger. 1996. SAS system for mixed models. SAS Inst., Cary, NC.

Mahmood, B., J.M. Russell, and G.L. Wall. 2002. Field-scale nitrate simulation. Trans. ASAE 45:1835–1842.

Mohtar, R.H., D. Buckmaster, and S.L. Fales. 1997a. A grazing simulation model: GRASIM A: Model development. Trans. ASAE 40:1483–1493.

Mohtar, R.H., J.D. Jabro, and D. Buckmaster. 1997b. Field testing of the grazing simulation model, GRASIM. Trans. ASAE 40:1495–1500.

McCuen, R.H. 1973. The role of sensitivity analysis in hydrologic modeling. J. Hydrol. 18:37–53.

Molina, J.A.E., and K. Richards. 1984. Simulation models of the nitrogen and carbon cycle in the soil-water-plant system, NCSWAP; Guide for the preparation of input data files and execution of NCSWAP. Soil Ser. 116. Dep. of Soil Sci., Univ. of Minnesota, St. Paul.

Ng, H.Y.F., C.F. Drury, V.K. Serem, C.S. Tan, and J.D. Gaynor. 2000. Modeling and testing of the effect of tillage, cropping and water management practices on nitrate leaching data in clay loam soil. Agric. Water Manage. 43:111–131.

Nielsen, S.N. 1994. Modeling structural dynamical changes in a Danish shallow lake. Ecol. Model. 73:13–30.

Pennell, K.D., A.G. Hornsby, R.E. Jessup, and P.S.C. Rao. 1990. Estimation of five simulation models for predicting aldicarb and bromide behavior under field conditions. Water Resour. Res. 26:2679–2693.

Shaffer, M.J., A.D. Halvorson, and F.J. Pierce. 1991. Nitrate Leaching and Economic Analysis Package (NLEAP): Model description and application. p. 285–322. *In* R.F. Follett et al. (ed.) Managing nitrogen for ground-water quality and farm profitability. SSSA, Madison, WI.

Smith, J., P. Smith, and T. Addiscott. 1996. Quantitative methods to evaluate and compare soil organic matter (SOM) models. p. 181–200. *In* D.S. Powlson et al. (ed.) Evaluation of soil organic matter models. NATO ASI Ser. I, Vol. 38. Springer-Verlag, Heidelberg.

Sogbedji, J.M., H.M. van Es, and J.L. Hutson. 2001. N fate and transport under variable cropping history and fertilizer rate on loamy sand and clay loam soils: I. Calibration of the LEACHM model. Plant Sci. 229:57–70.

Sourisseau, S., A. Basseres, F. Pere, and T. Caquest. 2008. Calibration, validation and sensitivity analysis of an ecosystem model applied to artificial streams. Water Res. 43:1167–1181.

USEPA. 1979. Methods for chemical analysis of water and wastes. USEPA, Cincinnati, OH.

van Genuchten, M.Th. 1980. A closed-form equation for predicting the hydraulic conductivity of unsaturated soils. Soil Sci. Soc. Am. J. 44:892–898.

Van Nes, E.H., and M. Scheffer. 2005. A strategy to improve the contribution of complex simulation models to ecological theory. Ecol. Model. 85:153–164.

Chapter 4

Special Features of the SPACSYS Modeling Package and Procedures for Parameterization and Validation

L. Wu
A. Shepherd

Abstract

Process modeling and simulation are commonly used tools in both practical and research-oriented activities in agriculture. Recently released models have been developed to simulate integrated agricultural systems in one way or another, an activity which involves many processes and needs a large number of parameters and input variables. After a model is implemented and verified and before it is applied to a real scenario or used to test a hypothesis, a sensitivity analysis, calibration, and validation of the model should be performed to make sure it is workable and reliable. The SPACSYS model is an ongoing developmental, field-scale, weather-driven, and daily-step dynamic process-based simulation model. One feature of the model is its integration of the interactions among below- and aboveground plant growth with development stages, nitrogen and carbon cycling, and water and heat movements in the plant–atmosphere–soil continuum into a single package, which can avoid shortcomings of some published models that ignore the effects of one or more parts of the system. This chapter will describe the main features of the process-based SPACSYS model briefly and illustrate the procedures to parameterize and validate some state variables in it. It is concluded that data quantity and quality are limits to the parameterization of the model, especially those describing the development of root architecture of plants and the definition of organic matter pools in the soil. It is necessary to use advanced technology to determine these parameters in the future.

L. Wu (Lianhai.Wu@bbsrc.ac.uk) and A. Shepherd, Sustainable Soils and Grassland Systems Dep., Rothamsted Research, North Wyke, Okehampton, Devon EX20 2SB, UK.

doi:10.2134/advagricsystmodel2.c4

Methods of introducing system models into agricultural research. L.R. Ahuja and L. Ma (ed.)
Advances in Agricultural Systems Modeling Series 2.

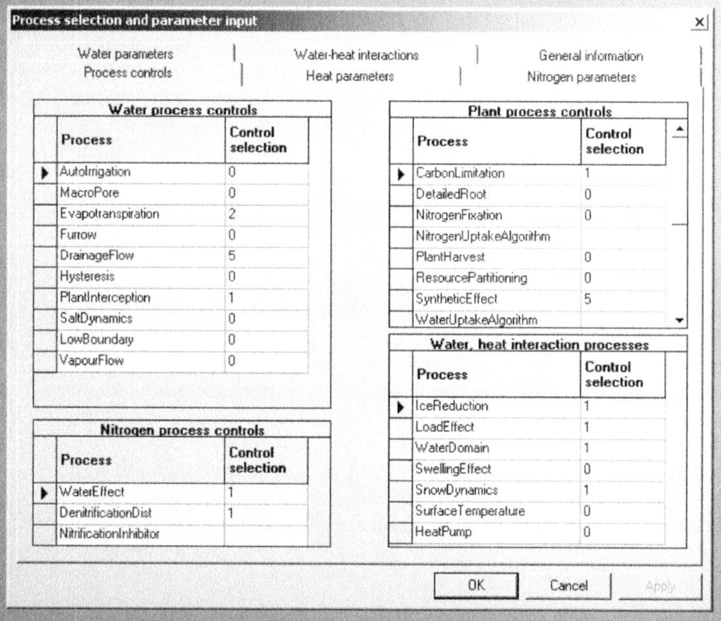
Process selection and parameter input
Water parameters
Water-heat interactions
General information
Process controls
Heat parameters
Nitrogen parameters
Water process controls
Process
Control selection
AutoIrrigation 0
MacroPore 0
Evapotranspiration 2
Furrow 0
DrainageFlow 5
Hysteresis 0
PlantInterception 1
SaltDynamics 0
LowBoundary 0
VapourFlow 0
Nitrogen process controls
Process
Control selection
WaterEffect 1
DenitrificationDist 1
NitrificationInhibitor
Plant process controls
Process
Control selection
CarbonLimitation 1
DetailedRoot 0
NitrogenFixation 0
NitrogenUptakeAlgorithm
PlantHarvest 0
ResourcePartitioning 0
SyntheticEffect 5
WaterUptakeAlgorithm
Water, heat interaction processes
Process
Control selection
IceReduction 1
LoadEffect 1
WaterDomain 1
SwellingEffect 0
SnowDynamics 1
SurfaceTemperature 0
HeatPump 0
OK
Cancel
Apply

Process modeling and simulation are commonly used tools in both practical and research-oriented activities in agriculture. Recently released models have been developed to simulate integrated agricultural systems, involving many processes and requiring a large number of parameters and input variables. Since a complex model contains more parameters, increasing errors and uncertainty in the model, it does not follow that all complex models are more accurate than more simple ones, even if necessary measures are taken. The development of a process-based model is a time-consuming, knowledge-assembling, and interdisciplinary collation process. Therefore, a general procedure for model development should be followed. The procedure has been recommended from different research fields (Foss, 1990; Jørgensen, 1995; Hangos and Cameron, 2001). Ideally, after a model is implemented and verified and before it is applied to a real scenario or used to test a hypothesis, a sensitivity analysis, calibration, and validation of the model should be performed to make sure it is workable and reliable (Fig. 4–1). A successful modeling procedure for modern models simulating a realistic system may require feedback and dynamic changes, since a model cannot necessarily be satisfactory in a single sweep through the procedure (Jørgensen, 1995).

A sensitivity analysis on a model determines the most sensitive parameters or processes. The sensitivity of a change in parameter is the percentage response in the state variable with time, divided by the percentage change in the forcing parameter over the same time. The responses to several stepped increases or decreases are noted. If the difference between including a submodel or bypassing it is compared, the sensitivity of that submodel will be obtained.

Following sensitivity analysis, parameterization, which aims to improve individual parameter estimation, and calibration, which assesses the performance of this improved individual parameter as part of the whole model parameters collectively, are introduced almost simultaneously. Both procedures are usually based on a comparison of simulated and observed data, but calibration of the model chooses an output that incorporates all of the most important or sensitive parameters and processes. Initial parameter estimates are often taken from literature. A set of parameters is often steadily varied at a time until the best comparison between simulated and observed data occurs. The process is iterated several times to arrive at the best set of parameters. Automatic calibration varies the parameter set until

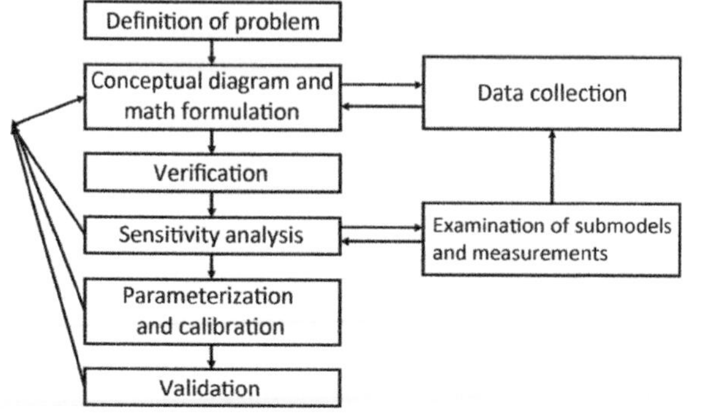

Fig. 4–1. A general guide for modeling procedure (modified from Jørgensen, 1995).

the occurrence of the lowest standard deviation (or other criteria) between simulated and observed values. Results may be fed back at this point into recalculating sensitivity, and recalibrating the effect of the whole model.

To validate the model, simulations are tested against observed data using conditions or time periods different from those used for calibration. Regression analysis often features in validation procedure; however, there are inconsistencies in published regression results. Toy et al. (1993) recommended that results should give, as a minimum standard, an illustrated scatter diagram, the standard error of the estimate and the sample size, to have a comparative indication of regression. The *R*-square of the regression is not adequate; a root mean square error around a 1:1 line is a better indicator of model predictability.

Methodologies for parameterization, calibration, and validation vary to compare simulated and observed data. There is no standard way to analyze, only what is most appropriate for the application. The objective of this chapter is to describe the main features of the process-based SPACSYS model (Wu et al., 2007, 2008) briefly and to illustrate the procedures to parameterize and validate some state variables in it.

Model Description

The SPACSYS model is an ongoing developmental, field-scale, weather-driven, daily-step dynamic simulation model. One feature of the model is to integrate the interactions among below- and aboveground plant growth with development stages, N and C cycling, water and heat movements in the plant–atmosphere–soil continuum into a single package, which can avoid shortcomings of some published models that ignore the effects of one or more parts of the system. Its framework is designed in such a way that other existing or newly developed components, such as P cycling, can be easily plugged in. The assembly of the model

Fig. 4–2. The interface acquiring information to connect databases in SPACSYS. Data collection in this interface is a starting point to operate the model.

ensures a flexibility of which components and their processes will be chosen for a simulation. Inputs and outputs of all components are organized as a database operated by MySQL 5.0, Microsoft SQL Server 2000, or Access 2000 and stored in tables within it. Before operating any functions of the model, the user will be asked to connect to a database in which the inputs and outputs will be physically stored (Fig. 4–2).

This is a generic model that includes a plant growth and development component with detailed representation of the root system, in addition to components for C and N cycling in the soil with links to the plant, a soil water component, and a heat transfer component. Since its original publication, the structure of the model has been modified to minimize information flows among the components, and more representation in root growth has been added. The components of aboveground plant C and N have been combined so that the model now consists of six modular components: plant aboveground, root system, soil C, soil N, soil water, and heat. Parallel to the detailed representation of the root system, a simplified root system that is represented by rooting depth, the vertical distribution of root length density and root biomass has been implemented in the modified version (Wu et al., 2008). Within a component, there may be embedded small components to represent specific processes with various methods to describe them.

The component of plant growth and development is generic. It is applicable not only for a monoculture crop, but also for intercropping that two crops can be grown in the field simultaneously. In terms of an intercropping system, extra processes describing competition between species for resources acquisition from below and aboveground have been implemented into relevant components. Live plants are virtually divided into four components: leaf, stem, seed, and root. Dead materials from the components contribute to the aboveground litter pool and the root litter pool. The main processes concerning plant growth in the model are plant development, assimilation, respiration, and partition of photosynthate and absorbed N, plus N fixation from a legume plant to various organs. Carbon flows from leaves to stems and roots, while N absorbed by root is allocated to green leaves and stems. During the reproductive stage, C and N from leaves and stems translocate to seeds. Partitioning coefficients of assimilated C and absorbed N to various components are dependent on plant development stages that are calculated using accumulated temperatures, air temperature, and daylength, and can be estimated with default portioning curves embedded in the model or set by the user.

Nitrogen cycling coupled with C cycling in the SPACSYS model covers the transformation processes for organic matter (OM) and inorganic N. Organic matter is further divided into fresh OM, dissolved OM (DOM), a litter pool, a humus pool, and a microbial biomass pool, and inorganic N includes a nitrate pool and an ammonium pool. Fluxes between the pools occur in a particular way according to physical and biological conditions in the source and destination pools. Fresh litter or manure and DOM in the uppermost soil layer may be lost if surface runoff occurs. The fresh litter, DOM, and humus pools receive contributions from aboveground litter fall and below-ground root litter. Decomposed OM from litter, DOM, and humus are partitioned to the different pools. The main processes and transformations causing size changes to inorganic N pools are mineralization–immobilization, nitrification, denitrification, and plant N uptake. External input such as chemical fertilizer, livestock manure, and atmospheric deposition is also included. Most of the transformations are dependent on soil water content and temperature. Nitrate, ammonium, and DOM are transported through the soil profile and into field drains or deep groundwater with water movement.

Both soil water and heat components are inherited from the SOIL model (Johnsson and Jansson, 1991). Assumptions are made about flows as the result of gradients in water potential with Darcy's law, which is formulated as Richards' equation (Richards, 1931) and in temperature with Fourier's law, together with the laws of conservation of mass and energy. Water is stored in three pools in the simulated system, in the soil profile, on the soil surface, and on the canopy. Incoming water from precipitation and/or irrigation will partly flow to the soil surface and partly remain stored temporarily on the canopy if one exists. Water on the surface may be subdivided into two parts, that contributing to the surface water pool and that infiltrating into the soil water store. Intercepted water can evaporate directly back to the atmosphere. Surface water may cause surface runoff if the quantity exceeds a certain threshold value set by the user. Soil water is depleted by plant uptake, soil evaporation, deep percolation, and horizontal flows to field drains (if present). Water in the soil profile is held mainly in soil matrix pores, but if the water content in a layer rises above a specified value a proportion is held in macropores where rapid water movements take place due to gravitational forces alone.

Input Requirements

The model needs information about the soil profile, field management, and weather to run a simulation. It could be easier to obtain the record of field management and weather data than the information on soil profile. Spatial variability is the main obstacle in characterizing the soil profile at field scale. The model assumes that soil characteristics are homogenous on a horizontal level, while it may be heterogeneous through the profile. It is necessary that the soil profile is divided into layers, each of which is considered uniform in terms of its characteristics. The number of the layers to be set in a profile is unlimited. All information required for a simulation is linked by a simulation identity. Soil properties for a soil profile and weather data at a site may be used for multiple simulations. Avoiding the duplication of the information, the table names are asked when editing parameters for a simulation (Fig. 4–3).

Daily Weather Data

Daily weather data as driving variables for the model are required before running any simulation. Daily maximum and minimum air temperatures (°C), precipitation (mm), wind speed (m s^{-1}) and relative humidity (%) must be available. It is desirable to have global radiation intensity (J m^{-2} d^{-1}) and net radiation (J m^{-2} d^{-1}) to be included. If these variables are unavailable, they are estimated in the model. If daily sunshine hours are available, the Ångström method (Ångström, 1924) is used to estimate radiation intensity. Otherwise, the Bristow–Campbell method (Bristow and Campbell, 1984) is used. When the estimation of daily net radiation is required, the Brunt's method is used to calculate long-wave radiation (Brunt, 1932).

Soil Properties

Properties to describe the soil profile for a specific soil type must be predefined. Either the Brooks and Corey equation (Brooks and Corey, 1964) or van Genuchten

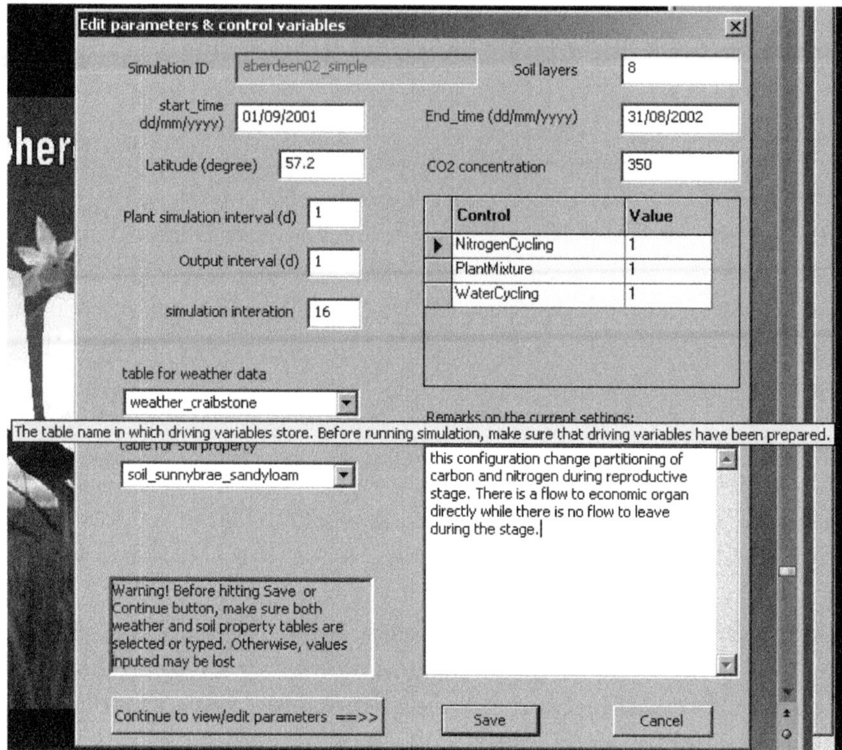

Fig. 4–3. Interface to collect general information to edit parameters and control variables in a simulation. When the mouse is hovering on a subwindow in the interface, a brief explanation of the parameter will be displayed automatically.

equation (van Genuchten, 1980) can be chosen to describe the water retention curve, that is, the relationship between water content (θ) and water potential (Ψ).

To determine soil properties by independent measurements would be time-consuming and very labor intensive, especially since some of these properties show substantial spatial heterogeneity. The model has a tool to derive some of the properties from pedotransfer functions (PTFs) using the amount of sand, clay, and silt, as well as organic matter content in soil horizons. The parameters in the Brooks and Corey equation plus water content at wilting point are estimated with the equations by Saxton et al. (1986). Saturated matrix conductivity is calculated with a published equation (Cosby et al., 1984; Liu and Dickinson, 2003). Macropore volume and macropore saturated conductivity, part of saturated total conductivity, are estimated with the procedure suggested by Rawls et al. (1996). The data input requirements for soil properties used in the model are summarized in Table 4–1. The data should be supplied for each soil horizon in the profile.

Table 4–1. Data requirement on soil properties for the SPACSYS model.

Required parameter		Symbol	Unit	Measured	Derived using soil texture information
Soil bulk density		ρ_a	g cm^{-3}	✓	
Water retention curve	residue water content	θ_r	m^3 m^{-3}		✓
	saturated water content	θ_{sat}	m^3 m^{-3}		✓
	wilt point	θ_{wilt}	m^3 m^{-3}		✓
	pore-size distribution index	λ	–		✓
	air-entry pressure	Ψ_a	kPa		✓
Hydraulic conductivity	saturated matrix conductivity	k_{mat}	mm d^{-1}		✓
	saturated total conductivity	k_{sat}	mm d^{-1}	✓	
pH value		pH	–	✓	
Brooks and Corey equation	macropore volume	θ_{mac}	m^3 m^{-3}	✓	
	tortuosity coefficient	τ	mm d	✓	

Field Management

The model will incorporate field management into relevant components in which some processes will be affected. The managements include: (i) plant management (sowing and harvest or cutting), (ii) soil management, (iii) irrigation events, and (iv) fertilizer and manure applications. Information for each of the categories is stored in a separate table in the database. All relevant inputs for field management are listed in Table 4–2.

Plant Management

The dates of plant sowing and/or harvest (cutting) are recorded accordingly. For a specific planting date, the information of seed rate (g m^{-2}) and sowing depth (m) are required. In the event of harvest or cutting, the model will make sure that the sum of removed leaf fraction and alive leaf fraction in stubble should be less than or equal to the unity. The same rule is applied to stems.

When a three-dimensional root system is simulated, information on root axis appearance should be supplied. For the purpose of simulation, root axis has a broad definition here. An axis can be either a seminal axis (appearing from the coleorhiza of the seed) or a nodal axis appearing from the coleoptile and stem nodes. To record each cohort of root axis emergence, appeared location (expressed as x, y, and z coordinates), axes number, and weight factor for geotropism of the cohort are required. Because of sparse available information on root systems, it is anticipated that it would be difficult to get accurate data. It is suggested that a one-dimensional root system be adopted in a simulation unless the user has

Table 4–2. Data requirement for field management events in the SPACSYS model. Each event will be logged with date and time.

Plant management				Fertilizer application	
Sowing	Harvest/cutting	Soil management	Irrigation	Fertilizer	Livestock manure
seed rate	fraction of removed leaf	plowing/cultivation depth	irrigation intensity	application method	
sowing depth	fraction of removed stem			amount of N applied	
	fraction of living leaf in the stubble			application depth (if any)	
	fraction of living stem in the stubble			fraction of ammonium	
	fraction of living root after harvest/cutting			fraction of nitrate	
					fraction of DON
					C/N for DOM
					C/N for fresh OM

a specific interest with root architecture and has collected enough information required to describe the system.

Soil Management

All events for field plowing or cultivation are recorded, and the information is used to redistribute C and N pools and water content in the soil profile. Each record represents one event that indicates plowing or cultivation depth (m). If plowing occurs during the simulation, all remaining living leaves and stems, roots down to the depth given by the parameter, and all aboveground residues are evenly mixed in the litter pool down to the depth. The roots below the plowing depth are incorporated into the corresponding litter pools.

Irrigation Events

Any irrigation event during a simulation is recorded with irrigation rate (mm d^{-1}). Only one event occurs within a day. The content of N in irrigated water is ignored in the current version of the model.

Fertilizer Application

There are two forms of external N sources: commercial fertilizer and livestock manure (farm yard manure or slurry). Commercial fertilizer N consists of two

Table 4–3. Initialization of state variables in the SPACSYS model.

	Soil profile	Water and heat	Soil carbon	Soil nitrogen	Plant status
Requirement for component	mandate	water and heat	soil C and N	soil C and N	plant
Defined for each soil layer	✓	✓	✓	✓	
Information collected	thickness	water content	microbial C	microbial N	accu. temperatures
		temperature	humus C	humus N	N in leaf
			DOC	DON	N in stem
			C in fresh OM (up to 6 pools)	N in fresh OM (up to 6 pools)	N in root
			C in litter (up to 2 pools)	N in litter (up to 2 pools)	leaf dry matter
				nitrate content	stem dry matter
				ammonium content	root dry matter

parts: ammonium and nitrate. Livestock manure is split up into three components: inorganic N, fresh organic matter, and soluble organic matter.

Different methods of application affect its distribution of N sources in the soil and some N and C transforming processes. The application methods considered include slurry injection to a certain depth, surface broadcasting, and incorporation within the soil (e.g., plowing) to a certain depth.

Initialization of State Variables

Most state variables need initial values at the start point for a simulation, which will depend on which components and what processes in the component will be involved in the simulation. Usually soil water contents, soil temperatures, soil C and N pools, and plant status are initialized (Table 4–3).

If soil water cycling and heat transformation are included in a simulation, soil water content and temperature in each soil layer must be initialized.

If the N cycle is coupled with the C cycle within a simulation, N and C contents in the soil profile must be initialized. The information about fresh organic matter pools depends on how many pools are defined in the system. The number of pools defined here should be consistent with the number used in the parameter configuration. For example, if two pools of fresh organic matter are defined in the configuration, C and N contents in soil layers of these pools should be initialized here. It is difficult to determine the number of OM pools. For simplicity of initialization, there should be two pools: one humus and one fresh OM pool. Again, the separation between the pools is still subjective. It is suggested that the user pre-run a simulation for 2 to 3 yr to have stabilized pools.

If a plant is involved in a simulation and the start point of the simulation is in the middle of the growing period, initial values of plant organs and growing stage, expressed in accumulated air temperature (degree days) since sowing are required.

If a three-dimensional root system simulation is included, and the start point of the simulation is in the middle of the growing period, the root structure of standing plants to describe root system and the growing root tips of standing plants (if roots still have potential for growth at the start of the simulation) are required. Constructing a root system manually would not be recommended because the size of the system is huge. In this case, it is worth simulating the system at the beginning of the growth period. Alternatively, one may use the results from a previous simulation that includes a root system recorded in output documents.

Parameterization and Description of Processes

Parameterization is the process of deciding and defining the parameters necessary for a complete or relevant specification of a model. The parameterization of a complex model is difficult because some of the parameters are hard to measure. Shaffer et al. (2001) itemized the reasons why parameterization is difficult in C and N models; similar reasons exist as for the parameterization of plant growth and development models. SPACSYS is a complicated process-based model with more than 400 parameters used in total. Although not all parameters are critical in the model, there should be a reference or default value for each one. The reference values are derived mainly from literature, with a few directly from laboratory or field experimental data. Meanwhile, water and heat components of the model are inherited from the SOIL model (Johnsson and Jansson, 1991). Therefore, the values for the parameters involved in these two components are brought in from the original source or published parameterization. The parameters to simulate plant growth and development are plant-specific. To run the model for a specific species of plant, a parameter set for the plant has to be predefined. During parameterization, each parameter will give a default value, the normal range of its value, and a short explanation for it through the help function, and the model will automatically store the edited information in the tables included in the connected database.

Interfaces of Parameterization

Parameterization for a simulation is done through interfaces designed in the model. When a new simulation identity is created, general information will be requested in the system, as shown in Fig. 4–3. It is important to give the right number of soil layers for the soil profile to be considered, as this will affect all process occurred in the soil. The thickness of each soil layer will be collected in a table. All inputs and outputs linked to the soil profiles will be identified with soil layer number rather than soil depth or thickness of the soil layers. When the simulation is running, the system will match the thickness definition with the layer number. During the parameterization, some processes involved in the simulation are chosen through control variables in the interface (Fig. 4–4). These choices should be considered carefully as they will affect parameter requirements. If a process is excluded from the simulation, the parameters for the process are not required in the interfaces.

Plant Parameters

Characteristics of a plant species are less dependant on geographic sites. They may be applied to a region where the climatic conditions are similar. The parameter values can be determined before running the model for various scenarios at a specific site. They will be stored in tables in the connected database (Fig. 4–5). Parameters to be considered in this section will depend on plant species. We will describe the requirements in general terms.

Plant Development

Plant development is estimated based on its requirement for degree days, and threshold temperatures during different stages and represented by development index (DI). The growing cycle of a plant is divided into three periods: sowing to emergence (DI defined between 0 and 1), emergence to heading (DI between 1 and 2) and heading to maturity (DI between 2 and 3). During the vegetative stage, photoperiod effect and vernalization (if a species is an over-winter plant) will be considered. For an over-winter plant, the vernalization process must occur first, and then the photoperiod will be incorporated in the development. All parameters required for plant development are listed in Table 4–4. Most of them are species-specific. They should be derived from field experiments for that species directly.

The development index for a given period is calculated as

$$\mathrm{DI} = \sum \frac{T_{\mathrm{add}}}{\mathrm{ADD}} \qquad [1]$$

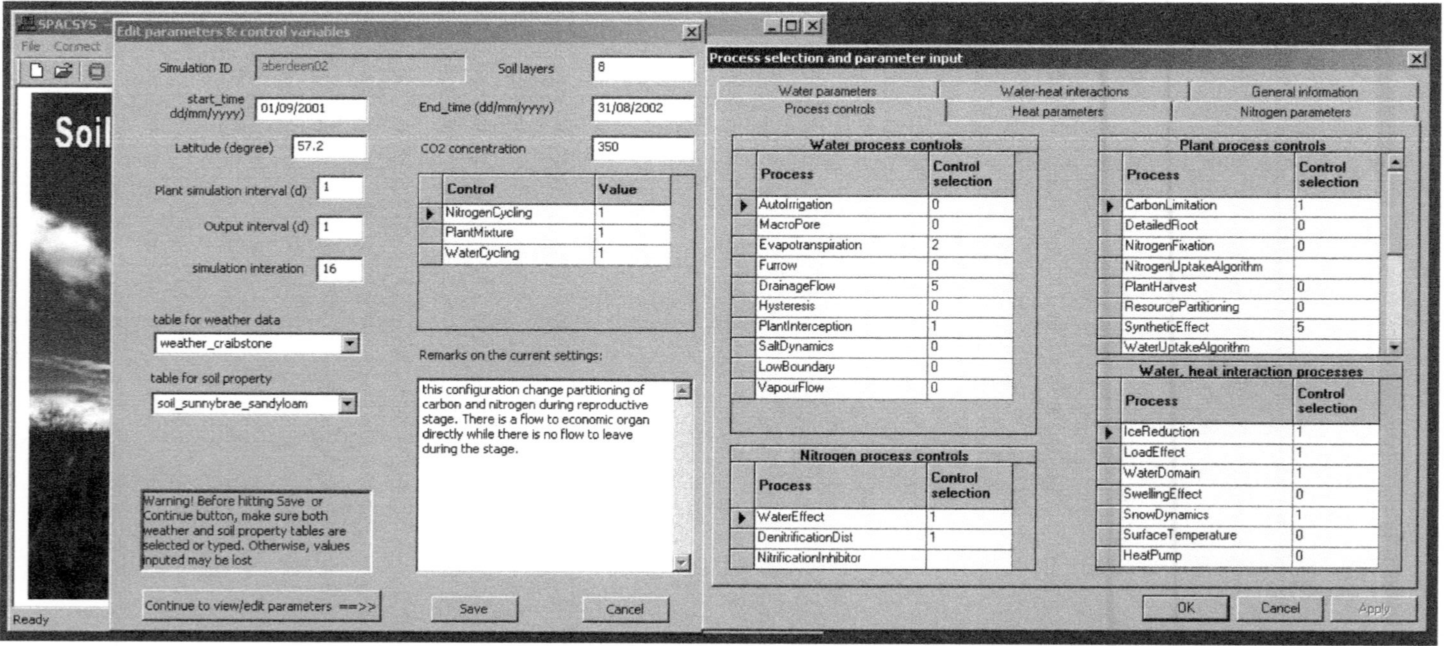

Fig. 4–4. Interface to choose processes involved in a simulation during parameterization for the simulation.

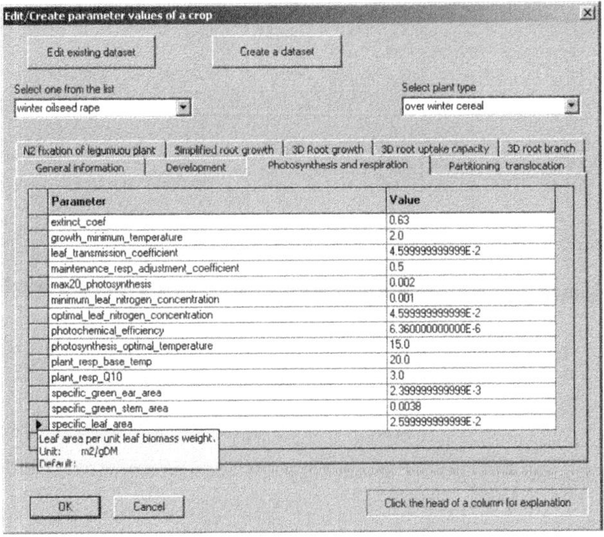

Fig. 4–5. Interface to create a new dataset or edit an exiting dataset for a crop. The parameters to describe a crop are split into nine categories. If there is no three-dimensional root system involved on the crop at all, the parameters under the three-dimensional categories could be ignored. If the crop is nonleguminous, the parameters for nitrogen fixation are ignored.

Table 4–4. Parameters in plant development and suggested values for oilseed rape.

Parameter	Unit	Default value	Source
Accumulated air temperature required from sow to emergence	°C d	120	Gabrielle et al. (1998)
Accumulated air temperature required from emergence to heading	°C d	1280	Malagoli et al. (2005)
Accumulated air temperature required from heading to maturity	°C d	550	Malagoli et al. (2005)
Accumulated air temperature required from emergence to flag leaf appearance	°C d	1280	Malagoli et al. (2005)
The critical photoperiod for vegetative stage below or over which plant development will not be affected by light	h	14.8	Habekotté (1997a)
Threshold temperature for emergence	°C	0.3	Habekotté (1997a)
Minimum air temperature below which vernalisation response function is zero	°C	−3.7	Habekotté (1997a)
Maximum air temperature above which vernalisation response function is zero	°C	17.2	Habekotté (1997a)
Optimum air temperature at which vernalisation response function is maximized	°C	3.0	Habekotté (1997a)
The critical photoperiod for vegetative stage below or over which plant development will stop	h	5.74	Habekotté (1997a)
Coefficient in the photo period response function	–	0.301	Eckersten and Jansson (1991)
Threshold temperature during reproductive stage	°C	4.9	Habekotté (1997a)
Threshold temperature for vegetative development	°C	0.5	Habekotté (1997a)

where ADD is an required accumulated degree days for the given period, and T_{add} is the daily effective temperature to be accumulated:

$$T_{\text{add}} = \begin{cases} T\left[1 - e^{\text{kp}(\text{Dayl}-\text{photop})}\right] & (T \geq T_c) \\ 0 & \end{cases} \qquad [2]$$

where kp is a photoperiod response control parameter, Dayl is daylength (h) in the day, photop is critical photoperiod during vegetative stage (h), and T_c is threshold temperature for a given development stage below which there is no temperature accumulated. For short-day plants that flower only after being exposed to light periods shorter than a certain critical length, T_{add} is zero when daylength on the day is greater than a critical photoperiod (pp_c). For long-day plants that require long hours of sunlight to set bloom, it is zero when daylength on the day is shorter than pp_c.

The sigmoid function described by Streck et al. (2003) is implemented for vernalization to avoid introducing more parameters into the model:

$$f(\text{VD}) = \frac{\text{VD}^5}{22.5^5 + \text{VD}^5} \qquad [3]$$

where VD is cumulated vernalization days since emergence of an over-winter plant. Because of the nature of the sigmoid curve, the function approaches its maximum value only as VD approaches infinity, which will cause a plant to have an indefinite period for the process. For simplification, we assume the vernalization process finishes when the function is greater than 0.95 (Chun, 1993). Daily contribution to the cumulated vernalization is estimated as

$$f_{\text{VD}} = \begin{cases} 0.0 & (T_i \leq T_{\text{vmin}} \text{ or } T \geq T_{\text{vmax}}) \\ \dfrac{T - T_{\text{vmin}}}{T_{\text{vopt}} - T_{\text{vmin}}} & (T_{\text{vmin}} < T \leq T_{\text{vopt}}) \\ \dfrac{T_{\text{vmax}} - T}{T_{\text{vmax}} - T_{\text{vopt}}} & (T_{\text{vopt}} < T < T_{\text{vopt}}) \end{cases} \qquad [4]$$

where T is daily average temperature, and T_{vmin}, T_{vopt}, and T_{vmax} are minimum, optimum, and maximum temperatures that control the response function to air temperature, respectively.

Assimilation and Respiration

The actual net photosynthetic rate of a plant is considered as the potential growth (temperature-limited and photosynthetic active radiation-limited) reduced by leaf N concentration and relative transpiration (the ratio of actual transpiration

Table 4–5. Plant parameters in assimilation and respiration and suggested values for oilseed rape.

Parameter	Unit	Default value	Source
Extinct coefficient	–	0.63	Tang et al. (2009)
Temperature at which photosynthesis ceases	°C	4.3	Tang et al. (2009)
Leaf transmission coefficient	–	0.046	Tang et al. (2009)
Leaf photosynthesis rate at saturating light levels, optimal temperature, water and N conditions	g CO_2 m^{-2} s^{-1}	0.002	Jensen et al. (1996)
Leaf N concentration at which the photosynthesis ceases	g N g^{-1} DM	0.002	assumed
Leaf N concentration at which the effect on photosynthesis is in unity	g N g^{-1} DM	0.0463	Habekotté (1997b)
Photochemical efficiency at optimal temperature, water and N conditions	g CO_2 MJ^{-1}	5.203	based on Dreccer et al. (2000)
Photosynthesis optimal temperature	°C	20	Paul et al. (1990)
Temperature when the Q_{10} function is in unity	°C	25	Tang et al. (2009)
Q_{10} value for plant maintenance respiration	–	2	Tang et al. (2009)
Specific green leaf area	m^2 g^{-1} DM	0.026	Jensen et al. (1996)
Specific green ear area	m^2 g^{-1} DM	0.0024	Jensen et al. (1996)
Specific stem area	m^2 g^{-1} DM	0.0038	Jensen et al. (1996)

to potential transpiration). Plant respiration is divided into two parts: growth respiration and maintenance respiration. The first part has been taken into account when daily net photosynthesis is estimated. The latter is estimated with the Q_{10} temperature coefficient, the rate of change of the respiration rate as a consequence of increasing the temperature by 10°C. All parameters used for the processes are summarized in Table 4–5.

The effect of leaf N concentration on net photosynthesis is expressed as a ratio of actual N concentration to optimum N concentration, above a minimum threshold:

$$f(N)=\frac{N_{\mathrm{con}}-N_{\mathrm{min}}}{N_{\mathrm{opt}}-N_{\mathrm{min}}} \qquad [5]$$

where N_{con} is N concentration in leaves, N_{opt} is leaf N concentration at which the effect on photosynthesis is in unity, and N_{min} is leaf N concentration at which the photosynthesis ceases.

Daily potential net photosynthesis is calculated based on the Hurley Pasture Model (Thornley, 1998):

$$\mathrm{Photo}=\frac{1}{2\Theta}\int_0^L\left[\frac{\alpha kI_0}{1-m}e^{-kl}+P_{\max}-\sqrt{\left(\frac{\alpha kI_0}{1-m}e^{-kl}+P_{\max}\right)^2-4\Theta P_{\max}\frac{\alpha kI_0}{1-m}e^{-kl}}\right]dl \qquad [6]$$

$$P_{max} = P^0_{max}\left[1-\varepsilon\left(1-e^{-kl}\right)\right]\frac{T-T_0}{T_{opt}-T_0}H \qquad [7]$$

where Θ is leaf photosynthesis parameter, α is photochemical efficiency, k is extinction coefficient, I_0 is photosynthetic active radiation, m is leaf transmission coefficient, l is cumulative leaf area index, $P^0{}_{max}$ is maximum rate of leaf photosynthesis, ε is the rate of decline of $P^0{}_{max}$ with irradiance, T_0 is the temperature at which photosynthesis ceases, T_{opt} is optimal temperature for photosynthesis, H is effective daylength, and L is green leaf area index, which has a broad meaning for the purpose of photosynthesis estimation:

$$L = W_{leaf}\,SLA + W_{stem}\,SSA + W_{grain}\,SGA \qquad [8]$$

where SLA, SSA, and SGA are specific leaf area, specific green stem area, and specific green grain area, respectively, and W_{leaf}, W_{stem}, and W_{grain} are biomass of leaf, stem, and grain, respectively.

Water and Nitrogen Uptake

Plant water and N uptake are critical and complicated processes to quantify the N cycling in the system accurately. The mechanisms of the processes are still unclear. Inevitably, there is space for model improvement in estimating water and N uptake. Here we briefly describe the implementation of them in the model and required parameters (Table 4–6).

Root water uptake is simulated by the macroscopic-scale approach in the current model in an attempt to avoid a high computational cost. Following the method used in the Clausnitzer and Hopmans (1994) model, water uptake rate at a given time is the product of a normalized potential root water uptake distribution, a potential transpiration rate, and an extraction function (f_e) introduced by van Genuchten (1987) to consider the combined effects of matric and osmotic potential on the root water uptake:

$$f_e = \frac{1}{\left[1+\left(\frac{h}{h_{50}}\right)^{pv_1}\right]\left[1+\left(\frac{p}{p_{50}}\right)^{pv_2}\right]} \qquad [9]$$

where h_{50} and p_{50} are the soil water pressure head and the osmotic head at which the uptake rate is reduced by 50%, respectively, and pv_1 and pv_2 are fitting parameters.

The model considers both passive or active N uptake by roots, but whether these two uptake processes work in parallel or one process triggers another is

Table 4–6. Parameters for water and nitrogen uptake and suggested values for spring barley.

Parameter	Unit	Default value
Critical inorganic N concentration in root zone below which that plant will take up organic N	g N m^{-3}	0.4
Maximum N uptake rate in Michaelis–Menten equation	g N m^{-2} root surface d^{-1}	0.027
Available N concentration in the root zone at which N uptake ceases	g N m^{-3}	0.08
Root ammonium uptake preference index	–	0.5
Passive N uptake fraction	–	0.2
Michaelis–Menten constant	g N m^{-3}	5.38
Fitting parameter in the effect of osmotic head on water uptake rate	–	3
Fitting parameter in the effect of pressure head on water uptake rate	–	3
Soil water osmotic head at half water uptake rate	cm	200
Soil water pressure head at half water uptake rate	cm water	400

poorly understood. We assume that both active and passive uptake are functioning simultaneously and the contribution of each process to total uptake is controlled by a user-specified partitioning fraction:

$$N_u = f_p N_p + (1 - f_p) N_a \quad [10]$$

where f_p is a user-specified partitioning fraction for passive uptake, N_p is passive N uptake rate through water uptake, and N_a describes the rate of active uptake represented by a Michaelis–Menten expression.

Plants may have preference regarding uptake of N in the form of ammonium or nitrate. There is a user-defined parameter to adjust the proportion of ammonium uptake (f_{amm}). The total inorganic N uptake rate (N_{uinn}) is expressed as:

$$N_{uinn} = f_{amm} N_{uamm} + (1 - f_{amm}) N_{unit} \quad [11]$$

where N_{uamm} and N_{unit} are plant uptake rates of ammonium and nitrate, separately.

To consider the contribution of organic N to plant N uptake and N flows in soil N cycling, the algorithm is implemented in the model. When mineral N concentration reaches a critical level, it will trigger plant uptake of organic N.

Nitrogen Fixation

Nitrogen fixation, a complex process, is estimated as a potential fixation rate taking into account the effect of mineral N concentration in the root zone, temperature, and soil moisture. The general procedure is to estimate potential fixation rate based on belowground root biomass first, and then adjust it by the environmental factors. Soil water function is assumed to have the same effect

Table 4–7. Parameters for biological N fixation of white clover.

Parameter	Unit	Value	Source
The minimum temperature below which N fixation ceases	°C	9	Frame and Newbould (1986)
The lower threshold of optimal temperature for N fixation	°C	13	Halliday and Pate (1976)
The upper threshold of optimal temperature for N fixation	°C	26	Halliday and Pate (1976)
The maximum temperature over which N fixation stops	°C	30	Lie (1971)
Potential N fixation rate	mg N × d^{-1} × g^{-1} nodule d.w.	110.6	Macduff et al. (1996)

as photosynthesis (the relative transpiration). The parameter values are highly dependent on legume species (Table 4–7).

Partitioning, Translocation, and Senescence

Roots receive photosynthate with the highest priority, then leaves and stems in order. If the root growth demand for carbohydrate is greater than the partitioned photosynthate available and the demand-driven mechanism for root growth is activated, photosynthate will be allocated to root. Partitions to leaves and stems will be adjusted. If there is extra photosynthate after the allocation to roots, it will be partitioned to leaves. How much photosynthate goes to leaves will depend on plant type, plant species, and its development stage. The model implements general partitioning patterns for over-winter plants and spring plants. Alternatively, the user can define DI-based partitioning coefficients, which will be used in a simulation to replace default curves embedded in the model.

Fixed N or N uptake will be partitioned to roots first, then leaves and stems in that order. The allocation to roots, leaves, and stems depends on the N demand of the correspondent component, which is estimated by optimal N concentration of the organ and allocated photosynthate to it. Alternatively, the user can define partitioning coefficients with plant development (Table 4–8).

When the plant is in the reproductive stage, dry matter and N in leaves and stems will translocate to economic organs (seeds). The translocation rates are the product of daily translocation rate and biomass or N content of leaves or stems. The parameters required for translocation are shown in Table 4–9.

Live plant components will die, and dead materials contribute to aboveground residue or root litter pool. Further, residue is partly converted into OM. The contribution of dead material to the residue pool is estimated by the product of stem biomass or N and its loss rate as litter. Gross leaffall is con-

Table 4–8. Partitioning fraction of N taken up with development index for oilseed rape (*Brassica napus* L.) derived from Malagoli et al. (2005).

Development index	Root proportion	Leaf proportion	Stem proportion
1.70	0.06	0.59	0.35
1.73	0.08	0.51	0.42
1.76	0.03	0.47	0.50
1.79	0.07	0.42	0.51
1.83	0.08	0.38	0.54
1.85	0.07	0.33	0.60
1.87	0.07	0.26	0.67
1.91	0.05	0.23	0.73

Table 4–9. Parameters for translocation and senescence of spring barley.

Parameter	Unit	Default value
Fraction of leaf biomass translocated to economic organ during reproductive stage	d^{-1}	0.004
Fraction of stem biomass translocated to economic organ during reproductive stage	d^{-1}	0.02
Fraction of leaf N translocated to economic organ during reproductive stage	d^{-1}	0.01
Fraction of stem N translocated to economic organ during reproductive stage	d^{-1}	0.04
Fraction of dry matter withdrawal from leaf to stem before leaf abscission	–	0.1
Fraction of N withdrawal from leaf to stem before leaf abscission	d^{-1}	0.4
Withdrawal fraction of N from matured roots to living roots	d^{-1}	0.0
Optimum stem N concentration	g N g^{-1} DM	0.02
Leaf age	°C d	600
Root biomass loss rate as litter	d^{-1}	0.03
Stem biomass loss rate as litter	d^{-1}	0.007

trolled by leaf age, expressed as accumulated temperature since the emergence of a leaf cohort (leaf area increment in a time-step). Before the dedicated dead leaves become aboveground residue, part of the dead leaf mass is translocated to live stem, the amount of which is a "first-order rate" of the gross leaffall. The remainder of the leaffall then goes to the aboveground residue pool.

Root Systems

One-Dimensional Root System

A root system is represented by rooting depth, the vertical distribution of root length density, and root biomass. Downward progress of rooting depth is estimated by using a genotype-specific maximum root extension rate, modified by soil temperature. A linear temperature response function is used when the average soil temperature within the root zone is greater than a critical temperature.

The distribution of root length density with depth is estimated by the well-cited Gerwitz and Page (1974) equation:

$$\frac{dP}{dx} = e^{-fx+C} \tag{12}$$

where dP is the percentage of root within a horizon of thickness dx at a depth of x cm, and f and C are constants. In the process of calculating the root length density, the parameter "root length per unit root dry matter" (RLS) will be used to estimate the density. It is assumed that the estimation of root length density should never be lower than a minimum value. To estimate root surface area in the root zone, the parameter "average root radius" should be given. The parameters required for estimating the root system are shown in Table 4–10. Note that those parameters are relatively easy to retrieve for a specific plant in literature or field experiments.

Three-Dimensional Root System

Root architecture and developmental processes are used to describe the dynamics of a root system. The architecture is described by root elongation, root growing direction, branching position on a root, branching orientation, and senescence. Developmental schemes categorize roots into several types. A root is called an axis when it has developed from the seed (seminal axis) or the stem (nodal axis). Roots arising from the axis are designated first-order laterals, and those branching from first-order laterals are designated second-order laterals, etc. Because the root architecture is a genetic characteristic of a plant, the number of orders is species-based and will be defined by the user.

The growth of a root system is expressed as the elongation rates of various root orders that are estimated with maximum growth rates for various root categories, reduced by response functions of soil temperature and soil strength. It is assumed that each segment produced in one time step has the same radius at both ends. The radius and surface area of a root segment are estimated with root dry weight per unit volume of root and geometry calculation. The growth direction of a root segment is computed using three vector components: the initial direction of the root at the previous time step, a vertical vector representing geotropism, and a vector representing mechanical constraint. The length of the first vector is set at 1, and the last two are expressed as the product of the elongation during the current time step and weighting factors. The parameters used for root growth and direction can be found in Table 4–11.

Insertion angle (vertical angle or branch angle) is used to locate root branching orientation and is specified as a parameter and modified from an initial

Table 4–10. Parameters for one-dimensional root system for oilseed rape (*Brassica napus* L.).

Parameter	Unit	Default value	Source
Maximum depth where root can penetrate to	m	1.8	Barraclough (1989)
Minimal root density that will never be below, if there is enough root mass	m m^{-3}	100	Svendsen et al. (1995)
Root density at the potential rooting depth	m m^{-3}	1000	Svendsen et al. (1995)
Potential root penetration rate	cm °C^{-1} d^{-1}	0.08	Gabrielle et al. (1998)
Threshold temperature below which root stop growing	°C	4	Malagoli et al. (2004)
Average root radius	cm	0.0085 (13°C)	Macduff et al. (1986)
Root length per unit root dry matter	m g^{-1} DM	585.4 (15°C)	Macduff et al. (1986)

Table 4–11. Parameters for three-dimensional root systems.

Parameter	Unit	Default value
Maximum executive period without carbohydrate supply that root can survive	d	15
Soil strength at which root growth ceases	Pa	6
Lower threshold temperature below which the response of root elongation is zero	°C	0
Optimum temperature at which the temperature function of root elongation is unity	°C	20
Reference gradient for soil strength component of growth direction vector	Pa cm^{-1}	0.5
Root dry weight per unit volume of root	g DM m^{-3} root	0.055
Upper threshold temperature above which the response of root elongation is zero	°C	40
Weighting factor of the geotropism component of the growth direction vector for the roots with first order	–	0

value by a random variation within a predefined range. We assume that distance between adjacent laterals or interbranch distance is a constant within the same lateral order. There are two zones that lack lateral roots: near the apex and in some instances near the root base (Table 4–12).

All parameters used in the root architecture are very much species dependant and would not be easy to determine. Unless the user has a specific interest in root architecture and makes sure that all parameters are available with a certain confidence in the values, it is advised to use the one-dimensional root system as representative of below-ground growth for a plant.

Parameters in Nitrogen and Carbon Cycling

The SPACSYS model includes as many processes in the C and N cycles as possible; inevitably more parameters are required (Table 4–13). The processes that are

Table 4–12. Parameters in three-dimensional root systems for the first-order roots of spring barley.

Parameter	Unit	Default value
Minimum biomass per root length	g DM cm^{-1}	0.00005
Potential root elongation rate	cm d^{-1}	0.48
Maximum root growth length	cm	25
Branch orientation angle	°	45
Basal non-branching distance	cm	8
Apical non-branching distance	cm	3.84
Distance between two adjacent branches with same order oriented from a same parental root	cm	0.5
Weighting factor for soil strength	–	1

influenced by soil temperature and water contents in the soil profile have unique response functions, except the water response function for denitrification. The temperature effect is expressed as a Q_{10} expression, with different Q_{10} values and temperatures when the function is unity. The effect of soil water content is based on the assumption that the chemical activity decreases on either side of an optimal range of soil water contents.

Deposition and Dissolution

Dry and wet depositions are considered one source of external mineral N input. In the parameter settings, dry deposition rate, N concentration in precipitation, and ammonium fractions in dry and wet depositions should be considered. Specific dissolution rate of chemical fertilizer is also parameterized. All the parameters are site specific and could be collected from the literature.

Residue Transformation

When surface runoff occurs, residues on the soil surface and litter/OM and DOM in the top soil compartment will be taken away from the system with a specific loss rate. Residues are transferred into a temporary litter C or N pool with a transferring fraction rate. Further, the transferred C or N will be partially partitioned into the DOC or DON pool. The transformation between pools is difficult to quantify. Therefore, the parameter values are arbitrary and should be optimized until values stabilize.

Decomposition of Organic Matter and Mineralization–Immobilization

The decomposition rate of organic matter is generally expressed as a specific decomposition rate for a given OM pool with further reduction due to soil temperature and soil moisture. The potential rates of fresh OM or litter pools are

Table 4–13. Default parameter values for soil C and N cycles in the SPACSYS model.

Parameter	Unit	Default value
Fraction of N and C in aboveground residues that are daily transformed to litter pool	–	0.05
Fraction of residue/litter/OM in uppermost layer lost per unit runoff	–	–
Specific dissolution rate of commercial fertilizer	d^{-1}	0.15
Dry deposition mineral N to soil surface	$g\ N\ m^{-2}\ d^{-1}$	–
Precipitation N concentration	$g\ N\ m^{-3}$	–
Ammonium fraction in dry deposition	–	–
Ammonium fraction in wet deposition	–	–
Weighting factor of acidity	–	–
Decomposition, mineralization/immobilization		
Synthesis efficiency constant	–	0.48
Humus fraction from other pools of organic matter	–	–
Potential decomposition rate for humus (sandy loam soil)	d^{-1}	0.000172
Potential decomposition rates for various pools of organic matter	d^{-1}	–
Fraction of ammonium content for immobilization	–	0.1
Critical C/N ratio favorable to immobilization	–	10
Nitrification		
Nitrate/ammonium ratio	–	8
Potential nitrification rate	d^{-1}	0.086
Denitrification		
Potential denitrification rate	$g\ N\ m^{-2}\ d^{-1}$	0.06
Nitrate concentration with half activity at optimum nitrate concentrations	$g\ N\ m^{-3}$	10
Coefficient in denitrification response to water	-	0.5
Water response function		
Coefficient in water function	–	1
Difference in water contents between wilting point and the lowest water content at which the water content response function has a value of unity	% (v/v)	13
Relative activity at saturated water content	–	0.6
Difference in water contents between saturation and the highest water content at which the water content response function has a value of unity	% (v/v)	8
Water content range in function for soil water content/aeration effect on denitrification	% (v/v)	17
Temperature effect		
Base temperature when the function is to unity for decomposition	°C	20
Base temperature when the function is to unity for nitrification	°C	20
Base temperature when the function is to unity for denitrification	°C	20
Q_{10} value for denitrification	–	2.58
Q_{10} value for decomposition	–	2.58
Q_{10} value for nitrification	–	2.58

dependant on OM quality (e.g., C/N or lignin/N) and can be determined from incubation experiments.

Decomposed OM has different destinations for different OM pools, which are determined by a synthesis efficiency constant, humification fraction, and the fraction partitioned to DOM. As with the residue transformation parameters, the fractions are empirical and would be difficult to estimate accurately.

Mineralization/immobilization depends on the C/N ratio in the microbial biomass. The critical C/N ratio above which the immobilization occurs should be defined. The default value is 8.

Nitrification

The nitrification rate is estimated based on a specific nitrification rate with the modifications of ammonium and nitrate contents in the soil, soil temperature and moisture, and soil acidity, which is a parameter and expressed as a weight factor. The response to ammonium and nitrate contents is estimated as

$$f_{\mathrm{NH_4/NO_3}} = \begin{cases} \dfrac{N_{\mathrm{NH_4}} - N_{\mathrm{NO_3}}}{r_{\mathrm{NO_3/NH_4}}} & \left(N_{\mathrm{NH_4}} > N_{\mathrm{NO_3}}\right) \\ 0 & \left(N_{\mathrm{NH_4}} \le N_{\mathrm{NO_3}}\right) \end{cases} \qquad [13]$$

where $r_{\mathrm{NO3/NH4}}$ is the nitrate/ammonium ratio.

The potential nitrification rate is affected by soil type. Therefore, the value should be different for different soil types. The parameter can be determined by incubation experiments for various soil types, or it can be derived from literature. It can be adjusted by calibrating nitrate and ammonium contents in the soil profile while having reasonable estimation on mineralization, plant uptake, and possible losses of nitrate and ammonium.

Denitrification

The method to estimate denitrification rate is similar to nitrification, but considering different environmental factors (soil temperature, water content, and nitrate concentration) or the type of response functions. The effect of N content on the process is introduced:

$$f_{\mathrm{NO_3}} = \frac{N_{\mathrm{NO_3}}}{N_{\mathrm{NO_3}} + c_{\mathrm{half}}} \qquad [14]$$

where c_{half} is nitrate concentration (g N m^{-3}) when the function is equal to 0.5.

There are three options to represent the contributions of different soil layers to nitrogenous gases emission: even distribution with a certain depth,

Table 4–14. Minimum parameter requirement for soil water cycles.

Parameter	Unit	Default value
Albedo of soil and canopy	%	23
Amount of water on the soil surface that could cover the whole soil surface completely	mm	50
Maximum water amount stored on the soil surface without causing any surface runoff	mm	0.01
Temperature threshold above which precipitation is regarded as pure rain	°C	2.0
Temperature threshold below which precipitation is considered as pure snow	°C	−2.0
Runoff first-order rate coefficient	d^{-1}	0.5

linear decrease distribution with the given soil depth, and exponential decrease distribution.

Water Cycle and Heat Transformation

Water cycle and heat transformation have been well parameterized through various applications (McGechan et al., 1997; Jansson and Karlberg, 2004; Alvenäs and Jansson, 1997; Jabro et al., 2001). The parameters used in the different model components and their values are listed only. There is minimum parameter requirement when no special processes are considered in a simulation (Table 4–14). If the special processes are included in the simulation, extra parameters are required based on which process is included and how it is to be considered. There are three options offered to estimate water flow to a drainage pipe when the groundwater table is above the bottom level of the pipe: linear drainage equation, Hooghoudt drainage equation (Hooghoudt, 1940), and Ernst drainage equation (Ernst, 1956). The parameters for the process are shown in Table 4–15. Similarly, the estimation of evapotranspiration rate with Penman–Monteith formula requires more parameters to calculate boundary resistances.

Snow dynamics is an option in the model to interact between heat and water components (Table 4–16). The equations and suggested parameter values can be found from the literature (Jansson and Karlberg, 2004).

Validation

Validation is the process of gaining confidence that the model fulfills its purpose. Rykiel (1996) stated that validation means a model is acceptable for its intended use because it meets its performance criteria. Validation is not a testing of scientific theory, nor is it a requirement of every modeling project. Validation means basically "acceptable for use" with respect to its intended application and conditions, not that it is the best model available.

Table 4–15. Parameter requirement with various process options for soil water cycles.

Parameter	Unit	Default value
With groundwater flow		
Distance between drainpipes	m	7
Drain pipe level	m	0.9
With groundwater (Hooghoudt or Ernst method applied)		
Diameter of the drain pipe	m	0.15
Thickness of the layer below the drain pipes and above a vertical impermeable horizon	m	–
With evapotranspiration		
Vapor pressure deficit causing 50% reduction of stomata conductance in the Lohammar equation (v_{half})*	MPa	100
Global radiation intensity that represents half-light saturation in the Lohammar equation (R_{half})	J m^{-2} d^{-1}	5×10^6
Maximum conductance of fully opened stomata in the Lohammar equation (r_{max})	m s^{-1}	0.02
Minimum roughness length	m	0.01

† The Lohammar equation is used to estimate surface resistance in the Penman–Monteith equation:

$r_s = \frac{1}{\max(\mathrm{LAI}\,r_l, 0.001)}$ and $r_l = \frac{R_s}{R_s + R_{half}} \frac{r_{max}}{1 + \frac{e_x - e_a}{v_{half}}}$ where R_s is global radiation, which is a driving variable.

Table 4–16. Parameters used in the model to consider snow dynamics.

Parameter	Unit	Default value
Old snow radiation melting factor	–	2.5
Precipitation thermal quality limit	–	0.9
Radiation coefficient in melt function	kg J^{-1}	10^7
Refreezing efficiency constant	m	0.1
Snow age coefficient	d^{-1}	0.1
Mean snow thickness that corresponds to a complete cover of the soil	m	0.03
Snowfall limit in snow age	kg m^{-2} d^{-1}	5
Temperature coefficient in melt function	kg °C^{-1} m^{-2} d^{-1}	3

The SPACSYS model is so complicated that it would be impossible to find or provide a dataset to validate the model as a whole. However, we may validate specific state variables in detail while controlling the quality of other variables in general. The soil water cycle is a critical component for other estimations in the system because plant growth and development and soil C and N cycling are affected by soil water status. Therefore, we must be sure that the processes and state variables of the component are estimated correctly. Then soil C and N should be tested in turn, and finally the plant growth component.

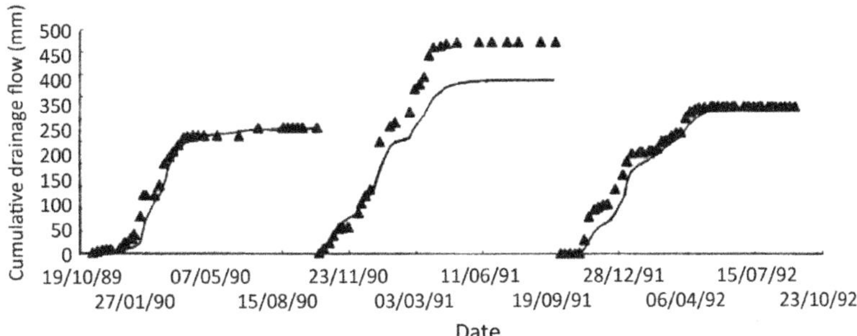

Fig. 4–6. Comparison between measured (triangles) and simulated (line) cumulative drainage flows on arable land (from Wu et al. [1998], with permission).

Validation of Soil Water

We used the field measurements of drainage water flows at Bush Estate near Edinburgh, UK to validate water drainage flows (Wu et al., 1998). The soil type is a sandy loam soil derived from partially sorted glacial till (Vinten et al., 1994). The cropping was typical of that in upland Britain where spring barley (*Hordeum vulgare* L.) is the main arable crop, and the experiment was performed over 4 yr. Parameter selection of soil water and heat transformation in the validation was reported by McGechan et al. (1997). The results showed that cumulative simulated drain flows from 1 September in each year are good fits to measured data for the winter period in every year, with slight discrepancies emerging during some years during spring and summer only (Fig. 4–6).

Validation of Soil Nitrogen

The experimental data was from an organic grazing system located at Craibstone, near Aberdeen, UK. It was sited on a Countesswells series freely drained sandy loam soil and was used to calibrate N gas emissions (Wu et al., 2006). The experimental treatments consisted of four systems of establishment for spring barley. These were (i) no grazing after late autumn, early plowed in January (abbreviated as GX/PJ); (ii) early grazed until January, early plowed in January (GJ/PJ); (iii) early grazed until January, late plowed in March (GJ/PM); and (iv) late grazed until March and late plowed in March (GM/PM). Parameter values related to plants are mainly from Wu et al. (1998), and those for water cycling and soil heat transformation are from McGechan et al. (1997) for this modeling exercise.

Measured N_2O emissions were reported by Ball et al. (2007). Simulated rates of N emissions and measured N_2O emission rates during the experimental period are shown in Fig. 4–7. Both have a similar pattern during the whole period, except for spring 2002 and 2003, when simulated rates exceeded measured rates. Because

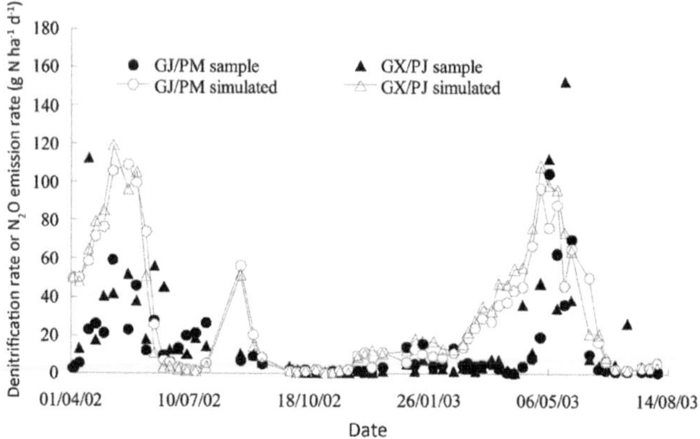

Fig. 4–7. Daily gas emissions during the whole experimental periods from the GJ/PM and GX/PJ treatments. Sampled data are N_2O emissions and simulated figures are N_2 plus N_2O emissions from denitrification processes. (GJ/PM represents the treatment of early grazed until January and late plowed in March; GX/PJ is that of no grazing after late autumn and early plowed in January).

the simulated emission rate is the summation of N_2O and N_2 emissions while the measured values are only for N_2O, theoretically the values of the simulation output should be higher than the corresponding sampled ones. Proportions of N_2O and N_2 in the emitted gases depend considerably on soil type, land use, climatic condition, and other environmental factors (Schindlbacher et al., 2004). The ratio of N_2/N_2O gas flux during denitrification is extremely variable and difficult to estimate (Riley and Matson, 2000; Parton et al., 1996; Elmi et al., 2003; Mathieu et al., 2006). If we assume that N_2O emissions account for about 60% of the total denitrification, the discrepancies between the sampled and estimated cumulative N_2O emissions become small (Table 4–17).

Validation of Biomass and Nitrogen Offtake

The field experiment was conducted at Gourdie farm, Dundee, UK. The soil was a freely drained fine sandy loam overlying loamy sand of the Buchanyhill series. Sixteen spring barley genotypes, selected to cover a range of dates of commercial introduction from 1930 to 2005, were sown in 6- by 1.5-m plots at a seed rate of 360 seeds m^{-2} and a row spacing of 15 cm. Each genotype was grown with and without N fertilizer. The model with three-dimensional root system simulated the period from 1 Jan. 2006 to 10 Sept. 2006, starting 3 mo ahead of the collection of experimental data to reduce the effect of errors in assuming initial soil N and water contents. As samplings of dry matter accumulation with time series were not taken, we used average values of all genotypes and compared the results of dry matter and N contents in plants at harvest. Many of the parameter values

Table 4–17. Comparison of corrected simulated and measured cumulative N_2O emissions for various treatments between April 2002 and July 2003.†

Treatment	GX/PJ	GJ/PJ	GJ/PM	GM/PM
	kg N ha^{-1}			
Measured N_2O emissions	7.4	5.1	5.4	8.0
Simulated N_2O emissions	7.9	7.3	6.3	8.2

† GX/PJ: no grazing after late autumn, early plowed in January; GJ/PJ: early grazed until January, early plowed in January; GJ/PM: early grazed until January, late plowed in March; and GM/PM: late grazed until March and late plowed in March.

Table 4–18. Comparison of simulated and measured dry matter for spring barley.

Treatment		Grain yield	Relative error	Aboveground	Relative error
		t ha^{-1}	%	t ha^{-1}	%
No fertilizer	simulated with a detailed root system	1.97	21.54	3.42	22.68
	sampled†	1.62 (0.26)		2.79 (0.35)	
Fertilizer applied	simulated with a detailed root system	3.47	−9.02	5.89	−11.05
	sampled	3.81 (0.53)		6.62 (0.66)	

† Sampled values are average of all varieties and the values in parentheses are the standard deviation.

used in this study, such as plant growth rates and rate coefficients for transformation processes in the soil, were based on a previous study (Wu et al., 1998). Parameter values related to the processes of soil water movement and heat transformation are from McGechan et al. (1997). Parameter values that describe root growth are shown in Tables 4–11 and 4–12, respectively.

Both experimental and simulation results show that grain yield and aboveground dry matter with no external fertilizer application during the growing season are significantly lower than those grown with chemical fertilizer (Table 4–18). It indicates that N status in the soil profile at this site was a major constraint to crop growth. The model overestimated both grain yield and aboveground dry matter for the nonfertilizer treatment, but underestimated both for the fertilizer treatment.

The measured offtakes of N in both grain and total aboveground biomass for the fertilizer treatment were three times greater than those for the nonfertilizer treatment (Table 4–19). The results of the simulation predicted N offtakes well, especially for the nonfertilizer treatment where the relative error was less than 1%. In the fertilizer treatment, the model underestimated grain N offtake by 4% and total offtake by 10%.

Table 4–19. Comparison of simulated and measured N content in grain and aboveground dry matter for spring barley.

Treatment		Grain	Relative error	Aboveground biomass	Relative error
		kg N ha^{-1}	%	kg N ha^{-1}	%
No fertilizer	simulated with a detailed root system	23.37	−0.70	29.29	−0.69
	sampled†	23.54 (2.44)		29.49 (2.61)	
Fertilizer applied	simulated with a detailed root system	64.22	−3.78	75.29	−9.72
	sampled	66.74 (7.04)		83.39 (6.98)	

† Sampled values are averages of all varieties, and the values in parentheses are the standard deviation.

Sensitivity Analysis

The sensitivity analysis determines the most important parameters and forcing functions, and the effects of different processes on specific state variables. It is also used to explore resource use efficiency by crops. The design of root systems that are more efficient in their capture of N is an important research goal because of the energy and greenhouse gas costs associated with the use of inorganic N fertilizer. To do this exercise, a published dataset for winter wheat (*Triticum aestivum* L.) is used, grown in 1974–1975 in Nottinghamshire, UK, published by Gregory et al. (1978a,b). A simple and convenient description of a root system, the distribution of root length density (RLD) within the soil profile, is included in the publication and is used to validate the performance of the model. It predicts the vertical distribution of the root system down the soil profile at various growth stages. Soil properties, field management, and the production of root axes, root depths, and distribution of root length with depth at intervals during the growing season were taken from the publications. Daily meteorological data for the site in the years of 1974 and 1975 were extracted from the DABC database (UK Meteorological Office, 2006).

A sensitivity analysis was performed to examine the relative effect of changing values of two root architectural parameters on RLD distribution: the interbranch distance and the orientation angle of lateral roots. Four scenarios were designed to conduct the sensitivity analysis: (i) short interbranch distance giving a high density of laterals, (ii) large interbranch distance giving a low lateral root density, (iii) small branching angle (relative to the parent root), and (iv) large branching angle. For each parameter, the normal baseline (default) value was close to the midpoint between the two extremes used (Table 4–20). When values of one parameter were varied, those of the other parameter were held at its

Table 4–20. Parameter values used in the sensitivity analysis of winter wheat.

	Root orders	Dense	Sparse	Default
Interbranch distance, cm	first-order lateral	0.1	0.5	0.25
	second-order lateral	0.1	0.5	0.25
	third-order lateral	0.2	0.7	0.45
		small angle	large angle	default
Branching angle, °	first-order lateral	20	70	45
	second-order lateral	20	70	45
	third-order lateral	20	70	45

default value. The root mean square error (RMSE) procedure was used to quantify the deviance of the simulated RLD for each of the designed scenarios from the baseline:

$$\mathrm{RMSE} = \sqrt{\frac{1}{N}\sum_{i=1}^{N}\left(V_{bi} - V_{si}\right)^2} \qquad [15]$$

where N is the number of samplings, V_{bi} is root length density in ith soil layer calculated from baseline parameter values, V_{si} is root length density in the same layer calculated from one of the designed scenarios.

A comparison of the measured RLD with soil depth and the outcome from the baseline simulation is shown in Fig. 4–8. With the exception of the top 10 cm of the soil, the model predicted the overall distribution of RLD reasonably well on 10 June. There was a tendency for the model to underestimate RLD in the topsoil, but overestimate it between the 50- and 80-cm soil depth. Thus the model predicted a more uniform distribution with depth than was found in practice. This discrepancy tended to be greater earlier in the season. Varying the interbranch distance had a pronounced effect on the size and relative distribution of the root system (Fig. 4–9a). A small spacing increased RLD in the top 80 cm of soil, while a large spacing

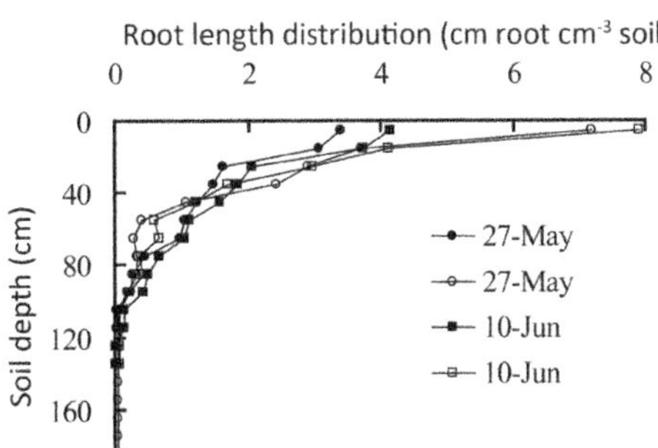

Fig. 4–8. Comparison of measured root length density (RLD) distribution with depth (open symbols) and predicted RLD (solid symbols) from simulations using the default parameter values for winter wheat.

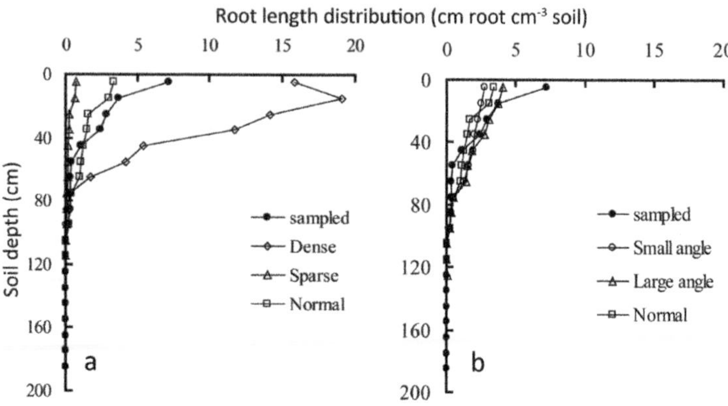

Fig. 4–9. Root length density distribution in the soil profile on 27 May and the predicted effects of varying interbranch distance and branch angle.

decreased it. By comparison varying the branch angle had a much smaller effect (Fig. 4–9b). This is reflected in the greater RMSE values for interbranch distance compared to branch angle scenarios (Table 4–21). The interbranch distance, and hence spacing of lateral roots, has a profound effect on the total length and distribution of root system produced compared with branch angle.

Discussion

SPACSYS is a complex systems model. It would be impossible to parameterize it with one dataset and validate it thoroughly using another—field experiments cannot carry out all the measurements simultaneously that the model requires. Therefore, data quantity is a limit to the parameterization of the model. A feasible way to meet the challenge is to collect parameter values from literature that included experiments with similar environments and treatments.

Data quality is another limitation to parameterization. In practice only the most sensitive parameters need calibrating. It is important to focus data and frequency of collection for parameterization during the most dynamic period when the state variable is likely to change rapidly, for example, plant nitrogen uptake in spring. It is important to use independent data for calibration that have not been used in any other test of the model.

Determining pools and transformation rates among the pools are critical issues in soil C and N cycles because they affect many parameters within the component and other relevant components. Although the model divided organic matter into several pools and the subpools of fresh OM, these are arbitrary pools that are distinguished on the basis of their resistance to microbial action rather

Table 4–21. RMSE values for root parameters of winter wheat on selected dates.

Dates	Interbranch distance		Branch angle	
	Dense	Sparse	Small	Large
Combined†	7.12	0.93	0.27	0.38
25 Feb.	3.29	0.33	0.17	0.04
06 May	8.08	0.95	0.21	0.29
27 May	7.66	1.26	0.42	0.66

† Combined analysis of all outputs from different soil layers and all dates.

than their chemical characteristics. In the future, we may use chemical characteristics to define OM pools and trace them with some organic compounds with the stable isotope technique and other advanced technology.

Root systems are central to the acquisition of water and nutrients by plants and a major pathway for the input of C and nutrients to soil. It is necessary to integrate temporal and spatial root development into the model. With such a root structure, we can describe soil–root water and N transfers, as well as C and N cycling involving the root system within a soil profile. Although a simulation model with three-dimensional root architecture may help us to understand better the process of root N uptake, the disadvantage is that more parameters to describe root architecture and growth are needed. This may introduce greater uncertainty into the simulation results.

Acknowledgments

The authors thank Dr. D. Chadwick for comments on the chapter. North Wyke is part of Rothamsted Research and supported by the Biotechnology and Biological Sciences Research Council of the UK.

References

Alvenäs, G., and P.E. Jansson. 1997. Model for evaporation, moisture and temperature of bare soil: Calibration and sensitivity analysis. Agric. For. Meteorol. 88:47–56.

Ångström, A. 1924. Solar and terrestrial radiation. Q. J. R. Meteorol. Soc. 50:121–126.

Ball, B.C., C.A. Watson, and I. Crichton. 2007. Nitrous oxide emissions, cereal growth, N recovery and soil nitrogen status after ploughing organically managed grass/clover swards. Soil Use Manage. 23:145–155.

Barraclough, P.B. 1989. Root growth, macro-nutrient uptake dynamics and soil fertility requirements of a high-yielding winter oilseed rape crop. Plant Soil 119:59–70.

Bristow, K.L., and G.S. Campbell. 1984. On the relationship between incoming solar radiation and daily maximum and minimum temperature. Agric. For. Meteorol. 31:159–166.

Brooks, R.H., and A.T. Corey. 1964. Hydraulic properties of porous media. Hydrology Paper 3. Colorado State University, Fort Collins.

Brunt, D. 1932. Notes on radiation in the atmosphere. Q. J. R. Meteorol. Soc. 58:389–418.

Chun, J.U. 1993. Variations in rate of leaf emergence, initiation of ear primordium, stem elongation and heading time as affected by vernalization duration of barley with differing growth habits. Field Crops Res. 32:159–172.

Clausnitzer, V., and J.W. Hopmans. 1994. Simultaneous modeling of transient three-dimensional root growth and soil water flow. Plant Soil 164:299–314.

Cosby, B.J., G.M. Hornberger, R.B. Clapp, and T.R. Ginn. 1984. A statistical exploration of the relationships of soil moisture characteristics to the physical properties of soils. Water Resour. Res. 20:682–690.

Dreccer, M.F., A.H.C.M. Schapendonk, M. van Oijen, C.S. Pot, and R. Rabninge. 2000. Radiation and nitrogen use at the leaf and canopy level by wheat and oilseed rape during the critical period for grain number definition. Aust. J. Plant Physiol. 27:899–910.

Eckersten, H., and P.-E. Jansson. 1991. Modelling water flow, nitrogen uptake and production for wheat. Fert. Res. 27:313–329.

Elmi, A.A., C. Madramootoo, C. Hamel, and A. Liu. 2003. Denitrification and nitrous oxide to nitrous oxide plus dinitrogen ratios in the soil profile under three tillage systems. Biol. Fertil. Soils 38:340–348.

Ernst, L.F. 1956. Calculation of the steady flow of ground water in vertical cross sections. Neth. J. Agric. Sci. 4:126–131.

Foss, A.M. 1990. Towards a general procedure for dynamic model development. Trans. Inst. Meas. Contr. 12:174–177.

Frame, J., and P. Newbould. 1986. Agronomy of white clover. Adv. Agron. 40:1–88.

Gabrielle, B., P. Denoroy, G. Gosse, E. Justes, and M.N. Andersen. 1998. Development and evaluation of a CERES-type model for winter oilseed rape. Field Crops Res. 57:95–111.

Gerwitz, A., and E.R. Page. 1974. An empirical mathematical model to describe plant root systems. J. Appl. Ecol. 11:773–781.

Gregory, P.J., M. McGowan, and P.V. Biscoe. 1978a. Water relations of winter wheat. 2. soil water relations. J. Agric. Sci. 91:103–116.

Gregory, P.J., M. McGowan, P.V. Biscoe, and B. Hunter. 1978b. Water relations of winter wheat. 1. growth of the root system. J. Agric. Sci. 91:91–102.

Habekotté, B. 1997a. A model of the phenological development of winter oilseed rape (*Brassica napus* L.). Field Crops Res. 54:127–136.

Habekotté, B. 1997b. Evaluation of seed yield determining factors of winter oilseed rape (*Brassica napus* L.) by means of crop growth modelling. Field Crops Res. 54:137–151.

Halliday, J., and J.S. Pate. 1976. The acetylene reduction assay as a means of studying nitrogen fixation in white clover under sward and laboratory conditions. Grass Forage Sci. 31:29–35.

Hangos, K., and I. Cameron. 2001. Process modelling and model analysis. Academic Press, London.

Hooghoudt, S.B. 1940. Contributions to the knowledge of some physical characteristics of soil. 7. General considerations of the problem of drainage and infiltration by means of parallel drains, furrows and canals. (In Dutch.) Versl Landbouwkd Onderz 46:515–707.

Jabro, J.D., W.L. Stout, S.L. Fales, and R.H. Fox. 2001. SOIL-SOILN simulations of water drainage and nitrate nitrogen transport from soil core lysimeters. J. Environ. Qual. 30:584–589.

Jansson, P.-E., and L. Karlberg. 2004. Coupled heat and mass transfer model for soil–plant–atmosphere systems. Royal Inst. of Technology, Stockholm.

Jensen, C.R., V.R. Mogensen, G. Mortensen, M.N. Andersen, J.K. Schjoerring, J.H. Thage, and J. Koribidis. 1996. Leaf photosynthesis and drought adaptation in field-grown oilseed rape (*Brassica napus* L.). Aust. J. Plant Physiol. 23:631–644.

Johnsson, H., and P.-E. Jansson. 1991. Water balance and soil moisture dynamics of field plots with barley and grass ley. J. Hydrol. 129:149–173.

Jørgensen, S.E. 1995. State of the art of ecological modelling in limnology. Ecol. Model. 78:101–115.

Lie, T.A. 1971. Symbiotic nitrogen fixation under stress conditions. Plant Soil 35:117–127.

Liu, Q., and R.E. Dickinson. 2003. Use of a two-mode soil pore size distribution to estimate soil water transport in a land surface model. Geophys. Res. Lett. 30:1331, doi:10.1029/2002GL016562.

Macduff, J.H., S.C. Jarvis, and I.A. Davidson. 1996. Inhibition of N_2 fixation by white clover (*Trifolium repens* L.) at low concentrations of NO_3^- in flowing solution culture. Plant Soil 180:287–295.

Macduff, J.H., A. Wild, M.J. Hopper, and M.S. Dhanoa. 1986. Effects of temperature on parameters of root growth relevant to nutrient uptake: Measurements on oilseed rape and barley grown in flowing nutrient solution. Plant Soil 94:321–332.

Malagoli, P., P. Lainé, E. le Deunff, L. Rossato, B. Ney, and A. Ourry. 2004. Modeling nitrogen uptake in oilseed rape cv capitol during a growth cycle using influx kinetics of root nitrate transport systems and field experimental data. Plant Physiol. 134:388–400.

Malagoli, P., P. Laine, L. Rossato, and A. Ourry. 2005. Dynamics of nitrogen uptake and mobilization in field-grown winter oilseed rape (*Brassica napus*) from stem extension to harvest. I. Global N flows between vegetative and reproductive tissues in relation to leaf fall and their residual N. Ann. Bot. (Lond.) 95:853–861.

Mathieu, O., J. Lévêque, C. Hénault, M.-J. Milloux, F. Bizouard, and F. Andreux. 2006. Emissions and spatial variability of N_2O, N_2 and nitrous oxide mole fraction at the field scale, revealed with ^{15}N isotopic techniques. Soil Biol. Biochem. 38:941–951.

McGechan, M.B., R. Graham, A.J.A. Vinten, J.T. Douglas, and P.S. Hooda. 1997. Parameter selection and testing the soil water model SOIL. J. Hydrol. 195:312–334.

Parton, W.J., A.R. Mosier, D.S. Ojima, D.W. Valentine, D.S. Schimel, K. Weier, and A.E. Kulmala. 1996. Generalized model for N_2 and N_2O production from nitrification and denitrification. Global Biogeochem. Cycles 10:401–412.

Paul, M.J., D.W. Lawlor, and S.P. Driscoll. 1990. The effect of temperature on photosynthesis and carbon fluxes in sunflower and rape. J. Exp. Bot. 41:547–555.

Rawls, W.J., D.L. Brakensiek, and S.D. Logsdon. 1996. Estimation of macropore properties for no-till soils. Trans. ASAE 39:91–95.

Richards, L.A. 1931. Capillary conduction of liquids through porous mediums. Physics 1:318–333.

Riley, W.J., and P.A. Matson. 2000. NLOSS: A mechanistic model of denitrified N_2O and N_2 evolution. Soil Sci. 165:237–249.

Rykiel, E.J. 1996. Testing ecological models: The meaning of validation. Ecol. Model. 90:229–244.

Saxton, K.E., W.J. Rawls, J.S. Romberger, and R.I. Papendick. 1986. Estimating generalized soil-water characteristics from texture. Soil Sci. Soc. Am. J. 50:1031–1036.

Schindlbacher, A., S. Zechmeister-Boltenstern, and K. Butterbach-Bahl. 2004. Effects of soil moisture and temperature on NO, NO_2 and N_2O emissions from European forest soils. J. Geophys. Res. 109:D17302.

Shaffer, M.J., L. Ma, and S. Hansen. 2001. Introduction to simulation of carbon and nitrogen dynamics in soils. p. 1–10. *In* M.J. Shaffer et al. (ed.) Modeling carbon and nitrogen dynamics for soil management. Lewis Publ., Boca Raton, FL.

Streck, N.A., A. Weiss, and P.S. Baenziger. 2003. A generalized vernalization response function for winter wheat. Agron. J. 95:155–159.

Svendsen, H., S. Hansen, and H.E. Jensen. 1995. Simulation of crop production, water and nitrogen balances in two German agro-ecosystems using the DAISY model. Ecol. Model. 81:197–212.

Tang, L., Y. Zhu, X. Liu, Y. Tian, X. Yao, and W. Cao. 2009. Simulation model for photosynthetic production in oilseed rape. Pedosphere 19:700–710.

Thornley, J.H.M. 1998. Grassland dynamics—An ecosystem simulation model. CAB International, Cambridge, UK.

Toy, T.J., W.R. Osterkamp, and K.G. Renard. 1993. Prediction by regression and intrarange data scatter in surface-process studies. Environ. Geol. 22:121–128.

UK Meteorological Office. 2006. MIDAS Land Surface Stations data (1853-current). Available at http://badc.nerc.ac.uk/data/ukmo-midas (verified 23 Mar. 2011). British Atmospheric Data Centre.

van Genuchten, M.Th. 1980. A closed form equation for predicting the hydraulic conductivity of unsaturated soils. Soil Sci. Soc. Am. J. 44:892–898.

van Genuchten, M.Th. 1987. A numerical model for water and solute movement in and below the root zone. Res. Rep. 121. USDA-ARS, U.S. Salinity Lab, Riverside, CA.

Vinten, A.J.A., B.J. Vivian, F. Wright, and R.S. Howard. 1994. A comparative study of nitrate leaching from soils of differing textures under similar climatic and cropping conditions. J. Hydrol. 159:197–213.

Wu, L., B.C. Ball, J.A. Baddeley, and C.A. Watson. 2006. Simulation of gaseous emissions of nitrogen from organic ley-arable systems under different management regimes. p. 101–108. *In* M. Andrews et al. (ed.) Delivering sustainability within profitable farming system—Is it possible? Aspects of Applied Biology 80. Studley Castle, Warwickshire, UK. 14–15 Dec. 2006. Assoc. of Applied Biologists, Warwick, England.

Wu, L., I.J. Bingham, J.A. Baddeley, and C.A. Watson. 2008. Modeling plant nitrogen uptake using three-dimensional and one-dimensional root architecture. p. 197–218. *In* L. Ma et al. (ed.) Quantifying and understanding plant nitrogen uptake systems modeling. CRC Press, Boca Raton, FL.

Wu, L., M.B. McGechan, D.R. Lewis, P.S. Hooda, and A.J.A. Vinten. 1998. Parameter selection and testing the soil nitrogen dynamics model SOILN. Soil Use Manage. 14:170–181.

Wu, L., M.B. McGechan, N. McRoberts, J.A. Baddeley, and C.A. Watson. 2007. SPACSYS: Integration of a 3D root architecture component to carbon, nitrogen and water cycling—Model description. Ecol. Model. 200:343–359.

Chapter 5

Special Features of the DayCent Modeling Package and Additional Procedures for Parameterization, Calibration, Validation, and Applications

S.J. Del Grosso
W.J. Parton
C.A. Keough
M. Reyes-Fox

Abstract

DayCent (Daily CENTURY) is a biogeochemical model of intermediate complexity used to simulate flows of carbon and nutrients for crop, grassland, forest, and savanna ecosystems. Required model inputs are soil texture, current and historical land use, vegetation cover, daily maximum and minimum temperature, and daily precipitation. For calibrating the model, we recommend testing model performance in the following order: soil water content, crop yield and plant growth, changes in soil organic matter levels, and N loss vectors. Different statistics should be used when evaluating model performance, including correlation coefficients, root mean square error, and mean error. For vectors that are highly variable in time (e.g., N_2O emissions), the model can represent treatment impacts on seasonal emissions correctly, but not necessarily the timing at the daily scale. In addition to comparing model outputs with field observations, comparisons with alternative models are advocated to more fully evaluate model performance. Some of the most uncertain model outputs include N_2, NH_3, and NO_x losses because these vectors are rarely measured in field experiments. Web accessible databases that include comprehensive model driver and testing data are needed to facilitate model comparisons and evaluation. DayCent has been used to simulate the impacts of climate and land use change on various crop, grassland, and forest systems around the world and is currently used to estimate soil N_2O emissions from cropped and grazed lands for the annual U.S. inventory of greenhouse gases compiled by the USEPA.

S.J. Del Grosso (delgro@nrel.colostate.edu) and M. Reyes-Fox, USDA-ARS, Fort Collins, CO 80523; S.J. Del Grosso, W.J. Parton, and C.A. Keough, Natural Resource Ecology Lab., Colorado State Univ., Fort Collins, CO 80523.

doi:10.2134/advagricsystmodel2.c5

Methods of introducing system models into agricultural research. L.R. Ahuja and L. Ma (ed.)
Advances in Agricultural Systems Modeling Series 2.

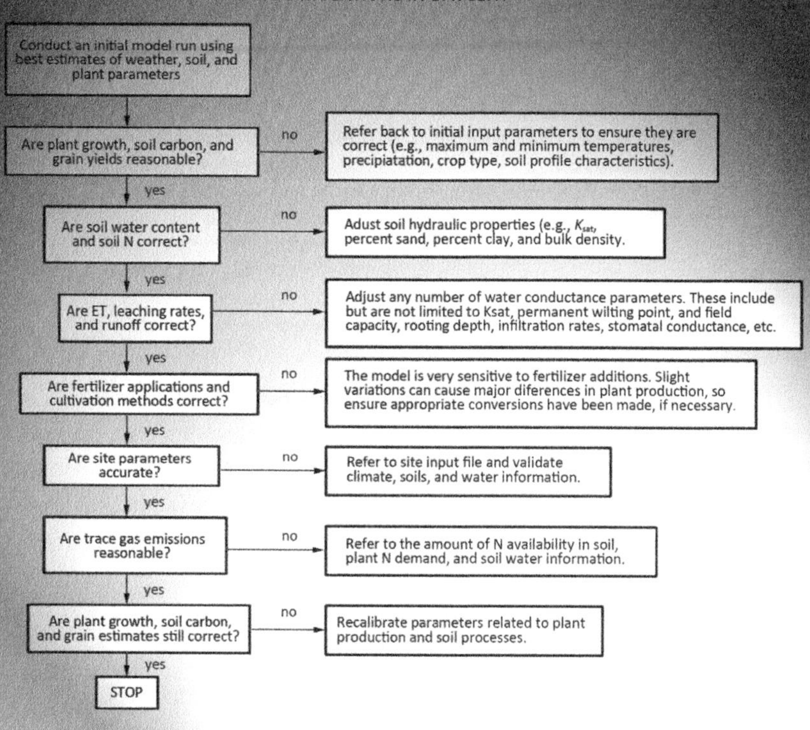
RUNNING SIMULATIONS IN DAYCENT
Conduct an initial model run using best estimates of weather, soil, and plant parameters
Are plant growth, soil carbon, and grain yields reasonable?
no
Refer back to initial input parameters to ensure they are correct (e.g., maximum and minimum temperatures, precipiatation, crop type, soil profile characteristics).
yes
Are soil water content and soil N correct?
no
Adust soil hydraulic properties (e.g., K_{sat}, percent sand, percent clay, and bulk density.
yes
Are ET, leaching rates, and runoff correct?
no
Adjust any number of water conductance parameters. These include but are not limited to Ksat, permanent wilting point, and field capacity, rooting depth, infiltration rates, stomatal conductance, etc.
yes
Are fertilizer applications and cultivation methods correct?
no
The model is very sensitive to fertilizer additions. Slight variations can cause major diferences in plant production, so ensure appropriate conversions have been made, if necessary.
yes
Are site parameters accurate?
no
Refer to site input file and validate climate, soils, and water information.
yes
Are trace gas emissions reasonable?
no
Refer to the amount of N availability in soil, plant N demand, and soil water information.
yes
Are plant growth, soil carbon, and grain estimates still correct?
no
Recalibrate parameters related to plant production and soil processes.
yes
STOP

This chapter follows the same general format as Chapter 1 (Ma et al., 2011, this volume), which describes the operation, parameterization, and calibration of the Root Zone Water Quality Model (RZWQM) model in detail. We first summarize the DayCent model and how it operates, then outline the steps recommended to parameterize and calibrate the model. In general, RZWQM is more detailed and mechanistic than DayCent. For example, RZWQM requires more meteorological inputs than DayCent and has a pesticide submodel, which DayCent does not. We also discuss issues regarding evaluation of model performance.

Overview of DayCent

Model Description

DayCent (Kelly et al., 2000; Parton et al., 1998; Del Grosso et al., 2001) is the daily time step version of the CENTURY model. CENTURY (Parton et al., 1994) operates on a monthly time step and was originally developed in the 1970s to simulate changes in soil organic matter (SOM), plant productivity, nutrient availability, and other ecosystem parameters in response to changes in land management and climate. CENTURY has been widely applied and has been shown to reliably simulate plant growth and changes in soil organic matter for most terrestrial ecosystems (Yari and Parton, 2005; Parton et al., 1995, 2011; Pepper et al., 2005). However, finer time-scale resolution is required to accurately simulate N gas emissions from soils because the processes that result in N gas fluxes respond nonlinearly to important controls, such as soil water content. Increased interest in performing complete greenhouse gas analyses, including soil nitrous oxide (N_2O) emissions, led to the development of DayCent. In terms of processes, the major difference between DayCent and CENTURY is that DayCent explicitly represents the processes (nitrification and denitrification) that lead to N_2O, NO_x, and N_2 emissions, whereas CENTURY assumes that a constant proportion of available N in a time step is lost as N gas, but the different N gas species are not distinguished. The SOM decomposition and plant growth submodels are similar, but the different time steps dictate that the equations representing the controls are different. The soil water and temperature submodels are also more resolved both spatially and temporally in DayCent.

DayCent simulates exchanges of carbon, nutrients, and trace gases among the atmosphere, soil, and plants as well as events and management practices such as fire, grazing, cultivation, and organic matter or fertilizer additions. Required model inputs are: soil texture, current and historical land use, and daily maximum and minimum temperature and precipitation data.

DayCent includes submodels for plant productivity, decomposition of dead plant material and SOM, soil water and temperature dynamics, and N gas fluxes (Fig. 5–1). Flows of C and nutrients are controlled by the amount of C in the various pools, the nutrient concentrations of the pools, abiotic temperature and soil water factors, and soil physical properties related to texture. Net primary productivity (NPP) is a function of nutrient availability, soil water and temperature, shading, vegetation type, and plant phenology (Metherell et al., 1993). Net primary productivity is divided among leafy, woody, and root compartments on the basis of plant type and phenology. The root/shoot ratio of NPP allocation is a function of soil water content and nutrient availability. The death rate of plant compartments is controlled by soil water, temperature, season, and plant-specific senescence parameters. Germination for annuals and beginning of the growing season for perennials can be defined by the user or made a function of soil temperature. Harvest dates and the end of the growing season can similarly be defined by the user or made a function of accumulated growing degree days. Management and disturbance events (e.g., grazing, fire, cultivation, irrigation, fertilization, and clear cuts) can easily be implemented.

DayCent can grow only one type of grass, crop, or tree at a time, but does simulate competition between trees and grasses (i.e., a savanna system). The savanna system is simulated using the existing tree and grassland–crop submodels, with the two subsystems interacting through shading effects and nutrient competition (Metherell et al., 1993). The fraction of the nutrient pools that are available for plant growth is a function of root biomass. The savanna model modifies maximum grass production by a shade modifier that is a function of tree leaf biomass and canopy cover. Additional constraints on plant production due to nutrient allocation between trees and grasses decrease maximum production rates for the grasses.

Soil organic matter is simulated for the top 20-cm soil layer and is divided into three pools (active, slow, and passive) on the basis of decomposition rates (Parton et al., 1993, 1994). Decomposed detrital material that has a low C/N ratio flows to the active SOM pool, which includes microbial biomass and the highly labile byproducts of decomposition that turn over in approximately 1 yr or less. The products of detrital decomposition that have a higher C/N ratio, flow to the slow SOM pool, which includes the relatively resistant (10–50 yr turnover rate)

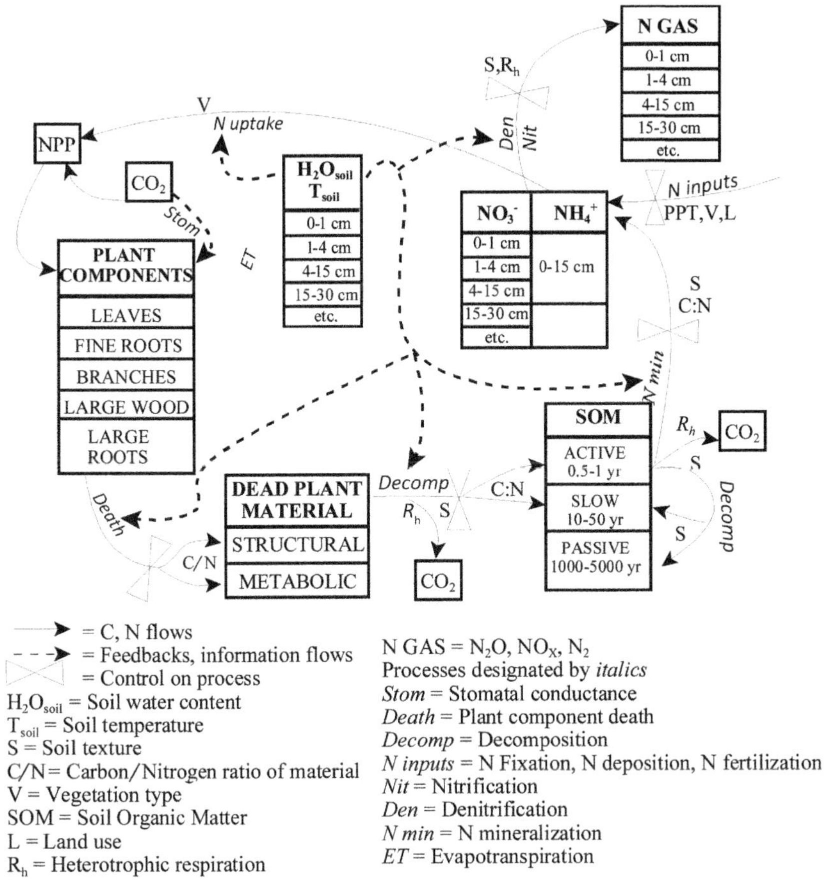

Fig. 5–1. Conceptual diagram of the DayCent ecosystem model (Del Grosso et al., 2001).

byproducts of decomposition. The passive SOM pool consists of humus that is extremely resistant to further decomposition. As soils become finer textured a lower portion of SOM is respired as CO_2 and more SOM is retained in stable form due to physical and chemical protection.

Decomposition of SOM and external nutrient additions supply the nutrient pool, which is available for plant growth and microbial processes that result in trace gas fluxes. Ammonium (NH_4^+) is modeled for the top 15 cm, while nitrate (NO_3^-), P, and S are distributed throughout the soil profile. The land surface submodel of DAYCENT simulates water flow through the plant canopy, litter, and soil profile, as well as soil temperature throughout the profile (Parton et al., 1998; Eitzinger et al., 2000). Saturated water flow is simulated down the soil profile on days that it rains, snow melts, or an irrigation event occurs. Unsaturated flow is simulated on all days that do not have water inputs sufficient to saturate the profile and can be up or down the profile depending on matric and gravitational potentials. If the water input rate exceeds the saturated hydraulic conductivity, excess water will not enter the profile and is assumed to be runoff. Soil water and dissolved NO_3, P, or S that exit the deepest soil layer simulated are assumed to be leached into groundwater or the subsoil, but the model does not simulate lateral transfer of water or nutrients.

The N gas submodel of DayCent simulates soil N_2O and NO_x gas emissions from nitrification and denitrification as well as N_2 emissions from denitrification. Nitrifying microbes oxidize NH_4^+ to NO_3^-, with some N_2O and NO_x released during the intermediate steps. Nitrogen gas flux from nitrification is assumed to be a function of soil NH_4^+ concentration, water content, temperature, and pH (Parton et al., 1996, 2001). Nitrification rates increase linearly with soil NH_4^+ concentration and a maximum of 10% of soil NH_4^+ can be nitrified in a day. Nitrification is limited by moisture stress on biological activity when soil water-filled pore space (WFPS = % relative saturation) is too low and by O_2 availability when WFPS is too high. Nitrification increases exponentially with temperature and stabilizes when soil temperature exceeds the site-specific average high temperature for the warmest month of the year. Nitrification is not limited when pH is greater than neutral, but decreases exponentially as soils become acidic.

Denitrification is a function of soil NO_3^- (e^- acceptor) concentration, labile C (e^- donor) availability, WFPS, and soil physical properties related to texture that influence gas diffusivity (Parton et al., 1996; Del Grosso et al., 2000a). Soil heterotrophic respiration is used as a proxy for labile C availability. No denitrification is assumed to occur until WFPS values exceed 50 to 60%, then denitrification increases exponentially until WFPS reaches 70 to 80%, and it stabilizes as soil water content approaches saturation. The model calculates $N_2 + N_2O$ emissions from denitrification by assuming that the process is controlled by the input (NO_3^-, respiration, WFPS) that is most limiting. Maximum daily denitrification rates range from close to 100% of soil NO_3^- at low soil NO_3^- concentration to 15% or less of soil NO_3^- at high soil NO_3^- concentration. Nitrous oxide emissions are calculated from $N_2 + N_2O$ gas emissions and an N_2/N_2O ratio function. The ratio of N_2/N_2O gases emitted due to denitrification ranges from 1 to 23 and is assumed to increase as the ratio of e^- acceptor (NO_3^-) to e^- donor (labile C) decreases and as soil gas diffusivity and O_2 availability decrease (Del Grosso et al., 2000a). Limited model sensitivity analyses showed that more than 2% and approximately 0.75% of N inputs were lost as N_2O from clay and sandy soils, respectively, on an annual basis (Del Grosso et al., 2006).

Nitrogen oxide gas emissions from soil are a function of total N_2O emissions, a NO_x/N_2O ratio equation, and a precipitation initiated pulse multiplier (Parton et al., 2001). Simulated N_2O gas emissions from nitrification and denitrification are summed to obtain total N_2O flux. The NO_x/N_2O ratio is high (maximum of ~20) when soil gas diffusivity is high and decreases to a minimum of ~1 as diffusivity decreases. The base NO_x emission rate may be modified by a pulse multiplier. Large pulses of NO_x are often initiated when precipitation falls on soils that were previously dry (Smart et al., 1999; Martin et al., 1998; Hutchinson et al., 1993). To account for these pulses the model incorporates the pulse multiplier submodel described by Yienger and Levy (1995). The magnitude of the multiplier is proportional to the amount of precipitation and the number of days since the latest precipitation event, with a maximum multiplier of 10.

Ammonia volatilization is simulated less mechanistically than the other N gas species. A soil texture specific portion of N excreted from animals is assumed to be volatilized (with more volatilization as soils become coarser), and a plant-specific portion of harvested or senesced biomass N is assumed to be volatilized.

DayCent Model Philosophy and Operation

Model complexity is a function of intended generality, availability of input data, purpose and scale of application, and other considerations. DayCent represents most of the processes that are responsible for C and nutrient fluxes, but not all of these are represented mechanistically. For example, soil gas diffusivity is an important control on N gas emissions because it influences O_2 availability, and hence, nitrification and denitrification rates. The model does not simulate O_2 diffusion into soil nor N gas diffusion from soil. However, the model does account for the impacts of gas diffusion by calculating a relative diffusion index and assuming that as total pore space and mean pore size decrease, relative gas diffusivity decreases. Relative gas diffusivity, in turn, influences emissions of the different N gas species (NO_x, N_2O, N_2).

All models have biases. As a biogeochemical model, nutrient availability is the major control on plant growth rates in DayCent. The current version of DayCent that is available for distribution does not simulate photosynthesis, but does account for the impacts of radiation as influenced by daylength and cloud cover. Daylength is a function of latitude and day of year while cloud cover is inferred by comparing diurnal maximum and minimum temperatures. A version of the model that explicitly simulates photosynthesis is currently being tested.

DayCent is a mixed language program written in FORTRAN and C and can be operated from a DOS window on a PC or on a UNIX/Linux platform. Input file names in DayCent have the following conventions. Input files with a ".100" extension are also used in monthly CENTURY. Input files unique to DayCent have an ".in" extension, except for weather files, which have a ".wth" extension. Output files with a ".lis" extension are common with monthly CENTURY, while unique DayCent output files have a ".out" extension. Software is available to convert between CENTURY and DayCent formatted input files. More information and downloads can be found at www.nrel.colostate.edu/projects/daycent/and www.nrel.colostate.edu/projects/century/.

Characterizing the "System": Input Data Requirements

DayCent model inputs can be divided into four categories: (i) weather information, (ii) soil information, (iii) plant information, and (iv) management information. Table 5–1 lists the basic information needed to define a "system" in DayCent.

Weather Information

DayCent has two options for weather inputs (Table 5–1). Users can run the model using only daily weather (maximum and minimum temperature [°C] and precipitation [cm] values), or they can choose to use these plus three extra drivers (solar radiation [ly d^{-1}], wind speed [mph], and relative humidity [%]). In addition to creating a daily weather file, users must also generate weather statistics using built in software. Weather statistics are used to calculate mean actual evapotranspiration (AET), which is used to calculate N inputs from atmospheric deposition and asymbiotic fixation. When running the model at the plot level, local weather station data should be used. But when running at regional or larger scales, datasets such as DAYMET (available at http://www.daymet.org [verified 30 Mar. 2011]; Thornton et al., 2000, 1997; Thornton and Running, 1999) can be used.

Site Information

Soil properties (Table 5–1) are probably the most important site level data needed because they influence plant growth, water and nutrient flows, and decomposition processes. Users specify soil horizonation (i.e., layer thickness and number of layers to bedrock or water table). Required soil properties for each layer include texture, bulk density, wilting point, field capacity, the extent to which water content can drop below the wilting point, root fraction, organic matter fraction, saturated hydraulic conductivity (K_{sat}), and pH. Soil texture, depth to bedrock, and pH are typically available for experimental plots and can also be derived from soil map units. If time series data for soil water content are available, then field capacity and wilting point can be estimated by observing maximum soil water contents a day or two after rainfall events and minimum soil water contents during dry down periods. Field capacity and wilting point, as well as K_{sat}, can be

Table 5–1. Data required to operate the DayCent model.

Date type	Data required
Daily meteorology data	Minimum and maximum air temperature, precipitation Extra drivers: wind speed, solar radiation, and relative humidity
Site specifications	Latitude, weather statistics, number of elements simulated The following can be changed through time: ambient CO_2 levels, daily temperature and precipitation values, soil pH, and soil temperature, soil organic matter eroded or deposited, nitrogen added or deposited, ^{14}C or ^{13}C labeling
Soil properties	Soil horizon delineation, texture, bulk density, field capacity, wilting point, saturated hydraulic conductivity, pH, rooting distribution
Plant	Specifying a crop cultivar from the "crop.100" file or create a new cultivar
Management practices	Day of year and amount or intensity for events such as cultivation, irrigation, planting/harvest, fertilization, burning, grazing

derived based on soil texture. We recommend using the algorithm developed by Saxton and Rawls (2006; available at http://hydrolab.arsusda.gov/soilwater/Index.htm [verified 30 Mar. 2011]). Default values for organic matter fraction are usually used, but root fraction will depend on the suite of vegetation represented during the simulation period. A water table can be simulated by specifying water table depth on a monthly basis. The water table can be a source of water flowing up the profile via unsaturated flow. When a water table is implemented, drainage out the bottom of the soil profile does not occur, so rainfall events can saturate soil layers above the water table. When running the model at the plot level, site-specific soil characteristics should be used. However, when running at regional or larger scales, datasets such STATSGO (http://soils.usda.gov/products/datasets/statsgo/, verified 14 Apr. 2011) can be used.

Users also specify site latitude, which is used in the plant growth submodel to calculate solar radiation, as well as the number of elements in addition to C, to be simulated. Users can select to include only N cycling, N and P cycling, or N, P, and S cycling. Changes in some environmental conditions can be simulated through time. Specifically, users can choose to linearly increase or decrease ambient CO_2 levels, daily temperature and precipitation values, soil pH, and soil temperature by specified amounts for a specified time period. The model does not explicitly account for topography, but erosion and deposition of soil organic matter can be simulated by specifying amounts on a monthly basis. The model can also simulate ^{14}C or ^{13}C labeling, incubation experiments, and microcosms (e.g., buried litter bag experiments).

Crop Information

DayCent has been parameterized to grow all the major U.S. commodity crops—corn (*Zea mays* L.), soybean [*Glycine max* (L.) Merr.], wheat (*Triticum aestivum* L.), hay, sorghum [*Sorghum bicolor* (L.) Moench, and cotton (*Gossypium hirsutum* L.)—as well as some other crops—barley (*Hordeum vulgare* L.), potato (*Solanum tuberosum* L.), pearl millet [*Pennisetum glaucum* (L.) R. Br.], tomato (*Solanum lycopersicum* L.), and sweet corn). The model also includes parameterizations to represent various grassland, forest, and savanna biomes (e.g., tallgrass prairie, shortgrass steppe, boreal forest, arid shrubland, tropical savanna). If users are growing a crop or vegetation mix that is not represented in our crop library, then we recommend modifying an existing parameterization that most closely resembles the new crop of interest. Data or estimates for the following are typically needed for each plant component (e.g., roots, shoots) to parameterize a new crop: C/nutrient ratio, C allocation fraction, death rate, and lignin concentration. In addition, maximum growth rate, growth response to temperature, harvest index, soil temperature for germination, and accumulated growing degree day impacts on phenology are required.

When growing vegetation, users need to specify phenology options. In addition to designating whether the vegetation is a perennial or an annual, users choose if the beginning and end of the growing season are specified or generated by the model. Specifically, users can designate the calendar day when growth for perennials or germination for annuals begins, and the day when harvest or senescence occurs at the end of the growing season. Alternatively, the beginning of the growing season can be made a function of soil temperature and the end of the growing season a function of accumulated growing degree days. For grain-filling annuals, users have the further option of using accumulated growing degree days to initiate anthesis (flowering), as well as harvest.

Management Information

In DayCent, it is easy to represent common management options (e.g., tillage, irrigation, fertilization, planting, harvesting, burning, grazing). Automatic fertilization and irrigation options are also available. In general, users specify the calendar day on which the management or disturbance event occurs, and parameters controlling which soil, vegetation, and litter pools are impacted, and the magnitude of impact on the different pools. Regarding automatic fertilization, the user specifies the degree to which fertilizer application will meet or exceed nutrient demand. When using the automatic irrigation option, the user specifies either to irrigate to field capacity, field capacity plus potential evapotranspiration, or to apply a specified amount of water.

Users should be aware of caveats when using soil temperature and accumulated growing degree days to control plant phenology. For example, if the soil temperature is cool in spring, then a fertilizer event scheduled on the typical planting day could occur before crop germination. Similarly, if accumulated growing degree days are used to control crop harvest, then a tillage event scheduled shortly after the typical harvest date could occur before harvest when the growing season has cooler temperatures than normal.

Initialization Options

Users can choose to enter initial values for C and nutrient quantities in the different soil organic matter and vegetation component pools or initial values can be generated using long-term simulations. We advocate using the latter because C and nutrient values are rarely known for all of the pools and using long-term simulations ensures that the model is close to equilibrium. Specifically, we recommend that users simulate at least 1000 yr of native vegetation (spin up simulation) followed by plow out and historical land use (base simulation). Before implementing the base simulation, users should verify that the spin up simulation was of sufficient duration to result in relatively stable intermediate and slow SOM pools. If local information regarding presettlement vegetation is not available, then the potential native vegetation map developed by Kuchler (1993) can be used to derive an appropriate vegetation mix to simulate during the spin up. For the base simulation, users should use their best guesses regarding when plow out occurred and what the land use history was likely to have been. Fortunately, model results are more sensitive to recent land management, which is likely to be known with greater confidence than earlier land management.

Model Calibration

It is desirable to have a general model that can be applied across different cropping systems without changing parameters other than site-specific ones relating to weather, soils, vegetation, and land management. However, use of generalized parameters will likely lead to poor model performance for at least some site–treatment combinations, and model results can always be improved by tuning the model to better represent particular conditions. We do not recommend arbitrarily changing parameters to improve model fit without explicitly stating which parameter values were changed and quantifying how well model performance was improved. Identifying sensitive parameters can lead to model improvements, especially if a relationship based on already required site information can be derived to predict how a parameter value should change across sites. For example, weather statistics can be used to predict if an early-, medium-, or late-maturing cultivar should be simulated at a particular site. Users should not use parameter values that do not make sense biologically, even if they improve model performance.

There are more than 1000 parameters in the DayCent model. Fortunately, users typically only need to calibrate a small subset of these. After input files have been formatted and checked for accuracy, the next step is to run the model and evaluate if the results are reasonable. We advocate verifying model outputs in the following order: soil water content, crop yields and plant growth rates, soil organic C levels, and N loss vectors (Fig. 5–2). Soil water content should be looked at first because this is a primary control on most of the key processes represented in the model. Plant growth rates and crop yields should be verified next because yield data are almost always available and if growth and harvest are not simulated correctly then other variables of interest are also likely to be simulated incorrectly. It is important to verify that SOC levels appear reasonable because SOC integrates many processes in the model (e.g., NPP, decomposition) that are sensitive to both environmental conditions and management decisions. Lastly, N flows should be verified. We elaborate on these steps below.

Model Inputs

Verify that the weather file is in the proper units and investigate any outliers, such as precipitation events that appear anomalously large. Crop rotations and amounts and timing of nutrient addition should also be verified.

Calibrating Soil Water Parameters

Soil water content and flows are crucial in DayCent, and key processes and parameters controlling water balance are summarized in Table 5–2. If time series observations of soil water content are available, compare with simulated values and adjust the field capacity and wilting point in the "soils.in" file so that maximum and minimum simulated soil water content values are not biased, high or low. If surface runoff does not seem reasonable, adjust the duration of precipitation events in the "sitepar.in" file and/or the K_{sat} of the surface layer in the "soils.in" file. Look at other water outputs in the "watrbal.out" file and see if water leaching, evaporation, transpiration, and sublimation seem reasonable. If transpiration seems incorrect, root distribution control in the "crop.100" or "soil.in" files may need adjustment. If leaching appears incorrect, adjust hydraulic water potential and/or saturated hydraulic conductivity of the deep storage layer in the "sitepar.in" file. Limitations that inhibit how well the soil water submodel works include: not accounting for topography or lateral transfer of water and the inability to precisely simulate water inputs from snow melt due to snow drifting, which is not accounted for in the model.

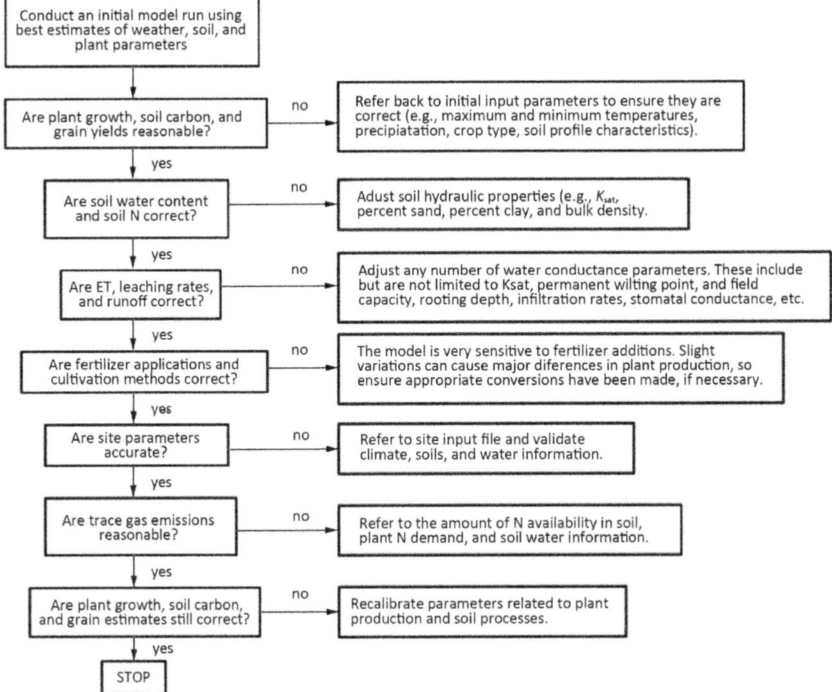

Fig. 5–2. Running simulations in DayCent.

Table 5–2. Data required to calibrate soil water balance components of DayCent.

Water balance component	Adjustment
Precipitation	No adjustment-driving variable (but ensure input file is correct)
Infiltration and runoff	Adjust assumed duration (hours) of precipitation events in "sitepar.in" file. Change the saturated conductivity of surface soil layer in "soils.in" file.
Evapotranspiration	Verify that the rooting depth of the crop is reasonable in both "crop.100" and "soils.in" files so that water extraction from soil layers is correct.
Leaching	Adjust hydraulic water potential and/or saturated hydraulic conductivity of deep storage layer in "sitepar.in" file.
Soil water content	Adjust runoff, evapotranspiration, field capacity, wilting point in "soils.in" file If using water table submodel, verify values in "sitepar.in" file

Calibrating Plant Parameters

In DayCent, plant growth is most sensitive to phenology, nutrient availability, environmental conditions, and maximum potential growth rate (Table 5–3).

Phenology

Look at monthly outputs for live biomass to ensure that phenology is reasonable. If the user has specified when a plant begins growing, is harvested, and senesces, check day of year values for these parameters in the schedule file. If using the growing degree submodel to control phenology, then the parameter controlling the minimum soil temperature for germination and/or the parameters controlling the required number of growing degree days for anthesis, harvests, and senescence may need adjustment. Growing degree days can be modified using the "crop.100" file.

Net Primary Productivity

If nutrients are not limiting, the potential growth rate and/or carbon to nutrient ratios of biomass components in the "crop.100" file may need adjustment. If nutrients are limiting, then parameters controlling external nutrient input rates in the "site.100" file and/or parameters controlling nutrient losses from leaching in the "fix.100" file may need adjustment. Check that nutrient inputs from fertilization events are correct. Plant temperature response may need to be adjusted based on crop cultivar.

Biomass

Assuming NPP is reasonable, parameters controlling harvest index in the "crop.100" file and/or biomass removal amounts in the "harv.100" file and/or parameters controlling biomass component senescence rates and/or carbon allocation in the "crop.100" file may need adjustment. If the plant growth looks correct but the amount of harvested material is not representative of the crop grown, check the timing of the harvest event in relation to senescence and/or cultivation events.

Soil Organic Matter Parameters

DayCent simulates SOM (C, N, P, and S) in the top 20 cm of the soil. For most model applications, it is more important that changes in SOM through time and among treatments are correctly represented than absolute values of SOM. Assuming plant growth and senescence rates which control carbon inputs are correct, then look at parameters controlling decomposition rates (Table 5–4). The intensity and duration of cultivation events may need adjustment. The default assumption is that cultivation events increase decomposition rates for 1 mo, but intensive plowing can disrupt soil aggregates and increase decomposition rates for subsequent months. Consequently, users may need to include additional cultivation events in the schedule file following the initial soil disturbance event to increase decomposition rates while not disturbing the soil or plant material.

Table 5–3. To calibrate crop growth in DayCent.

Steps	Related model parameters
1. Crop life cycle	Look at monthly outputs for biomass and check if phenology is reasonable. If the user has specified when a plant begins growing, is harvested, and senesces, check day of year values for these parameters in the schedule file. If using the growing degree submodel to control phenology, then the parameter controlling the minimum soil temperature for germination and/or the parameters controlling required number of growing degree days in the "crop.100" file may need adjustment.
2. Net primary production (NPP)	If nutrients are not limiting, the potential growth rate and/or carbon to nutrient ratios of biomass components in the "crop.100" file may need adjustment. If nutrients are limiting, then parameters controlling external nutrient input rates in the "site.100" file and/or parameters controlling nutrient losses from leaching in the "fix.100" file may need adjustment. Check that nutrient inputs from fertilization events are correct. Plant temperature response may need to be adjusted based on crop cultivar.
3. Standing biomass	Assuming NPP is reasonable, parameters controlling harvest index in "crop.100" file and/or biomass removal amounts in "harv.100" file and/or parameters controlling biomass component senescence rates and/or carbon allocation in the "crop.100" file may need adjustment.
4. Verify that both NPP and seasonality of standing biomass for above and below ground pools are reasonable.	
5. Conduct a sensitivity analysis to see how the calibrated model responds to water and nutrient inputs.	If there is not a large range of treatments in the study, create a few management scenarios to evaluate model response to water and nutrient applications; model response to air temperature can also be tested using different weather files.

Improper simulation of site history also influences present day SOC levels. Verify that the simulation of pre-crop vegetation is correct and of sufficient duration to achieve relatively stable intermediate and slow SOM pools and that the recent site history is correct (e.g., when plow-out occurred, historical rotations, nutrient inputs). In rare circumstances, decomposition rate constants or C/nutrient ratios for SOM pools in the "fix.100" file may need adjustment.

The fast, intermediate, and slow SOM pools in DayCent roughly correspond with microbial biomass and labile SOM (fast), resistant plant material and soil-stabilized microbial products (slow), and physically and chemically stabilized SOM (passive). However, efforts to compare these conceptual pools in the model with SOM data based on laboratory fractionations often do not correspond very well. Consequently, we believe it is most important that the model shows the correct response to measured changes in total SOM.

Table 5–4. To calibrate soil organic matter components of DayCent.

Soil organic matter component	Adjustment
Soil organic C	Assuming plant growth and senescence rates that control C inputs are correct, look at parameters controlling decomposition rates. The intensity and duration of cultivation events may need adjustment. The default assumption is that cultivation events increase decomposition rates for 1 mo, but intensive plowing can disrupt soil aggregates and increase decomposition rates for subsequent months. Consequently, users may need to include additional cultivation events in the schedule file following the initial cultivation soil disturbance event that increase decomposition rates but do not disturb the soil or plant material. Improper simulation of site history also influences preset day soil organic C levels. Check that simulation of pre-crop vegetation is correct and of sufficient duration to achieve relatively stable intermediate and slow SOM pools and that the recent site history is correct, e.g., when plow-out occurred, historical rotations, and nutrient inputs. Verify that soil temperature and water content are correct. Check if standing surface residue pool amount is correct. In rare circumstances, decomposition rate constants for soil organic matter pools in "fix.100" file may need adjustment.
Soil organic N	In rare circumstances, C/N ratios of soil organic matter pools in "fix.100" file may need adjustment.

Soil Nitrogen Parameters

Soil Nitrogen Distribution and NO_3 Leaching

Ammonium is assumed to be confined to the top 15 cm of soil, while nitrate is mobile throughout the profile. Assuming water leaching is correct but NO_3 leaching appears to be incorrect, then the parameters controlling the fraction of N available for plants and/or the fraction of NO_3 that leaches to the next soil layer during saturated water flow in the "fix.100" file may need adjustment (Table 5–5).

Nitrous Oxide Emissions from Nitrification and Denitrification

Plant N demand and N losses from leaching impact N availability for nitrification and denitrification. Ensure that plant phenology, NO_3 leaching losses, and soil water content look reasonable. Often, mean N_2O emissions agree reasonably well with field data, but the timing of emissions peaks on a daily basis do not compare favorably. This is to be expected. Nitrous oxide emissions are notoriously variable in time and space, and some of the factors that control emissions are not represented in the model (e.g., topography, microbial community, spatial heterogeneity in nutrient availability). If mean N_2O emissions are incorrect, then the portion of nitrified N that is assumed to be lost as N_2O in the "sitepar.in" file can be adjusted. Adjustment of this parameter should be noted, and we do not recommend arbitrary

Table 5–5. To calibrate nitrogen components of DayCent.

Nitrogen component	Adjustment
N removed via disturbance events	C/N ratios of plant components and/or fraction of N that goes to grain in "crop.100" file may need adjustment. Biomass removal amounts during harvest, grazing, and burning events may need adjustment. N return amounts during grazing and burning events may need adjustment.
Nitrate leaching	Modify the fraction of N available for plants. Change the fraction of NO_3 that leaches to the next soil layer during saturated water flow.
N gas losses	Adjust portion of nitrified N that is assumed to be lost as N_2O in "sitepar.in" file Ensure that plant phenology related to N uptake is correct Ensure that soil water content is simulated correctly

tuning this parameter just to improve model fit. Using N_2O emission data from irrigated corn cropping in Colorado, this parameter was calibrated to 1% (Del Grosso et al., 2008a) and this value applied generally as the model performed well at simulating N_2O emissions observed at various sites (Del Grosso et al., 2008b, 2010). We also recommend that users look at N_2O emissions from nitrification and denitrification separately to help diagnose problems.

Model Evaluation

Comparisons with Observations and Other Models

When evaluating model performance, different statistics should be used (r^2, RMSE, mean error) because, for example, r^2 is not influenced by model bias. In addition to comparing model outputs with field data, we also advocate comparing outputs among different models. One reason for this is to evaluate model performance relative to standard, commonly used models. For example, model performance when compared solely with observations may be judged as poor (e.g., low correlation coefficient), but when another model is included in the comparison, the relative performance of the model of interest could be more favorable.

Another reason to compare results from different models is that field data are always incomplete, and some important variables, such as N_2 and NH_3 losses from soils, are almost never measured. If models that use different assumptions agree, then this increases confidence in model performance. Conversely, disagreement suggests that understanding is limited. For example, a comparison of six models, including DayCent, showed good agreement with one another and with observations for crop yields, stream flow, and NO_3^- leaching for a watershed in Illinois where most of the land is used for tile drained corn–soybean cropping (David et al., 2009). However, the model results diverged sharply for denitrification rates and N_2 emissions, partially because these variables are rarely measured, as was the case in this study. Consequently, more observational data are needed to properly evaluate the ability of models to represent denitrification rates.

Because the different N loss vectors interact, we advocate complete N cycle evaluation. Unfortunately, few field studies measure all the important N loss vectors. Even if complete field data are lacking, N loss vectors for competing models can still be compared.

Model evaluation is dependent on the variable of interest, the scale of measured data, and the intended model application. For SOM, simulating the SOM differences between different treatments correctly is usually more important than simulating the absolute SOM values correctly. For grain yields, both absolute values and treatment differences should be simulated correctly. The model should be expected to simulate absolute N losses correctly, as well as treatment differences when the results are aggregated temporally. The model should not necessarily be expected to accurately predict N losses on a daily basis. If a primary model application is to predict how annual N losses vary across climate zones, soil types, and management options, then correctly simulating mean losses is acceptable. But, if the model is intended to investigate the small-scale mechanisms that control N cycling, then agreement with measured daily loss rates is more important.

Model Testing Databases

The developers of DayCent have compiled N_2O flux, crop yield, NO_3 leaching, and model driver data from about a dozen sites in North America (Del Grosso et al., 2005, 2010). Web accessible databases that include model driver and testing data from various experimental plots that span geographical regions are particularly useful. Access to such data enables rigorous model testing and facilitates model comparisons. One example is the U.S. Trace Gas Network, or TRAGNET (http://www.nrel.colostate.edu/projects/tragnet/ [verified 30 Mar. 2011]), which contains N_2O, CH_4, and CO_2 flux data, as well as some environmental and soil variables from about a dozen sites in North America and Europe (Ojima et al., 2000). However, TRAGNET is limited because land management information is incomplete, weather data are not included, and the types of data reported are not always consistent across sites. Nonetheless, TRAGNET helped develop and test some commonly used models for trace gas fluxes (Bogner et al., 2000; Davidson and Verchot, 2000; Del Grosso et al., 2000a,b; Groffman et al., 2000). A more recent example is GRACEnet (Greenhouse gas Reduction through Agricultural Carbon Enhancement network: www.ars.usda.gov/research/programs/programs.htm?np_code=204&docid=17271&page=1 [verified 30 Mar. 2011]), a USDA-ARS cross location project designed to quantify the impacts of conventional and improved management on crop yields, greenhouse gas emissions, and other environmental factors for cropped and grazed systems (Jawson et al., 2005). The GRACEnet database will include extensive model driver and testing data from more than 30 research sites across the United States; see the web address above for a map of locations.

DayCent Model Applications

DayCent has been used to simulate the impacts of climate and land use change on various crop, grassland, forest, and savanna systems around the world (Pepper et al., 2005; Parton et al., 2007, 2011; Conant et al., 2005; Morgan et al., 2004; Hartman et al., 2007). Beginning in 2005, county level DayCent simulations have been used to estimate N_2O emissions from cropped and grazed systems for the Inventory of U.S. Greenhouse Gas Emissions and Sinks compiled annually by the USEPA (Del Grosso et al., 2006, 2010; USEPA 2010). DayCent has been used to estimate soil greenhouse gas mitigation for corn, soybean, and wheat at the global scale (Del Grosso et al., 2009). The model has also been used to calculate crop yields and soil greenhouse gas emissions for different biofuel cropping systems (Adler et al., 2007; Davis et al., 2010, Robertson et al., 2010).

Conclusions

There is no perfect method to quantify many agroecosystem variables of interest, and all measurement methods and models have strengths and weaknesses. Ground-based measurements are limited because snapshots of state variables at small spatial and temporal scales, (e.g., standing biomass, chamber trace gas concentration) are used to infer plant growth and gas flux rates at larger scales. Models are a simplification of reality but have the advantage of offering full spatial and temporal coverage. The accuracy and utility of estimates based on both models and measurements vary across spatial and temporal scales. For example, at smaller scales, process-based models usually agree more closely with N flux observations than simple empirical models, but as scale increases, estimates based on different modeling and measuring methodologies tend to converge (Del Grosso et al., 2008b).

When evaluating models, the following must be kept in mind. There is uncertainty in inputs used to drive models and measurements used to test model outputs. The reliability of some measurements (e.g., grain yields) is greater than others (e.g., annual soil surface N_2O flux). For these reasons, model results should be put in perspective by comparing outputs not only with measurements but also with outputs from other models (e.g., Frolking et al., 1998; Cramer and Field 1999; David et al., 2009). When testing models, results are almost always mixed, with different outputs comparing favorably with measurements under some conditions but not others. In addition to evaluating models based on comparisons with measurements and outputs from other models, how well faulty model behavior can be explained, in terms of, for example, reliability of input data, should also be considered.

References

Adler, P.R., S.J. Del Grosso, and W.J. Parton. 2007. Life cycle assessment of net greenhouse gas flux for bioenergy cropping systems. Ecol. Appl. 17:675–691Bogner, J.E., and R.L. Sass, and B.P. Walter. 2000. Model comparisons of methane oxidation across a management gradient: Wetlands, rice production systems, and landfill. Global Biogeochem. Cycles 14(4), doi:10.1029/1999GB001224.

Conant, R.T., K. Paustian, S.J. Del Grosso, and W.J. Parton. 2005. Nitrogen pools and fluxes in grassland soils sequestering carbon. Nutr. Cycl. Agroecosyst. 71:239–248.

Cramer, W., and C.B. Field. 1999. Comparing global models of terrestrial net primary production (NPP): Overview and key results. Glob. Change Biol. 5(Suppl. 1):1–15.

David, M.B., S.J. Del Grosso, X. Hu, G.F. McIsaac, W.J. Parton, E.P. Marshall, C. Tonitto, and M.A. Youssef. 2009. Modeling denitrification in a tile-drained, corn and soybean agroecosystem of Illinois, USA. Biogeochemistry 93:7–30.

Davidson, E., and L. Verchot. 2000. Testing the hole-in-the-pipe model of nitric and nitrous oxide emissions from soils using the TRAGNET database. Global Biogeochem. Cycles 14:1035–1043.

Davis, S.C., W.J. Parton, F.G. Dohleman, N.R. Gottel, C.M. Smith, S.J. Del Grosso, A.D. Kent, and E.H. DeLucia. 2010. Comparative biogeochemical cycles of bioenergy crops reveal nitrogen-fixation and low greenhouse gas emissions in a *Miscanthus* × *giganteus* agro-ecosystem. Ecosystems 13:144–156, doi:10.1007/s10021-009-9306-9.

Del Grosso, S.J., A.R. Mosier, W.J. Parton, and D.S. Ojima. 2005. DAYCENT model analysis of past and contemporary soil N_2O and net greenhouse gas flux for major crops in the USA. Soil Tillage Res. 83:9–24, doi:10.1016/j.still.2005.02.007.

Del Grosso, S.J., S.M. Ogle, W.J. Parton, and F.J. Breidt. 2010. Estimating uncertainty in N_2O emissions from U.S. cropland soils. Global Biogeochem. Cycl. 24:GB1009, doi:10.1029/2009GB003544.

Del Grosso, S.J., D.S. Ojima, W.J. Parton, E. Stehfest, M. Heistemann, B. Deangelo, and S. Rose. 2009. Global Scale DAYCENT Model Analysis of Greenhouse Gas Mitigation Strategies for Cropped Soils. Global Planet. Change 67:44–50, doi:10.1016/j.gloplacha.2008.12.006.

Del Grosso, S.J., W.J. Parton, A.R. Mosier, M.D. Hartman, J. Brenner, D.S. Ojima, and D.S. Schimel. 2001. Simulated interaction of carbon dynamics and nitrogen trace gas fluxes using the DAYCENT model. p. 303–332. *In* M. Schaffer et al. (ed.) Modeling carbon and nitrogen dynamics for soil management. CRC Press, Boca Raton, FL.

Del Grosso, S.J., W.J. Parton, A.R. Mosier, D.S. Ojima, A.E. Kulmala, and S. Phongpan. 2000a. General model for N_2O and N_2 gas emissions from soils due to denitrification. Global Biogeochem. Cycles 14:1045–1060.

Del Grosso, S.J., W.J. Parton, A.R. Mosier, D.S. Ojima, C.S. Potter, W. Borken, R. Brumme, K. Butterbach-Bahl, P.M. Crill, K. Dobbie, and K.A. Smith. 2000b. General CH_4 oxidation model and comparisons of CH_4 oxidation in natural and managed systems. Global Biogeochem. Cycl. 14:999–1019.

Del Grosso, S.J., W.J. Parton, A.R. Mosier, M.K. Walsh, D.S. Ojima, and P.E. Thornton. 2006. DAYCENT national scale simulations of N_2O emissions from cropped soils in the USA. J. Environ. Qual. 35:1451–1460 doi:10.2134/jeq2005.0160.

Del Grosso, S.J., W.J. Parton, D.S. Ojima, C.A. Keough, T.H. Riley, and A.R. Mosier. 2008a. DAYCENT Simulated effects of land use and climate on county level N loss vectors in the USA. p. 571–

595. *In* R.F. Follett and J.L. Hatfield (ed.) Nitrogen in the environment: Sources, problems, and management. 2nd ed. Elsevier Science Publishers, New York.

Del Grosso, S.J., T. Wirth, S.M. Ogle, and W.J. Parton. 2008b. Estimating agricultural nitrous oxide emissions. EOS 89:529–530.

Eitzinger, J., W.J. Parton, and M.D. Hartman. 2000. Improvement and validation of a daily soil temperature submodel for freezing/thawing periods. Soil Sci. 165:525–534.

Frolking, S.E., A.R. Mosier, D.S. Ojima, C. Li, W.J. Parton, C.S. Potter, E. Priesack, R. Stenger, C. Haberbosch, P. Dörsch, H. Flessa, and K.A. Smith. 1998. Comparison of N_2O emissions from soils at three temperate agricultural sites: Simulations of year round measurements by four models. Nutr. Cycl. Agroecosyst. 52:77–105.

Groffman, P., R. Brumme, K. Butterbach-Bahl, K. Dobbie, A. Mosier, D. Ojima, H. Papen, W. Parton, K. Smith, and C. Wagner-Riddle. 2000. Evaluating annual nitrous oxide fluxes at the ecosystem scale. Global Biogeochem. Cycl. 14:1061–1070.

Hartman, M.D., J.S. Baron, and D.S. Ojima. 2007. Application of a coupled ecosystem–chemical equilibrium model, DayCent-Chem, to stream and soil chemistry in a Rocky Mountain watershed. Ecol. Model. 200:493–510, doi:10.1016/j.ecolmodel.2006.09.001.

Hutchinson, G.L., G.P. Livingston, and E.A. Brams. 1993. Nitric and nitrous oxide evolution from managed subtropical grassland. p. 290–316. *In* R.S. Oremland (ed.) Biogeochemistry of global change: Radiatively active trace gases. Chapman and Hall, New York.

Jawson, M.D., S.R. Shafer, A.J. Franzluebbers, T.B. Parkin, and R.F. Follet. 2005. GRACEnet: Greenhouse gas reduction through agricultural carbon enhancement network. Soil Tillage Res. 83:167–172, doi:10.1016/j.still.2005.02.015.

Kelly, R.H., W.J. Parton, M.D. Hartman, L.K. Stretch, D.S. Ojima, and D.S. Schimel. 2000. Intra and interannual variability of ecosystem processes in shortgrass steppe. J. Geophys. Res. 105:20,093–20,100, doi:10.1029/2000JD900259

Kuchler, A.W. 1993. Potential natural vegetation of the conterminous United States. Digital vector data in an Albers Equal Area Conic polygon network and derived raster data on a 5 km by 5 km Albers Equal Area 590×940 grid. *In* Global Ecosystems Database Version 2.0. NOAA National Geophysical Data Center, Boulder, CO.

Ma, L., L.R. Ahuja, S.A. Saseendran, R.W. Malone, T.R. Green, B.T. Nolan, P.N.S. Bartling, G.N. Flerchinger, K.J. Boote, and G. Hoogenboom. 2011. A protocol for parameterization and calibration of RZWQM2 in field research. p. 1–64. *In* L.R. Ahuja and L. Ma (ed.) Methods of introducing system models into agricultural research. Advances in Agricultural Systems Modeling 2. ASA, CSSA, and SSSA, Madison, WI.

Martin, R.E., M.C. Scholes, A.R. Mosier, D.S. Ojima, E.A. Holland, and W.J. Parton. 1998. Controls on annual emissions of nitric oxide from soils of the Colorado shortgrass steppe. Global Biogeochem. Cycles 12:81–91.

Metherell, A.K., L.A. Harding, C.V. Cole, and W.J. Parton. 1993. CENTURY soil organic matter model environment, Technical documentation, Agroecosystem version 4.0. Great Plains System Research Unit Technical Rep. 4. USDA-ARS, Fort Collins, CO.

Morgan, J.A., D.E. Pataki, C. Körner, H. Clark, S.J. Del Grosso, J.M. Grünzweig, A.K. Knapp, A.R. Mosier, P.C.D. Newton, P.A. Niklaus, J.B. Nippert, R.S. Nowak, W.J. Parton, H.W. Polley, and M.R. Shaw. 2004. Water relations in grassland and desert ecosystems exposed to elevated atmospheric CO_2. Oecologia 140:11–25.

Ojima, D., A. Mosier, S. Del Grosso, and W. Parton. 2000. TRAGNET analysis and synthesis of trace gas fluxes. Global Biogeochem. Cycl. 14:995–997.

Parton, W.J., M. Hartman, D.S. Ojima, and D.S. Schimel. 1998. DAYCENT: Its land surface submodel: Description and testing. Global Planet. Change 19:35–48.

Parton, W.J., E.A. Holland, S.J. Del Grosso, M.D. Hartman, R.E. Martin, A.R. Mosier, D.S. Ojima, and D.S. Schimel. 2001. Generalized model for NO_x and N_2O emissions from soils. J. Geophys. Res. 106(D15):17403–17420.

Parton, W.J., J.A. Morgan, G. Wang, and S.J. DelGrosso. 2007. Projected ecosystem impact of the prairie heating and CO_2 enrichment experiment. New Phytol. 174:823–834, doi:10.1111/j.1469–8137.2007.02052.x.

Parton, W.J., A.R. Mosier, D.S. Ojima, D.W. Valentine, D.S. Schimel, K. Weier, and K.E. Kulmala. 1996. Generalized model for N_2 and N_2O production from nitrification and denitrification. Global Biogeochem. Cycl. 10:401–412.

Parton, W.J., D.S. Ojima, C.V. Cole, and D.S. Schimel. 1994. A general model for soil organic matter dynamics: Sensitivity to litter chemistry, texture and management. p. 147–167. *In* R.B. Bryant and R.W. Arnold (ed.) Quantitative modeling of soil forming processes. SSSA Spec. Publ. 39. SSSA, Madison, WI.

Parton, W.J., R.J. Scholes, K. Day, J. Carter, and R. Kelly. 2011. CENTURY-SAVANNA model for tree–grass ecosystems. p. 443–462. *In* M.J. Hill and N.P. Hanan (ed.) Ecosystem function in savanahs: Measurement and modeling at landscape to global scales. CRC Press, Boca Raton, FL.

Parton, W.J., J.M.O. Scurlock, D.S. Ojima, T.G. Gilmanov, R.J. Scholes, D.S. Schimel, T. Kirchner, J.C. Menaut, T. Seastedt. E.G. Moya, A. Kamnalrut, and J.L. Kinyamario. 1993. Observations and modeling of biomass and soil organic matter dynamics for the grassland biome worldwide. Global Biogeochem. Cycl. 7:785–809.

Parton, W.J., J.M.O. Scurlock, D.S. Ojima, D.S. Schimel, and D.O. Hall. 1995. Impact of climate change on grassland production and soil carbon worldwide. Glob. Change Biol. 1:13–22, doi:10.1111/j.1365-2486.1995.tb00002.x.

Pepper, D.A., S.J. Del Grosso, R.E. McMurtrie, and W.J. Parton. 2005. Simulated carbon sink response of shortgrass steppe, tallgrass prairie and forest ecosystems to rising [CO_2], temperature and nitrogen input. Global Biogeochem. Cycl. 19:GB1004, doi:10.1029/2004GB002226.

Robertson, G.P., S.K. Hamilton, S.J. Del Grosso, and W.J. Parton. 2010. The biogeochemistry of bioenergy landscapes: Carbon, nitrogen, and water considerations. Ecol. Appl. doi:10.1890/09-0456.1.

Saxton, K.E., and W.J. Rawls. 2006. Soil water characteristic estimates by texture and organic matter for hydrologic solutions. Soil Sci. Soc. Am. J. 70:1569–1578, doi:10.2136/sssaj2005.0117.

Smart, D.R., J.M. Stark, and V. Diego. 1999. Resource limitation to nitric oxide emissions from a sagebrush-steppe ecosystem. Biogeochemistry 47:63–86.

Thornton, P.E., H. Hasenauer, and M.A. White. 2000. Simultaneous estimation of daily solar radiation and humidity from observed temperature and precipitation: An application over complex terrain in Austria. Agric. For. Meteorol. 104:255–271.

Thornton, P.E., and S.W. Running. 1999. An improved algorithm for estimating incident daily solar radiation from measurements of temperature, humidity, and precipitation. Agric. For. Meteorol. 93:211–228.

Thornton, P.E., S.W. Running, and M.A. White. 1997. Generating surfaces of daily meteorology variables over large regions of complex terrain. J. Hydrol. 190:214–251.

USEPA. 2010. Inventory of U.S. greenhouse gas emissions and sinks: 1990–2008. USEPA, Washington, DC.

Yari, J., and B. Parton. 2005. Potential changes in carbon dynamics due to climate change measured in the past two decades. Can. J. For. Res. 35:2258–2267.

Yienger, J.J., and H. Levy. 1995. Empirical model of global soil biogenic NO_x emissions. J. Geophys. Res. 100:11447–11464.

Chapter 6

Special Features of the EPIC and APEX Modeling Package and Procedures for Parameterization, Calibration, Validation, and Applications

Xiuying Wang
Armen R. Kemanian
Jimmy R. Williams

Abstract

Numerous hydrologic and environmental models, with a range of different levels of approach and complexity, have been developed to strengthen and complement field studies. The Environmental Policy Impact Climate (EPIC) and the Agricultural Policy Environmental eXtender (APEX) models are flexible and dynamic tools that were developed to evaluate a wide array of management strategies applied to crop, pasture, and grazing lands. They are capable of estimating long-term sustainability of land management in respect to erosion (wind, sheet, and channel), economics, water supply, water quality, soil quality, plant competition, weather, and pests for cropland, as well as grazing and pasture land. This chapter describes the main features of the EPIC and APEX models and reports on the parameterization, calibration, and validation of the models. As a case study, the APEX model was applied to the North Bosque watershed in central Texas, where the major land use/land cover is native pastures/range and improved pasture, with dairies and milk sheds. Influential model parameters were adjusted during the model calibration period on the basis of the monthly stream flow, sediment yield, and nutrient losses at the Hico monitoring station. The model performance was reasonable, and this application shows that the APEX model is a powerful tool for simulating feedlot water quality.

X. Wang (swang@brc.tamus.edu) and J.R. Williams (jwilliams@brc.tamus.edu), Blackland Research and Extension Center, Texas AgriLIFE Research, 720 East Blackland Rd., Temple, TX 76502; A.R. Kemanian, Dep. of Crop and Soil Sciences, The Pennsylvania State Univ., 116 Agricultural Sciences & Industries Bldg., University Park, PA 16802 (akemanian@psu.edu).

doi:10.2134/advagricsystmodel2.c6

Methods of introducing system models into agricultural research. L.R. Ahuja and L. Ma (ed.)
Advances in Agricultural Systems Modeling Series 2.

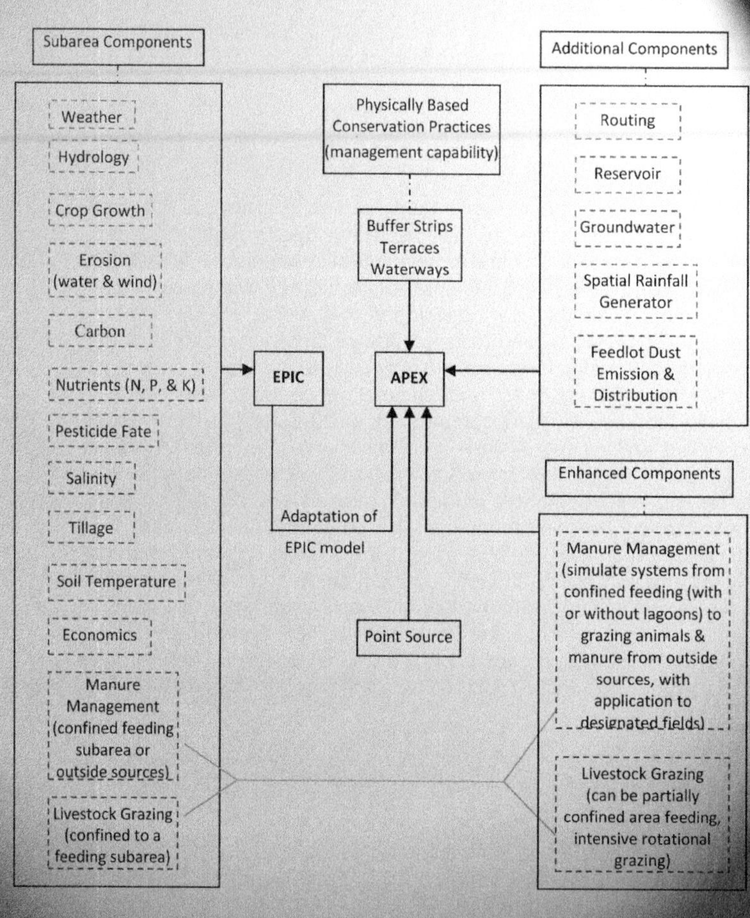
Subarea Components
Weather
Hydrology
Crop Growth
Erosion (water & wind)
Carbon
Nutrients (N, P, & K)
Pesticide Fate
Salinity
Tillage
Soil Temperature
Economics
Manure Management (confined feeding subarea or outside sources)
Livestock Grazing (confined to a feeding subarea)
EPIC
Adaptation of EPIC model
Physically Based Conservation Practices (management capability)
Buffer Strips
Terraces
Waterways
APEX
Point Source
Additional Components
Routing
Reservoir
Groundwater
Spatial Rainfall Generator
Feedlot Dust Emission & Distribution
Enhanced Components
Manure Management (simulate systems from confined feeding (with or without lagoons) to grazing animals & manure from outside sources, with application to designated fields)
Livestock Grazing (can be partially confined area feeding, intensive rotational grazing)

The Environmental Policy Impact Climate (EPIC) model (Williams et al., 1984; Williams, 1990, 1995) was developed in the early 1980s to assess the effect of soil erosion on soil productivity. It can be used to determine the effect of management strategies on agricultural production and soil and water resources. The drainage area considered by EPIC is generally a field-sized area of up to 100 ha, where weather, soils, and management systems are assumed to be homogeneous. The major components in EPIC are hydrology, erosion–sedimentation, nutrient cycling, pesticide fate, plant growth, soil temperature, tillage, economics, and plant environment control. It also includes a weather generator (Fig. 6–1). The Agricultural Policy Environmental eXtender (APEX) model (Williams et al., 1995; Williams, 2002; Williams and Izaurralde, 2006) is an extended and expanded version of the EPIC model. It was developed to facilitate multiple fields' scenarios in whole-farm or watershed management. The farm or watershed may be subdivided based on fields, soil types, landscape positions, or subwatersheds. Therefore, APEX can provide a consistent approach for evaluating various land management strategies at scales ranging from field to farm to watershed. Both EPIC and APEX run on a daily time-step. In APEX modeling, the individual field simulation uses the functions originally developed in EPIC (Fig. 6–1). In addition to EPIC functions, APEX has components for routing water, sediment, nutrients, and pesticides across landscapes through channels, floodplains, and reservoirs to the watershed outlet. The APEX groundwater component partitions flow between deep percolation and return flow. APEX also has a grazing component that provides flexibility to simulate confined or partially confined feeding area, intensive rotational grazing, cropland grazing after harvest, and other options (Fig. 6–1). Both EPIC and APEX models have been widely applied (Gassman et al., 2005, 2010).

The EPIC and APEX models require a relatively large number of input parameters. Like any other model, proper parameterization of EPIC and APEX is essential for making use of their modeling capabilities. For example, while the EPIC model was used for crop rotations in Southern France (Cabelguenne et al., 1990), the crop parameters including biomass energy ratio (WA), harvest index (HI), optimum air temperature (T_0), base temperature (T_b), and maximum leaf area index (LAI_{mx}) were adjusted to obtain a more satisfactory simulation of the seasonal development of leaf area index and grain yield. Gaiser et al. (2010)

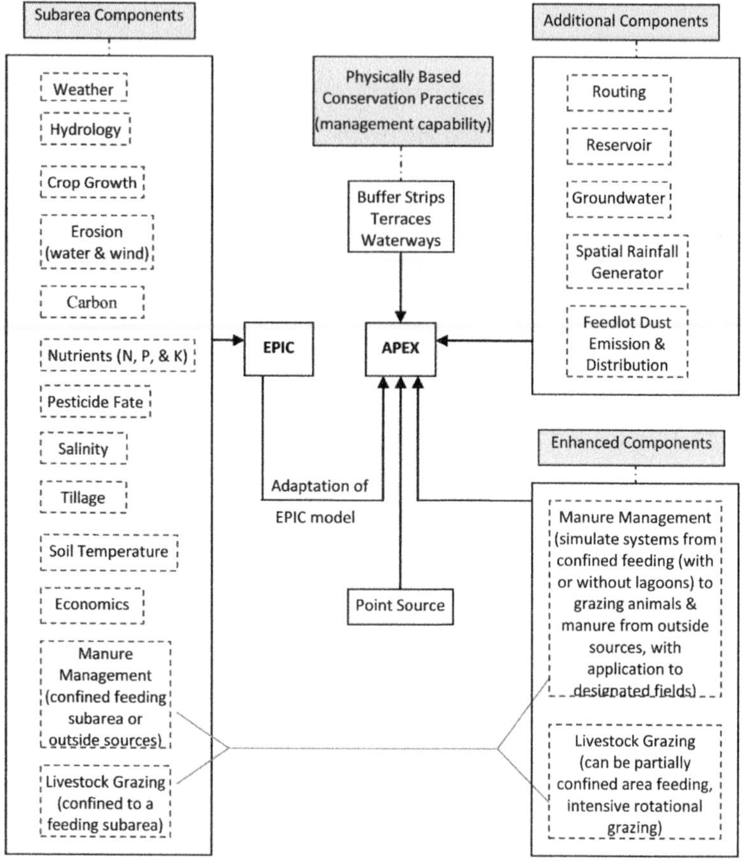

Fig. 6–1. The EPIC and APEX models.

indicated that crop growth models in tropical regions require extended testing under a range of climatic, soil, and management conditions and they used multi-location data to calibrate and validate the EPIC model for maize production in western Africa and northeastern Brazil. Wang et al. (2005) tested the EPIC crop growth and soil organic C components and conducted a sensitivity analysis for nine EPIC parameters using a variance-based sensitivity analysis method—the extended Fourier amplitude sensitivity test (FAST) (Saltelli et al., 1999, 2000). In Wang et al. (2006c), the sensitivities of 15 key APEX parameters involving hydrology, water and wind erosion, nutrient losses, soil organic C, and crop growth were analyzed for 159 locations across the conterminous United States. The analysis covers a wide range of soil types and climatic conditions, includes cropping systems of corn (*Zea mays* L.), soybean [*Glycine max* (L.) Merr.], and wheat (*Triticum aestivum* L.), and tillage systems of no-till, mulch, and conventional till. Yin et al. (2009) performed a sensitivity analysis for 13 key parameters affecting surface runoff and sediment loss using the FAST method before the model testing phase. Collectively, the USDA-NRCS curve number (CN), curve number index coefficient (P42), and the Hargreaves potential evapotranspiration (PET) equation (Hargreaves and Samani, 1985) exponent (P34) (if the Hargreaves method is used) are influential for runoff- and water-related output variables, such as soil loss by water and N and P losses in runoff. The revised universal soil loss equation (RUSLE) C factor coefficients (P46 and P47) are influential for erosion, sediment yield, and N and P losses in sediment. The available soil water capacity (the difference of soil water contents at field capacity and wilting point), potential heat units (PHU), biomass/energy ratio (WA), and harvest index are influential for the crop growth component (Table 6–1). The microbial decay rate coefficient and fraction of humus in passive pool are influential for the soil organic C component. The nitrate leaching ratio (nitrate concentration in surface runoff/nitrate concentration in percolate) and soluble P sorption coefficient (P concentration in sediment divided by that of the water) are influential to P loss (Table 6–1). Although numerous studies have been conducted using the EPIC and APEX models, as reviewed by Gassman et al. (2005; 2010), there is a need to summarize the step by step procedures for the models' parameter estimation and calibration. The objectives of this chapter are to describe the special features of the EPIC and APEX modeling package, to discuss parameterization, calibration and validation issues, and to demonstrate an APEX model's application.

Special Features of the EPIC and APEX Modeling Package

The EPIC and APEX models have been actively maintained since their inception. The models are refined, expanded, and improved constantly (Williams et al., 2008), a process mainly driven by various users' interaction, feedback, and interests for specific applications. Consequently, the models have a wide range of applications (Gassman et al., 2005, 2010) and also contain multiple options for simulating processes, such as hydrology and soil erosion–sedimentation (Table 6–2). Therefore, users can prefer one over the other depending on data availability and geographic or agricultural settings.

The models dynamically represent multiple soil properties, including depth and bulk density for each soil layer. The models remove eroded soil, attached organic nutrients, pesticides, and C from the soil profile as part of the emphasis on erosion productivity. Soil hydraulic properties and bulk density are estimated from soil texture and soil organic C using the equation described in Rawls and Brakensiek (1985). The models also consider the effect of changes in atmospheric

Table 6–1. Review of influential EPIC and APEX parameters.

Reference	Study location	Model component	Influential parameter†
Cabelguenne et al. (1990)	Southern France	crop grow	WA; HI; T_0; T_b; LAI_{mx}; RDMX; DLAI; DLAP1; DLAP2; RLAD; CNY; WSYF; BN1; BN2; BN3
Chung et al. (2001, 2002)	Lamberton, MN Nashua, IA	hydrology; N cycling	CN2; N fixation coefficient (P7)
Wang et al. (2005)	Plots in WI Arlington, Agricultural Research Station in south-central Wisconsin	crop grow; carbon	Soil water capacity (FC-FW); potential heat units (PHU); WA; HI; P70 (used to be P20 in old EPIC version); FHP
Wang et al. (2006a)	Zi-Fang-Gully watershed in northwestern China	hydrology; erosion/ sedimentation	soil saturated conductivity (SATC); floodplain saturated conductivity (FPSC); P42; P46; P47
Wang et al. (2006b)	Riesel, TX	hydrology; erosion–sedimentation; nutrients	CN2; P42; P46 (used to be P23 in old version); PEC; phosphorus sorption ratio (PSP)
Wang et al. (2006c)	159 locations in the United States	hydrology; crop grow; erosion–sedimentation; nutrients; carbon	P2; P5; P8; P11; P34; P42; P46; P52; PHU; UXP; FHP; return flow ratio (RFP)
Wang et al. (2008)	Treynor watersheds in Iowa	hydrology; carbon; erosion–sedimentation	CN2; P42; P46; P47; P70
Yin et al. (2009)	Middle Huaihe River watershed, China	hydrology; erosion–sedimentation	Peak runoff rate-rainfall energy adjustment factor (APM); CN2; P42; PEC
Wang et al. (2009)	Shoal Creek watershed, central Texas	hydrology; erosion–sedimentation	CN2; P18; P42; P46; RSRR
Gaiser et al. (2010)	Tropical subhumid West Africa and semiarid Brazil	crop grow	WA; HI, WSYF; LAI_{mx}; PPLP1; PPLP2; Aluminum tolerance index (ALT); Critical aeration factor (CAF)
Mudgal et al. (2010)	Goodwater Creek watershed in north central Missouri	hydrology; pesticide fate	P3; P5; P16; P17; P24; P34; P38; P42; P44; P63; PHLS

Continued.

Table 6–1. Continued.

Reference	Study location	Model component	Influential parameter†
Wang et al. (2010)	subwatershed within the Bosque River watershed in central Texas	hydrology; nutrients; erosion–sedimentation	CN2; P8; P14; P18; P19; P29; P31; P35; P42; P46; P59; P72

† BN1–3: normal fraction of N in crop biomass at emergence, mid-season, and maturity, respectively; CN2: Initial input of condition 2 curve number; CNY: normal fraction N in yield; DLAP1 & DLAP2: first and second point on leaf development curve, respectively; DLAI: fraction of growing season when LAI declines; HI: harvest index; LAI_{mx}: maximum leaf area index; P2: root growth soil strength; P3: water stress harvest index; P5: soil water lower limit in the top 0.5 m soil depth; P8: soluble P runoff coefficient; P11: moisture fraction required for seed germination; P14: nitrate leaching ratio, which is the ratio of nitrate concentration in surface runoff to that in the percolate; P16: expands CN retention parameter; P17: soil evaporation plant cover factor; P18: sediment routing exponent; P24: pesticide leaching ratio; P29: biological mixing efficiency simulates mixing in top soil by, e.g., earthworms; P31: maximum depth for biological mixing; P34: Hargreaves PET equation exponent; P35: denitrification soil-water threshold, which is the fraction of field capacity soil water storage to trigger denitrification; P38: water stress weighting coefficient; P42: NRCS curve number index coefficient; P44: upper limit of CN retention parameter; P46 & P47: RUSLE C factor coefficients; P52: exponential coefficient of tillage effect on residue decay rate; P59: P upward movement (by evaporation) coefficient; P63: pesticide loss coefficient; P70: microbial decay rate coefficient; P72: volatilization/nitrification partitioning coefficient; PEC: erosion control practice factor; PHLS: pesticide half-life in soils; PPLP1 & PPLP2: plant density_LAI parameters; RDMX: Maximum root depth; RLAD: leaf area index decline rate parameter; RSRR: average time to release flood storage through the rock dams in day; T_0: optimum air temperature; T_b: base temperature; WA: energy conversion ratio; WSYF: lower limit of harvest index; FHP: fraction of humus in passive pool; UXP: power parameter of modified exponential distribution of wind speed.

CO_2 on growth and transpiration and can incorporate trends in technology change through changes in the harvest index.

The models enable simulation of mixed stands with plant competition for light, water, and nutrients. An extensive array of tillage systems can be simulated. Tillage mixes soil layers, changes standing residue to flat residues, accelerates mineralization, and changes bulk density. The soil bulk density gradually increases after tillage based on the amount of water infiltrating through the corresponding layer. The difference in bulk density between a settled and tilled condition controls the effect of tillage in the mineralization rate. Both the soil bulk density settling and its effect on the mineralization rate are represented by empirically derived equations. The models enable wind erosion. Manure can also be eroded from feeding areas and manure application fields. Depending on the amount of manure cover of the soil the erosion varies from essentially all manure to a combination of manure and soil. Since manure is considered residue, a heavy cover in a feedlot may completely eliminate soil erosion but may create the potential for severe manure erosion. Soil erosion potential is also very low in manure application fields with a good grass cover, but manure erosion can be high. The models have manure management with automatic application from stockpile or lagoon, and a hail occurrence–damage component developed for use in crop insurance. The southern oscillation index is considered in the weather simulator for better rainfall definition during drought and for real time simulation.

Table 6–2. Major components and options available in the EPIC and APEX models.

Component	Method or principle
Weather	Weather generator options, if not input: First-order Markov chain model for precipitation (Nicks and Lane, 1989) Multivariate generation model for air temperature/solar radiation (Richardson, 1981) Daily wind speed distribution and dominant direction model (Richardson and Wright, 1984) Relative humidity model for daily average relative humidity
Hydrology	
Potential evapotranspiration	Penman–Monteith (Monteith, 1965) Penman (Penman, 1948) Priestly–Taylor (Priestly and Taylor, 1972) Hargreaves (Hargreaves and Samani, 1985) Baier–Robertson (Baier and Robertson, 1965)
Surface runoff	Green and Ampt method (Green and Ampt, 1911) for infiltration and rainfall excess rate Curve number (CN) (Mockus, 1969; USDA-NRCS, 2004) Options for daily CN adjustment: Variable daily CN nonlinear CN/SW with depth soil water weighting Variable daily CN nonlinear CN/SW no depth weighting Variable daily CN linear CN/SW no depth weighting Nonvarying CN–CN2 used for all storms Variable daily CN soil moisture index
Peak runoff rate	Modified rational formula (Williams, 1995) SCS TR55 Peak Rate estimate (USDA-SCS, 1986)
Subsurface flow	Storage routing and pipe flow equations
Crop growth	Potential growth Water use; nutrient uptake (N, N fixation, P) Growth constraints (water stress, temperature stress, nutrient stress, aeration stress) Biomass; root growth; crop yield Plant competition Winter dormancy
Available soil water capacity	Options for estimate field capacity and wilting point (if no good data): Estimated using the Rawls method (dynamic) Estimated using the Rawls method (static) Inputted using the Rawls method (dynamic) Estimated using nearest neighbor (dynamic) Estimated using nearest neighbor (static)
Soil erosion	
Water	Water erosion equations: Universal Soil Loss Equation (USLE) (Wischmeier and Smith, 1978) Onstad- Foster modification of the USLE (AOF) (Onstad and Foster, 1975) Revised USLE (RUSLE) (Renard et al., 1997) Modified USLE (MUSLE) (Williams, 1975) MUSLE variation 1 (MUST) (Williams, 1995) MUSLE variation 2 (MUSS) (Williams, 1995) MUSLE with input coefficients (MUSI) RUSLE2 (Foster, 2005) Slope length and steepness factor: RUSLE slope length and steepness factor (preferred for steep slopes >20%) MUSLE slope length and steepness factor

Continued.

Table 6–2. Continued.

Component	Method or principle
Wind	Wind erosion continuous simulation Surface roughness factor (Potter et al., 1990, Potter and Zobeck, 1990) Roughness factor Unsheltered distance factor (Stout, 1990) Wind erosion rate (Skidmore, 1986)
Nutrients	Nitrogen (organic N, denitrification, N cycling, nitrification, volatilization, rainfall N) Phosphorus (soluble P loss, P transport by sediment, mineralization, mineral P cycling)
Plant environmental control	Drainage, irrigation, fertilization, liming, pests, furrow diking

Additional Features of APEX

As stated above, APEX extended EPIC's ability to simulate complex farm or watershed landscape processes and management practices. APEX can simulate detailed management practices related to farm animal productions, such as confined or partially confined area feeding, intensive rotational grazing, and cropland grazing after harvest. APEX has a spatial rainfall simulator for larger watersheds that provides daily rainfall distributed among subareas considering distance from the storm center and east–west or north–south gradients. The APEX routing component provides for physically based simulation of conservation practices like filter strips, terraces, and waterways. The routing mechanisms route water, sediment, nutrients, and pesticides across landscapes through channels, floodplains, and reservoirs to the watershed outlet. The routing feature of APEX enables the configuration of novel land management strategies. Soils are usually uniquely related to landscape position. Allowing the consideration of upper and lower landscape positions and representation of spatial locations of the various land uses and cultural practices can greatly improve modeling accuracy and allow alternative treatment such as best management practices in different landscape positions for maximum efficiency.

Input Data and Parameterization

The EPIC and APEX models require inputs of daily weather, soil attributes by layer, field management (e.g., planting crop, plant population, harvesting, tillage, fertilizer application rate and depth, pesticide application rate, irrigation), site information (e.g., latitude, longitude, elevation, CO_2 concentration in atmosphere, average concentration of N in rainfall), catchment characteristics (e.g., drainage area, average upland slope, average upland slope length), and channel geometry of routing reaches. The routing mechanism needs to be set up for an APEX run if a farm or watershed study involves multiple fields and subareas.

Climate Inputs

The weather variables required for EPIC and APEX simulations are daily precipitation, maximum and minimum air temperature, and solar radiation. Wind speed and relative humidity are also required if the Penman methods (Penman, 1948) are used to estimate potential evaporation. Wind speed is also needed when wind-induced erosion or dust emission and distribution are to be simulated. These weather variables can be all entered by the user or generated at run time. Both models also have options for using different combinations of both measured and generated weather variables. If any variables are user input (measured), then the daily precipitation data must be user input too.

For weather generation and other uses, monthly weather statistics are required. These include average monthly maximum and minimum air temperatures, monthly average standard deviations of daily maximum and minimum temperatures, average monthly precipitation, monthly standard deviation of daily precipitation, monthly skew coefficient for daily precipitation, monthly probabilities of wet day after dry day and wet day after wet day, average number days of rain per month, monthly maximum 0.5-h rainfall, average monthly solar radiation, and monthly average relative humidity. These variables are used to estimate missing daily data or generate daily weather variables (precipitation, maximum and minimum air temperatures, solar radiation, and relative humidity). Other uses include the monthly temperatures for calculating potential heat units (PHU, thermal time) for tree crops, the maximum 0.5-h rainfall for estimating peak runoff rate, and rainfall energy. The monthly wind weather statistics include average monthly wind speed and monthly wind percentage from 16 wind directions, which are critical if simulating wind erosion, as well as dust distribution and air quality from feedlots. These variables are used to generate daily wind speed and direction.

The monthly weather data can be prepared using historical daily weather data with a stand-alone weather generating program WXPM (Williams et al., 2006). The models' database contains monthly weather statistics for about 1050 weather stations in the United States. If a long-term daily weather series is not available to generate the monthly weather statistics, the EPIC and APEX programs can reference the latitude and longitude of the simulation site to find the closest weather station in the database and use its monthly weather statistics.

Soils

The required soil data include soil albedo, hydrologic soil group, soil attributes by layer, including depth, bulk density, texture, organic carbon, and pH. The models have options to estimate missing soil attributes if soil inputs provide insufficient information. For example, the soil water contents at wilting point and field capacity can be estimated using either the Rawls method (Rawls and Brakensiek, 1985) or the nearest-neighbor method (Nemes et al., 2006).

Databases

The models have built-in databases for crops, tillage tools, fertilizers, and pesticides (Williams et al., 2006). The crop database contains information needed for EPIC and APEX to simulate the growth of a particular crop. The growth parameters in the plant growth database define plant growth under ideal conditions and quantify the impact of some stresses on plant growth. There are more than 100 plant species included in the crop database. When compiling the list of plants in the default database, an attempt was made to include the most economically important plants as well as those that are widely distributed in the landscape. However, the list is by no means exhaustive, and users may need to add plants to the list. When adding a new plant or land cover to the database, a review of existing literature should provide most of the parameter values needed to simulate plant growth. The tillage database contains information needed for EPIC and APEX to simulate the redistribution of nutrients and pesticides that occurs in a tillage operation and the surface roughness and bulk density changes caused by tillage operation. The common tillage management activities in agricultural land use are included. The fertilizer database summarizes nutrient fractions for various fertilizers and types of manures. The pesticide database contains properties of the active ingredients in pesticides. These databases are not comprehensive but have been designed so that users can easily append new information regarding crops, tillage tools, fertilizers, and pesticides.

APEX Routing Instruction

In APEX, there are only three variables, which need to be specified for each subarea to determine the routing structure of a watershed. In case of extreme areas (subareas that receive no flow from another subarea) channel length (CHL) and routing reach channel length (RCHL) are identical (CHL = RCHL). For subareas that contain a routing reach (i.e., receive inflow from one or more subareas) CHL is greater or equal to RCHL. A negative watershed area (WSA) indicates that there is output from another subarea in storage that must be added to the subarea output. Therefore, it is possible to manually construct APEX's routing structure (Williams et al., 2006). For a larger watershed or a complex landscape, geographic information system (GIS) tools are available to build APEX inputs. For example, the ArcAPEX tool is designed to use ArcGIS 9.x GIS platform to prepare APEX inputs (Tuppad et al., 2009). The GIS layer of a digital elevation model (DEM) is required to define the catchment outer boundary and also to delineate the stream networks within the catchment. The subarea in APEX is defined using the resulting stream networks from the DEM delineation such that each subarea is draining the area upstream of a certain location in the stream networks. User-defined streams and subarea boundaries, which are closely associated with specific agricultural field boundaries, can be used in place of the DEM-based delineation in ArcAPEX. User defined subareas could be delineated from readily available aerial imagery, e.g., United States Department of Agriculture-National Agriculture Imagery Program (USDA, 2009) or other remote sensing datasets.

Calibration and Validation

Comprehensive models like EPIC and APEX contain a small number of parameters that are subject to calibration by the user and a large number of parameters that are contained in the algorithms and are inaccessible to the users. For that reason, the calibration process, being either manual or automated, requires careful inspection of outputs, good judgment, and above all a disclosure of the parameters adjusted to obtain good agreement between observed and simulated data. Model calibration is to adjust influential model parameters or inputs within their reasonable ranges so that the modeling results (see major outputs in Table 6–3) are comparable with observed data. Normally some variables fit measured data better than others because they are relatively independent. For example, sediment yield, nutrients, and pesticides depend on runoff so errors in simulating runoff can propagate in those variables. Corrective measures can be taken in the calibration by identifying the actual source of the disagreement.

The first step before any calibration is to examine the EPIC or APEX output file "*.out" from a preliminary model run. This output file provides comprehensive information about the simulation (inputs; daily, monthly, or annual simulation results; and summary). Therefore, it is the most important file for identifying possible causes for poor simulation results. A basic check is to see if the total watershed area obtained by adding all subareas during APEX simulation is correct. Incorrect total watershed area indicates mishandling of the routing instructions in the input files. This may occur if the routing instructions are prepared manually. Water balance for the entire watershed should be checked. Average annual surface runoff, water yield, sediment yields and nutrient yields should be checked. If they are not reasonable, then check individual subarea contributions to narrow down possible causes. The user needs to decide in this case what a reasonable outcome is. It is extremely important to use actual information to judge model outcomes, and to avoid imposing our expected outcome by adjusting calibration parameters ("drawing the output").

Table 6–3. The EPIC and APEX models' input files likely to be updated during calibration and major model output files. See user manuals in Williams et al. (1990, 2006) for detail.

File		Short description
Input		
Model control	*cont.dat	Includes a variety of data parameters, e.g., submodel options can be indicated in this file
Model parameter	P*.dat	Contains definitions of s-curve used to describe the behavior of many processes and P1-P83 mainly for defining equation coefficients
Soil attributes	*.sol	Lists soil attributes, such as hydrologic soil group, years of cultivation before simulation, and other soil attributes by layer
Field management	*.opc or *.ops	Lists operation schedule by date, such as planting (crop ID, potential heat units, land use number or CN2, plant population), harvesting, tillage, fertilizer application (fertilizer ID, application rate, depth), pesticide application (pesticide ID, application rate)
Subarea information	*.sub	Contains subarea information, such as soils and operation schedules used for each subarea, catchment characteristics, and channel geometry of routing reaches
Crop database	*crop.dat	Lists plant species and plant growth parameters that define plant growth under ideal conditions and quantify the impact of some stresses on plant growth
Output		
Detailed simulator	*. out	Provides comprehensive information about the simulation (inputs/options used; daily, monthly, or annual simulation results; and summary)
Annual crop yield	*.acy	Annual crop grain, fiber, and forage yield, biomass; stress days of drought, N, P, temperature, and water; nutrients in the root zone for each subarea
Annual subarea	*.asa	Annual evapotranspiration, surface runoff, subsurface flow, percolation, drainage flow, C loss with sediment, erosion–sedimentation, soluble N and P losses, organic N and P losses for each subarea
Annual watershed	*.aws	Annual upland and watershed flow, sediment yield, C loss with sediment, soluble N and P losses, organic N and P losses
Daily subarea pesticide	*.dps	Pesticide name; amount applied; losses from runoff, percolation (root zone), subsurface flow, sediment, and drainage flow; pesticide degradation from foliage and soil
Annual soil organic C and nitrogen	*.acn	Bulk density; C pools of biomass, slow humus, passive humus, and total C pool; structural litter N; metabolic litter N; N pools of biomass, slow humus, passive humus, and total N pool for each soil layer and a total across layers
Daily subarea hydrology	*.dhy	Daily surface runoff, subsurface flow, quick return flow, water yield, peak flow rate, root zone soil water, groundwater storage
Monthly watershed file	*.mws	Watershed flow, soil loss from erosion, soluble N and P losses, organic N and P losses
Reach output	*.rch	Area of subarea or reach outlet; daily, monthly or annual reach in and out variables of flow, sediment, soluble N and P loads; organic N and P loads

After making sure that the fertilizer/manure rates, water and nutrient dynamics are reasonable, limited adjustment can be performed based on the comparison between predicted and observed data. After the preliminary investigation, the influential model parameters should be identified based on sensitivity analysis, experience, or literature. Useful guides can be found in Wang et al. (2005, 2006c) and Yin et al. (2009). Sensitivities are dynamic in both temporal and spatial dimensions; therefore, a site-specific sensitivity analysis is encouraged.

Automatic calibration procedures with Monte Carlo simulations and multi-objective functions have been successfully used in EPIC (Wang et al., 2005) and APEX (Yin et al., 2009) calibration. Zhang et al. (2010) developed a multi-objective optimization algorithm coupled with EPIC for exploring the trade-offs between biofuel energy production and unintended ecosystem-service responses. The algorithm could be adopted for parameter optimization where measured data are available. However, model users should use good judgment when analyzing outputs from automatic and manual calibration considering errors in measured variables and inputs. The widespread argument that a model may be tuned for a right answer but with a wrong reason, is valid. Therefore, it is always a good practice to look into various outputs during interactive manual calibration.

The calibration should first consider a correct water balance, including all components, and a correct simulation of erosion, crop growth, and nutrient cycling. These processes are intimately intertwined, and the calibration requires interpretation. A description of a standard procedure for calibrating and testing EPIC and APEX is provided in the following sections. Given the large number of processes simulated, the procedure is generic and needs to be tailored to the specific simulation under consideration. Table 6–3 lists the model input files where some inputs and parameters are likely to be adjusted during calibration.

Calibration of Hydrologic Balance

The hydrologic balance component encompasses the key processes that occur in the hydrologic cycle. Initially, rainfall may be intercepted by plant canopy or fall directly on the soil surface. Incoming precipitation, snowmelt, or irrigation is partitioned between surface runoff and infiltration. Infiltrated water stored in the soil profile is subject to subsurface flow and evapotranspiration (ET). The vertical subsurface flow may leave the root zone and enter the groundwater storage, where it is partitioned into deep percolation from the system and return flow to channel. Horizontal flow within the root zone is partitioned into lateral and quick return flow. The algorithms to represent these processes have a mix of mechanistic equations and empirically derived equations and associated parameters to simulate mass transport.

Potential Evapotranspiration

The five PET methods included in the models (Table 6–2) vary in the amount of required weather inputs. The Hargreaves method (Hargreaves and Samani, 1985) provides a simple and relatively robust method for estimating PET that only requires daily maximum and minimum temperature. The Penman–Monteith method is generally considered the most accurate but requires solar radiation, air temperature, relative humidity, and wind speed, which are not always available from measurements. The prediction of PET by Penman–Monteith can be adjusted through the stomatal conductance coefficient P1 (Table 6–4). The Priestley–Taylor method is a simplification of the Penman–Monteith method. It requires solar radiation and air temperature. Like the Hargreaves method the Baier–Robertson method requires minimal data; however, it was developed in Canada and is provided mainly as an alternative for use in cold climates. Because the Hargreaves PET equation requires minimal inputs and is commonly used in EPIC–APEX applications, only this method is described in this chapter.

$$\mathrm{PET} = \mathrm{P23}\left(\frac{\mathrm{RAMX}}{2.5 - 0.0022T_{\mathrm{x}}}\right)\left(T_{\mathrm{a}} + 17.8\right)\left(T_{\mathrm{x}} - T_{\mathrm{n}}\right)^{\mathrm{P34}} \quad [1]$$

where P23 (range 0.0023–0.0032) and P34 (range 0.5–0.6) are empirical coefficients, RAMX is the clear sky daily radiation at the site elevation (MJ m^{-2} d^{-1}), T_a is the mean daily air temperature (°C), T_x and T_n are the daily maximum and minimum air temperatures (°C), Increasing P34 increases PET and decreases soil water and crop yield because transpiration and growth are decoupled in these models when using the radiation use efficiency to simulate growth.

Surface Runoff

The USDA-NRCS runoff curve number (CN) as modified for continuous simulation is a reliable method for estimating surface runoff given daily precipitation. The models vary CN continuously as soil water changes through the retention parameter, *S*. The models have four different methods of linking *S* (CN) and soil water plus a constant CN option (Table 6–2). Generally the soil moisture index (Option 5, see Table 6–2) is relatively robust and reliable. Kannan et al. (2008) has demonstrated that this method can produce realistic runoff estimates over a wide range of soils. Because this method is commonly used in EPIC–APEX applications, all the other methods also have no provision for calibration. Therefore, only this method in which two parameters (P42 and P44) can be calibrated, is described in this chapter. This method calculates the daily retention parameter *S* accounting for plant evapotranspiration using the equation:

$$S = S_{\text{prev}} + \text{PET}\exp\left(-\text{P42}\frac{S_{\text{prev}}}{S_{\max}}\right) - P_{\text{prev}} + Q_{\text{prev}} + Q_{\text{return}} + \text{SSF} + \text{PRK}; \qquad S < \text{P44}S_{\max} \quad [2]$$

where S is the retention parameter for a given day (mm), S_{prev} is the retention parameter on the previous day (mm), PET is the potential evapotranspiration for the day (mm d^{-1}), P42 is the CN index coefficient, P44 is the upper limit of CN retention parameter, S_{max} is the maximum value the retention parameter can achieve (mm), which is associated with CN_1 (the curve number for moisture condition 1 or dry condition), P_{prev} is the rainfall reaching the ground surface after plant interception on the previous day (mm), and Q_{prev} is the runoff on the previous day (mm), Q_{return} is quick return flow on the previous day (mm), SSF is the lateral subsurface flow on the previous day (mm), and PRK is percolation on the previous day (mm).

Considerable runoff can be infiltrated into floodplains depending on travel time and floodplain saturated conductivity. Flow through a routing reach can be increased by return flow from groundwater and quick return flow. Groundwater storage must exceed a threshold value before return flow occurs. Therefore, the floodplain saturated conductivity, groundwater storage threshold, maximum groundwater storage, groundwater residence time, and ratio of return flow to return flow plus deep percolation can be used to adjust runoff loss or gain during routing (Table 6–4).

Calibration of Crop Grain Yield

The models are capable of simulating growth for both annual and perennial crops. Phenological development of a crop is based on daily heat unit accumulation. Annual crops grow from planting to harvest or until the accumulated thermal time equals the potential heat units for the crop. Perennial crops maintain their root systems throughout the year, although they may become dormant after frost. The model is also capable of simulating mixed stands of up to 10 crops or other plants in a competitive environment.

If crop yields are available, they can be calibrated together with runoff calibration or after runoff calibration. Realistic PET and proper crop growing season are important to both runoff and crop yield. Unreasonable ET are usually associated with a crop not emerging or stressed due to lack of water and/or nutrients. Excessive PET or runoff estimates can reduce crop yields because it can increase crop water stress. Plant-available water is another important limitation that causes water stress. Erroneous estimates of plant-available water occur when

Table 6–4. Calibration of major EPIC and APEX outputs.

Output	Influential process	Influential input or parameter	Range	Input file
Runoff	PET equations (indicating option in *cont.dat)	If Hargreaves, P34 If Penman-Monteith, P1	0.5–0.6 1.0–2.0	P*.dat
	Crop growing season	Potential heat units (PHU) planting and harvest dates If automatic heat unit schedule (HUSC)	1.0–1.2	*.opc or *.ops
	Runoff equation– curve number equations (indicating option in *cont.dat)	CN2 or land use number (LUN)		*.opc or *.ops
		If variable daily soil moisture index, P42	0.5–2.5	P*.dat
	Irrigation (if used)	Irrigation runoff ratio (EFI)		*.sub or *.opc
	Routing	Floodplain saturated conductivity (FPSC) Maximum groundwater storage (GWSO) Groundwater residence time (RFTO) Return flow/(return flow + deep percolation) (RFPO)		*cont.dat (values overridden by values > 0. in *.sol)
		Groundwater storage threshold: P40	0.001–1.0	P*.dat
Crop yield	Crop growing season	See above		
	Growth constraints (check stress days in *.out) Root growth stress	High bulk density (BD) or aluminum saturation caused by low pH < 5 (Applying deep tillage and lime to lower BD and pH, respectively.)		*.sol (if correcting inputs)
	Water stress: excessive PET or runoff low plant available water Nutrient stress: low mineralization rates inadequate fertilization excessive leaching of N	See above for related parameters Difference between field capacity (FC) and wilting point (WP)		*.sol (if correcting inputs) *cont.dat (estimation option)
		Organic N, P, and C Years of cultivation at simulation (RTN1) Fertilization rates; or if auto fertilization: Auto-fertilization application rate (FNP4) Max. annual N application for a crop (FMX) Soil plant-available water (FC- WP) saturated conductivity	C/N near 10 N/P near 8	*.sol *.sub

Continued.

Table 6–4. Continued.

Output	Influential process	Influential input or parameter	Range	Input file
Crop yield, continued				
		Plant population		*.opc or *.ops
		Winter dormancy, P6	0–1	*P.dat
		Maximum potential leaf area index (DMLA) Harvest index (HI) Biomass-energy ratio (WA)		cropcom.dat
Erosion–sedimentation	Erosion/sedimentation equations (indicating option in *cont.dat)	RUSLE C factor coefficient: P46 P47	0.5–1.5 0.01–3.0	P*.dat
		Erosion control factor (PEC)	0.0–1.0	*.sub
		Slope length and steepness factor equations		*cont.dat
	Routing	Sediment routing exponent, P18 Sediment concentration at 1.0 m s^{-1} velocity, P19	1.0–1.5 0.005–0.25	P*.dat
		Channel and floodplain geometry Normally unknown, if let the model estimate, then input: channel capacity flow rate in mm h^{-1} (QG)	about 2 yr frequency flow rate	*cont.dat
		Channel cover factor (RCHC) Soil erodibility factor (RCHK)	0.001–1.0 0.001–1.0	*.sub
Nutrient	Nitrogen cycling Carbon cycling Phosphorus cycling	Biological mixing efficiency, P29 Soluble P runoff coefficient, P8 P upward movement by evaporation coefficient, P59 Nitrate leaching ratio, P14 Denitrification soil_water threshold, P35 N fixation coefficient, P7 Volatilization/nitrification partitioning coefficient, P72 Microbial decay rate coefficient, P70 Fraction of humus in passive pool, FHP	0.1–0.5 10–20 1–30 0.1–1.0 0.9–1.1 0.0–1.0 0.05– 0.5 0.05– 1.5 0.3– 0.9	P*.dat *.sol

field capacity (FC) or wilting point (WP) are incorrect. Soil water storage is particularly important in dry climates. Soil depth may also curtail crop yield. Plant population and thermal time from planting to harvest impact crop grain yield as well. Crop parameters contained in the crop database may need to be adjusted, including biomass energy ratio (WA), base temperature (T_b), harvest index (HI), and maximum leaf area index (LAI_{mx}) (Table 6–1).

Root growth stresses due to bulk density (BD) or aluminum toxicity (ALSAT) can reduce crop yields greatly. Bulk density can be lowered by deep tillage or the addition of crop residue. Lime may be applied automatically by the models to increase pH and reduce aluminum toxicity stress.

Nitrogen and P stress may be caused by low mineralization rates, inadequate fertilizer, gaseous losses, or excessive leaching of N (Table 6–4). The soil organic C/N should be near 10, and N/P should be near 8. The mineralization rate can be increased by decreasing the number of years of cultivation at the beginning of simulation (RTN1). High N leaching (PRKN) relative to annual N fertilizer results from high percolation (PRK), which may result from low ET or runoff, low soil plant-available water storage (FC–WP), or high saturated conductivity values.

Calibration of Erosion–Sedimentation

Runoff must be realistic before calibrating erosion–sedimentation. Proper cover and residue are critical; therefore, the crop growth must also be realistic. For watershed analyses, the water erosion equations of Modified Universal Soil Loss Equation (MUSLE) and two variations of MUSLE, MUSS and MUST, are recommended. MUSS is the small watershed version of MUSLE, and MUST is the theoretically developed version of MUSLE. The three equations usually give similar results and are appropriate for estimating sediment yield from small watersheds up to about 250 km^2. The universal soil loss equation (USLE) is an erosion equation that is useful in studies assessing the effect of erosion on productivity. These equations are in the following form:

$$Y = X \times \mathrm{EK} \times \mathrm{CVF} \times \mathrm{PEC} \times \mathrm{SL} \times \mathrm{ROKF} \qquad [3]$$

USLE and RUSLE:

$$X = \mathrm{EI} \qquad [4]$$

MUSLE:

$$X = 1.586\left(Qq_{\mathrm{p}}\right)^{0.56} \mathrm{WSA}^{0.12} \qquad [5]$$

MUST:

$$X = 2.5\left(Qq_{\mathrm{p}}\right)^{0.5} \qquad [6]$$

MUSS:

$$X = 0.79\left(Qq_p\right)^{0.65} \mathrm{WSA}^{0.009} \qquad [7]$$

where Y is the sediment yield (Mg ha^{-1}), X is the rainfall and runoff factor, EI is rainfall energy factor for USLE, EK is the soil erodibility factor, CVF is the crop management factor, PEC is the erosion control practice factor, SL is the slope length and steepness factor, ROKF is the coarse fragment factor, Q is the runoff volume (mm), q_p is the peak runoff rate (mm h^{-1}), and WSA is the watershed area (ha).

Because the plant cover varies during the growth cycle, the CVF factor is internally updated for all days when runoff occurs. It is calculated as a function of crop residue, crop height, standing live biomass of the crop, and soil surface random roughness using the following equations:

$$\mathrm{CVF} = \exp\left[-0.026(\mathrm{RRUF} - 6.1)\right]\mathrm{FRSD}\ \mathrm{FBIO} \qquad [8]$$

where RRUF is the soil surface random roughness (mm), FRSD is the crop residue factor, and FBIO is the growing biomass factor. The FRSD and FBIO factors are calculated using the following equations:

$$\mathrm{FRSD} = \exp(-\mathrm{P46}\ \mathrm{CVRS}) \qquad [9]$$

$$\mathrm{FBIO} = 1.0 - \exp(-\mathrm{P47}\ \mathrm{CPHT})\frac{\mathrm{STL}}{\mathrm{STL} + \exp(1.175 - 1.748\mathrm{STL})} \qquad [10]$$

where CVRS is the aboveground crop residue (Mg ha^{-1}), P46 and P47 are coefficients in the exponential functions, CPHT is the crop height (m), and STL is the standing live biomass of the crop (Mg ha^{-1}).

Increasing P46 decreases FRSD and CVF, which leads to lower sediment yield estimates. The effect of crop height on the CVF factor is governed by the exponential coefficient P47. Increasing P47 diminishes the effect of ground cover (represented by STL in Eq. [10]), increases FBIO and CVF, which increases sediment yield estimates. The two parameters can be adjusted for sediment yield calibration.

The channel vegetative cover factor (RCHC) ranges from near 0.0 for a channel with extremely good cover, like a grass waterway, to about 0.8 for a channel essentially bare of vegetation. The soil erodibility factor (RCHK) ranges from near 0.0 for channels with extremely stable side slopes and bottom to about 0.5 for shallow channels in cropped fields. P19 is the sediment concentration at a flow velocity of 1.0 m s^{-1}. The concentration (t m^{-3}) should normally range from 0.01 to 0.05. High values cause overestimation of sediment transport. These parameters and the sediment routing exponent (P18) may be adjusted for sediment yield calibration.

Calibration of Nutrients

The first item to check is the application rate of fertilizer/manure; then check the individual components (denitrification, volatilization, leaching, runoff, crop uptake, and N fixation). Parameters that may be adjusted to improve results include P29 (biological mixing efficiency), P14 (the ratio on N concentration in the runoff to that of the percolate), P72 (partitions between volatilization and nitrification), P7 (N fixation coefficient ranges from 0.0 for the highest estimate to 1.0 for the lowest, P8 (P linear adsorption coefficient—inversely related to soluble P in the runoff), P70 (microbial decay rate coefficient), and FHP (fraction of humus in passive pool) (Table 6–4).

Application

The APEX model is used in an ongoing Special Emphasis Conservation Effects Assessment Project (CEAP) watershed study for the North Bosque watershed (322,700 ha) in central Texas (Plate 6–1; see color insert section). In 2000, the North Bosque River was listed as an impaired water body in the Texas Water Quality Inventory for concerns of elevated levels of bacteria, chlorophyll a, and nutrients entering the segment from tributary watersheds. Upper North Bosque was also placed on the 303(d) list for elevated levels of sediment, nitrogen, phosphorus, chloride, sulfate, and chlorophyll a. These impairments have mainly been associated with the dairy industry in the northern part of the watershed. The objective of the CEAP watershed study is to identify conservation practices to reduce the phosphorous loading levels flowing from the North Bosque watershed into Lake Waco by testing alternative practices and evaluating the timing, location and suites of practices that trap, filter, or otherwise mitigate phosphorus loss. APEX has been calibrated at Hico (Plate 6–1) on the basis of available historical data. The following sections mainly focus on the calibration and validation of the APEX model for monthly flow, sediment, nitrogen, phosphorous stream loadings from the North Bosque watershed above Hico.

Watershed Description and Input Data

The North Bosque originates in Erath County and flowing southeast through Bosque and McLennan Counties where it is impounded to create Lake Waco. The lake provides 75% of the water supply for the city of Waco, TX. In the early 2000s the study area has the largest concentration of dairy animals in the state of Texas. There are an estimated 9300 ha of manure application fields. In the CEAP watershed study, it was assumed there are 61 permitted dairy farms (Plate 6–2) with 39,825 confined dairy cows. Some estimates indicate the number of dairy cows may be around 40,000. However, no effort was made to actually quantify the number of cows in each of the individual dairies; because of variations by season and by years depending on many external factors, the number cannot be verified. We used a percentage of the permitted number of cows to distribute the cows among the individual dairy farms.

The watershed area of the North Bosque watershed above Hico is 92,650 ha. Hico is located in the very center of the North Bosque watershed (Plate 6–1). This subwatershed contains 41 of the 61 dairies in the North Bosque (Plate 6–2). Waste application fields identified in the Texas Institute of Applied Environmental Research (TIAER) and the Texas Commission on Environmental Quality (TCEQ) database were used to locate the waste application fields within the subwatersheds. We did not have an accurate account of which waste application fields were used by which dairy, so we assigned the fields to the nearest dairy. This assumption, however, is likely to cause some misrepresentation of the actual conditions. Unless we had specific information to assign land use, all waste application fields were assumed to have coastal bermudagrass [*Cynodon dactylon* (L.) Pers.] as the land cover. These assumptions, however, unavoidably contribute to the uncertainty of model inputs.

Daily weather data were obtained from historical weather records of the National Oceanic and Atmospheric Administration (NOAA) National Weather Service. Ten stations (three above Hico) were chosen that had both temperature and precipitation records from 1965 to 2004 for the CEAP watershed study. The long-term average annual precipitation for this watershed is 842 mm. The average monthly maximum temperature is 25°C, and the average monthly minimum temperature is 11°C. The 40-yr period was used for various scenario analyses in the CEAP study. However, observed data of stream flow, sediment, and nutrient losses are only available at the Hico monitoring station for 1993 through July 1998. In this study, the simulated monthly data were compared with the available monthly measured data at Hico.

A new land use–land cover inventory was taken for the CEAP watershed study. Unsupervised classification was used to stratify 300 points into 25 categories each using 2002 Thematic Mapper satellite data. Of these 300 points, a total of 260 points could be reached and classified by ground observation. These points were used to classify the entire watershed for land use–land cover as of 2002. The land use was 63.5% native pastures–Range, 14.2% improved pasture(assumed to be improved Bermuda grass), 10.4% misquote–cedar mix, 6.4% deciduous trees (including wetlands), 4.3% cropland, 1% Urban, and 0.2% Dairies and milk sheds. Plate 6–3 shows the land use–land cover and the Soil Survey Geographic (SSURGO) soils data for the watershed above Hico.

The 10-m elevation data from Texas Natural Resource and Information Systems (TNRIS) were used to develop the detailed watershed data. The watershed attributes required for APEX include delineation of the subarea boundaries, estimation of the drainage system, slope and slope length, channel lengths, and subbasin dimensions (length and width). The philosophy of the study design was to divide the Bosque basin into small enough subunits so as to let one subbasin represent one field or pasture. After some experimentation and generation of subbasins using GIS tools and viewing these with the land use and the National Agriculture Imagery Program (NAIP) data, the North Bosque watershed was divided into a total of 7544 subbasins. These subbasins were again divided into upper and lower landscape positions to make a total of 15,088 subbasins, among which 3475 subareas are above Hico. Plate 6–4 shows the upper and lower landscape positions within one 12-digit watershed (delineated by U.S. Geological Survey using a nationwide system based on surface hydrologic features) in the study watershed. This separates the lower position areas in the landscape to allow alternative treatment placements. The separation of the landscape into two positions was accomplished using a GIS algorithm packaged in the Field Hydro Tool (Kemanian et al., 2009) that performed several functions:

1. The algorithm calculated the position of each 10 by 10 m cell by assigning a percentage distance between the stream line and the ridge line of each of the subbasin.
2. It then calculated the change in slope for each of the cells.
3. The next step was to identify the cells that had the most rapid change in slope found in the lower 40% of the subbasin.
4. The algorithm finally assigned the cells below the maximum slope change to the lower landscape position and above that line to the upper landscape position. The set of rules required that all upper landscape positions flow into or through a lower landscape position before entering the stream channel. The lower landscape position was assumed to have a drainage channel receiving water from the immediately adjacent upper landscape position.

Cropping systems were chosen that represented crops and management practices in the area.

Coastal bermudagrass was used for all waste application areas. This constituted the planning of the coastal bermudagrass and harvesting of hay from the waste application area four times a year. The pasture and range areas identified from the land cover classifications were assigned various combinations of grasses and shrubs, including Mesquite trees [*Prosopis chilensis* (Molina) Stuntz], little bluestem grass (*Dichanthium* Willem.), and side oats grama grass [*Bouteloua curti-*

pendula (Michx.) Torr.]. The grass on this land cover was harvested three times a year to simulate livestock grazing. Heavy woodlands were simulated by a cover of deciduous trees with no harvesting of biomass was removed from these areas. Improved pasture was simulated either as coastal bermudagrass or buffalograss [*Bouteloua dactyloides* (Nutt.) J.T. Columbus], with commercial fertilizer used. A simple mix of primary crops was chosen for croplands. Three crops—corn, sorghum [*Sorhum bicolor* (L.) Moench], and wheat—were placed into three rotations for the study. These rotations were (i) continuous corn, (ii) corn–sorghum, and (iii) wheat–sorghum. The automatic fertilizer (nitrogen application) option was used to provide the nutrient for these rotations.

There are two primary types of reservoirs found in the study area. First are those constructed by the USDA-NRCS as flood prevention reservoirs. Most of these reservoirs were constructed in the 1950s and 1960s. The Bosque watershed has 41 such reservoirs located in the entire basin above Lake Waco. The second type of reservoir includes private, water supply reservoirs and those built by other organizations. We identified 34 of these reservoirs to be included for the CEAP watershed study. A total of 41 reservoirs were within the watershed above Hico. The data for both types dams and reservoirs are found in the national inventory of dams (http://geo.usace.army.mil/pgis/f?p=397:12:1816408540398851 [verified 1 Apr. 2011]). This inventory provides sufficient information to allow us to parameterize each of these individual dams found in the watershed.

The ArcGIS toolbox and the APEX interface, WinAPEX (Magre et al., 2006), were used to process and manage the databases and model runs. WinAPEX has built-in algorithms that extracts the data from a MS Access database and formats the properties into the respective files needed by APEX.

APEX Calibration and Validation

The monthly record of stream flow, sediment yield, mineral N, mineral P, organic N, and organic P available at the Hico monitoring station were used to calibrate and validate the APEX model. About one-half of the observed data (July 1995–July 1998, i.e., 3 yr) were chosen for calibration because this period includes the biggest and smallest rainfall events. The remaining data (January 1993–June 1995) were used for validation. The model options chosen for this study were the USDA-NRCS curve number (CN) method for runoff estimation, variable daily CN soil moisture index method to estimate daily CN, modified rational equation to calculate peak flow, the Hargreaves method to calculate potential evapotranspiration, a variation of the modified Universal Soil Loss Equation—the MUST equation (Williams, 1995)—to calculate erosion/sedimentation (Table 6–2), and the GLEAMS enrichment ratio method (Leonard et al., 1987) for P transport and transformation.

The parameters adjusted for flow calibration included P42 (1.0), P40 (0.1), and RFPO (0.5). The parameters adjusted for sediment yield included P46 (0.62), P47 (1.0), P18 (1.0), and P19 (0.007). The calibration of N and P losses were conducted by adjusting P8 (10), P14 (0.09), P29 (0.2), P35 (1.1), and P59 (10). These parameters were described in Table 6–4. The values within the parentheses denote the actual calibrated values. Statistical measures including mean, standard deviation, R^2, Nash–Sutcliffe efficiency (NSE) (Nash and Sutcliffe, 1970), percentage bias (PBIAS), and the ratio of the root mean square error to the standard deviation of observed data (RSR) were used to evaluate the model performance based on criteria suggested by Moriasi et al. (2007).

Results and Discussion

The simulated monthly stream flow, sediment yield, and nutrient losses compared well with observed values for the calibration period, as evidenced by the values of NSE (0.69–0.85), R^2 (0.70–0.92), PBIAS (−18–8%), and RSR (0.39–0.56) in Table 6–5. Based on the statistical criteria (established based on values of NSE, RSR, and PBIAS) for evaluating water quality model performance proposed by Moriasi et al. (2007), the model performance is good for monthly flow, sediment, and nutrient losses during the calibration period (July 1995–July 1998). During the validation period (January 1993–June 1995) the model performance was good for monthly flow, mineral P, and Total P, for which NSE values ranged from 0.74 to 0.78 and RSR values from 0.49 to 0.56 and satisfactory for monthly sediment, organic P, and total N, for which NSE values ranged from 0.54 to 0.62 and RSR values from 0.68 to 0.70. However, the model performance was unsatisfactory for organic N and mineral N losses during the validation period. This could partly be attributed to the uncertainty in model input data in the N fertilizer application with the assumption that commercial fertilizer was used for improved pasture and the automatic N fertilizer option in APEX was used for cropland, and that no commercial fertilizer was applied for the remaining land uses. Nutrient loads were relatively lower during the validation period than in the calibration period. With recognized uncertainty in measured data (Harmel et al., 2006; Harmel and Smith, 2007), low loads might contribute to the relatively weaker comparison during the validation period because low loads may be difficult to detect therefore with higher uncertainty. Harmel et al. (2006) indicated that model results within 10 to 31% of the measured values are within the average uncertainty range of water quality data measured with a typical quality assurance–quality control effort. Moriasi et al. (2007) also proposed that the values of PBIAS (percentage error or bias) for simulated average monthly N or P within ±25 to ±40% of

Table 6–5. Summary statistics of monthly calibration and validation results for flow (m^3 s^{-1}), sediment (Mg ha^{-1}), and nutrient (kg ha^{-1}) at Hico monitoring station.†

		Observed		Simulated		NSE	R^2	PBIAS	RSR
		Mean	SD	Mean	SD				
								%	
Calibration (July 1995– July 1998) N = 36	Flow	5.07	6.05	5.10	7.28	0.85	0.92	0.7	0.39
	Sediment	0.05	0.12	0.06	0.08	0.81	0.84	8.0	0.42
	Mineral P	0.03	0.05	0.03	0.04	0.69	0.70	−13.6	0.56
	Organic P	0.04	0.07	0.04	0.05	0.77	0.79	−5.1	0.47
	Mineral N	0.10	0.12	0.10	0.13	0.70	0.73	5.5	0.55
	Organic N	0.24	0.36	0.20	0.26	0.77	0.83	−18.1	0.49
	Total P	0.08	0.12	0.07	0.09	0.77	0.79	−8.8	0.47
	Total N	0.34	0.47	0.30	0.38	0.79	0.82	−11.2	0.46
Validation (January 1993– June 1995) N = 31	Flow	3.54	4.17	3.54	4.20	0.74	0.76	0.0	0.54
	Sediment	0.04	0.07	0.04	0.05	0.62	0.64	−10.6	0.70
	Mineral P	0.02	0.03	0.02	0.04	0.78	0.84	28.6	0.49
	Organic P	0.03	0.04	0.04	0.04	0.61	0.62	5.1	0.70
	Mineral N	0.08	0.09	0.09	0.11	0.17	0.52	8.3	0.94
	Organic N	0.20	0.28	0.16	0.21	0.46	0.48	−21.7	0.73
	Total P	0.05	0.07	0.06	0.08	0.74	0.76	13.4	0.56
	Total N	0.28	0.35	0.24	0.30	0.54	0.57	−13.2	0.68

† NSE, Nash–Sutcliffe efficiency; PBIAS, percentage bias; RSR, ratio of the root mean square error to the standard deviation of observed data.

observed values are acceptable. The PBIAS values in this study were within ±20%, except for mineral P (28.6%) and organic N (−21.7%) (Table 6–5).

Simulated monthly flow, sediment yield, total N, and total P at the Hico monitoring station followed, in general, the patterns of observed values during both the calibration and the validation periods (Fig. 6–2 through 6–4). Total N and total P losses were clearly underestimated in April 1993 and February 1997. However, APEX overestimated the total N and total P losses in September 1993 and August 1996 (Fig. 6–4 and 6–5). One possible reason for the discrepancies between the observed and simulated monthly nutrient losses could be that the actual number of cows in each of the individual dairies, as well as seasonal or annual changes to their counts, were unknown. The actual waste application field for each of the individual dairies was also unknown. Further calibration by adjusting the number of cows or seasonal changes in each dairy or adjusting the assignment of waste application fields for each dairy may improve the results. Plate 6–5 shows the total P loss from each subarea in the study watershed. The high total P losses were from the waste application fields, especially where the waste application areas were inadequate to support the waste application from nearby dairies.

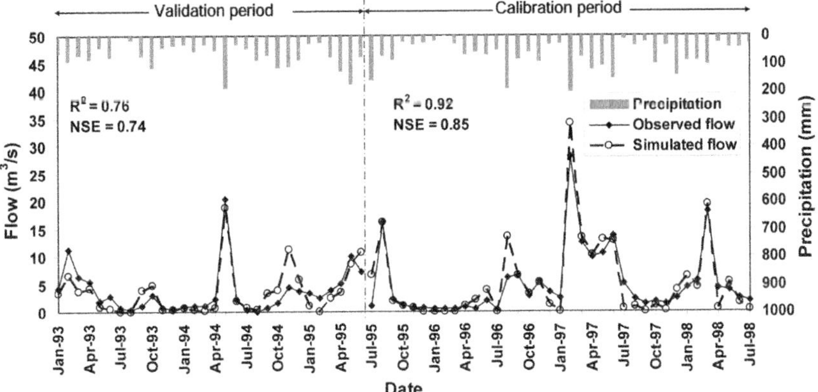

Fig. 6–2. Observed and simulated flow for the calibration and the validation periods at the Hico monitoring station.

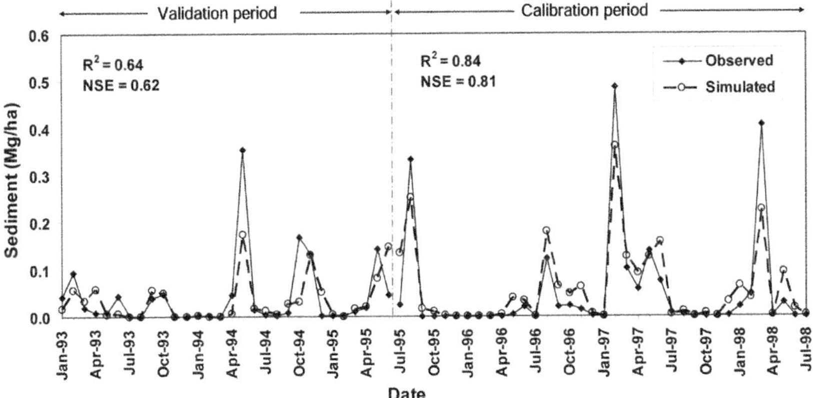

Fig. 6–3. Observed and simulated sediment yield for the calibration and the validation periods at the Hico monitoring station.

Conclusions

Both EPIC and APEX are comprehensive terrestrial ecosystem models capable of simulating the basic biological, chemical, hydrologic, and meteorological processes of farming systems and their interactions. The models dynamically represent multiple soil properties including depth and bulk density for each soil layer during simulations. They remove eroded soil, attached organic nutrients, pesticides, and carbon from the soil profile as part of the emphasis on erosion productivity. The models also consider the effect of changes in atmospheric CO_2 on growth and transpiration. They can simulate mixed stands with plant competition for light, water, and nutrients. The multi-subarea capabilities of the

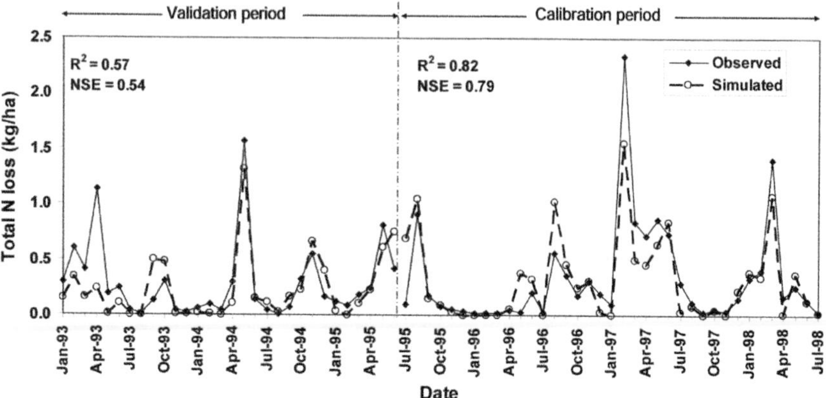

Fig. 6–4. Observed and simulated total N loss for the calibration and the validation periods at the Hico monitoring station.

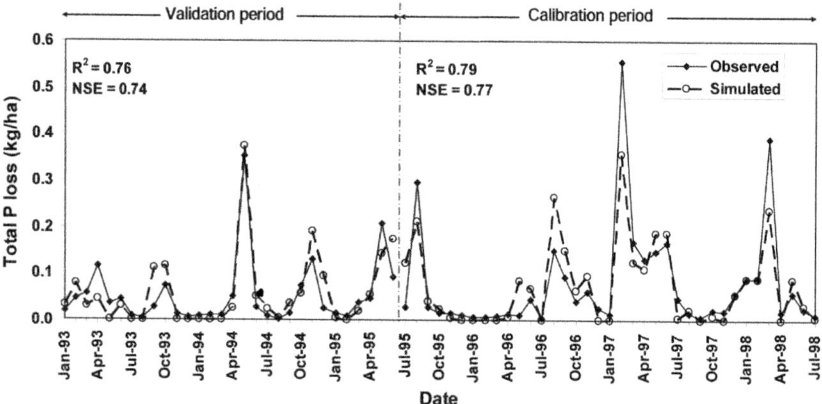

Fig. 6–5. Observed and simulated total P loss for the calibration and the validation periods at the Hico monitoring station.

APEX model greatly expand the simulation strengths in EPIC. APEX can simulate detailed management practices related to farm animal productions, such as confined or partially confined area feeding, intensive rotational grazing, and cropland grazing after harvest. APEX enables the physically based simulation of conservation practices like filter strips, terraces, and waterways.

The EPIC and APEX models require daily weather data (precipitation, maximum and minimum air temperature, solar radiation, wind speed, and relative humidity). They can be user inputs or generated within the models based on long-term monthly weather statistics. Other main inputs required are soil properties and attributes by layer, land use and field operation schedule, and/or subarea delineation (for APEX).

The model calibration should start with the hydrology component by investigating the PET, ET, crop growth season, and base flow proportion and by comparing simulated flow with observed values. The user can choose to use a different PET equation or curve number equation, and adjust the initial CN2 values for different land use to improve model performance. The crop growth should be realistic to represent proper cover and residue; therefore, if biomass or crop grain yield data are available, they can be calibrated before the calibration of erosion–sedimentation. The user can choose different erosion equations. Calibrations of nutrients and pesticides follow after flow and sediment calibrations. However, because the processes of hydrology, erosion, crop growth, nutrient cycling, and pesticide fate are intimately intertwined, back and forth refining and adjustment may be needed for achieving an overall satisfactory model performance.

In this study the APEX model was manually calibrated and validated for monthly stream flow, sediment yield, and nutrient losses from the North Bosque Watershed at the Hico monitoring station (1993–1998), Texas. The watershed area above Hico is 92,650 ha. It contains 41 dairies. Influential model parameters were adjusted during model calibration. The NSE values ranged from 0.69 to 0.85 for the calibration period. The NSE values ranged from 0.61 to 0.78 for flow, sediment, mineral P, organic P, and total P losses for the validation period. The model performance was relatively poor for organic N and mineral N losses during the validation period. This could partly be attributed to the uncertainty in model input data in the N fertilizer application with the assumption that commercial fertilizer was used for improved pasture (same rate for all the simulation years) and the automatic N fertilizer option was used for cropland, and that no commercial fertilizer was applied for the remaining land uses. No effort was made to adjust model inputs in regard to numbers of cows and the assignment of waste application fields for individual dairies. However, in general, the model's predictions were reasonable based on statistical criteria for establishing satisfactory water quality model performance proposed by Moriasi et al. (2007). This study shows that the APEX model is a powerful tool for simulating feedlot water quality.

Experience has taught us that combining modeling with data collection programs advances research much more rapidly than either effort alone. The data collection provides essential information for use in developing and testing models, and models can provide feedback to the data collection program that can be used to enhance the experimental design and expand the temporal and spatial scales. In reality, models are often used where data have been collected without knowing the actual model input requirement or where it is not possible to collect all the required inputs, especially for larger scale watershed modeling. Assumptions and best judgment have to be made. With uncertainties associated with model inputs, model parameters, model structures, and observations, it should be expected that model predictions cannot fully replicate observations. Users should not be encouraged to over-tune model inputs.

Acknowledgments

The authors gratefully acknowledge Paul Duckworth for his work in developing and using the Field Hydro Tool for delineation of the watershed and Paul Dyke for providing data used in this exercise. The North Bosque Watershed is a Special Emphasis Watershed study in the Conservation Effects Assessment Project (CEAP) sponsored by USDA conservation programs.

References

Baier, W., and G.W. Robertson. 1965. Estimation of latent evaporation from simple weather observations. Can. J. Plant Sci. 45:276–284.

Cabelguenne, M., C.A. Jones, J.R. Marty, P.T. Dyke, and J.R. Williams. 1990. Calibration and validation of EPIC for crop rotations in southern France. Agric. Syst. 33:153–171.

Chung, S.W., P.W. Gassman, R. Gu, and R.S. Kanwar. 2002. Evaluation of EPIC for assessing tile flow and nitrogen losses for alternative agricultural management systems. Trans. ASAE 45:1135–1146.

Chung, S.W., P.W. Gassman, D.R. Huggins, and G.W. Randall. 2001. EPIC tile flow and tile nitrate losses for three Minnesota cropping systems. J. Environ. Qual. 30:822–830.

Foster, G.P. 2005. Science documentation. Revised universal soil loss equation, Version 2 (RUSLE 2). National Sedimentation Lab., USDA-ARS, Oxford, MS.

Gassman, P.W., J.R. Williams, V.W. Benson, R.C. Izaurralde, L. Hauck, C.A. Jones, J.D. Atwood, J. Kiniry, and J.D. Flowers. 2005. Historical development and applications of the EPIC and APEX models. Working Paper 05-WP 397. Available at www.card.iastate.edu/publications/synopsis.aspx?id=763 (accessed 5 June 2009, verified 30 Mar. 2011). Center for Agricultural and Rural Development, Iowa State Univ., Ames.

Gassman, P.W., J.R. Williams, X. Wang, A. Saleh, E. Osei, L. Hauck, C. Izaurralde, and J. Flowers. 2010. The Agricultural Policy Environmental Extender (APEX) model: An emerging tool for landscape and watershed environmental analyses. Trans. ASAE 53:711–740.

Green, W.H., and G.A. Ampt. 1911. Studies on soil physics: 1. Flow of air and water through soils. J. Agric. Sci. 4:1–24.

Gaiser, T., I. de Barros, F. Sereke, and F.M. Lange. 2010. Validation and reliability of the EPIC model to simulate maize production in small-holder farming systems in tropical sub-humid West Africa and semi-arid Brazil. Agric. Ecosyst. Environ. 135:318–327.

Hargreaves, G.H., and Z.A. Samani. 1985. Reference crop evapotranspiration from temperature. Appl. Eng. Agric. 1:96–99.

Harmel, R.D., R.J. Cooper, R.M. Slade, R.L. Haney, and J.G. Arnold. 2006. Cumulative uncertainty in measured streamflow and water quality data for small watersheds. Trans. ASABE 49:689–701.

Harmel, R.D., and P.K. Smith. 2007. Consideration of measurement uncertainty in the evaluation of goodness-of-fit in hydrologic and water quality modeling. J. Hydrol. 337:326–336.

Kannan, N., C. Santhi, J.R. Williams, and J.G. Arnold. 2008. Development of a continuous soil moisture accounting procedure for curve number methodology and its behaviour with different evapotranspiration methods. Hydrol. Processes 22:2114–2121.

Kemanian, A.R., P. Duckworth, and J. Williams. 2009. A spatially distributed modeling approach for precision conservation and agroecosystem design. p. 153–154. *In* Proc. of the Int. Farming Systems Design Conf., Monterey, CA. Available at http://ww.iemss.org/farmsys09/uploads/2009_FSD_Proceedings.pdf (accessed 15 Apr. 2011).

Leonard, R.A., W.G. Knisel, and D.A. Still. 1987. GLEAMS: Groundwater loading effects on agricultural management systems. Trans. ASAE 30:1403–1428.

Magre, M., J.R. Williams, W.L. Harman, T. Gerik, J. Greiner, L. Francis, E. Steglich, and A. Meinardus. 2006. WinAPEX users guide. Available at http://winapex.brc.tamus.edu/media/17085/winapexv0604_nov2006.pdf (verified 30 Mar. 2011). Blackland Research and Extension Center, Temple, TX.

Mockus, V. 1969. Hydrologic soil-cover complexes. p. 10.1–10.24. *In* SCS national engineering handbook. Section 4. Hydrology. USDA-SCS, Washington, DC.

Moriasi, D.N., J.G. Arnold, M.W. Van Liew, R.L. Binger, R.D. Harmel, and T. Veith. 2007. Model evaluation guidelines for systematic quantification of accuracy in watershed simulations. Trans. ASABE 50:885–900.

Monteith, J.L. 1965. Evaporation and environment. Symp. Soc. Exp. Biol. 19:205–234.

Mudgal, A., C. Baffaut, S.H. Anderson, E.J. Sadler, and A.L. Thompson. 2010. APEX model assessment of variable landscapes on runoff and dissolved herbicides. Trans. ASAE 53:1047–1058.

Nash, J.E., and J.V. Sutcliffe. 1970. River flow forecasting through conceptual models. Part 1: A discussion of principles. J. Hydrol. 10:282–290.

Nemes, A., W.J. Rawls, and Ya.A. Pachepsky. 2006. Use of a non-parametric nearest-neighbor technique to estimate soil water retention. Soil Sci. Soc. Am. J. 70:327–336.

Nicks, A.D., and L.J. Lane. 1989. Weather benerator. p. 2.1–2.19. Profile model documentation. USDA-Water Erosion Prediction Project: Hillslope Profile Version. NSERL Rep. 2. USDA-ARS National Soil Erosion Research Lab., West Lafayette, IN.

Onstad, C.A., and G.R. Foster. 1975. Erosion modeling on a watershed. Trans. ASAE 18:288–292.

Penman, H.L. 1948. Natural evaporation from open water, bare soil, and grass. Proc. R. Soc. Lond. A 193:120–146.

Potter, K.N., and T.M. Zobeck. 1990. Estimation of soil microrelief. Trans. ASAE 33:156–161.

Potter, K.N., T.M. Zobeck, and L.J. Hagen. 1990. A microrelief index to estimate soil erodibility by wind. Trans. ASAE 33:151–155.

Priestly, C.H.B., and R.J. Taylor. 1972. On the assessment of surface heat flux and evaporation using large-scale parameters. Mon. Weather Rev. 100:81–92.

Rawls, W.J., and D.L. Brakensiek. 1985. Prediction of soil water properties for hydrologic modeling. p. 293–299. *In* E.B. Jones and T.J. Ward (ed.) Watershed management in the Eighties. American Society of Civil Engineers, New York.

Renard, K.G., G.R. Foster, G.A. Weesies, D.K. McCool, and D.C. Yoder. 1997. Predicting soil erosion by water: A guide to conservation planning with the revised universal soil loss equation (RUSLE). Agric. Handb. 703. USDA-ARS, Washington, DC.

Richardson, C.W. 1981. Stochastic simulation of daily precipitation, temperature, and solar radiation. Water Resour. Res. 17:182–190.

Richardson, C.W., and D.A. Wright. 1984. WGEN: A model for generating daily weather variables. ARS-8. USDA-ARS, Washington, DC.

Saltelli, A., K. Chan, and E.M. Scott (ed.) 2000. Sensitivity analysis. John Wiley and Sons, Chichester, UK.

Saltelli, A., S. Tarantola, and K. Chan. 1999. A quantitative, model-independent method for global sensitivity analysis of model output. Technometrics 41:39–56.

Skidmore, E.L. 1986. Wind-erosion climatic erosivity. Clim. Change 9:195–208.

Stout, J.E. 1990. Wind erosion with a simple field. Trans. ASAE 33:1597–1600.

Tuppad, P., M. Winchell, R. Srinivasan, and J.R. Williams. 2009. ArcAPEX: ArcGIS interface for Agricultural Policy Environmental eXtender (APEX) hydrology/water quality model. Int. Agric. Eng. J. 18:59–71.

USDA. 2009. National Agriculture Imagery Program. Available at http://www.fsa.usda.gov/FSA/apfoapp?area=home&subject=prog&topic=nai (verified 30 Mar. 2011) USDA, Washington, DC.

USDA-NRCS. 2004. Estimation of direct runoff from storm rainfall. p. 10.1–10.22 *In* NRCS national engineering handbook, Part 630: Hydrology. Available at http://directives.sc.egov.usda.gov/viewerFS.aspx?hid=21422 (accessed 5 June 2009, verified 8 Apr. 2011). USDA-NRCS, Washington, D.C.

USDA-SCS. 1986. Urban hydrology for small watersheds. Tech. Release 55. USDA-SCS, Washington, DC.

Wang, E., C. Xin, J.R. Williams, and C. Xu. 2006a. Predicting soil erosion for alternative land uses. J. Environ. Qual. 35:459–467.

Wang, X., P.W. Gassman, J.R. Williams, S. Potter, and A.R. Kemanian. 2008. Modeling the impacts of soil management practices on runoff, sediment yield, maize productivity, and soil organic carbon using APEX. Soil Tillage Res. 101:78–88.

Wang, X., X. He, J.R. Williams, R.C. Izaurralde, and J.D. Atwood. 2005. Sensitivity and uncertainty analyses of crop yields and soil organic carbon simulated with EPIC. Trans. ASAE 48:1041–1054.

Wang, X., R.D. Harmel, J.R. Williams, and W.L. Harman. 2006b. Evaluation of EPIC for assessing crop yield, sediment and nutrient losses in runoff from watersheds with poultry litter fertilization. Trans. ASABE 49:47–59.

Wang, X., D.W. Hoffman, J.E. Wolfe, J.R. Williams, and W.E. Fox. 2009. Modeling the effectiveness of conservation practices at Shoal Creek watershed, TX, using APEX. Trans. ASABE 52:1181–1192.

Wang, X., S.R. Potter, J.R. Williams, J.D. Atwood, and T. Pitts. 2006c. Sensitivity analysis of APEX for national assessment. Trans. ASAE 49:679–688.

Wang, X., P. Tuppad, and J.R. Williams. 2010. Modeling agricultural management systems with APEX. *In* Soil hydrology, land use and agriculture: Measurement and modelling. CABI Press, Wallingford, UK.

Williams, J.R. 1975. Sediment yield prediction with universal equation using runoff energy factor. ARS-S-40. USDA-ARS, Washington, DC.

Williams, J.R. 1990. The erosion productivity impact calculator (EPIC) model: A case history. Philos. Trans. R. Soc. Lond. B Biol. Sci. 329:421–428.

Williams, J.R. 1995. The EPIC model. p. 909–1000. *In* V.P. Singh (ed.) Computer models of watershed hydrology. Water Resources Publ., Highlands Ranch, CO.

Williams, J.R. 2002. The APEX manure management component. p. 44–51. *In* A. Saleh (ed.) Proc. of the Total Maximum Daily Load (TMDL) Environmental Regulations, Fort Worth, TX. 11–13 Mar. ASAE, St. Joseph, MI.

Williams, J.R., J.G. Arnold, J.R. Kiniry, P.W. Gassman, and C.H. Green. 2008. History of model development at Temple, Texas. Hydrol. Sci. J. 53:948–960.

Williams, J.R., P.T. Dyke, W.W. Fuchs, V.W. Benson, O.W. Rice, and E.D. Taylor. 1990. EPIC—Erosion/productivity impact calculator: 1. User manual. A.N. Sharpley and J.R. Williams (ed.) USDA Tech. Bull. 1768. USDA, Washington, DC.

Williams, J.R., and R.C. Izaurralde. 2006. The APEX model. p. 437–482. *In* V.P. Singh and D.K. Frevert (ed.) Watershed models. CRC Press/Taylor & Francis, Boca Raton, FL.

Williams, J.R., C.A. Jones, and P.T. Dyke. 1984. A modeling approach to determining the relationship between erosion and soil productivity. Trans. ASAE 27:129–144.

Williams, J.R., C.A. Jones, P.W. Gassman, and L.M. Hauck. 1995. Simulation of animal waste management with APEX. p. 22–26. *In* Proc. Innovations and New Horizons in Livestock and Poultry Manure Management, Austin, TX. 6–7 Sept.

Williams, J.R., E. Wang, A. Meinardus, W.L. Harman, M. Siemers, and J.D. Atwood. 2006. APEX users guide. V. 2110. Texas A&M Univ., Texas Agricultural Extension Service, Texas Agricultural Exp. Stn., Blackland Research Center, Temple, TX.

Wischmeier, W.H., and D.D. Smith. 1978. Predicting rainfall erosion losses, a guide to conservation planning. USDA Agric. Handb. 537. USDA, Washington, DC.

Yin, L., X. Wang, J. Pan, and P.W. Gassman. 2009. Evaluation of APEX for daily runoff and sediment yield from three plots in the middle and upland Huaihe River watershed, China. Trans. ASAE 52:1833–1845.

Zhang, X., R.C. Izaurralde, D. Manowitz, T.O. West, W.M. Post, A.M. Thomson, V.P. Bandaru, J. Nichols, and J.R. Williams. 2010. An integrative modeling framework to evaluate the productivity and sustainability of biofuel crop production systems. Global Change Biol. Bioenergy 2:258–277.

Chapter 7

Parameterization of the GPFARM-Range Model for Simulating Rangeland Productivity

S.G.K. Adiku
L.R. Ahuja
G.H. Dunn
J.D. Derner
A.A. Andales
L. Garcia
P.N.S. Bartling

Abstract

One of the major limitations to rangeland model usage Is the lack of parameter values appropriate for reliable simulations at different locations and times. In this chapter we seek to show how the GPFARM-Range, a rangeland model, which has been previously parameterized, tested, and validated for the central locations of the Great Plains, could be reparameterized to extend its domain of application to other locations of the Plains. Two main parameter determination methods are proposed: (i) manual adjustment of default parameter values and (ii) direct empirical parameter determination from some experimental data. It was recommended that in view of the level of information and expertise required for the second method, at this point the users should follow the first method—comparing the simulated output with observed data in a statistical sense such that the sum of squared deviations between the simulated and observed is minimized. It is noteworthy, however, that the published literature is a major source of data that could be used for parameterizing models. A number of texts are suggested for consultation to assist in the empirical determination of parameter values. It is shown that by varying the default values of only a few key parameters, the GPFARM-Range model could simulate forage growth under varying weather and grazing conditions at Miles City, located in northern part of the Great Plains.

S.G.K. Adiku (sadiku@mail.colostate.edu) and L. Garcia (luis.garcia@colostate.edu), Dep. of Civil and Environmental Engineering, Colorado State Univ., Fort Collins, CO 80523; L.R. Ahuja (laj.ahuja@ars.usda.gov), G.H. Dunn (gale.dunn@ars.usda.gov), and P.N.S. Bartling (pat.bartling@ars.usda.gov), USDA-ARS, Agricultural Systems Research Unit, Fort Collins, CO 80526; J.D. Derner (justin.derner@ars.usda.gov), USDA-ARS, High Plains Grasslands Research Station 8408 Hildreth Road Cheyenne, WY 82009-8899; A.A. Andales (allan.andales@colostate.edu), Dep. of Soil and Crop Sciences, Colorado State Univ., Fort Collins, CO 80523.

doi:10.2134/advagricsystmodel2.c7

Methods of introducing system models into agricultural research. L.R. Ahuja and L. Ma (ed.)
Advances in Agricultural Systems Modeling Series 2.

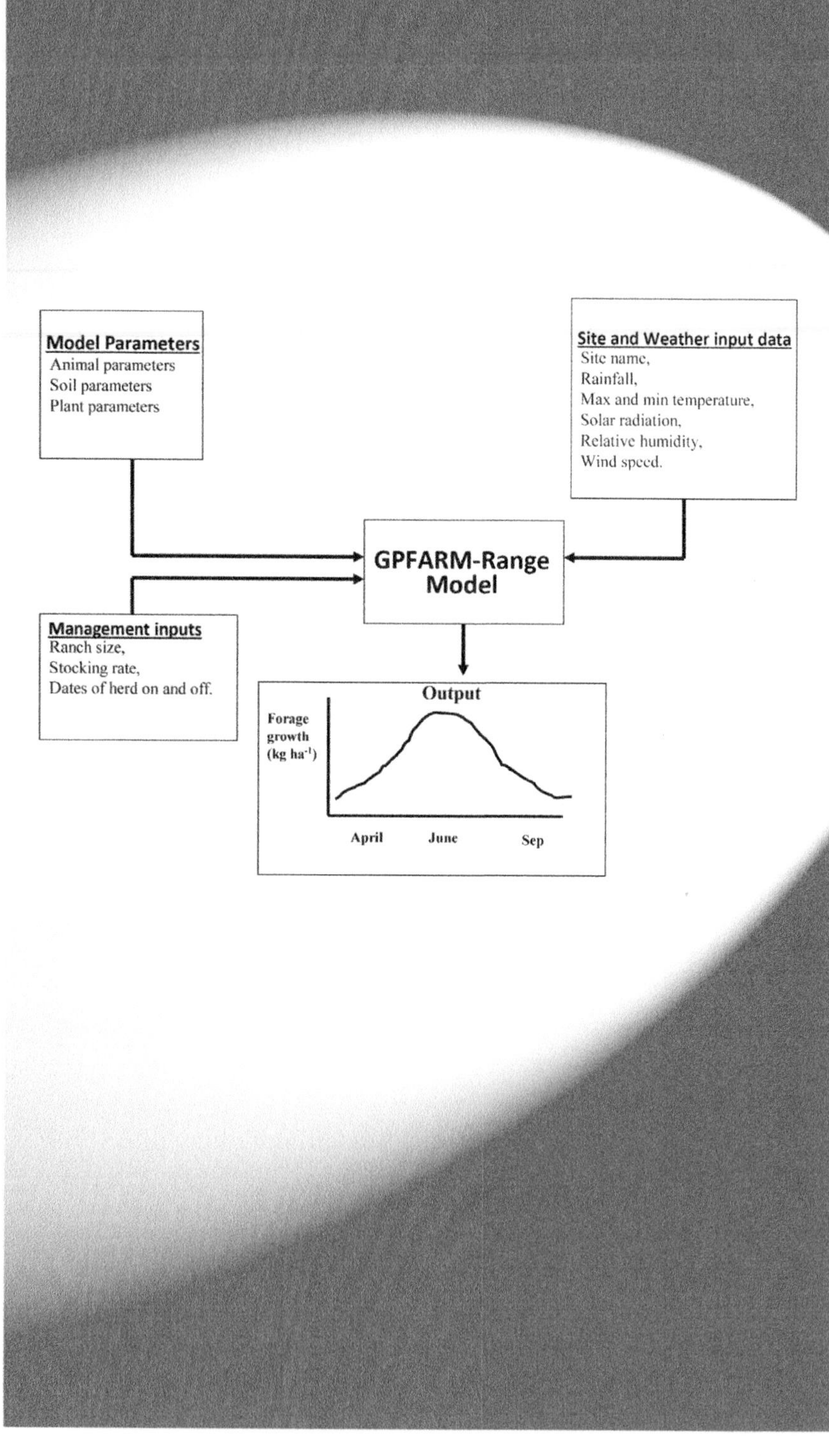
Model Parameters
Animal parameters
Soil parameters
Plant parameters
Site and Weather input data
Site name,
Rainfall,
Max and min temperature,
Solar radiation,
Relative humidity,
Wind speed.
GPFARM-Range
Model
Management inputs
Ranch size,
Stocking rate,
Dates of herd on and off.
Output
Forage
growth
(kg ha-1)
April
June
Sep

Rangelands comprise about 50% of the world's land area. They occur largely in low rainfall zones and on marginal lands and have diverse vegetation types, including (i) natural grasslands, (ii) deserts shrublands, (iii) savanna woodlands, (iv) forests, and (v) tundras. These vegetation types comprise mixtures of C3 and C4 species. Economically, rangelands are primarily used for grazing ruminants. However, rangelands also offer a range of ecosystem goods and services, such as wildlife habitat, water supply, and conservation of nature.

Most rangelands have been studied to understand the ecology of the system. Studies have focused on vegetation composition and its dynamics, hydrology, and landscape processes. Since the maintenance of rangeland productivity is at the heart of the ranch and animal husbandry industry, it is of interest to understand what factors affect it and how management could be used to sustain this productivity. Therefore, by far, the rangelands have been studied with economic motives, as the source of forage for domesticated animals, which form a vital component of human nutrition. The major aspects of rangelands currently being studied by grazing experimental stations include range species and primary production (Derner and Hart, 2007), pest infestation, invasion of poisonous plants (Blumenthal et al., 2005), and soil fertility and water relations (Heitschmidt et al., 1999, 2005).

Traditionally, field observations involving destructive biomass harvests from quadrats or along transects and soil sampling are used to determine rangeland productivity. Increasingly, nondestructive methods have been developed that estimate biomass from leaf area (determined using remote sensors), canopy height measurement, or from weighted plate measurements (Ganguli et al., 2000). Over the years, the use of nondestructive methods such as moisture meters for soil measurement has also increased compared with the traditional soil augering.

However, due to the high labor demands, increasing cost, and time and location specificity of field observations, modeling approaches are now emerging as additional methods for the assessment of rangeland productivity. In particular cases, where it is desired to predict rangeland productivity in space and time to gain a foreknowledge of the range condition under varying management to match stocking rates to forage availability, or to assess the effect of future changes in climate on rangeland productivity, models have been found to be more appropriate tools (National Resource Models in the Rangelands, 2004). Several models have now been developed for rangeland management studies. One important range model is the Simulation of Production and Utilization of Rangelands, or SPUR (Hanson et al., 1987), and this model has been used by many researchers to assess forage growth and cattle production in the Great Plains of the United States. The model was further enhanced into SPUR II (Hanson et al., 1992). The SPUR models provided a good foundation for the development of a broader agricultural production model for cultivated crops and forage and animal production, referred to as the Great Plains Framework for Agricultural Resource Management (GPFARM) Decision Support System (DSS) (Ascough et al., 2007; Shaffer et al., 2000). Later, the components of GPFARM DSS that simulated rangeland forage growth and cattle production were extracted to form the basis of a new model now called the GPFARM-Range model. Details of the new GPFARM-Range model can be found in Andales et al. (2005, 2006). Information on the soil properties, potential evapotranspiration, water balance, and chemical transport modules are similar to those of GPFARM DSS publications (see Andales et al., 2003). The GPFARM-Range has been validated for some central locations of the Great Plains of the United States (Andales et al., 2005, 2006)

The purpose of this chapter is to present the GPFARM-Range model with a focus on how it could be parameterized and applied to simulate rangeland productivity at other locations of interest within the Great Plains of the USA or elsewhere.

The GPFARM-Range Model

Model Description

The GPFARM-Range is an object-oriented model written in Java. It comprises several modules that describe the growth of up to five forage functional groups and animal types (e.g., cattle). The model requires information about the sites, animal types, weather variables, and management as inputs and several animal–soil–plant parameters for execution (Fig. 7–1). Parameters are constants that describe the behavior of a system under varying environmental conditions. Their values can be either hard-coded in the models or read from parameter files. For example, the relative growth rate, RGR, which is the relative increase in mass per unit mass per unit time (kg kg^{-1} d^{-1}), is a simple growth parameter that is often used to describe the growth of a range plant. More detailed physiological models would require parameters such as the radiation use efficiency (RUE) or the photosynthetic efficiency rate. Parameter values are derived from detailed experimental data, and a host of them have been published in the literature. Input variables, on the other hand, are external to the model and include environmental variables (e.g., temperature, radiation, rainfall, and initial soil and plant conditions) and

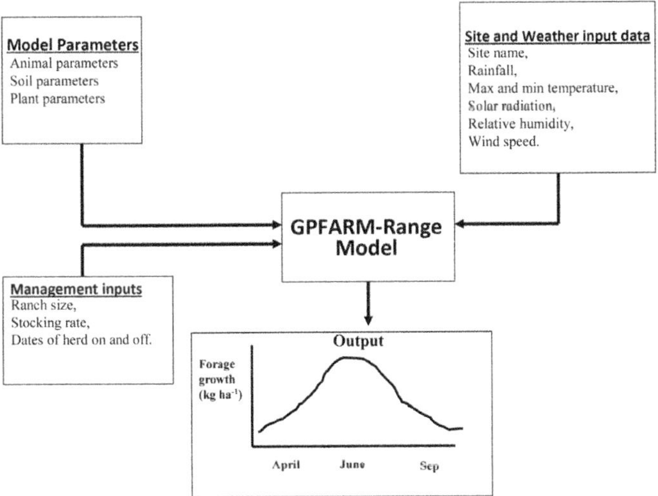

Fig. 7–1. Schematic diagram of the GPFARM-Range model.

management factors (e.g., stocking rates) that interact with the parameters and process descriptions to predict behavior of the system.

The GPFARM-Range model requirements are grouped into four classes: (i) site and weather inputs, (ii) animal parameters, (iii) soil parameters, and (iv) plant parameters. Each of these classes is discussed in greater detail below.

Inputs and Parameters

Site and Weather Inputs

The site information required for running the GPFARM-Range model include the site name and the Cartesian coordinates, namely latitude and longitude. The weather input variables constitute the external drivers of the GPFARM-Range model. Daily input data are required for rainfall amount (mm) and duration (h), maximum and minimum temperature (°C), solar radiation (Langleys), wind velocity (m s^{-1}), and relative humidity (%). The weather data can be historical or forecast and must be available for all the years for which the simulations are to be run. The site coordinates are helpful to estimate some weather variables, such as solar radiation, if measured data are not available

Animal Parameters and Input Variables

Animal parameters enable the prediction of animal growth and weight gain during the season. The parameters relate to the animal types currently on the range. For example, for cattle, parameters include the various groups such as open cows, pregnant cows, calves, heifers, steers and bulls, as well as their initial weight, daily forage intake rate, and forage utilization efficiency. Input variables for the animal component include stocking rate, the forage use criterion ("useCrit"), which specifies the fraction of total forage available for grazing. Setting the "useCrit" to zero implies no grazing whereas a fraction of 0.5 would represent the "take half leave half" rule. Further input variables include the details of the grazing events, that is, the dates when the animals were on and off the rangeland. Table 7–1 summarizes the major parameters of the animal module.

Soil Parameters

The soil parameters enable the simulation of the soil water balance and other soil processes. As with the site description, it is useful to specify the general classification (according to U.S. soil taxonomy) of the soil, although this information is not directly used in the model execution. There are two sets of parameters: the hydrologic group, which relates to the soil surface condition (e.g., crusting, soil albedo, and residue cover) and affects the simulation of evaporation, and the soil profile group, which comprises the detailed description of the properties for each soil layer. As shown on Table 7–2, the parameters include soil layers and layer thickness, sand and clay percentages, bulk density, saturated soil water content, saturated hydraulic conductivity, the air-entry value, and the pore-size distribution index for each layer. The latter soil water related parameters are estimated from soil texture and bulk density if measured data are not provided. These parameters are used to simulate the distribution of soil water with time and depth, using the Darcy descriptions for soil water flow simulation.

Plant Parameters

Plant parameters enable the simulation of forage development and growth in rangelands. As indicated earlier, rangelands have many plant species growing in a mixture, and it is not practicable to simulate the development and growth of each individual species. For convenience, species are classified into functional groups, based on their major physiological similarities. In the GPFARM-Range model, all the C4 graminoids are grouped into warm season grasses (WSG), and the C3 ginoids into cool season grasses (CSG). Three other recognized functional groups are the legumes, the forbs, and the shrubs. Each of these groups has distinctive parameters for development and growth (Table 7–3).

Plant development is expressed in terms of physiological, rather than chronological, time. The main development stages recognized are (i) emergence or green up, (ii) anthesis, (iii) senescence, and (iv) maturity, and the duration of each stage is expressed as growing degree days (GDD). The calculation of GDD requires knowledge of functional-group specific cardinal temperatures, namely

Table 7–1. Some animal parameter values for Central Plains Experiment Research Station, Nunn, CO.

Parameter	Cow	Calf	Heifer	Bull
Mature weight, kg	544	–	–	500
Daily feed requirement, kg head^{-1}	7.7	3.8	7.7	7.7
Weight gain rate, kg d^{-1}	1.4	0.5	0.5	1.4
Feed utilization efficiency, kg kg^{-1}	0.6	0.6	0.6	0.6

Table 7–2. Example of soil parameters for Central Plains Experiment Research Station, Nunn, CO.

Depth	Bulk density	Clay	Sand	Saturated soil water content	Saturated hydraulic conductivity	Air-entry value	Soil pore-size distribution index
cm	g cm^{-3}	——— % ———		cm^3 cm^{-3}	cm h^{-1}	cm	
7	1.45	15.00	71.05	0.45	2.59	14.66	0.32
15	1.45	15.00	71.05	0.45	2.59	14.66	0.32
41	1.44	24.29	54.04	0.39	0.43	28.08	0.25
56	1.45	22.50	53.84	0.39	0.43	28.08	0.25
86	1.48	11.50	71.79	0.45	2.59	14.66	0.32
116	1.48	11.50	71.79	0.45	2.59	14.66	0.32
152	1.48	11.50	71.79	0.45	2.59	14.66	0.32

the base, optimum, and maximum temperatures. Because rangeland forage is not sown, there is no definite sowing date. For modeling convenience, the GDD value is calculated from January 1 until the occurrence of a particular stage for a given functional group.

Forage growth is controlled by the functional group-specific relative growth rate (RGR), temperature, and soil water availability. Growth is initiated by translocation of root biomass reserves to shoots at the beginning of the growing season. The proportion of the various functional groups in the mixture governs the contribution of each functional group to the overall forage growth. The senescence of both the above- and below-ground biomass adds to soil carbon and nitrogen and determines the fertility status of the range.

Running the Model using Default Parameters

To facilitate model execution, the GPFARM-Range model is equipped with a Microsoft Office Excel interface for input and output. Three Excel worksheets enable the user to input location specific information for running the model. These are (i) the "Input" sheet, (ii) the "Weather" sheet, and (iii) the "Events" sheet. The "Input" sheet provides all the default parameters for each model component: site, animal, soil, and plant (Fig. 7–2 and 7–3). The "Weather" sheet provides the daily weather variables for the location of interest. The "Event" sheet provides informa-

Table 7–3. Example of plant parameters for Central Plains Experiment Research Station, Nunn, CO.

Parameter	Description	Warm season grasses	Cool season grasses	Legumes	Shrubs	Forbs
emergGDD, °C d	Growing degree days from January 1 to green up	50	80	105	89	105
senGDD, °C d	Growing degree days from January 1 to senescence	1200	1200	1335	1877	1188
matureGDD, °C d	Growing degree days from January 1 to maturity	1600	2200	1855	2300	1865
maxGrowthRate, kg kg^{-1} d^{-1}	Maximum relative growth rate	0.26	0.25	0.17	0.17	0.17
propPop	Proportion of functional group in forage	0.64	0.23	0.00	0.04	0.07
respRate, kg kg^{-1} d^{-1}	Respiration rate	0.04	0.04	0.04	0.04	0.04
senRate, kg kg^{-1} d^{-1}	Senescence rate	0.018	0.013	0.005	0.001	0.001
rootBiomass, kg ha^{-1}	Initial root biomass	7168	2576	0	672	784
foragePref, 0–1	Preference for forage by grazing animals	0.4	0.9	0.9	0.2	0.1
waterStresssSen, 0–1	Sensitivity to water stress	0.15	0.30	0.6	0.45	0.18
tempBase, °C	Base temperature	8	3	3	4	3
tempOpt, °C	Optimal temperature	27	22	20	21	23
tempMax, °C	Maximum temperature	41	36	35	36	35

tion on grazing events, such as the dates of herd on and off the rangeland. Data from these sheets can be edited by the user and are used by the Java program to simulate several animal, soil, and forage growth attributes.

The results from the model simulations are also output to the same Excel interface. The animal weight gain, soil water distribution with time and depth, and forage growth during the simulation period are output on the Excel sheets "herd.out," "water.out," and "plant.out," respectively. A typical output of the forage growth at Central Plains Experiment Research Station (CPER) at Nunn, CO for four functional groups is shown in Fig. 7–4.

An important question relates to how well a model performs in simulating the behavior of the soil–plant–animal system. This is often determined by comparing simulated results with observations. The agreement is based on statistical procedure. There are several statistical criteria for judgment, but the most com-

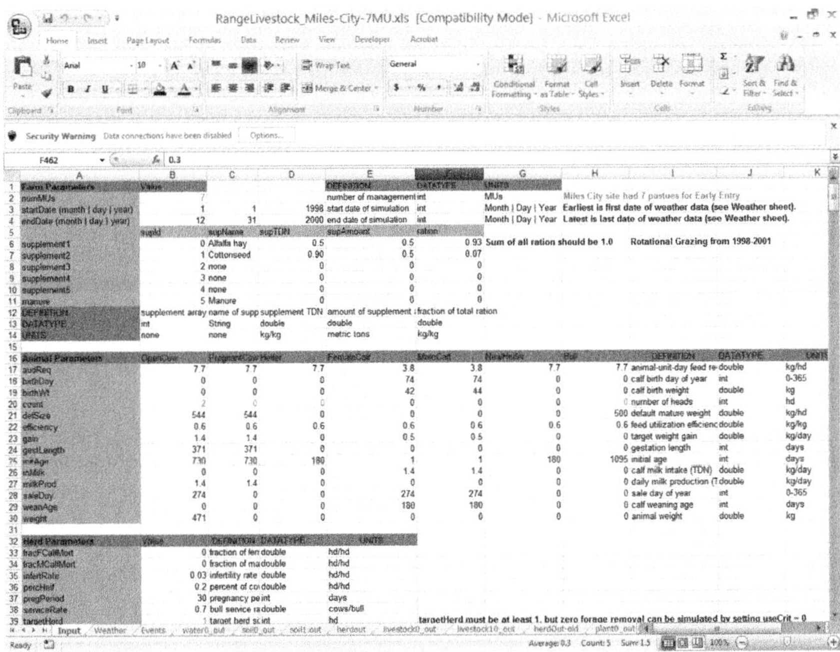

Fig. 7–2. Excel interface focused on the animal input spreadsheet for the Miles City, MT location.

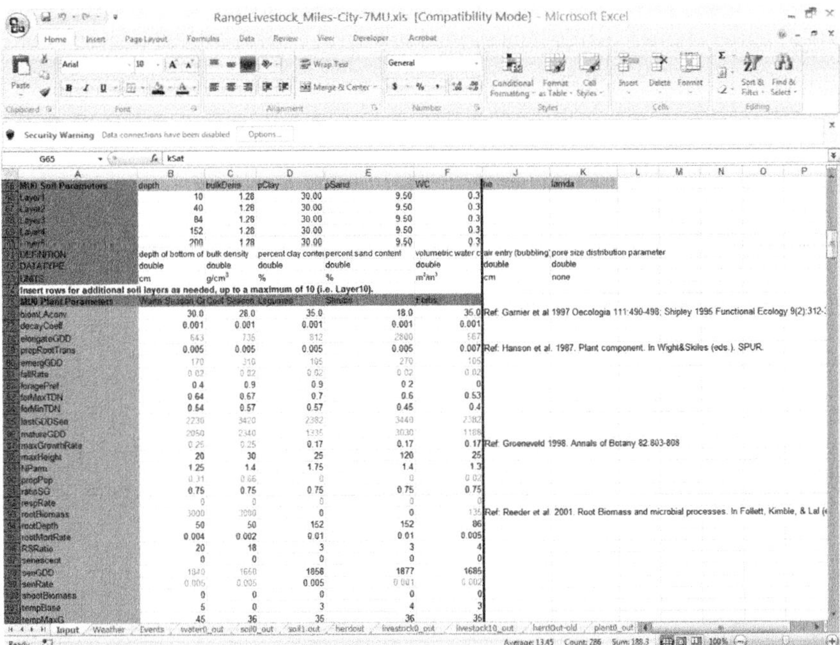

Fig. 7–3. Excel interface focused on the plant parameter input spreadsheet for the Miles City, MT location.

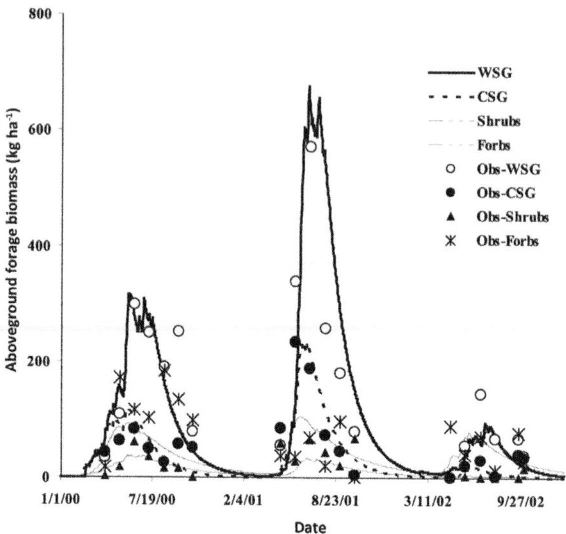

Fig. 7–4. Simulated growth of the various forage functional groups at the Central Plains Experiment Station (CPER), Nunn, CO. (Source: Andales et al., 2005).

monly used are (i) the coefficient of determination (R^2), (ii) the root mean square error (RMSE), and (iii) the Willmott (1981) *d* index of modeling efficiency (EF). The equations for the last two statistics are as follows:

$$\mathrm{RMSE} = \sqrt{\frac{\sum_{i=1}^{i=n} (p_i - o_i)^2}{n}} \qquad [1]$$

$$d = 1 - \left[\frac{\sum_{i=1}^{i=n} (p_i - o_i)^2}{\sum \left(|p_i'| + |o_i'| \right)^2} \right] \qquad 0 \le d \le 1 \qquad [2]$$

where p_i and o_i are the predicted and observed forage, $p_i' = p_i - \bar{o}$, $o_i' = o_i - \bar{o}$, with $\bar{o}$ being the observed mean. A perfect agreement between the predicted and the observed is indicated by a *d* index of 1. In general, the RMSE must be as small as possible, and a *d* value approaching unity is desired.

Step-by-Step Parameterization of the GPFARM-Range Model

More often, users would want to apply the model to their own observations and validate the model for their locations or other locations of interest. The default parameter values currently used in the model were drawn from the published literature and experience with using the model for rangeland experiment stations in the U.S. Great Plains. The Excel interface cites references for parameters taken from literature, and the user may customize the entry and citations when the field is changed (Fig. 7–3). As indicated, the model was originally parameterized and validated for Nunn, CO and Cheyenne, WY, both of which are located in the more central locations of the Great Plains. However, as one moves away toward either the more northern or southern parts of the plains, vegetation types and soils change. Hence, several parameter values may differ from the default values currently in GPFARM-Range model. In such situations, the direct application of the model based on the default parameters could lead to incorrect simulations. Therefore, models need to be "re-parameterized," recalibrated, and validated for different locations. Parameterization, calibration, and validation are often very tedious and time-consuming procedures. There are two ways to parameterize the model. The first is to manually adjust the default values such that the simulation output matches the observed growth data. The second is to directly determine the parameter values empirically (from experimental data) for the specific animal, soil, and forage type in question.

Manual Adjustment of Default Parameter Values

Because the number of model parameters is often large and they span a wide range of scientific disciplines, it is often not practical to determine all the parameters for every location and time. The first step in model parameterization is therefore to indentify the key process determinants and adjust their parameter values in a way that will minimize the deviations from the observed, using Eq. [1] and [2]. It is often practical to begin by setting the soil parameters, followed by plant parameters, and then the animal parameters. For the soil component, information on texture is required, and this is often available from USDA-NRCS database (http://soils.usda.gov/survey/geography/ssurgo/, verified 4 Apr. 2011) or from other Soil Survey manuals. In the absence of detailed soil hydraulic properties, the GPFARM-Range provides pedotransfer algorithms that use only texture and bulk density data for their estimation. The user must specify the use of these algorithms by setting the "soilPropOpt" option in the hydrologic group to 0.

Having specified the parameters for the soil, the next step is to parameterize the plant module. It is recommended to begin with a non-grazing situation so that the potential forage growth can be first evaluated. This is achieved by setting the "useCrit" = 0. For the plant component, the key determinants are the GDDs accumulated for the various development stages, the RGR, forage respiration and senescence rate, and the initial root biomass. The plant parameters for each functional group are adjusted upward or downward and the model executed using the weather data for the location in question. Based on Eq. [1] and [2], a good agreement between the model output and the observed soil water content distribution, evaporation, transpiration, and forage growth indicates that the adjusted parameter values are adequate for the location in question. Several iterations are usually needed to achieve the desirable agreement.

Once acceptable simulation results for soil water balance and potential forage growth are attained, the animal component can be activated. The key animal parameters include the initial weight, forage intake rate, weight gain rate, and feed conversion efficiency. By varying these values and with the grazing event schedule input, the model is executed and validated once more for the animal parameters as well. Table 7–4 shows the adjustments to parameters from original values from cattle grazing at Nunn, CO to sheep grazing at Miles City, MT.

As noted, the manual adjustment of parameter values is often tedious and time-consuming and requires painstaking efforts to ensure reliable parameterization. Automated parameter estimation methods based on concepts such as the generalized likelihood uncertainty estimation, GLUE (Beven and Binley, 1992) are now being adapted and introduced into crop modeling (He et al., 2010). However, these are not yet available for rangeland modeling.

It is worth noting that the manual or automated methods of parameter estimation are indirect or inverse procedures, using an end product value (e.g., biomass) to determine the value of an input variables (e.g., RGR). In doing so, model users need to ensure that the input parameter values derived lie within physiologically accepted ranges.

Empirical Determination of Parameter Values

The direct empirical parameter determination is also often a daunting task; therefore, before resources are spent on any determinations, users are encouraged to first consult the literature for published data availability for their locations. Procedures and protocols for data collection are available in field and laboratory manuals that are available to consult for guidance.

The animal growth and intake module allows small, medium, and large animal body types. The default body type for cattle studies is large, and mature animal weight can be obtained from the animal breed characteristics. The daily feed requirement is the amount necessary to obtain the daily weight gain rate goal. The amount of available range forage subject to the "UseCrit" utilization efficiency, in addition to any supplemental feed, is used to meet the daily requirement. When demand is not met, the weight gain goal is not attained. Parameterizing the animal

Table 7–4. Adjustment in parameter values from cattle grazing at Nunn, CO to sheep grazing at Miles City, MT.†

	Previous parameter values			New parameter values		
Animal module	**Cow**		**Calf**	**Ewe**		**Lamb**
Mature weight, kg	540			75		25
Daily forage requirement, kg head^{-1}	7.7		3.8	1.5		0.6
Plant module	**Warm season grass**		**Cool season grass**	**Warm season grass**		**Cool season grass**
tempBase, °C	8.0		3.0	5.0		0.0
matureGDD, °C d	1600		2200	2045		2160
maxGrowthRate, kg kg^{-1} d^{-1}	0.26		0.25	0.16		0.10
respRate, kg kg^{-1} d^{-1}	0.04		0.04	0.005		0.005
senRate, kg kg^{-1} d^{-1}	0.018		0.013	0.01		0.01
propPop	0.64		0.23	0.35		0.60
Soil module	**Clay**	**Sand**	**Bulk density**	**Clay**	**Sand**	**Bulk density**
	——— % ———		g cm^{-3}			g cm^{-3}
Layer						
10 cm	15.00	71.05	1.45	22.50	9.50	1.33
40 cm	24.29	54.04	1.43	26.50	9.00	1.31
80 cm	11.50	71.79	1.48	30.00	6.80	1.28
152 cm	11.50	71.79	1.48	30.00	6.80	1.28

† Parameters whose values remain unchanged are not listed.

module entails feeding experiments in relation to the metabolic weight of the animals. Walker (1993) provided details for such experiments. As noted, the minimum determinations must include the animal weight and weight gain rate, forage intake and utilization efficiency, dietary preferential grazing of functional groups, among others. In the absence of detailed feeding trials, breed characteristics from literature can be used to obtain initial parameter estimates.

For the soil module, the determination of texture and bulk density is simple and straight forward, and these are often routinely determined in most laboratories. However, the hydraulic properties (e.g., saturated hydraulic conductivity K_{sat}, pore-size distribution index λ, air-entry value h_e) are not normally available for most locations and must be determined for individual situations following standard procedures. The pressure plate method is commonly used to establish the soil moisture retention or characteristic curve, while the saturated hydraulic conductivity can be determined in the laboratory using the constant-head permeameter setup. For field conditions, simple methods such as the single ring infiltrometer (Wu et al., 1999) may suffice to determine the saturated hydraulic conductivity, although more sophisticated and more accurate methods such as the Guelph permeameter (Elrick and Reynolds, 1992) are also available. Given data on the soil bulk density and water content at field capacity (33 kPa suction), the air-entry value and the pore-size distribution index can be derived using the approaches such as the one-parameter for soil moisture characteristic (Williams and Ahuja 2003) and saturated hydraulic conductivity from effective porosity (Ahuja et al., 1989). Texts such as the *Methods of Soil Analysis* series (Klute, 1986; Dane and Topp, 2002) also provide method descriptions.

For the forage plant growth module, both parameters affecting the plant's phenology and productivity of tissue require data. Several plant ecophysiological and modeling texts, such as Charles-Edwards et al. (1986), provide useful directions for determining plant parameters. Recordings of the calendar dates for green up, anthesis, senescence, and maturity for each functional group and the daily temperature would provide the necessary data to parameterize the phenology aspects of the plant module using the GDD formula:

$$\mathrm{GDD} = \sum_{t=1}^{t=n} \left(T_{\mathrm{av}} - T_{\mathrm{b}} \right) t \qquad [3]$$

where t is time (days), n is the number of days for a given development stage, T_{av} is average daily temperature, and T_b is the base temperature, a threshold for development. The model describes daily growth rate of each forage group i by:

$$\frac{\mathrm{d}W_i}{\mathrm{d}t} = W_i \,\mathrm{RGR}\,\mathrm{EVP}_i \qquad [4]$$

where W_i is the biomass of group i, (g m^{-2} or kg ha^{-1}), RGR is the potential relative growth rate (g g^{-1} d^{-1}), t is time (d), and EVP_i is an environmental fitness factor that combines the temperature and water stress effects on growth (0–1) of class i. The RGR is the major plant growth parameter and can be determined from sequential forage biomass clippings. This can be determined as (South, 1995):

$$\mathrm{RGR} = \frac{\mathrm{Ln}(W_2) - \mathrm{Ln}(W_1)}{t_2 - t_1} \qquad [5]$$

where W_1 and W_2 are any two biomass harvests at times t_1 and t_2, and Ln is the natural logarithm. However, to translate measured RGR to the parameter maximum relative growth rate one would need to assume no grazing, water, or temperature stress, which is an unlikely occurrence in rangelands. Therefore, biomass clippings from grazing protected enclosures across years (so as to include wet climate years) at times before peak standing biomass is reached for each functional group would be most ideal.

Example of Parameterizing the GPFARM-Range for a New Location

As an example of parameterizing the GPRAFRM-Range model for a different location in the Great Plains, we simulated the rangeland studies conducted during sheep grazing at Fort Keogh Livestock and Range Research Laboratory located near Miles City, MT (46°22′ N, 105°5′ W). The site receives an annual rainfall of 340 mm with 60% falling from mid-April to mid-September. The mean daily temperature ranges from −10°C in the winter months to 24°C in the summer months. Soils range from silty clay loam to fine montmorillonitic Ustochrepts. The vegetation is a mixed grass with grama–needlegrass–wheatgrass (*Bouteloua* Lag.–*Stipa* spp.–*Agropyron* Gaertn.) as the dominant species (Küchler, 1964).

Two sets of studies were performed from 1993 to 1996 (Set 1) and from 1998 to 2001 (Set 2). During these studies, forage growth was determined under ambient rainfall (*A*) or imposed short-term drought (*D*) and grazing conditions. Eight dry matter harvests were performed in each year and each treatment. The imposition of drought occurred via the exclusion of rainfall using rainout shelters. In the Set 1 studies, the drought was imposed only in 1994, from late May to mid October. In the Set 2 studies, the drought was imposed in 1998 and 1999, and the periods were from 1 April to 30 June. Grazing, which we described as "flash" due to the high intensity and very short duration of a few hours a day, was implemented by allocating six ewes and their twin lambs to graze paddocks of size 50 m^2, removing more than 75% of the standing biomass. According to Heitschmidt et al. (1999, 2005), the grazing start dates were early June and July each year. For the simulation we set the grazing dates to 2 June and 3 July each year. Control studies in both Set 1 and Set 2 involved no drought and no grazing treatments. Details of this study can be found in Heitschmidt et al. (1999, 2005).

Using this information in addition to the daily weather data for the site during the study years, the GPFARM-Range model was re-parameterized for the Fort Keogh site as follows. First, the soil profile data, which included depth, texture, and bulk density of horizons, were obtained from the USDA-NRCS. The hydraulic parameters required for water balance simulation were estimated from texture and bulk density using the pedotransfer functions provided in the GPFARM-Range model, by setting the "soilPropOpt" = 0. Second, the model was adapted to sheep, which hitherto was not one of the animal types. To do so, the animal parameters were modified by substituting ewes and lambs for cows and calves, and setting the weights of the ewes and lambs to 65 and 25 kg, respectively. Further, the maximum daily dry matter intake by the ewes and lambs were also set to 1.5 and 0.6 kg, respectively (Table 7–4). Third, the forage growth component of the model was re-parameterized by manually adjusting the RGR values for each functional group. The meteorological input variables, except solar radiation, were obtained from the website of the Western Regional Climate Center (Western Regional Climate Center, 2009), for the years 1993 to 1996 and 1998 to 2001. The daily solar radiation was estimated from the daily maximum and minimum temperature and latitude, using algorithms of Hargreaves and Samani (1982).

As shown in Table 7–4, there were major differences between the CPER (the previous site of model parameterization) and Miles City (the new location of interest). First, the soil type at Miles City was more clayey compared with the more sandy loam at the CPER. Second, whereas the WSG dominated at CPER (64%), the reverse was the case at Miles City. Presumably, the species that comprised the WSG and CSG at Miles City were different from those at the much warmer central locations at CPER. Thus, in applying the model to Miles City conditions, the differences between sites must reflect in parameter values.

The results of the studies showed that forage growth under the control conditions (ambient rainfall and non-grazed) followed the rainfall patterns closely (Fig. 7–5). For non-grazed forage growth in the Set 1 studies, the high rainfall in 1993 (Fig. 7–5a) resulted in relatively high forage growth (Fig. 7–5b), whereas declining rainfall in the years 1994 to 1996 resulted in lower growth in those years. In the Set 2 studies when rainfall was comparatively lower than Set 1 (Fig. 7–5c), forage growth was reduced (Fig. 7–5d).

The model could mimic the observed trends of the peak standing crop (PSC), despite the overestimation in 1993 and 1998. The statistical comparison between the simulated and the observed gave an $r^2 = 0.72$, RMSE = 194.5, and a d index = 0.87, with a slight negative bias. Judging from the modeling statistics, it could be concluded that the adapted model satisfactorily simulated the forage growth under the range of conditions considered. Hence, the adjusted parameters could be assumed to be adequate for simulating potential forage growth at the Miles City site.

A further evaluation of the re-parameterized model involved applying it to situations involving drought and grazing. As shown in Fig. 7–6, the model generally captured the forage growth trends under varying drought and grazing conditions. For Set 1 studies, forage growth was largely determined by the ambient rainfall when there was no grazing or imposed drought in 1993 (Fig. 7–6a). In 1994, however, the imposed drought not only decreased the peak standing crop, but also there was a sharp decline of the post-grazing forage biomass. In 1995, the peak forage growth was still low despite the removal of the drought treatment. Forage regrowth after the grazing event was considerably low. Forage growth recovered in 1996 when both the drought and grazing effects were removed. The

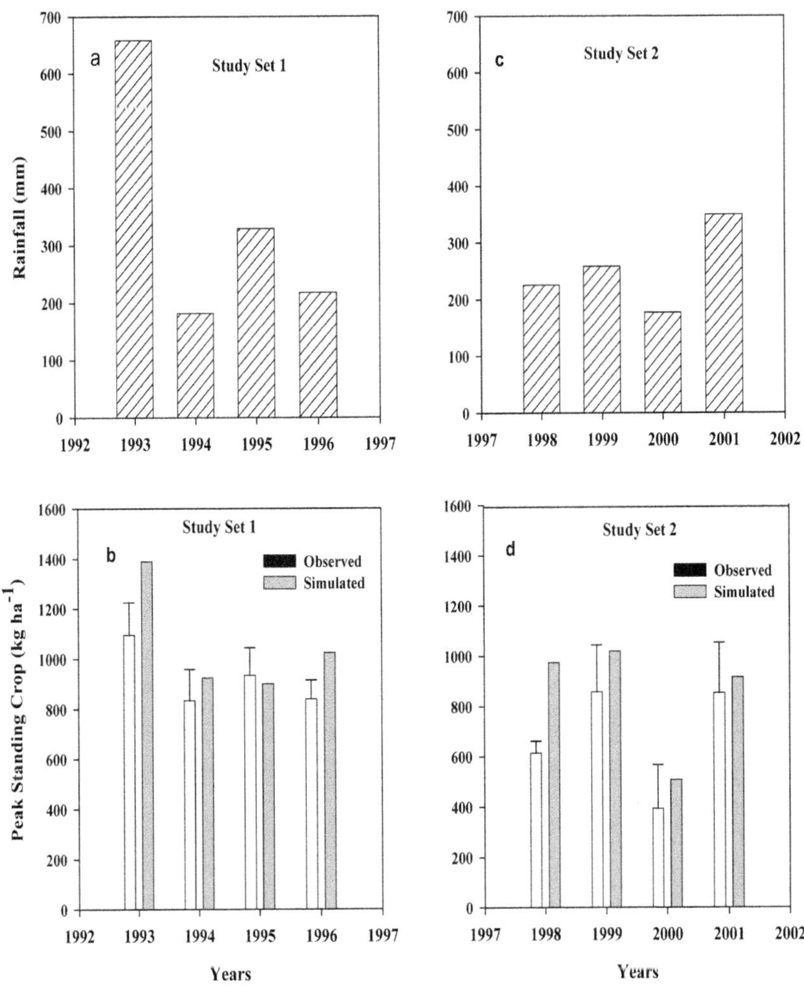

Fig. 7–5. Seasonal (April–September) (a, c) rainfall for the two sets of studies and (b, c) observed and simulated peak standing crop for non-drought and non-grazed conditions at Miles City, MT.

simulated forage growth followed the observed trends satisfactorily except for the overestimated growth under the ambient rainfall and non-grazed conditions in 1993 and 1996.

In the Set 2 studies (Fig. 7–6b), 2 yr of repeated drought phases and flash grazing events in 3 yr resulted in a drastic decline in forage growth during the first 3 yr, with the annual peak growth below 300 kg ha^{-1}. Although growth recovered in 2001 when both drought and grazing treatments were removed, the annual peak growth of 600 kg ha^{-1} was far less than that observed in the same year under ambient rainfall and non-grazed conditions. Thus, full recovery did

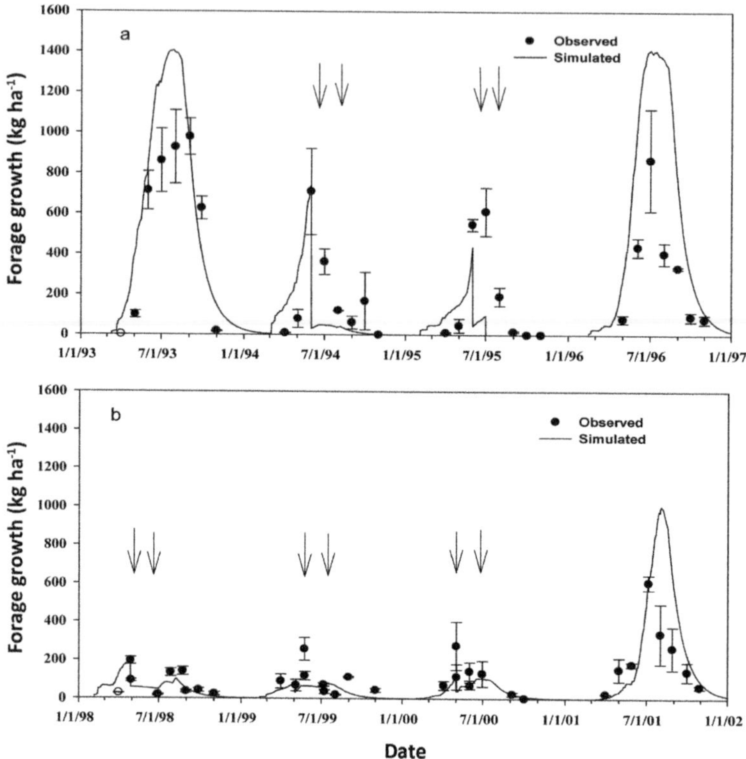

Fig. 7–6. Time course of observed (closed circles) and simulated (lines) forage growth at Miles City, MT. (a) Set 1 includes studies with drought imposed in 1994 and grazing in 1994 and 1995; (b) Set 2 includes studies with drought imposed in 1998 and 1999 with grazing during 1998 through 2000. Arrows indicate approximate grazing times.

not occur after the prolonged drought and grazing effects. The model captured the low growth trends during the first 3 yr (1998–2000) but overestimated the recovery in year 2001.

The comparison of the simulated and observed forage growth for treatments indicated that despite the variability of the observed data, the agreement was satisfactory ($R^2 = 0.68$, RMSE = 145 kg ha^{-1}, and Willmott's $d = 0.91$).

The flexibility of model to adapt it from cattle to sheep grazing systems expands the scope of the application of the GPFARM-Range model to many locations within the Great Plains. Further, the parameterization of the GPFARM-Range model resulted in acceptable model output. Herein lies the strength of models as tools for exploring soil–plant–animal–management interactions for a wide range of locations and time.

Further Improvements of the GPFARM-Range Model

The GPFARM-Range model is constantly improved and upgraded to address the many issues that determine rangeland productivity. In a recent paper, Adiku et al. (2010) proposed a framework for simulating the effect of soil compaction due to animal trampling on forage growth. Another effort is to simulate forage composition change due to grazing and other disturbances. In an ongoing work, efforts are being made to include the sensitivity of the model to global climate change by introducing functions that relate the stomatal conductance and RGR to changing atmospheric CO_2. In relation to this is the question of how much carbon rangelands sequester compared to other land use systems. To address this, efforts are under way to include components in the model that simulate soil carbon and nitrogen dynamics.

Conclusions

This chapter sought to briefly present the GPFARM-Range model and explore some of the practical ways of parameterizing the model for application at different locations and management purposes. Two main methods were proposed: (i) manual adjustment of default parameter values and (ii) direct empirical parameter determination from experimental data. It was recommended that in view of the level of information and expertise required for the second method, users could first adjust default values to minimize the difference between the simulated and observed. However, the published literature is a major source of data that could be used in parameterizing models. In the case where parameter values must be determined empirically, we recommend strict adherence to standard protocol for making measurements. A number of texts were suggested to assist in empirical parameter determination. It was also shown that by varying the values of some few key parameters, the GPFARM-Range model could simulate forage growth at new locations.

References

Adiku, S.G.K., G.H. Dunn, L.R. Ahuja, S. Gunter, J. Bradford, L. Garcia, and A. Andales. 2010. Simulation of forage growth under varying stocking rates. Rangeland Ecol. Manag. 63:546–552.

Ahuja, L.R., D.K. Cassel, R.R. Bruce, and B.B. Barnes. 1989. Evaluation of spatial distribution of hydraulic conductivity using effective porosity data. Soil Sci. 148:404–411.

Andales, A.A., L.R. Ahuja, and G.A. Peterson. 2003. Cropping systems: Evaluation of GPFARM for dryland cropping systems in eastern Colorado. Agron. J. 95:1510–1524.

Andales, A.A., J.D. Derner, L.R. Ahuja, and R.H. Hart. 2006. Strategic and tactical prediction of forage in northern mixed-grass prairie. Rangeland Ecol. Manag. 59:576–584.

Andales, A.A., J.D. Derner, P.N.S. Bartling, L.R. Ahuja, G.H. Dunn, R.H. Hart, and J.D. Hanson. 2005. Evaluation of GPFARM for simulation of forage production and cow-calf weights. Rangeland Ecol. Manag. 58:247–255.

Ascough, J.C., G.S. McMaster, A.A. Andales, N.C. Hansen, and L.A. Sherrod. 2007. Evaluating GPFARM crop growth, soil water and soil nitrogen components for Colorado dryland locations. Trans. ASABE 50:1565–1578.

Beven, K., and A. Binley. 1992. The future of distributed models: Model calibration and uncertainty prediction. Hydrol. Processes 6:279–298.

Blumenthal, D.M., N.R. Jordan, and E.L. Svenson. 2005. Effects of prairie restoration on weed invasions. Agric. Ecosyst. Environ. 107:221–230.

Charles-Edwards, D.A., D. Doley, and G.M. Rimmington. 1986. Modelling plant growth and development. Academic Press, Sydney, Australia.

Dane, J.H., and G.C. Topp (ed.) 2002. Methods of soil analysis. Part 4. Physical methods. SSSA Book Ser. 5. ASA and SSSA, Madison, WI.

Derner, J.D., and R.H. Hart. 2007. Grazing-induced modifications to peak standing crop in Northern mixed grass prairie. Rangeland Ecol. Manag. 60:270–276.

Elrick, D.E., and W.D. Reynolds. 1992. Infiltration from constant-head well permeameters and infiltrometers. p. 1–24. *In* G.C. Topp et al. (ed.) Advances in measurement of soil physical properties: Bringing theory into practice. SSSA Spec. Publ. 30. SSSA, Madison, WI.

Ganguli, A.C., L.T. Verneire, R.B. Mitchell, and M.C. Wallace. 2000. Comparison of four non-destructive techniques for estimating standing crop in shortgrass plains. Agron. J. 92:1211–1215.

Hanson, J.D., B.B. Baker, and R.M. Bourbon. 1992. SPUR II model description and user guide: GPSR Tech. Rep. 1. USDA-ARS, Great Plains Systems Research Unit, Ft. Collins, CO.

Hanson, J.D., J.W. Skiles, and W.J. Parton. 1987. Plant component. *In* J. Wight and J.W. Skiles (ed.) SPUR: Simulation of Production and Utilization of Rangelands. Documentation and user guide. USDA-ARS 63.

Hargreaves, G.H., and Z.A. Samani. 1982. Estimating potential evapotranspiration. J. Irrig. Drain. Eng. 108:223–230.

He, J., J.W. Jones, W.D. Graham, and M.D. Dukes. 2010. Influence of likelihood function choice for estimating crop model parameters using generalized likelihood uncertainty method. Agric. Syst. 103:256–264.

Heitschmidt, R.K., M.R. Haferkamp, M.G. Karl, and A.L. Hild. 1999. Drought and grazing: 1. Effects on quantity of forage produced. J. Range Manage. 52:440–446.

Heitschmidt, R.K., K.D. Klement, and M.R. Haferkamp. 2005. Interactive effects of drought and grazing on Northern Great Plains rangelands. Rangeland Ecol. Manag. 58:11–19.

Klute, A. (ed.) 1986. Methods of soil analysis. Part 1. Physical and mineralogical methods. SSSA Book Ser. 5. ASA and SSSA, Madison, WI.

Küchler, A. 1964. Potential natural vegetation of the conterminous United States. AGS Spec. Publ. 36. American Geographical Society, New York.

National Resource Models in the Rangelands. 2004. A Review Undertaken for the National Land and Water Resources Audit. Available at http://lwa.gov.au/files/products/national-land-and-water-resources-audit/er040822/er040822.pdf (verified 4 Apr. 2011). CSIRO, QLD, Australia.

Shaffer, M.J., P.N.S. Bartling, and J.C. Ascough II. 2000. Object-oriented simulation of integrated whole farms: GPFARM framework. Comput. Electron. Agric. 28:29–49.

South, D.B. 1995. Relative growth rates: A critique. South African For. J. 173:43–48.

Walker, J.W. 1993. Nutritional models for grazing animals. Icel. Agric. Sci. 7:45–57.

Western Regional Climate Center. 2009. Miles City FAA Airport, Montana (245690). Available at http://www.wrcc.dri.edu/cgi-bin/cliMAIN.pl?mt5690 (verified 4 Apr. 2011).

Williams, R.D., and L.R. Ahuja. 2003. Scaling and estimating the soil water characteristic using a one-parameter model. p. 35–48. *In* Y. Pahepsky et al. (ed.) Scaling methods in soil physics. CRC Press, Boca Raton, FL.

Willmott, C.J. 1981. On the validation of models. Phys. Geogr. 2:184–194.

Wu, L., L. Pan, J. Mitchell, and B. Sanden. 1999. Measuring saturated hydraulic conductivity using generalized solution for single-ring infiltrometers. Soil Sci. Soc. Am. J. 63:788–792.

Chapter 8

Spatial Relationships of Soil Properties, Crop Indices, and Nitrogen Application Pattern with Wheat Growth and Yield in a Field

Ole Wendroth
K. Christian Kersebaum
G. Schwab
L. Murdock

Abstract

Inherent spatial variability of agricultural fields causes spatial differences in crop growth and demand of resources. Efficient use of input and minimizing the risk of environmental hazard and economical losses require appropriate concepts and tools on which to base site-specific management decisions. The objective of this study was to identify how helpful site-specific soil and crop field sampling is for explaining crop growth and yield variability, how sensitive remotely sensed crop reflectance indices are to quantify site-specific crop N demand, and how sensitive an uncalibrated crop growth simulation model (DSSAT) is to site-specific soil input. A 27-m-wide and 645-m-long strip in a farmer's field in Kentucky was planted with winter wheat. Nitrogen fertilizer was applied in 43 cells each 27 m wide and 15 m long in a sinusoidal spatial pattern. Mineral soil N was sampled site-specifically as well as aboveground plant biomass. Optical reflectance indices were monitored with two active sensors, the GreenSeeker and the Yara-ALS. Crop indices derived from sensor measurements taken in the spring reflected the grain yield patterns very well. Thus, these integrated state variables are good choices to predict site-specific yield variability and as management decision aids. Cell-specific input was given to the crop growth simulation model DSSAT 4.0. Measured state variables were compared with simulated ones. On an absolute basis, model results deviated substantially from measurements. On a relative basis, among all simulation scenarios applied here, simulated grain yield based on average soil input resulted in the highest correlation with measured grain yield. Further research with an improved calibrated, validated crop model is needed to explore the advantages of site-specific inputs for modeling.

O. Wendroth (owendroth@uky.edu) and G. Schwab, Univ. of Kentucky, Dep. of Plant and Soil Sciences, Lexington, KY 40546; K.C. Kersebaum, Leibniz Centre for Agricultural Landscape Research (ZALF), Institute for Landscape System Analysis, Müncheberg, Germany; L. Murdock, Univ. of Kentucky, Dep. of Plant and Soil Sciences, Research and Education Center at Princeton, Princeton, KY 42445.

doi:10.2134/advagricsystmodel2.c8

Methods of introducing system models into agricultural research. L.R. Ahuja and L. Ma (ed.)
Advances in Agricultural Systems Modeling Series 2.

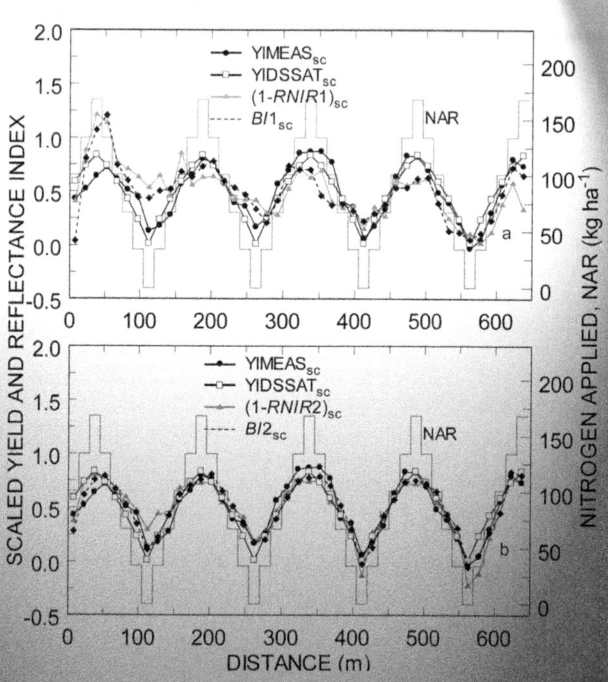

YIMEAS$_{sc}$
YIDSSAT$_{sc}$
(1-*RNIR*1)$_{sc}$
*BI*1$_{sc}$
NAR
a
YIMEAS$_{sc}$
YIDSSAT$_{sc}$
(1-*RNIR*2)$_{sc}$
*BI*2$_{sc}$
NAR
b
SCALED YIELD AND REFLECTANCE INDEX
NITROGEN APPLIED, NAR (kg ha^{-1})
DISTANCE (m)

Field soils exhibit spatial variability in their properties due to soil genesis, surface topography, land use history, and previous and current management (Trangmar et al., 1985; Pennock and Frick, 2001). These different factors cause characteristic spatial variation of soil water and nutrient status, crop performance, and yield at different scales. Moreover, the spatial variation of many important soil and vegetative properties is strongly affected by temporal dynamics of relevant state variables, such as soil water content and nutrient status, for example, N (Kersebaum, 1995; Wendroth et al., 2001). The interaction between temporal dynamics of soil water and its spatial variability results in a pronounced temporal change in the range of its spatial representativity (Greminger et al., 1985; Wendroth et al., 1999). Hence, at different times of the year with different soil water status, different characteristic spatial patterns of soil water determine crop growth patterns across the field. Depending on whether it is a rather dry or wet year, different underlying characteristics affect the spatial variability of crop production and grain yield (Timlin et al., 1998).

To adequately understand and describe processes in heterogeneous areas, effective average parameters need to be identified, as well as their necessary observation density. For example, Ahuja et al. (2010) compared different sets of effective average parameters in terms of their capacity to describe field average water infiltration and redistribution. In heterogeneous field soils grown with crops, relevant questions are: What are the important indicative state variables that support the site-specific description of crop growth processes and the final crop yield in a field? How can these indicators efficiently be sampled to make management decisions that help counteract unfavorable growing conditions or avoid unintended environmental hazard (Scharf et al., 2005; Hong et al., 2007) and economical damages caused by inappropriate N fertilizer application rates?

Two important aspects of these questions are (i) how representative, reliable, relevant, and supportive are field measurements for predictions in agricultural system models on the one hand and (ii) how sensitive is a modeling approach to site-specific soil and management input parameters for predicting crop growth and underlying soil processes and final yield development on the other hand? In other words, how sensitive is an existing crop growth model to site-specific input, and what site-specific input is worth sampling and at what spatial and temporal resolution?

Crop growth is an extremely complex process. Under field conditions, spatial differences of crop growth can be caused by locally specific soil textural composition, microclimatic conditions, soil moisture availability, or aeration stress at particular locations and over limited time periods and many other reasons. It is impossible to isolate the temporarily growth limiting factors and to come up with a comprehensive scheme for which variable to measure at which soil depth, at what time, and at what spatial and temporal resolution.

Crop growth models have been shown to capture field processes, including important soil water and nutrient dynamics, plant water and nutrient uptake, biomass development, and final crop yield in a more or less satisfying accuracy if they are fed with average soil property information representing an area (e.g., Kersebaum, 1989; Mirschel and Wenkel, 2007). Therefore, computer simulation models are one way to approach the complexity of crop growth processes by running scenarios with field-average or more or less site-specifically resolved input information.

As an alternative, indicative state variables can be monitored that integrate many relevant processes. If early in the growing period of a crop, local nutrient deficiency appears, it may persist throughout crop development and until harvest. Therefore, an indicator variable that reflects early nutrient deficiency would be useful for management decisions. Hatfield et al. (2008) reviewed various opportunities for applying canopy reflectance measurements to calculate several vegetative indices showing local variation of leaf area, crop cover, biomass, and nutrient status. Raun et al. (2001) used red and near-infrared reflectance measurements obtained early in the growing season to calculate the normalized difference vegetation index (NDVI) and applied it in an algorithm to predict potential grain yield of winter wheat (*Triticum aestivum* L.). Later, the N use efficiency of winter wheat could be increased through reflectance measurements at a very high spatial resolution (Raun et al., 2002). Several studies have shown the applicability of in-season crop sensing for detecting crop N status (e.g., Chua et al., 2003; Bronson et al., 2003, 2005; Kitchen et al., 2010) or to derive N requirements of the crop (e.g., Sripada et al., 2007, 2008; Scharf and Lory, 2009; Roberts et al., 2010). Zarco-Tejada et al. (2005) applied airborne narrow-band hyperspectral indices measured in a cotton (*Gossypium hirsutum* L.) field at high temporal and spatial resolution to diagnose crop development and yield variability, and Ritchie et al. (2010) identified simple indices to estimate crop growth during the growing season.

Agronomic experiments comparing the impact of different treatment levels on crop yield or another relevant outcome are usually designed as a randomized plot or block experiment. Most agricultural field soils exhibit an inherent spatial variability caused by the factors mentioned above. It would become a complicated task to separate treatment effects from spatial soil variation in a randomized design. Therefore, in this study, we follow the example of Robert L. Fox, University of Hawaii (cited in Nielsen and Wendroth, 2003), who was, to our knowledge, the first scientist who changed the application rate continuously across the landscape to capture effects of gradual soil variation. Later, Bazza et al. (1988) followed this scheme in their work on the relationship between surface soil temperature and salinity level of previous irrigation water. Recently, Shillito et al. (2009) adopted this approach when applying cyclically varying amounts N in a potato (*Solanum tuberosum* L.) field to detect spatially changing response of potato to N fertilization rate. The study by Shillito et al. (2009), the present attempts in this direction, and future efforts will have to show what investigators and farmers can understand from experimental designs in which treatments are imposed on top of gradual soil changes rather than dealing with variability as a random and aggravating or disturbing obstacle in the study. As described in Wendroth et al. (2011a), conceptually relying on Bazza et al. (1988), in this study, N fertilizer is applied in a sinusoidal pattern along a transect in a farmer's field to test the site-specific performance of the crop growth model DSSAT for winter wheat and to evaluate spectral reflectance measurements.

The objectives of this study were (i) to find out how helpful site-specific soil and crop measurements collected during the growing period are to explain spatial crop growth variability, (ii) to identify the sensitivity of spectral reflectance indices for identifying N deficiency, biomass status, and their relationship to grain yield, and (iii) to determine the sensitivity of the crop growth model DSSAT to site-specific model input in different input scenarios for describing site-specific crop growth, soil processes, and final grain yield. From the results of this study, a set of relevant and helpful measurements and an efficient experimental scheme was identified, as well as obvious strengths and deficiencies in both data collection strategy and modeling approach.

Materials and Methods

The experiment was conducted in a farmer's field in Princeton, Caldwell County, Kentucky (lat: 37.030909, long: −87.858079). The soil is classified as a Crider (fine-silty, mixed, active, mesic typic Paleudalfs) silt loam, is high in natural fertility, and is very high in moisture-supplying capacity (United States Department of Agriculture, 1966). The field is managed in a corn (*Zea mays* L.)–wheat–soybean [*Glycine max* (L.) Merr.] (double crop) rotation, and the soil is kept under no-tillage. The investigation was performed between October 2008 and June 2009. In the field, a strip 27 m wide by 645 m long was laid out as a transect. The entire strip was divided into 43 cells, each 27 m wide by 15 m long. In this study, a cell is the smallest considered spatial unit in the analysis. Notice, that the support size differs for the variables investigated, due to their specific instruments and approaches to monitor the individual variables.

On the day before planting of winter wheat, 30 kg N ha^{-1} and 77 kg P ha^{-1} were applied as diammonium phosphate fertilizer, in addition to 4.5 t ha barnyard manure with 2% N. On 3 Oct. 2008, winter wheat (cv. R35) was planted at a rate of 135 kg ha^{-1}. Two days before planting, soil samples were taken at the center of each cell with four Puerckhauer augers down to 90 cm. Augers were driven in with a jack hammer and lifted with a jack lift. The entire core in each auger was divided in four depth increments: 0 to 15, 15 to 30, 30 to 60, and 60 to 90 cm, and the respective compartment from each of the four replicate augers mixed to one bulk sample per depth increment. Two further campaigns for mineral soil N were on 22 January and 24 April 2009. Soil samples were frozen upon sampling and later analyzed for soil texture with sieving and the pipette method (Gee and Bauder, 1986), organic C content (Skjemstad and Baldock, 2008), and for mineral N content (N_{min}), i.e., ammonium concentration (Weatherburn, 1967) and nitrate concentration (Crutchfield and Grove, 2011). Moreover, a subsample from each of the depth compartments was separated for gravimetric soil water content determination.

In spring of 2009, mineral N fertilizer at several different levels was applied at two times. The first application occurred on 20 Feb. 2009 and the second on 27 Mar. 2009. On both of these days, half of the total amount of fertilizer N was applied. Applications occurred as urea ammonium nitrate (UAN) in liquid form. Six levels of N were in total: 0, 34, 67, 101, 134, and 168 kg ha^{-1}. These levels were distributed in a spatially continuous sinusoidal pattern, beginning with 101 kg ha^{-1} in Plot 1, then stepwise increasing to 168 kg ha^{-1} in Plot 3, and then decreasing, and again increasing resulting in a total of slightly more than four sinus waves (Fig. 8–1). With the sprayer boom being 9 m wide, the 27-m-wide experimental strip was covered in three parallel paths.

Two active vehicle-borne crop sensor systems were used to scan the wheat crop along the transect. One system was the so-called GreenSeeker (NTech Industries), which emits red and near-infrared light to the leaf surface, activates the chlorophyll, and registers the remitted signal. This system has been used often for quantifying the normalized difference vegetation index (NDVI). In this study, the red/near-infrared ratio (RNIR) was derived according to Scharf (2006)

$$\mathrm{RNIR} = \frac{\rho_{670}}{\rho_{780}} \qquad [1]$$

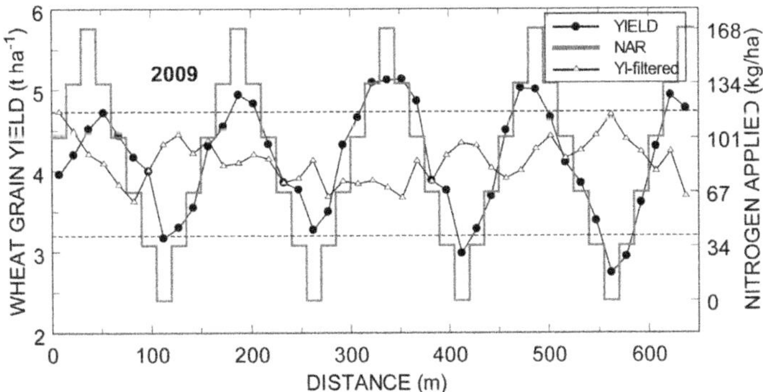

Fig. 8–1. Sinusoidal spatial distribution of mineral N fertilizer (NAR) applied in spring in two splits along a transect in a farmer's field in western Kentucky. Wheat grain yield and filtered grain yield (YI-filtered, see text) are shown. The levels of maximum and minimum yields obtained in the first sinewave are indicated with the dashed line for illustrating the spatial trend of the yield series.

with ρ_{670} and ρ_{780} denoting the red and near-infrared reflectance from the canopy, respectively. This index is expected to represent N deficiency of the crop. The vehicle used for the scanning was equipped with a 9-m-wide boom, carrying five green seeker sensor heads, i.e., one in the center, two 2.47 m, and two 4.22 m left and right from the center, respectively. The other system is an ALS (YARA) device that actively emits a light beam to the canopy surface and measures the reflection, resulting in a biomass index. It is derived from measurements at two wavelength bands, i.e., 730 and 760 nm (Reusch, 2005). This sensor measures in two 0.5-m-wide strips approximately 3.5 m left and right of the center of the machine. The result is a so-called biomass index *BI*, which is calculated as (Reusch, 2005)

$$BI = \ln(\rho_{760}) - \ln(\rho_{730}) \quad [2]$$

Both sensors were equipped with a differential global positioning system (DGPS). The 27-m-wide strip in the farmer's field was covered in the same three paths in which the N fertilizer was applied. Sensor data obtained with both the GreenSeeker and the ALS were aggregated using a self-written FORTRAN computer code over a rectangular area 20 m wide by 8 m long, based on the center northing and easting coordinates of each of the 43 cells.

Aboveground biomass (ABM) was harvested by hand twice on 24 April and 19 May 2009. A locally representative plant row was picked that was six or seven rows away from the northern traffic path and harvested for a length of 50 cm. Plants were cut 2 cm above the ground surface. Harvested plant material was dried at 60°C for 6 d, weighed, and processed for chemical analysis. Grain yield was determined with an automatic yield monitoring system (AgLeader) mounted to a Massey Ferguson combine harvester, type MF 8, equipped with a 2.0-m-wide cutting head and a DGPS. For calibration purposes, harvested grains were collected in a weigh wagon for each harvested path, and later the entire set of harvested yield data was calibrated. Machine harvested grain yield data were aggregated in the same way as described above for the remotely sensed indices.

Winter wheat growth and underlying soil processes were simulated using uncalibrated DSSAT (Decision Support System for Agrotechnology Transfer) version 4.0 (Hoogenboom et al., 2003). The DSSAT cropping system model captures growth and development of 16 different crops, while details on crop management (fertilization, pesticide application), layered soil profile properties, and daily weather data are provided to the computer algorithm (Jones et al., 2003). This computer simulation model has been widely applied on an international level for simulating the effects of management, for prediction of wheat yield (e.g., Bannayan et al., 2003) and in connection with the Root-Zone-Water-Quality-Model (RZWQM) for quantifying soil water processes and nitrate leaching (e.g., Fang et al., 2008). Since this study was focused on relative model behavior, and cultivar-specific parameters for the planted variety (R35) were not available, the variety Newton was used from the model database for simulations.

In the simulations for this study, each of the 43 soil profiles representing the cells was divided into four layers according to measurement depths. Soil textural composition was provided for each of the 43 cells, as well as initial values of mineral soil N concentration and soil water content. This soil textural information is intended to provide the basis for site-specifically calculated upper and lower drainage limit, saturated soil water content, and hydraulic conductivity (Jones et al., 2003). The soil module in DSSAT is configured to derive these parameters from a class pedotransfer function based on Saxton et al. (1986); i.e., for the 43 soil profiles in this study, the parameters would only vary across the four depths but not spatially among the 43 soil profiles all classified as silt loam soil type. In this configuration of DSSAT, the only site-specific input parameter varying for all 43 locations is the initial mineral soil N concentration. Individual calculation of upper and lower drainage limits outside of DSSAT based on Saxton et al. (1986) did not in all cases yield results that were consistent with saturated soil water content. Therefore, a continuous pedotransfer function of Wösten (1997) was used for calculating upper and lower drainage limits, saturated soil water content and hydraulic conductivity. In this case, the physical parameters differed for all 43 locations and across the four layers. In addition to these two pedotransfer function approaches, DSSAT was run for the six fertilization levels, while only average soil input was given, that is, only one profile with field average properties,

Table 8–1. Four model runs of this study. Site specific soil texture and initial soil mineral N (N_{min}) information is presented in Table 8–2.

DSSAT run	Soil texture	Initial N_{min}	PTF
CLASS-PTF-SSN	average profile, 4 layers	43 profiles, 4 layers	Saxton et al. (1986)
CLASS-PTF-AVG	average profile, 4 layers	average profile, 4 layers	Saxton et al. (1986)
CONT-PTF-SS	43 profiles, 4 layers	43 profiles, 4 layers	Wösten (1997)
CONT-PTF-AVG	average profile, 4 layers	average profile, 4 layers	Wösten (1997)

as well as field-average initial mineral soil N concentration. The different model runs with their respective source of information are listed in Table 8–1.

Comparison of model performance and remotely sensed crop indices with measured grain yield was based on ordinary correlation and on observed series scaled with respect to their mean and standard deviation by (Nielsen and Wendroth, 2003)

$$x_{i,\mathrm{sc}} = \frac{x_i - \left(\overline{x} - 2\sigma_x\right)}{4\sigma_x} \qquad [3]$$

where $x_{i,sc}$ is the scaled version of the variable x at location i, $\overline{x}$ and σ_x are the mean and the standard deviation of x_i, respectively. Through this simple transformation, each dataset's mean and standard deviation become 0.5 and 0.25, respectively. This way, only the relative magnitude of observations contributed to the comparison, and limitations due to lack of model and measurement calibration did not affect the comparisons. In an additional step before the transformation of RNIR data, their individual values are subtracted from 1. This way, all variables considered here are positively related, which facilitates the use of the sum of squared deviations as a unique quality criterion and visual comparison of the variables' behavior relative to each other.

Daily weather data collected from a nearby weather station consisting of precipitation, relative humidity, air temperature, global radiation, and wind speed were compiled for the simulations.

With the N fertilizer being applied in a sinusoidal spatial pattern, frequency-domain based statistical analysis is the predestinated tool, such as spectral and cross-spectral analysis for identifying spatial relationships and explaining variability patterns (Davis, 1986; Shumway, 1988; Bazza et al., 1988). The power spectrum integrates the autocorrelation function and is defined as (Shumway, 1988)

$$S(f) = 2\int_0^\infty r(h)\cos(2\pi fh)\,\mathrm{d}h \qquad [4]$$

where $S(f)$ is the variance spectrum as a function of frequency, and $r(h)$ is the autocorrelation behavior with lag distance h. The power spectrum depicts the intensity of variance sources repeating each other at a frequency f, which is the inverse of the wavelength.

In general, the variation of two series of observations may be caused by periodic variance sources that have the same frequency components. In this study, series of yield, crop sensor, clay and N content observations are analyzed for their common cyclic variance components. The tool to quantify common frequencies between the two series is the squared coherency spectrum defined as (Shumway, 1988)

$$\mathrm{Co}(f) = \frac{\mathrm{Co}^2(f) + Q^2(f)}{S_\mathrm{A}(f)\,S_\mathrm{B}(f)} \qquad [5]$$

with Co(f) and $Q(f)$ being the co- and quadspectra, respectively, and the $S_A(f)$ and $S_B(f)$ the individual power spectra of two series A and B. At each frequency, where the two series of consideration show common variance behavior, the squared coherency is close to 1, similar to the coefficient of determination in regression analysis. Small-scale behavior is depicted at high frequencies, that is, in the right-hand side of the spectrum; large-scale behavior is depicted at low frequencies.

Results and Discussion

Soil Measurements

Representing the soil input to the model (Table 8–2), two sets of the site-specific model inputs are illustrated in Fig. 8–2, soil clay content in the upper 30 cm layer and mineral soil N content measured across the 90-cm soil profile for each of the 43 cell center points. Notice that clay content and silt content are not only given for the 0- to 30-cm layer but specifically for the four depths of measurements. Moreover, mineral N is provided by depth as well. Clay content starts at about 20 g (100 g)$^{-1}$ for the first 450 m of the transect and then exhibits an increasing trend for approximately 100 m before it decreases again toward the end of the transect. Clay content increases in a zone of decreasing elevation. Probably, coarser textured soil material has been eroded, so that the underlying clay horizon appears at shallower soil depths. Mineral soil N levels in the first 150 m of the transect are slightly higher than in the remaining part. Only in the middle part (150 m) of the transect, does mineral soil N exhibit a relatively smooth spatial process, whereas

Table 8–2. Site- and layer-specific model input with silt and clay content, and initial mineral soil nitrogen concentration.

Profile	Depth	Silt	Clay	NH_4–N	NO_3–N	Profile	Depth	Silt	Clay	NH_4–N	NO_3–N
	cm	%		ppm			cm	%		ppm	
1	0–15	68	28	1.3	9.1	12	0–15	79	18	1.3	9.0
	15–30	63	30	0.3	1.7		15–30	84	14	0.6	2.0
	30–60	52	40	0.8	2.2		30–60	88	9	1.0	0.5
	60–90	38	53	0.6	1.6		60–90	78	20	0.4	0.2
2	0–15	78	18	2.2	14.0	13	0–15	80	17	1.6	4.3
	15–30	75	21	0.4	5.5		15–30	79	18	0.7	0.8
	30–60	57	36	0.4	4.6		30–60	69	28	0.4	0.3
	60–90	39	55	0.6	2.6		60–90	65	32	0.5	0.3
3	0–15	71	24	3.5	9.2	14	0–15	79	17	2.6	5.5
	15–30	63	32	0.5	3.6		15–30	73	24	1.0	0.8
	30–60	57	37	0.5	1.4		30–60	63	34	0.9	0.5
	60–90	49	43	0.3	0.6		60–90	65	29	0.5	0.5
4	0–15	78	20	1.4	8.2	15	0–15	78	19	5.1	6.6
	15–30	74	23	0.8	2.5		15–30	68	30	0.7	0.5
	30–60	61	34	0.5	0.8		30–60	67	31	1.1	0.4
	60–90	59	34	0.3	0.6		60–90	68	28	0.9	0.5
5	0–15	76	22	2.5	8.7	16	0–15	78	19	6.8	8.3
	15–30	73	26	0.9	2.8		15–30	69	28	1.1	0.9
	30–60	66	33	0.4	0.6		30–60	67	31	1.8	0.3
	60–90	67	29	0.9	1.1		60–90	68	28	0.7	0.6
6	0–15	78	20	2.5	15.0	17	0–15	75	23	2.6	4.1
	15–30	69	30	0.6	1.3		15–30	72	27	0.8	0.8
	30–60	66	32	0.5	0.7		30–60	68	30	1.6	0.5
	60–90	65	32	0.3	0.6		60–90	70	27	0.4	0.6
7	0–15	79	19	5.3	5.1	18	0–15	80	18	1.6	6.7
	15–30	74	25	0.6	0.6		15–30	75	24	1.2	1.2
	30–60	69	29	0.2	0.4		30–60	70	29	0.7	0.8
	60–90	67	30	0.1	0.5		60–90	70	28	0.4	0.3
8	0–15	77	21	4.0	9.6	19	0–15	81	16	1.5	6.7
	15–30	70	29	0.8	1.5		15–30	78	20	1.0	0.7
	30–60	66	32	0.5	1.0		30–60	68	31	1.0	0.4
	60–90	66	31	0.5	0.6		60–90	70	29	0.2	0.2
9	0–15	78	19	1.0	6.7	20	0–15	82	16	2.0	7.3
	15–30	74	23	0.5	1.3		15–30	78	20	0.8	0.7
	30–60	73	26	0.5	0.4		30–60	71	28	0.3	0.2
	60–90	69	28	0.3	0.6		60–90	69	29	1.5	0.5
10	0–15	78	19	0.6	6.5	21	0–15	81	17	2.0	6.8
	15–30	77	22	0.7	1.7		15–30	74	24	1.2	1.1
	30–60	68	30	0.9	0.5		30–60	68	31	0.4	0.5
	60–90	66	30	0.9	0.3		60–90	68	30	1.0	0.3
11	0–15	78	19	3.2	9.8	22	0–15	81	15	1.4	6.2
	15–30	81	16	0.6	1.9		15–30	75	24	1.3	1.2
	30–60	83	14	0.6	0.6		30–60	68	30	1.1	0.4
	60–90	68	29	1.0	0.3		60–90	70	28	0.8	0.4

Continued.

Table 8–2. Continued.

Profile	Depth	Silt	Clay	NH_4–N	NO_3–N	Profile	Depth	Silt	Clay	NH_4–N	NO_3–N
	cm	——% ——		——ppm ——			cm	——% ——		——ppm ——	
23	0–15	80	17	5.1	4.4	34	0–15	77	18	2.6	5.1
	15–30	74	23	0.9	0.8		15–30	77	20	1.0	1.1
	30–60	68	30	0.9	0.4		30–60	80	16	0.5	0.3
	60–90	69	30	1.1	0.2		60–90	76	19	0.1	0.3
24	0–15	81	15	1.4	5.8	35	0–15	77	18	1.6	6.4
	15–30	75	23	0.8	0.8		15–30	80	16	1.1	1.1
	30–60	69	30	1.5	0.6		30–60	77	20	0.7	0.5
	60–90	70	28	0.7	0.4		60–90	80	17	0.5	0.3
25	0–15	81	15	2.0	3.4	36	0–15	59	35	2.7	2.8
	15–30	74	24	1.0	0.6		15–30	55	38	1.1	1.5
	30–60	69	30	1.5	0.5		30–60	52	42	1.3	2.2
	60–90	69	29	1.1	0.7		60–90	52	41	0.9	4.3
26	0–15	82	15	1.7	3.9	37	0–15	64	32	1.5	4.0
	15–30	79	20	1.0	1.2		15–30	61	34	0.6	0.8
	30–60	70	29	0.9	0.6		30–60	57	37	0.0	0.5
	60–90	69	29	1.2	0.4		60–90	57	35	0.0	1.9
27	0–15	81	16	2.8	6.3	38	0–15	75	19	1.3	8.5
	15–30	77	22	1.8	1.8		15–30	79	19	0.7	1.2
	30–60	70	28	0.8	0.4		30–60	71	27	0.4	0.8
	60–90	70	28	1.2	0.5		60–90	64	32	0.6	0.4
28	0–15	87	9	1.4	5.5	39	0–15	74	23	1.3	4.0
	15–30	80	17	4.8	0.9		15–30	69	30	0.6	0.5
	30–60	75	23	0.0	0.3		30–60	66	32	1.2	0.5
	60–90	69	30	0.8	0.2		60–90	68	27	0.5	1.6
29	0–15	81	16	7.0	6.7	40	0–15	70	27	0.5	3.6
	15–30	84	14	0.8	0.8		15–30	68	30	1.2	0.8
	30–60	74	25	0.3	0.3		30–60	68	29	0.7	0.4
	60–90	72	27	0.3	0.4		60–90	67	29	0.6	1.9
30	0–15	80	18	1.6	1.5	41	0–15	70	27	1.6	5.4
	15–30	76	23	1.0	1.0		15–30	65	31	0.6	2.2
	30–60	71	27	1.3	0.5		30–60	60	34	0.7	0.5
	60–90	70	29	0.8	0.9		60–90	58	34	0.5	1.2
31	0–15	78	14	2.8	6.4	42	0–15	85	11	2.2	8.1
	15–30	83	14	1.4	1.1		15–30	69	28	0.8	1.6
	30–60	68	31	1.3	0.5		30–60	61	35	0.8	0.6
	60–90	68	30	0.9	0.3		60–90	56	37	0.4	1.0
32	0–15	80	17	10.2	3.7	43	0–15	74	20	0.7	8.6
	15–30	72	27	0.8	0.9		15–30	79	16	0.2	1.7
	30–60	68	31	1.4	0.5		30–60	76	22	0.7	0.6
	60–90	68	30	1.0	1.0		60–90	70	28	0.1	0.8
33	0–15	73	23	0.7	3.4						
	15–30	63	35	0.4	0.7						
	30–60	61	35	0.9	0.5						
	60–90	58	37	0.7	1.6						

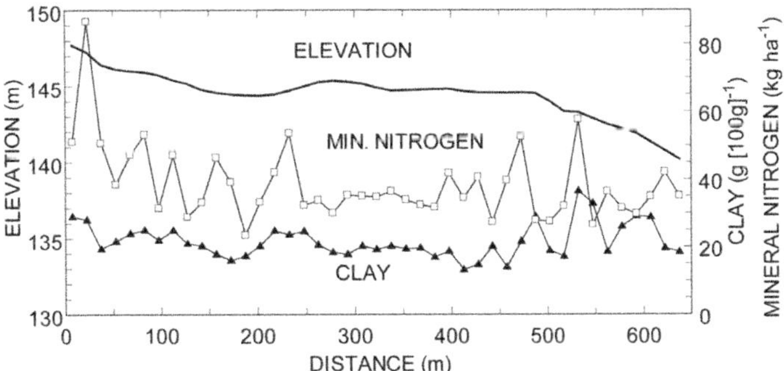

Fig. 8–2. Surface topography depicted by elevation, soil clay content in the upper 0- to 30-cm depth of the soil profile and profile mineral soil N content (N_{min}) at the 0- to 90-cm depth along the transect in a farmer's field in western Kentucky.

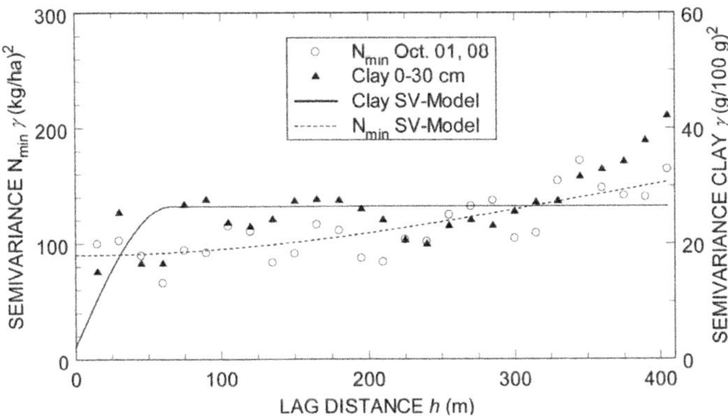

Fig. 8–3. Experimental semivariograms for the two data sets of Clay content (0–30 cm depth), and profile mineral soil nitrogen content (N_{min}) depicted in Fig. 8–2, and their semivariogram (SV) models.

in the first 250 m and the last 225 m its variation is more pronounced, and N_{min} fluctuates strongly from cell to cell. This behavior of mineral soil N variation is common (Wendroth et al., 2001; Giebel et al., 2006). For the first 400 m of the transect, the change in mineral soil N tends to coincide with surface topography, quantified by elevation data (Fig. 8–2).

The corresponding semivariograms (Isaaks and Srivastava, 1989) for both variables (clay content and mineral soil N) are presented in Fig. 8–3. Clay content exhibits a spatial range of influence of at least 30 m. The structured variation would not have been captured appropriately if sampling distances had been chosen further apart. The resolution of textural measurements should not be coarser. On the other hand, three of the four experimental semivariogram values for distances of up to 75 m are obviously smaller than the sill variance level. For long lag distances, a slight spatial trend of soil clay content is indicated by increasing experimental semivariogram values. Such a trend behavior is even more pronounced for the initial mineral soil N content. At a small scale, N_{min} varies strongly as indicated by the noisy experimental semivariogram behavior at short lag distances. The overall decreasing trend of N_{min} at 0 to 200 m (Fig. 8–2) affects the shape of the semivariogram obviously and is manifested by the increasing semivariance-lag distance relationship (Fig. 8–3). No small-scale variability structure can be identified through sampling at 15-m distances. In general, the spatial range of representativity for N_{min} can vary extremely within the same field, depending on the time of sampling, on the crop grown, and even on the scale of the sampling scheme (Giebel et al., 2006). Certainly, this complicated spatial variability behavior is partly due to the tremendous temporal dynamics of N_{min} caused by crop uptake, transformations in the soil, losses due to leaching, and gaseous emissions. Moreover, there exists a discrepancy between the scale at which we manage N—how we apply mineral nitrogen fertilizer—and the scale at which small-scale inhomogeneities in N transformations dominate the pattern of N_{min}, caused by, for example, inhomogeneous plant residue distribution during harvest. Whether or not a spatial variability structure for mineral soil N can be identified, depends, among other factors, on the time of the year and the level of N currently in the soil at the time of sampling (Springob et al., 1985; Stenger et al., 1998; Wendroth et al., 2001; Giebel et al., 2006).

Soil clay content being spatially correlated over relatively short distances can be considered an important variable for explaining other nonrandom processes. Questions relevant for the experimentalist and the modeler are: What other state variables coincide with the gradual changes of soil clay content? Is this gradual change of clay content, given to the model as input information, reflected in any of the model outputs? The spatial process of mineral soil N behaves different from soil clay content. As a result derived from semivariogram analysis, the change of mineral N along the transect follows a trend across the first 250 m; however, large point-to-point fluctuations cause a large nugget variance. Therefore, we conclude that with the sampling design and scale chosen, the small-scale spatial process of N_{min} may not precisely be delineated from cell to cell. Its behavior across several cells is rather relevant as an input to the model, and as a spatial covariate to measured and simulated grain yield, as well as its reflection in remotely sensed crop indices. It will be of interest to know whether its apparent spatial variability behavior affects the yield pattern.

Grain Yield

Measured wheat grain yield distribution along the transect is presented together with spring N fertilizer application rate in Fig. 8–1. Along the transect, the amplitude of wheat yield is generally increasing from the left to the right, obvious from the two dashed lines at the maximum and minimum yield level in the first wave. The cells receiving no N fertilizer always exhibit the lowest yields locally, and there is a decrease in these yield minima from the first to the fourth cycle. On the other hand, from the first to the third cycle, maximum yield in the highly fertilized plots increases, and then remains on a high level without further increase in Cycles 4 and 5 (Fig. 8–1). The maximum wheat yields are not always exactly observed in the 168 kg ha^{-1} fertilizer N cells, but there is a shift by one cell observed in Cycles 1, 4, and 5. Several factors could have caused this behavior. If errors in the GPS were the explanation, why would the same error not have occurred in those plots receiving no N fertilizer in spring? Hence, GPS error would only explain the spatial shift for the cells that have received high N rates. A delay in the electronic yield monitoring, known as the convolution effect (Pringle et al., 1999), could have affected the results, presumably here mainly occurring in the high-yield plots but not in the low-yield plots.

Despite some possible measurement errors that require further research to be resolved, it is obvious that with changing soil properties along the transect, the response to cyclically varying amounts of N fertilizer changes. In the right-hand half of the transect, larger yields were possible with high amounts of N than in the first half; however, yields were clearly more reduced in the right-hand side than left-hand side when N was lacking. Remember, that the growing season was a rather wet season. High clay contents in the right-hand half of the transect that would presumably cause a lower water availability in dry years than in the first half of the transect probably did not limit yield. On the other hand, lack of N associated with low N fertilizer application rates caused lower yields in the second half than in the first half of the transect. Moreover, a slower seed emergence and phenological development at early growth stages of wheat was observed in this zone with higher clay contents.

A series of filtered yield data is presented in Fig. 8–1 as well. In this series the effect of sinusoidal fertilizer application is separated from measured yields, as explained below, to compare the spatial behavior of filtered yield with soil variation.

Crop Reflectance Indices

After the first and the second N application, canopy reflectance was measured (i.e., on 21 March and 24 April, respectively). The resulting red/infrared ratio (RNIR) data are presented along the transect with grain yield and N application rate in Fig. 8–4. In this study, RNIR is not converted into N fertilizer demand; rather it is directly presented. It reflects fertilizer demand on a relative basis. From the first to the second crop scanning, RNIR generally decreases. At the first scanning, which occurred approximately 1 mo after the first N fertilizer and 6 d before the

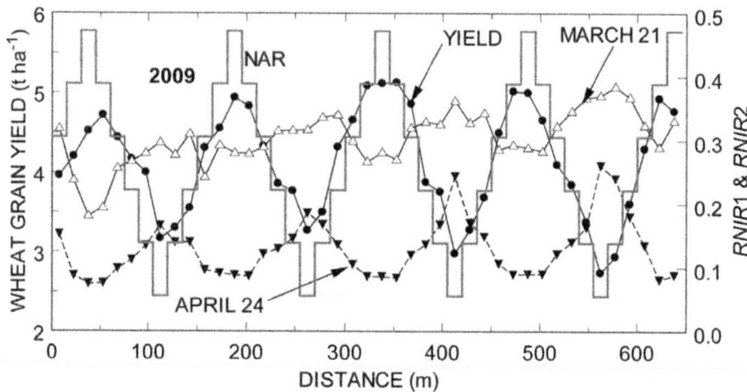

Fig. 8–4. Wheat grain yield and relative nitrogen demand derived from the red/near-infrared ratio measured at two times (RNIR1 and RNIR2) in spring 2009 in a farmer's field in Western Kentucky, and the sinusoidal spatial pattern of nitrogen application rate NAR.

second application, the cyclic pattern of applied N fertilizer was well reflected by sensor results. In cells of low N fertilizer application, RNIR was higher than in cells of high application and vice versa. Overall, an increasing tendency of RNIR with increasing distance is obvious. In other words, despite the implied locally varying NAR in cells, the general level of relative N fertilizer demand increases from left to right in the transect. This spatial trend in crop response to NAR may be due to the change in mineral N at the time of planting and to the slight increase in clay content across the last 300 m of the transect.

At the second scanning campaign (24 April), approximately 1 mo after the second N fertilizer application, the relative N demand was generally lower than at the first campaign. The cyclic pattern of RNIR is more pronounced; that is, it has a larger amplitude than during the first scanning, and the crop status obviously reflects the different N application pattern. Most of the mineralized N has been taken up by the crop, depleting the available N in the profile, as we will see from results presented later. Along the transect, the RNIR proceeds at a similar level for those cells that received high N fertilizer rates; however, RNIR obviously increased from left to right in plots that received no or only little fertilizer. This spatial trend behavior is inversely related to the behavior of grain yield. An increasing demand of N fertilizer is indicated, especially in the zone with slightly increased clay contents (450–640 m). In this zone, the mineralization or storage and supply of plant-available N probably has been lower than in the zone of 0 to 350 m. The sum of squared deviations between scaled measured grain yield and RNIR1 and RNIR2 is 2.2315 and 0.5426, and correlation coefficients are 0.59, and 0.90, respectively (Table 8–3), indicating that from the first to the second scanning date, the relationship between yield and crop index strongly increased.

Table 8–3. Spatial relationship between measured grain yield, crop indices, soil properties, and different crop growth model input scenarios. SSD between grain yield and respective variables is calculated based on their scaled values (Eq. [3]).

Variable/model scenario	SSD	*r*
1-RNIR1	2.2315	0.585
1-RNIR2	0.5426	0.899
BI1	1.7418	0.676
BI2	0.3112	0.942
CONT-PTF-SS (site-specific soil texture and initial N_{min})	0.6872	0.872
CONT-PTF-AVG (average soil texture and initial N_{min})	0.5199	0.903
CLASS-PTF-SSN (average soil texture, site-specific initial N_{min})	0.6407	0.881
CLASS-PTF-AVG (average soil texture and initial N_{min})	0.4791	0.911

Biomass index (BI) data obtained with the ALS are presented for the two scanning dates in Fig. 8–5. This index appears spatially inverse to RNIR. Biomass index proceeds at a lower level at the first scanning than at the second scanning, and there is an overall decreasing tendency of BI1 along the transect observed on 21 March. As expected, the BI2 results obtained on April 24 show an obviously cyclic pattern. The local variation of grain yield coincides closely with *RNIR* and BI. For both scanning dates, BI is closer related to measured grain yield than the respective results of RNIR, as indicated in Table 8–3.

We conclude, that RIRR and BI both indicate a strong inverse and positive relationship, respectively, with local and field-scale variation of grain yield. The BI2 yields the highest correlation with measured grain yield at all, and also

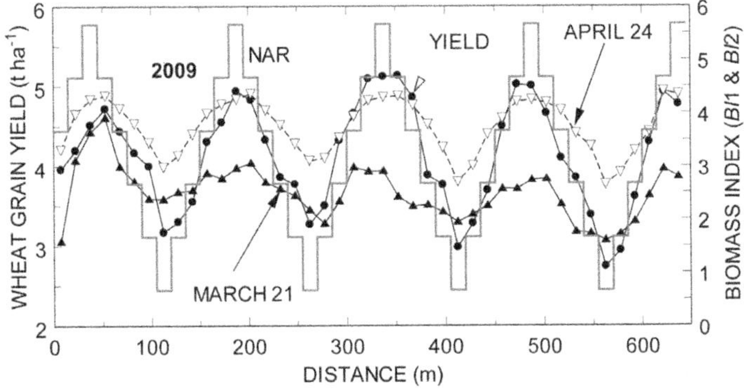

Fig. 8–5. Wheat grain yield and biomass index measured at two times (BI1 and BI2) in spring 2009 in a farmer's field in western Kentucky, and the sinusoidal spatial pattern of N application rate NAR.

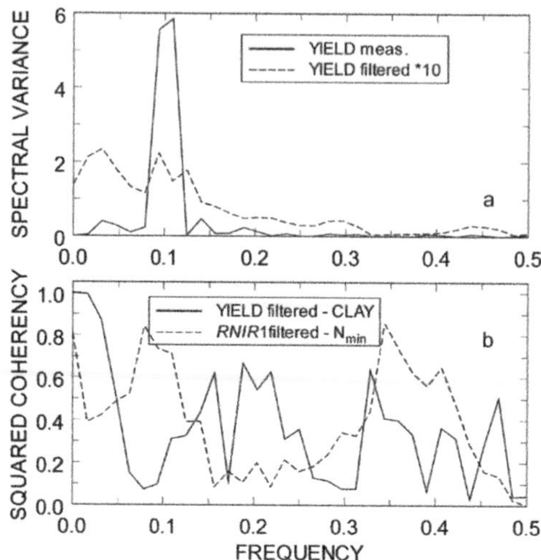

Fig. 8–6. (a) Power spectra for winter wheat grain yield and filtered yield (see text; for illustration, spectral variance of filtered yield is multiplied by 10) presented in Fig. 8–1, and (b) squared coherency spectra for filtered yield and clay, and for filtered RNIR1 and N_{min} (sampled in October 2008).

RNIR2 is highly correlated (Table 8–3). From this result, we conclude that due to their integrative characteristics, crop indices are very promising state variables for supporting site-specific crop yield prediction. The effort to collect crop sensor data is much lower than collecting site specific soil information, and the spatial resolution for crop indices can be much finer than for soil sampling.

Imposed Treatment and Underlying Soil Variation

The sinusoidal spatial variance behavior of grain yield as affected by the cyclic pattern of N application rate is depicted in the power spectrum in Fig. 8–6a. There is an obvious peak at a frequency of 0.1, reflecting a cyclic variation with a wavelength of 10 points. This signal corresponds to the cyclic pattern of N application rate because the wavelength of the sinusoidal application is 10 points. Since yield response is not uniform but changes along the transect, its variation is not only caused by N fertilizer but also by other underlying processes. The relationship between these and other processes shall be quantified based on the integrated autocovariance and cross-covariance function in spectral analysis. First, the sinusoidal variation of the grain yield series is filtered to study the remaining variation and its causes. For this purpose, each of the 43 cell-average wheat yield values yi_i is subtracted from its respective treatment average $\overline{yi}_t$. The total wheat yield average $\overline{yi}$ computed as the average of all 43 cells yield values is added afterward, resulting in a series of 43 filtered yield values $yi\,filt_i$

$$yi\,filt_i = \overline{yi}_t - yi_i + \overline{yi} \quad [6]$$

The filtered yield series is shown in Fig. 8–1. The spectrum of the filtered yield series depicted in Fig. 8–6a obviously shows a major portion of variance at small frequencies reflecting a large-scale source of variation. Kachanoski et al. (1985), Kachanoski and De Jong (1988), and Wendroth et al. (2011b) identified landscape features as the source of such large-scale variation. The reason that there still exists a variation source at the same frequency band as the fertilizer effect is that there still remains some variation at the wavelength imposed through the fertilizer application pattern because only the average treatment effect has been filtered, and the yield response to fertilizer varied along the transect.

The long-wave or small-frequency band of variation strongly coincides with the variation of soil clay content (Fig. 8–2), as is apparent from the coherency spectrum depicted in Fig. 8–6b. Although there are also some common frequencies in the range of small-scale variations (right-hand side of the spectrum), these peaks are based on small contributions to the variance of the filtered yield and the unfiltered clay content series.

Upon filtering the RNIR1 series shows common large-scale frequency bands with N_{min} between 0.07 and 0.10 and small-scale bands around 0.35 appear (Fig. 8–6b). The first indicates that there exists association in the variance of optical reflectance (i.e., crop N deficiency) and the pattern of mineral N in the soil. The latter shows common fluctuations at wavelengths of three points, that is, the small-scale scatter of observations. Therefore, on filtering crop observations, the impact of underlying spatial soil processes becomes apparent, emphasizing the importance of site-specific soil sampling or monitoring of integrative state variables, such as optical indices for understanding and predicting the yield variability pattern.

DSSAT Computer Simulations

With different levels of soil input to DSSAT, i.e., site specific soil texture and initial N_{min} (CONT-PTF-SS), field average soil texture and N_{min} (CONT-PTF-AVG), average soil texture but site-specific initial N_{min} (CLASS-PTF-SSN), or average soil texture and initial N_{min} (CLASS-PTF-AVG), simulated grain yield differs and so does its relationship to measured grain yield. In Fig. 8–7, the scenarios with the most and the least specific soil input are presented, CONT-PTF-SS and CLASS-PTF-AVG, respectively. Except for the lower two N application rates, the measured grain yield of wheat is below the level of simulated yields, and this discrepancy is larger at larger amounts of applied nitrogen fertilizer. Only for those plots that received no N fertilizer in spring at all, did simulated yields fit the measured ones almost perfectly for the CLASS-PTF-SSN scenario (Fig. 8–7). However, note that the model was not calibrated, and the objective is to identify how well the model captures the observed spatial change of yield to the N response.

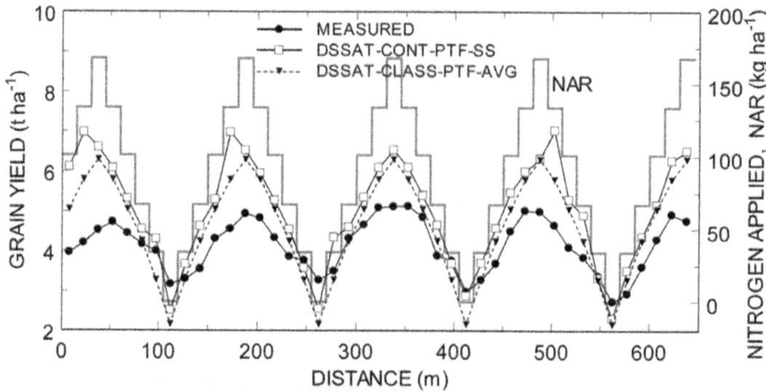

Fig. 8–7. Winter wheat grain yield, measured and simulated (DSSAT), and pattern of spring N fertilizer application rate (NAR) in a farmer's field in western Kentucky. Grain yield was simulated based on a continuous pedotransfer function and site-specific soil input (CONT-PTF-SS) for each of the 43 cells, and based on a class pedotransfer function and average soil input for each of the 43 cells (CLASS-PTF-AVG), respectively.

All simulation scenarios yielded correlation coefficients >0.87 with measured grain yield (Table 8–3). This result is probably dominated by the wide range of N fertilizer application rates. The highest level of specified soil input yields the lowest correlation coefficient to measured grain yield of all model scenarios (Table 8–3). On the other hand, the least specific input to the model, i.e., average soil textural profile for the entire transect and average initial N_{min} only, yields the largest correlation with scaled measured grain yield (Table 8–3). This result was not expected, and it shows that the modeling approach chosen including the underlying pedotransfer functions is supportive of calculating plant growth and soil processes in general, however, although the PTF by Wösten (1997) (Table 8–4) provided more consistent soil physical results than the one by Saxton et al. (1986), the model was not sufficiently parameterized to describe trends in the spatially varying response to N fertilizer more precisely.

The dynamics of soil water and nutrients were simulated together with crop growth. In Fig. 8–8, aboveground biomass (ABM) measured on 24 April and 19 May 2009, and the values simulated in the CLASS-PTF-SSN are shown for the same days together with the spatial pattern of NAR in spring. The hand-harvested ABM on 24 April reflects the cyclic behavior of the N application pattern. From 0 to approximately 550 m distance, the fluctuations behaved relatively uniformly. In the zone between 560 and 600 m, where low or no N fertilizer was applied and clay contents are relatively high (Fig. 8–2), the smallest amounts of aboveground biomass were observed. For ABM harvested on 24 April, no overall increasing or decreasing trends along the transect were obvious. In contrast, ABM sampled on 19 May shows a more pronounced spatial pattern. Probably due

Table 8–4. Upper (UDL) and lower (LDL) drainage limit for the 43 soil profiles, based on the continuous pedotransfer function by Wösten (1997), used in the calculations for CONT-PTF-SS. At the end, average UDL and LDL for all 43 profiles based on Wösten (1997) are given, used for CONT-PTF-AVG, and the average UDL and LDL based on Saxton et al. (1986), used in CLASS-PTF-SSN and CLASS-PTF-AVG.

Depth	Profile	UDL	LDL	Profile	UDL	LDL	Profile	UDL	LDL	Profile	UDL	LDL
cm												
0–15	1	0.316	0.204	13	0.255	0.133	25	0.239	0.117	37	0.332	0.224
15–30		0.304	0.187		0.246	0.118		0.279	0.155		0.319	0.205
30–60		0.332	0.219		0.300	0.172		0.306	0.181		0.325	0.208
60–90		0.202	0.103		0.137	0.050		0.125	0.042		0.148	0.058
0–15	2	0.297	0.181	14	0.255	0.133	26	0.239	0.117	38	0.269	0.148
15–30		0.312	0.196		0.279	0.155		0.258	0.131		0.252	0.124
30–60		0.325	0.208		0.317	0.197		0.303	0.177		0.297	0.168
60–90		0.174	0.078		0.125	0.042		0.125	0.042		0.137	0.050
0–15	3	0.262	0.141	15	0.269	0.148	27	0.247	0.125	39	0.292	0.175
15–30		0.263	0.137		0.304	0.187		0.269	0.144		0.304	0.187
30–60		0.322	0.205		0.309	0.185		0.300	0.172		0.312	0.190
60–90		0.207	0.108		0.121	0.039		0.121	0.039		0.117	0.037
0–15	4	0.275	0.155	16	0.269	0.148	28	0.173	0.059	40	0.311	0.198
15–30		0.274	0.150		0.296	0.177		0.239	0.111		0.304	0.187
30–60		0.317	0.197		0.309	0.185		0.282	0.148		0.303	0.177
60–90		0.144	0.055		0.121	0.039		0.129	0.044		0.125	0.042
0–15	5	0.287	0.169	17	0.292	0.175	29	0.247	0.125	41	0.311	0.198
15–30		0.288	0.166		0.292	0.172		0.216	0.088		0.308	0.191
30–60		0.315	0.194		0.306	0.181		0.290	0.158		0.317	0.197
60–90		0.125	0.042		0.117	0.037		0.117	0.037		0.144	0.055
0–15	6	0.275	0.155	18	0.262	0.141	30	0.262	0.141	42	0.200	0.079
15–30		0.304	0.187		0.279	0.155		0.274	0.150		0.296	0.177
30–60		0.312	0.190		0.303	0.177		0.297	0.168		0.320	0.201
60–90		0.137	0.050		0.121	0.039		0.125	0.042		0.155	0.063

Continued.

Table 8–4. Continued.

Depth	Profile	UDL	LDL	Profile	UDL	LDL	Profile	UDL	LDL	Profile	UDL	LDL
cm												
0–15	7	0.269	0.148	19	0.247	0.125	31	0.231	0.108	43	0.275	0.155
15–30		0.283	0.161		0.258	0.131		0.216	0.088		0.232	0.103
30–60		0.303	0.177		0.309	0.185		0.309	0.185		0.278	0.142
60–90		0.129	0.044		0.125	0.042		0.129	0.044		0.121	0.039
0–15	8	0.281	0.162	20	0.247	0.125	32	0.255	0.133			
15–30		0.300	0.182		0.258	0.131		0.292	0.172			
30–60		0.312	0.190		0.300	0.172		0.309	0.185	CONT-PTF-AVG		
60–90		0.133	0.047		0.125	0.042		0.129	0.044	Wösten (1997)		
0–15	9	0.269	0.148	21	0.255	0.133	33	0.292	0.175		0.269	0.148
15–30		0.274	0.150		0.279	0.155		0.322	0.209		0.279	0.155
30–60		0.293	0.163		0.309	0.185		0.320	0.201		0.303	0.177
60–90		0.121	0.039		0.129	0.044		0.155	0.063		0.133	0.047
0–15	10	0.269	0.148	22	0.239	0.117	34	0.262	0.141			
15–30		0.269	0.144		0.279	0.155		0.258	0.131	CLASS-PTF-SSN		
30–60		0.306	0.181		0.306	0.181		0.248	0.104	CLASS-PTF-AVG		
60–90		0.129	0.044		0.121	0.039		0.079	0.017	Saxton et al. (1986)		
0–15	11	0.269	0.148	23	0.255	0.133	35	0.262	0.141		0.381	0.142
15–30		0.232	0.103		0.274	0.150		0.232	0.103		0.329	0.115
30–60		0.234	0.089		0.306	0.181		0.269	0.130		0.373	0.175
60–90		0.125	0.042		0.129	0.044		0.068	0.013		0.325	0.202
0–15	12	0.262	0.141	24	0.239	0.117	36	0.343	0.238			
15–30		0.216	0.088		0.274	0.150		0.332	0.222			
30–60		0.185	0.047		0.306	0.181		0.336	0.225			
60–90		0.084	0.019		0.121	0.039		0.168	0.073			

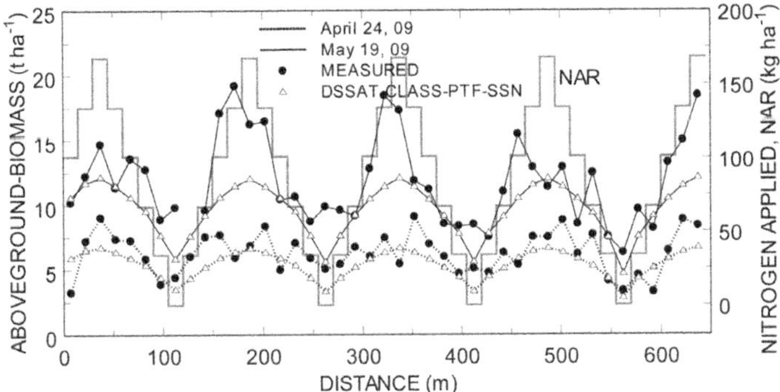

Fig. 8–8. Aboveground biomass (ABM) measured and simulated (CLASS-PTF-SSN) on 24 April and 19 May 2009 and pattern of spring N fertilizer application rate (NAR) in a farmer's field in western Kentucky.

to the huge difference in spatial sample support of harvested grain yield and ABM, the large spatial increasing trend of grain yield in cells that received high NAR is not reflected in ABM. The model scenario (CLASS-PTF-SSN) underestimates the magnitude of aboveground biomass, for both sampling dates, except for a few locations (Fig. 8–8). In general, the agreement between measured and simulated aboveground biomass is closer for the plots that have received low or no N fertilizer in spring than for the highly fertilized plots. The discrepancies between measurement and model result are much larger for the 19 May sampling than for that of 24 April. The spatial series of simulated data proceeds much smoother than the two measured series exhibiting considerable point-to-point fluctuations, which may be an effect of the small sample support area.

The results on measured and simulated (CLASS-PTF-SSN) mineral soil N content in the 0- to 90-cm soil profile are shown in Fig. 8–9. In general, the simulated N content on 22 January is approximately 10 to 20 kg ha^{-1} greater than the measured. However, for this day, the spatial patterns of measured and simulated N show similar fluctuations, especially at distances from 0 to 400 m. Apparently, the initial concentration of N_{min} has a considerable impact on the subsequent spatiotemporal soil processes of N_{min}. On 24 April, when the two N applications have occurred and substantial amounts of N have been taken up by the wheat, the measured N content is larger than the simulated. Moreover, simulations exhibit N depletion of the soil profile throughout the transect, whereas measurements still reveal relatively low but nonnegligible amounts of N in the profile. No spatial differentiation among the different treatment levels of N is reflected by simulation results, except for the two treatment levels receiving the highest and second-highest amount of N fertilizer. Notice that although simulated N levels in the soil are lower than measured ones, the simulated ABM is also lower than the measured one (Fig. 8–8).

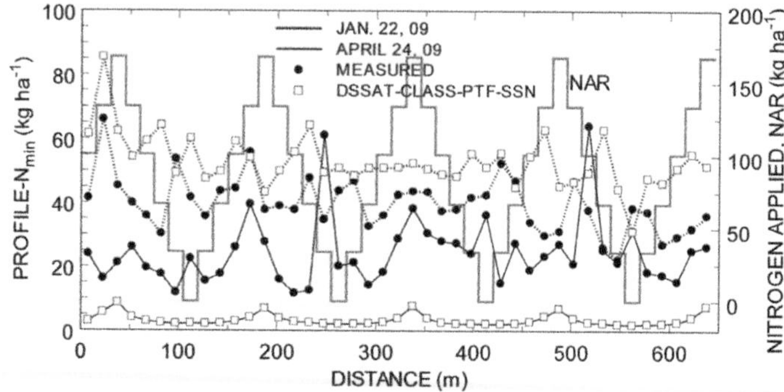

Fig. 8–9. Mineral soil N (N_{min}) in the profile from the 0- to 90-cm depth on two sampling dates (22 Jan. and 24 Apr. 2009), measured and simulated (CLASS-PTF-SSN), and pattern of spring N fertilizer application rate (NAR) in a farmer's field in western Kentucky.

In Fig. 8–10, time series of measured daily rainfall and average air temperature are displayed (Fig. 8–10a), together with measured and simulated (CLASS-PTF-SSN) time series of mineral soil N (Fig. 8–10b), aboveground biomass and grain yield (Fig. 8–10c). Data for three N levels presented in Fig. 8–10b and 8–10bc are based on averages among all those cells that received the same amount of spring NAR throughout the transect. Arrows in Fig. 8–10b indicate the times of N application, and an obvious increase in simulated mineral soil N can be noticed right after both N fertilizer applications. Soon after that increase, a rapid decrease of simulated N_{min} due to plant uptake occurs. We learn from Fig. 8–10b that for the first N_{min} sampling campaigns 1 Oct. 2008 and 22 Jan. 2009, levels of measured N_{min} averaged for the same treatment did not differ substantially. This result is probably due to the fact that the soil was treated uniformly thus far. However, treatment-averaged mineral soil N contents also did not differ substantially for the 24 April sampling (i.e., Day 204), although at this time, the three N_{min} treatment averages presented differ (Fig. 8–10b). For measuring any treatment-related N_{min} differences, the timing for this sampling was not well chosen. According to the simulation results, a sampling 10 d earlier would have been more informative and would have supported an evaluation of the model with respect to its capability to reflect site- and management-specific differences in N_{min}. At the actual day of sampling (24 April), the simulated amounts of N_{min} are below those being measured. However, the difference among the simulated results is almost as small as among the measured values.

As already obvious from Fig. 8–8, simulated (CLASS-PTF-SSN) aboveground biomass on 24 April fits measured biomass much better than on 19 May, when the model underestimated biomass more severely for all three treatment averages

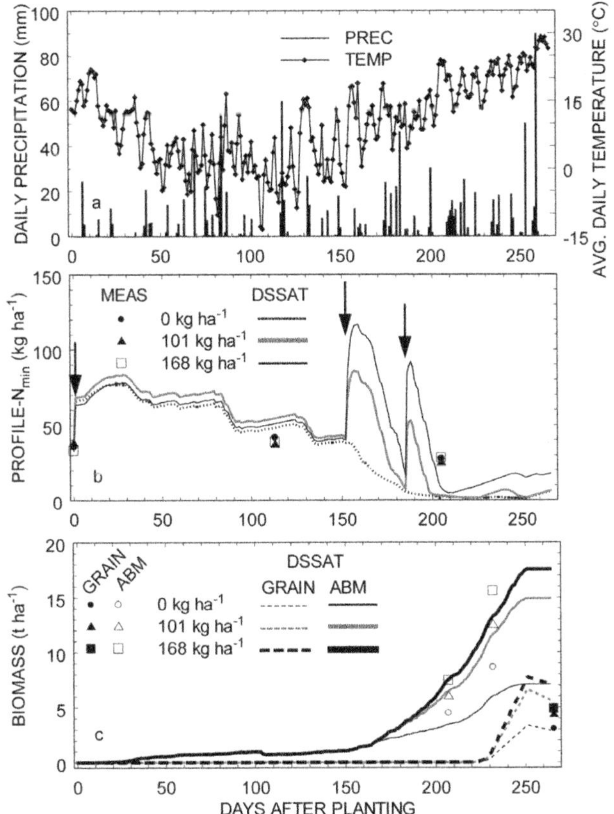

Fig. 8–10. (a) Daily precipitation and average air temperature, (b) measured and simulated (DSSAT, scenario: CLASS-PTF-SSN) mineral soil N in the profile (N_{min}), (c) aboveground measured and simulated biomass (ABM) and grain yield in a farmer's field in western Kentucky. In plot c, symbols denote measured grain yield (GRAIN) and aboveground biomass (ABM), respectively.

(Fig. 8–10c). Despite these discrepancies between measured and simulated ABM, astonishingly the measured grain yield averaged over all cells that received no N in spring is matched perfectly by the model. The other two treatment averages shown here are strongly overestimated by the model, as is also obvious for site-specific results in Fig. 8–8.

In general, the local magnitudes of grain yield measurements and simulations deviate from each other. However, a relevant aspect is how the spatial variation of measured grain yield compares with the simulated yield on a relative basis. In other words, if the model were calibrated with respect to the mean and the standard deviation of grain yield across the 43 cells, how close would it reflect the spatial behavior of measured yield on a relative basis? And in this

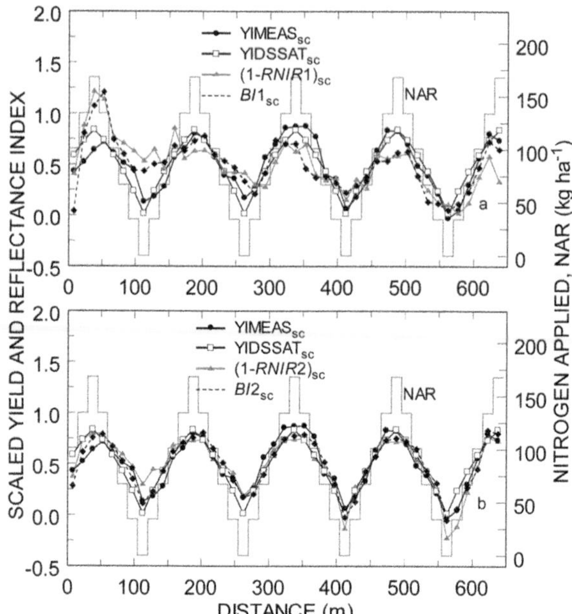

Fig. 8–11. Scaled yield measured (YIMEAS$_{sc}$) and simulated (YIDSSAT$_{sc}$, scenario CLASS-PTF-SSN) and scaled canopy indices from (a) campaign 1 [(1-RNIR1)$_{sc}$, BI1$_{sc}$] and (b) campaign 2 [(1-RNIR2)$_{sc}$, BI2$_{sc}$], respectively and pattern of spring N fertilizer application rate (NAR) in a farmer's field in western Kentucky.

context, how would scaled reflectance indices behave relative to measured and simulated yields?

In Fig. 8–11a, scaled measured and simulated (CLASS-PTF-SSN) yields are shown in coincidence with scaled reflectance indices, i.e., the scaled RNIR and BI index values measured at the first date ([1-RNIR1]$_{sc}$, and BI1$_{sc}$). From left to right, in the first three plots, the scaled simulated grain yield as well as both scaled reflectance indices proceed above the scaled measured yield. A possible explanation for this behavior could be that the higher level of initial mineral soil N observed in this part of the field has caused a larger N uptake and a larger abundance of biomass early in the season that is captured by the simulation model. A month later when the reflectance is measured again, the relative preference in reflectance of the first three plots in the field almost completely diminishes relative to the rest of the field (Fig. 8–11b).

Whereas scaled measured and simulated grain yields coincide strongly until approximately 250 m distance, optical indices are at a relatively higher level than both simulated and measured yields (scaled). In the zone between 250 and 360 m, measured yields exceed the simulated on a relative basis, except for one plot. Interestingly, in this zone, scaled reflectance indices are closer to scaled simulated than to scaled measured yields (Fig. 8–11a). Between 360 and 650 m, scaled measured and simulated yields concur closely, whereas both sensors provide higher relative index values in the zones of low fertilizer rates and lower relative values for cells that have received high fertilizer rates (Fig. 8–11a).

At distances >150 m and for the second scanning campaign, both sets of scaled indices show a consistent or close behavior with measured yield, while the $BI1_{sc}$ results are somewhat closer to scaled grain yields than $[1\text{-}RNIR1]_{sc}$ (Fig. 8–11b). In general, scaled sensor results obtained for the second campaign coincide much more closely with scaled measured and simulated grain yields than for the first campaign, as is obvious from correlation coefficients and SSD in Table 8–3.

Conclusions

How helpful is site-specific soil and crop sampling to explain crop growth and yield variability?

In general, any site-specific soil and crop information sampled is helpful, as long as it is spatially autocorrelated. Only if it is autocorrelated a variable is known to represent a certain domain, and it can be related to another variable, whose range of influence is also known (Nielsen and Wendroth, 2003; Wendroth et al., 2011a). In this study, information on soil texture was helpful to explain the trend in soil quality change and why measured crop yields tended to decrease in the last 250 m of the transect where clay contents increase. Overall, clay content change is relatively smooth throughout the field except for the zone from 450 to 640 m, where some considerable fluctuations between cells are observed. Inasmuch as clay contents do not vary strongly between neighboring locations, closer sampling intervals would probably not have decreased the nugget semivariance. Notice also that the overall semivariance across the domain investigated is not very large. It is not easy to derive the importance of knowing site-specific N_{min} from soil sampling. The main variation in crop growth and yield is caused by spatially varying N application in spring. At the first two times of sampling, spatial differences are not strongly pronounced for two reasons. First, in fall and winter, N fertilizer has not yet been applied at varying rates. Also, the April sampling campaign occurs too late to observe more clearly patterns that the site-specific fertilization has caused. This scheduling would not be very helpful to validate how well a plant growth model captures site-specific soil N dynamics. The spatial design of sampling N_{min} reveals its spatial trend and its long range of spatial structure along the transect and is therefore a helpful initial condition for the crop growth model. Measuring ABM in each cell reveals the sinusoidal pattern of vegetative growth caused by the spatial application scheme of NAR. In an experiment like this with NAR varying substantially, spatial differences in ABM can be well detected and correlated with other variables (results not shown).

How sensitive are remote sensors and canopy reflectance measurements to crop nitrogen demand and plant growth status?

Both crop sensor systems provided spectral reflectance indices that related very well to plant growth and N deficiency artificially induced by the way nitrogen was distributed in the space. Besides reflecting the sinusoidal pattern of NAR and plant growth status, the indices revealed spatial trends along the transect that appeared in the spatial grain yield behavior as well as in soil textural differences along the transect. The indices obtained from the crop sensors integrate the main spatial soil differences and processes as well as differences caused by management in a very comprehensive way, and their spatial behavior is closely related to grain yield even if spectral reflectance was measured early in the season. This result makes the crop sensors and indices tested here very promising tools for optimizing crop management systems, especially with respect to site-specific yield prediction and N application.

How sensitive is the crop growth simulation model to site-specific soil input?

The fact that the simulation scenario based on the least effort in terms of model input (i.e., average soil texture and average initial N_{min}) resulted in the closest correlation with measured grain yield on a relative basis leads to the conclusion that the model used here is not sufficiently sensitive to site-specific soil input. To what extent this result is due to the particular simulation of soil processes will have to be determined after further tests using other modeling approaches or at least other soil modules, and other more variable field soils.

Given the strong spatial relationships between canopy sensor results and measured grain yield, a more efficient and site-specific simulation model application could be based on incorporating sensor measurements as integrators of past crop development in a model calibration or updating step that still needs to be technically implemented. Unless better concepts for describing field soil processes in models are derived and used, integrative state variables such as the crop indices applied in this study are the better choices to predict site-specific crop yield variability and to derive management decisions.

Acknowledgments

This study was financially supported by the College of Agriculture Precision Resources Management Grant program and by the Kentucky Small Grain Growers' Association. The authors thank the farmer, Trevor Gilkey, Princeton, KY for his support and for allowing this research in his field. The excellent technical assistance of many individuals is highly appreciated: R. Jason Walton, for running the monitoring devices for crop scanner measurements, and for soil sampling and lab work; Danny Reid, Diane Hunter, Colleen Steele, and Paul Wilson in the UK Soils Regulatory Service Laboratory; Tammi Smith and Jim Crutchfield in the Soil and Plant Analysis Lab; and John James at the Research and Education Center at Princeton. Special thanks to Dr. Charles Dougherty for making his GreenSeeker sensors available and to James Dollarhide for machinery adjustments and excellent maintenance.

References

Ahuja, L.R., L. Ma, and T.R. Green. 2010. Effective soil properties of heterogeneous areas for modeling infiltration and redistribution. Soil Sci. Soc. Am. J. 74:1469–1482.

Bannayan, M., N.M.J. Crout, and G. Hoogenboom. 2003. Application of CERES-Wheat model for within-season prediction of winter wheat yield in the United Kingdom. Agron. J. 95:114–125.

Bazza, M., R.H. Shumway, and D.R. Nielsen. 1988. Two-dimensional spectral analyses of soil surface temperature. Hilgardia 56:1–28.

Bronson, K.F., J.D. Booker, J.W. Keeling, R.K. Boman, T.A. Wheeler, R.J. Lascano, and R.L. Nichols. 2005. Cotton canopy reflectance at landscape scale as affected by nitrogen fertilization. Agron. J. 97:654–660.

Bronson, K.F., T.T. Chua, J.D. Booker, J.W. Keeling, and R.J. Lascano. 2003. In-season nitrogen status sensing in irrigated cotton: II. Leaf nitrogen and biomass. Soil Sci. Soc. Am. J. 67:1439–1448.

Chua, T.T., K.F. Bronson, J.D. Booker, J.W. Keeling, A.R. Mosier, J.P. Bordovsky, R.J. Lascano, C.J. Green, and E. Segarra. 2003. In-season nitrogen status sensing in irrigated cotton: I. Yields and nitrogen-15 recovery. Soil Sci. Soc. Am. J. 67:1428–1438.

Crutchfield, J.D., and J.H. Grove. 2011. A new cadmium reduction device for the microplate determination of nitrate in water, soil, plant tissue, and physiological fluids. J. AOAC Intl. (in press.)

Davis, J.C. 1986. Statistics and data analysis in geology. 2nd ed. Wiley and Sons, New York.

Fang, Q., L. Ma, Q. Yu, R.W. Malone, S.A. Saseendran, and L.R. Ahuja. 2008. Modeling nitrogen and water management effects in a wheat–maize double-cropping system. J. Environ. Qual. 37:2232–2242.

Gee, G.W., and J.W. Bauder. 1986. Particle-size analysis. p. 383–423. *In* A. Klute (ed.) Methods of soil analysis. Part 1. Physical and mineralogical methods. 2nd ed. SSSA Book Ser. 5. ASA and SSSA, Madison, WI.

Giebel, A., O. Wendroth, H.I. Reuter, K.C. Kersebaum, and J. Schwarz. 2006. How representatively can we sample soil mineral nitrogen? J. Plant Nutr. Soil Sci. 169:52–59.

Greminger, P.J., K. Sud, and D.R. Nielsen. 1985. Spatial variability of field-measured soil-water characteristics. Soil Sci. Soc. Am. J. 49:1075–1082.

Hatfield, J.L., A.A. Gitelson, J.S. Schepers, and C.L. Walthall. 2008. Application of spectral remote sensing for agronomic decisions. Agron. J. 100:117–131.

Hong, N., P.C. Scharf, J.G. Davis, N.R. Kitchen, and K.A. Sudduth. 2007. Economically optimal nitrogen rate reduces soil residual nitrate. J. Environ. Qual. 36:354–362.

Hoogenboom, G., J.W. Jones, C.H. Porter, P.W. Wilkens, K.J. Boote, W.D. Batchelor, L.A. Hunt, and G.Y. Tsuji (ed.). 2003. Decision support system for agrotechnology transfer version 4.0. Volume 1: Overview. Univ. of Hawaii, Honolulu.

Isaaks, E.H., and R.M. Srivastava. 1989. Applied geostatistics. Oxford Univ. Press, New York.

Jones, J.W., G. Hoogenboom, C.H. Porter, K.J. Boote, W.D. Batchelor, L.A. Hunt, P.W. Wilkens, U. Singh, A.J. Gijsman, and J.T. Ritchie. 2003. The DSSAT cropping system model. Eur. J. Agron. 18:235–265.

Kachanoski, R.G., and E. De Jong. 1988. Scale dependence and the temporal persistence of spatial patterns of soil water storage. Water Resour. Res. 24:85–91.

Kachanoski, R.G., D.E. Rolston, and E. De Jong. 1985. Spatial and spectral relationships of soil properties and microtopography. I. Density and thickness of A horizon. Soil Sci. Soc. Am. J. 49:804–812.

Kersebaum, K.C. 1989. Die Simulation der Stickstoffdynamik in Ackerböden. Ph.D. diss. Univ. of Hannover, Germany.

Kersebaum, K.C. 1995. Application of a simple management model to simulate water and nitrogen dynamics. Ecol. Model. 81:145–156.

Kitchen, N.R., K.A. Sudduth, S.T. Drummond, P.C. Scharf, H.L. Palm, D.F. Roberts, and E.D. Vories. 2010. Ground-based canopy reflectance sensing for variable-rate nitrogen corn fertilization. Agron. J. 102:71–84.

Mirschel, W., and K.O. Wenkel. 2007. Modelling soil-crop interactions with AGROSIM model family. p. 59–73. *In* K.Ch. Kersebaum et al. (ed.) Modelling water and nutrient dynamics in soil-crop systems. Springer, Dordrecht, The Netherlands.

Nielsen, D.R., and O. Wendroth. 2003. Spatial and temporal statistics—Sampling field soils and their vegetation. Catena, Reiskirchen, Germany.

Pennock, D.J., and A.H. Frick. 2001. The role of field studies in landscape-scale applications of process models: An example of soil redistribution and soil organic carbon modeling using CENTURY. Soil Tillage Res. 58:183–191.

Pringle, M.J., B.M. Whelan, M.L. Adams, S.E. Cook, and G. Riethmuller. 1999. Yield deconvolution—A wetted grain pulse to estimate the grain flow transfer function. p. 1177–1184. *In* P.C. Robert et al. (ed.) Proc. 4th Int. Conf. Prec. Agric., St. Paul, MN. 1998. ASA, CSSA, SSSA, Madison, WI.

Raun, W.R., J.B. Solie, G.V. Johnson, M.L. Stone, E.V. Lukina, W.E. Thomason, and J.S. Schepers. 2001. In-season prediction of potential grain yield in winter wheat using canopy reflectance. Agron. J. 93:131–138.

Raun, W.R., J.B. Solie, G.V. Johnson, M.L. Stone, R.W. Mullen, K.W. Freeman, W.E. Thomason, and E.V. Lukina. 2002. Improving nitrogen use efficiency in cereal grain production with optical sensing and variable rate application. Agron. J. 94:815–820.

Reusch, S. 2005. Optimum waveband selection for determining the nitrogen uptake in winter wheat by active remote sensing. p. 261–266. *In* J.V. Stafford (ed.) Precision agriculture '05. Wageningen Academic Publishers, Wageningen, The Netherlands.

Ritchie, G.L., D.G. Sullivan, W.K. Vencill, C.W. Bednarz, and J.E. Hook. 2010. Sensitivities of normalized difference vegetation index and a green/red ratio index to cotton ground cover fraction. Crop Sci. 50:1000–1010.

Roberts, D.F., N.R. Kitchen, P.C. Scharf, and K.A. Sudduth. 2010. Will variable-rate nitrogen fertilization using corn canopy reflectance sensing deliver environmental benefits? Agron. J. 102:85–95.

Saxton, K.E., W.J. Rawls, J.S. Romberger, and R.I. Papendick. 1986. Estimating generalized soil-water characteristics from texture. Soil Sci. Soc. Am. J. 50:1031–1036.

Scharf, P. 2006. Variable-rate nitrogen fertilizer application in corn using in-field sensing of leaves or canopy. Available at http://www.mo.nrcs.usda.gov/technical/agronomy/technote35_2.html (verified 7 Apr. 2011).

Scharf, P.C., N.R. Kitchen, K.A. Sudduth, J.G. Davis, V.C. Hubbard, and J.A. Lory. 2005. Field-scale variability in optimal nitrogen fertilizer rate for corn. Agron. J. 97:452–461.

Scharf, P.C., and J.A. Lory. 2009. Calibrating reflectance measurements to predict optimal side-dress nitrogen rate for corn. Agron. J. 101:615–625.

Shillito, R.M., D.J. Timlin, D. Fleisher, V.R. Reddy, and B. Quebedeaux. 2009. Yield response of potato to spatially patterned nitrogen application. Agric. Ecosyst. Environ. 129:107–116.

Shumway, R.H. 1988. Applied statistical time series analysis. Prentice Hall, Englewood Cliffs, NJ.

Skjemstad, J.O., and J.A. Baldock. 2008. Total and organic carbon. p. 225–238. *In* M.R. Carter and E.G. Gregorich (ed.) Soil sampling and methods of analysis. 2nd ed. Canadian Society of Soil Science, CRC Press, Boca Raton, FL.

Springob, G., R. Anlauf, K.C. Kersebaum, and J. Richter. 1985. Räumliche Variabilität von Bodeneigenschaften und Nährstoffgehalten zweier Schläge auf Löss-Parabraunerden. Mitt. Dtsch. Bodenkdl. Ges. 43:691–696.

Sripada, R.P., D.C. Farrer, R. Weisz, R.W. Heiniger, and J.G. White. 2007. Aerial color infrared photography to optimize in-season nitrogen fertilizer recommendations in winter wheat. Agron. J. 99:1424–1435.

Sripada, R.P., J.P. Schmidt, A.E. Dellinger, and D.B. Beegle. 2008. Evaluating multiple indices from a canopy reflectance sensor to estimate corn N requirements. Agron. J. 100:1553–1561.

Stenger, R., E. Priesack, and F. Beese. 1998. Distribution of inorganic nitrogen in agricultural soils at different dates and scales. Nutr. Cycling Agroecosyst. 50:291–297.

Timlin, D.J., Ya. Pachepsky, V.A. Snyder, and R.B. Bryant. 1998. Spatial and temporal variability of corn grain yield on a hillslope. Soil Sci. Soc. Am. J. 62:764–773.

Trangmar, B.B., R.S. Yost, and G. Uehara. 1985. Application of geostatistics to spatial studies of soil properties. Adv. Agron. 38:45–94.

USDA. 1966. Soil Survey Report Caldwell County Kentucky. USDA-SCS in cooperation with the Kentucky Agricultural Experiment Station.

Weatherburn, M.W. 1967. Phenol-hypochlorite reaction for determination of ammonia. Anal. Chem. 39:971–974.

Wendroth, O., P. Jürschik, K.C. Kersebaum, H. Reuter, C. van Kessel, and D.R. Nielsen. 2001. Identifying, understanding, and describing spatial processes in agricultural landscapes- four case studies. Soil Tillage Res. 58:113–128.

Wendroth, O., S. Koszinski, and V. Vasquez. 2011a. Soil spatial variability. p. 10-1–10-22. *In* P.M. Huang et al. (ed.) Handbook of soil science, 2nd ed. CRC Press, Boca Raton, FL.

Wendroth, O., W. Pohl, S. Koszinski, H. Rogasik, C.J. Ritsema, and D.R. Nielsen. 1999. Spatio-temporal patterns and covariance structures of soil water status in two Northeast-German field sites. J. Hydrol. 215:38–58.

Wendroth, O., E.L. Ritchey, S. Nambuthiri, J.H. Grove, and R.C. Pearce. 2011b. Spatial variability of soil physical properties. p. 827–839. *In* J. Gliński et al. (ed.) Encyclopedia of agrophysics. Springer, Dordrecht, The Netherlands.

Wösten, J.H.M. 1997. Pedotransfer functions to evaluate soil quality. p. 221–245. *In* E.G. Gregorich and M.R. Carter (ed.) Soil quality for crop production and ecosystem health. Developments in Soil Science. Elsevier, Amsterdam, The Netherlands.

Zarco-Tejada, P.J., S.L. Ustin, and M.L. Whiting. 2005. Temporal and spatial relationships between within-field yield variability in cotton and high-spatial hyperspectral remote sensing imagery. Agron. J. 97:641–653.

Chapter 9

Parameterization of Energy Balance Components and Remote Sensing in Systems Modeling

Jerry L. Hatfield

ABSTRACT

Estimation of a number of parameters using simulation models has proven to be a valuable source of information from which we can assess the impact of scenarios that would be difficult to determine experimentally, or for which it would be difficult to conceptualize an appropriate experiment design. However, simulation models require extensive inputs that are not always easily found or exist at the spatial or temporal resolution needed for the models. Many simulation models require energy inputs that represent the energy balance of the surface, and there have been several attempts to derive different inputs. There have been various methods to estimate solar radiation from combinations of air temperature, altitude, and precipitation. Albedo has been estimated from several different methods using either combinations of reflectance or simple regression models. Long-wave radiation from the atmosphere has been estimated using regression models of vapor pressure and air temperature. Many of these parameterizations have been derived using locally available data, and efforts are needed for broader evaluation of these methods. Crop simulation models produce a variety of estimates for plant growth; among these are leaf area index, biomass, and ground cover. These parameters can be measured directly, often a laborious task and not at the scale needed for model evaluation, or they can be estimated from remotely sensed observations. This approach not only provides an independent measure of the crop parameters to compare with model simulations, but a potential feedback into the model simulation to help correct the model over time. Challenges remain in our efforts to improve models and provide the input necessary to further our ability to understand the complexities of the interactions in the soil–plant–atmosphere continuum.

J.L. Hatfield, USDA-ARS, National Laboratory for Agriculture and the Environment, 2110 University Blvd., Ames, IA 50011 (jerry.hatfield@ars.usda.gov).

doi:10.2134/advagricsystmodel2.c9

Methods of introducing system models into agricultural research. L.R. Ahuja and L. Ma (ed.)
Advances in Agricultural Systems Modeling Series 2.

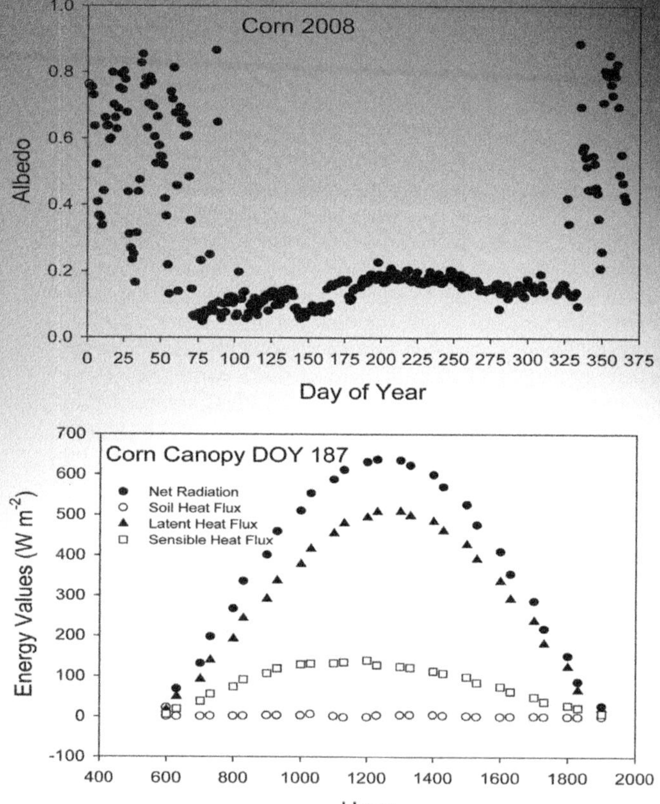
Corn 2008
Albedo
1.0
0.8
0.6
0.4
0.2
0.0
0 25 50 75 100 125 150 175 200 225 250 275 300 325 350 375
Day of Year
Corn Canopy DOY 187
Net Radiation
Soil Heat Flux
Latent Heat Flux
Sensible Heat Flux
Energy Values (W m^{-2})
700
600
500
400
300
200
100
0
-100
400 600 800 1000 1200 1400 1600 1800 2000
Hour

Simulation models provide a valuable tool for assessing the interactions among complex processes in the soil–plant–atmosphere continuum. There are extensive examples of models that have been developed that simulate these processes in detail; however, one of the major problems that remains is being able to parameterize these models with the necessary information required to allow for their full use across a range of spatial and temporal scales. All models require some type of input, and without the required data it may be necessary to either assume a value for the input data or estimate the required data from some other associated, often more readily available, parameter. The major dilemma that many researchers face in using different models or even testing them outside of the location in which they were developed is finding the required data as inputs into the models. Many of these inputs can't be ignored or eliminated because they are critical variables in the model.

When we consider the data required for effective application of models there is a need for inputs or surrogates for those inputs, enhanced spatial resolution of the input data so the models generate results that represent the spatial scale appropriate for the specific application, and finally, the proper relationship among the parameters so the results accurately represent the process being modeled and are not unrealistic or skewed. These issues do not represent unrealistic expectations for either experimental studies or simulation studies but represent the demands we need to place on how we view the parameters being evaluated. In this chapter, we will explore how energy balance parameters can be assessed for use in simulation models, with examples of energy balance components and how remote sensing data can be used to generate variables that are often simulated by these models as a method of providing an independent comparison of simulated versus estimated parameters.

One of the major difficulties is locating suitable input data for the various meteorological data needed for the model inputs or for parameterization methods. Some worldwide databases are maintained by the Food and Agriculture Organization (FAO) and are available online (www.fao.org/nr/climpag/data_5_en.asp, verified 8 Apr. 2011). This database covers monthly data for 28,100 stations and includes up to 14 observed and computed agroclimatic parameters, including evapotranspiration (mm/month), precipitation (mm/month), sunshine (monthly total), temperature (monthly mean, monthly mean maximum daily, monthly mean minimum daily), vapor pressure (mean monthly), wind speed (mean daily, monthly) On this website there are estimation routines to estimate local climate data for use at the monthly time scale. This dataset comprises long-term averages for the period from 1961 to 1990 and time series for rainfall and temperature. These data can be retrieved by geographic area, time period, and parameter and can be downloaded in different formats. The variables available in this database include maximum air temperature, minimum air temperature, mean air temperature, mean nighttime air temperature, mean daytime air temperature, total daily rainfall, dew point temperature, relative humidity, actual vapor pressure, potential evapotranspiration using Penman–Monteith equation, windspeed, global solar radiation, sunshine fraction, and sunshine hours. This is a rich database for the assessment of the meteorological resources.

As a beginning point, it is important to realize that all of these parameters have both a spatial and temporal component. Meteorological variables exhibit well-characterized, defined temporal variation both within a day and across the year. These defined patterns for specific parameters offer potential for the estimation of different values with some degree of confidence; however, the exact magnitude of a given parameter (e.g., temperature) would be dependent on a large number of variables that define the temperature at any given time and location. The spatial aspect of different meteorological variables is much different than the temporal variation. For example, the variation in solar radiation is dependent on cloud cover and frontal passage, which will add a degree of complexity to any attempt to both parameterize and compare modeled and observed results. Throughout this chapter the reader needs to be aware that both spatial and temporal variation exists in all of the different parameters, and there is no single accepted method for characterizing and quantifying either spatial or temporal variation.

There are several challenges that will be covered in this chapter. These include improved spatial and temporal inputs into simulation models, refinement of the methods used to parameterize models, evaluation of the feedback from models when parameters rather than actual data are incorporated into the model, and evaluation of the stability of the predictive methods. These are not impossible challenges and serve to provide a framework for how we should be viewing parameterization methods. These challenges may be addressed from the viewpoint of how energy is exchanged between the surface and the atmosphere and how we could use this framework to develop an understanding of model parameterization. The components within the energy balance approach provide examples of the different parameterization methods. These concepts are not specific to any one model but are described to help foster discussion and continued expansion of parameterization algorithms.

Energy Balance Components

The surface energy balance provides a representation of the energy exchanges and partitioning at the earth's surface and described in a generic form as follows:

$$R_n - G = H + \text{LE} \qquad [1]$$

where R_n is the net radiation (W m^{-2}), G the soil heat flux (W m^{-2}), H the sensible heat flux (W m^{-2}), and LE the latent heat flux (W m^{-2}). This form of the energy balance is a useful approximation of the energy exchanges but is not very useful in simulation models because the processes are not described in sufficient detail in terms of understanding the details that govern the process. For example, R_n, is more fully described as follows:

$$R_n = S_t(1 - \alpha_l) + L_d - \varepsilon\sigma T_d^4 \qquad [2]$$

where S_t is the incoming solar radiation (W m^{-2}), α_l the albedo of the surface, L_d the incoming long-wave radiation (W m^{-2}), and $\varepsilon\sigma T_d^4$ is the outgoing long-wave radiation as a direct function of surface temperature (T_s), ε is the emissivity of the surface, and σ the Stefan–Boltzmann constant of 5.67×10^{-8} W m^{-3} K^{-4}. Equation [2] allows for a detailed examination of parameterization of models because there are methods that have been used to estimate these various inputs from more easily derived variables. Likewise, we can expand the H and LE terms into more complex relationships; then there are other possible parameters that can be identified in these relationships. For example, we can expand H into

$$H = \rho C_p(T_s - T_a)R_{ah} \qquad [3]$$

where ρ is the density of air (kg m^{-3}), C_p the specific heat of air (J kg^{-1} °C^{-1}), T_s is the surface temperature (°C), T_a the air temperature (°C), and r_{ah} the aerodynamic resistance (s m^{-1}) for sensible heat transfer. This equation has been used in several different approaches for estimation of sensible heat. There are other aerodynamic forms that are often used to estimate H as well as the temperature difference method. If we express LE into an expanded form to estimate LE on the basis of surface to air differences then this form is expressed as

$$\text{LE} = \left(\frac{\rho\lambda m}{P}\right)(e_s - e_a)/(r_c + r_{av}) \qquad [4]$$

where λ is the latent heat of vaporization (J kg^{-1}), m the ratio of molecular weight of water vapor to air (0.622), P the barometric pressure (kPa), e_s the saturation vapor pressure (kPa) at the surface temperature, e_a the actual vapor pressure of the overlying air above the canopy, r_c the canopy resistance for water vapor transfer (s m^{-1}), and r_{av} the aerodynamic resistance (s m^{-1}) for water vapor transfer. As with H, there are alternative forms of estimating LE. Both Eq. [3] and [4] allow for a direct representation of the surface changes, which makes it easy to visualize how H and LE respond to changes in the surface conditions.

The energy balance equation also allows for a framework to describe different methods of parameterizing models because Eq. [1] can be separated into the components that illustrate different approaches that could be used.

Parameterizing Radiation Models

Solar Radiation

Solar radiation is one of the critical components of the energy balance model and all plant growth models since this provides the energy. Estimation of S_t values can be derived starting with fairly basic equations because of the known facts about the geometric relationships between the earth and the sun. These are based on the declination, equation of time, daylength, and daily extraterrestrial radiation values, which can then be used to estimate the instantaneous clear-sky irradiance for both the direct and diffuse components. This approach was provided by Ham (2005) as part of a summary of equations useful in micrometeorology. These approaches provide a very good estimation process for solar radiation in the direct and diffuse components under clear sky conditions. Clouds present a unique challenge because of their variability in thickness, time of day, and type. There have been several methods proposed to estimate S_t for cloudy conditions using empirical relationships (e.g., Bristow and Campbell, 1984). Bristow and Campbell (1984) used a relationship based on the difference between the daily maximum and minimum temperatures (ΔT) described as

$$S_t = a[1 - \exp(-b\Delta T^c)]S_e \qquad [5]$$

where S_e is the extraterrestrial solar radiation (W m^{-2}) and a, b, and c are empirically derived values. There are 16 different forms of empirical relationships, as shown in Table 9–1. These empirical models represent a range of different variables and equation forms. The advantage of these types of models is the use of more commonly available meteorological variables compared to a relatively sparse network of solar radiation sensors. Liu et al. (2009) compared these different models in China and found the Bristow and Campbell (1984) model correlated significantly with the common meteorological parameters. There is still a need for calibration of these models, but in their comparison this model was the most

Table 9–1. Empirical relationships for estimating solar radiations based on meteorological variables.†

Empirical relationship	Estimated parameters	Reference
$S_t = a[1 - \exp(-b\Delta T^c)]S_e$	*a, b, c*	Bristow and Campbell (1984)
$S_t = a\sqrt{\Delta T}S_e$	*a*	Hargreaves (1981)
$S_t = a\Delta T^b(1 + cP + dP^2)S_e$	*a, b, c, d*	DeJong and Stewart (1993)
$S_t = a\left[1 - \exp\left(-b\dfrac{\Delta T^c}{\Delta T^b}\right)\right]S_e$	*a, b, c*	Donatelli and Campbell (1998)
$S_t = a\{1 - \exp[-b(T_{avg})\Delta T^c)]\}S_e$ $f(T_{avg}) = 0.017\exp[\exp(-0.053T_{avg}\Delta T^c)]$	*a, b, c*	Donatelli and Campbell (1998)
$S_t = a\{1 - \exp[-bf(T_{avg})\Delta T^2 g(T_{min})]\}S_e$ $g(T_{min}) = \exp(T_{min}/T_{nc})$	*a, b,* T_{nc}	Donatelli and Campbell (1998)
$S_t = a\sqrt{\Delta T}S_e + b$	*a, b*	Hunt et al. (1998)
$S_t = a\sqrt{\Delta T}S_e + bT_{max} + cP + dP^2 + e$	*a, b, c, d, e*	Hunt et al. (1998)
$S_t = a\left[1 - \exp\left(-b\dfrac{\Delta T^c}{S_e}\right)\right]S_e$	*a, b, c*	Goodin et al. (1999)
$S_t = 0.75[1 - \exp(-b\Delta T^2)]S_e$	*b*	Meza and Varas (2000)
$S_t = 0.75\left[1 - \exp\left(-b\dfrac{\Delta T^2}{S_e}\right)\right]S_e$	*b*	Weiss et al. (2001)
$S_t = 0.75\{1 - \exp[-bf(T_{avg})\Delta T^2]\}S_e$	*b*	Weiss et al. (2001)
$S_t = a(1 + 2.7\times10^{-5}\mathrm{Alt})\sqrt{\Delta T}S_e$	*a*	Annandale et al. (2002)
$S_t = \left(a + \sqrt{\Delta T} + b\right)S_e$	*a, b*	Chen et al. (2004)
$S_t = a\{1 - \exp[-bf(T_{avg})\Delta T^2 g(T_{min})]\}S_e$	*b,* T_{nc}	Abraha and Savage (2008)
$S_t = 0.75\left[1 - \exp\left(-b\dfrac{\Delta T^c}{\Delta T_m}\right)\right]S_e$	*b, c*	Abraha and Savage (2008)

† ΔT, diurnal range of temperature (°C); T_{max}, daily maximum temperature (°C); T_{min}, daily minimum temperature (°C); T_{avg}, $(T_{max} + T_{min})/2$; ΔT_m, monthly mean temperature (°C); S_t, solar radiation (W m^{-2}); S_e, extraterrestrial radiation (W m^{-2}); Alt, altitude (m); *P*, precipitation (mm).

stable over the different times. These types of models provide valid methods for the estimation of solar radiation, and many have been developed specifically for use with crop simulation models. There needs to be an expansion of the comparison of these models over wider spatial areas and times of year to evaluate the robustness of the relationships.

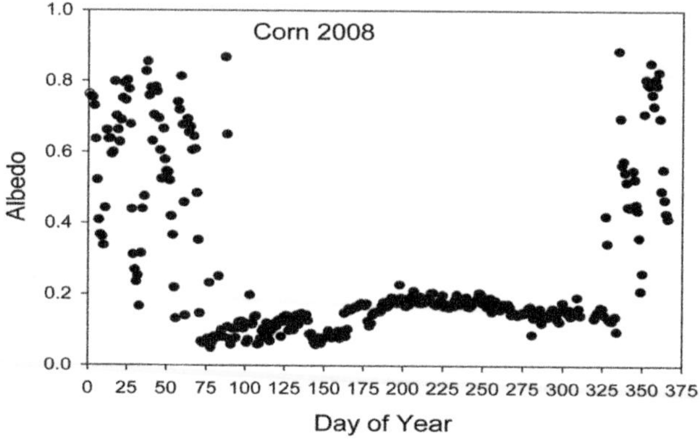

Fig. 9–1. Albedo changes during the course of a year for a corn field in central Iowa.

Albedo

Estimation of albedo is necessary if the energy balance is based on the radiation components and not a direct measure of net radiation. Albedo values for a cropped surface range from 0.22 to 0.32 over the course of a growing season and often a mean value of 0.27 is selected out of convenience. Albedo values vary as a function of soil, with light colored soils having albedo values near 0.30 and darker soils values near 0.15 or less. An example of the changes in albedo throughout the course of a year over a corn (*Zea mays* L.) field in central Iowa is shown in Fig. 9–1. Albedo values are very high during the winter when there is snow cover and change rapidly as the snow melts between snowfall events; during the early spring when there is no crop canopy present the albedo values are near 0.10. Albedo values increase to 0.2 as the canopy develops and then decrease again in the fall as the crop is harvested and the soil is tilled. Each year has different snowfall amounts, which affects the winter albedo, but the trends during the spring to fall are consistent in this area, with increasing albedo values because of the dark soils. In light colored soils, there would be a reversal of these patterns during the course of the year.

Wright (1982) proposed a fairly simple empirical relationship to predict albedo for a crop as a function of day of year (DOY):

$$\alpha = 0.29 + 0.06 \sin\left(\frac{\mathrm{DOY}+96}{57.3}\right) \qquad [6]$$

This is a simple relationship for an alfalfa (*Medicago sativa* L.) crop and is given as an illustration of how α could be quantified for a surface from measurements.

Albedo over a surface varies throughout the year, and in areas with snow cover the annual range in values can be between 0.2 and 0.8. Evaluation of different methods of estimating albedo would provide a valuable input into crop simulation and energy balance models.

Methods are available to obtain albedo from remote sensing measurements, and Carrer et al. (2008) showed that these provided reliable estimates of the spatial variation in albedo and suitable for weather forecasting. The errors in these models would be on the same order of magnitude as other measurements, with errors of less than 5% in the reported values. Reflectivity of a surface is easily obtained with remote sensing data, and for many of the studies on crop simulation modeling or energy balance model that represents a field scale, the use of high resolution models would be necessary. The same principles would apply across all of the different scales in terms of the types of algorithms to estimate albedo.

Long-Wave Radiation

Long-wave radiation is a large component of the total radiation budget and a factor that is often overlooked as to its importance in energy exchanges. As shown in Eq. [2] there are two components in the long-wave portion of net radiation. There are methods available to estimate these fractions of the net radiation balance.

Incoming

The incoming long-wave radiation is a function of the emission from the atmosphere. There have been several attempts to relate incoming long-wave to temperature and humidity in methods similar to those discussed for solar radiation. These different methods were summarized in Hatfield et al. (1983) and are shown in Table 9–2. The comparison of these models over several locations in the United States revealed that that original Brunt formula and the Brutsaert model with a coefficient of 0.575 predicted incoming long-wave radiation with errors less than 5%. These models didn't include a local correction for water vapor, which further increases the simplicity of the model.

Outgoing

The emission of long-wave radiation is a simple function of temperature as shown in Eq. [2], and σ is the Stefan–Boltzmann constant and ε the emissivity of the surface. To estimate the outgoing long-wave radiation requires a measure of the surface temperature or the assumption that the near surface air temperature and the surface temperature are equal or nearly equal. This is not often the case; however, it does provide a reasonable approximation. The other variable term is emissivity, which for most natural surfaces ranges from 0.92 to 0.98. One could assume a con-

Table 9–2. Empirical long-wave radiation estimation methods based on meteorological data.

Empirical relationship	Reference
$L_d = 0.51 + 0.66e_o^{1/2}$	Brunt (1932)
$L_d = 0.92 \times 10^{-5} e_o^{2}$	Swinbank (1963)
$L_d = 1 - 0.261[-7.704(273 - T_o)^2]$	Idso and Jackson (1969)
$L_d = 0.533e_o^{1/7}$	Brutsaert (1975)
$L_d = 0.575e_o^{1/7}$	Idso (1981)
$L_d = 0.179e_o^{1/7}\exp(350/T_o)$	Idso (1981)
$L_d = 0.70 + 5.95 \times 10^{-5} e_o \exp(1500/T_o)$	Idso (1981)

e_o, water vapor pressure (millibars); T_o, air temperature (K).

stant value of 0.95 or 0.96 and not create a large error in the estimation of long-wave radiation. For crop canopies the range of emissivity values is between 0.97 and 0.99, which further reduces the potential error in the ongoing long-wave calculation.

Soil Heat Flux

One of the more difficult parameters to obtain in the energy balance is soil heat flux, *G* (Eq. [1]). Values for *G* can be estimated in simulation models by using the change in spoil temperature with time and depth and estimating the amount of energy required to cause the change in soil temperature assuming the water content, heat capacity of the soil, and bulk density of the soil. These are complex equations that require several inputs, and the approach for the Root Zone Water Quality Model (RZWQM) was described by Flerchinger et al. (2000). Throughout the growing season, *G* can vary from nearly 30% of net radiation to less than 5% under a dense canopy. Because soil heat flux is related to the development of the plant canopy in annual crops, there have been several approaches to estimate values for *G*. Clothier et al. (1986) observed during the regrowth cycles in alfalfa that and found that the ratio of near-infrared to red (NIR/R) was a linearly decreasing function relative to G values. They suggested that this ratio could be used to accurately estimate G in alfalfa crops. Kustas and Daughtry (1990) found that the NIR/R ratio worked equally well in cotton (*Gossypium hirsutum* L.) canopies compared to alfalfa canopies and the standard error of estimate was 0.03 in the ratio of G/R_n. This would be acceptable in energy balance studies. The use of these simple ratios to estimate values for G would be useful in crop simulation models since values of G are rarely observed except in detailed energy balance studies. Bastiaanssen et al. (1998) derived all of the components of the surface energy balance to estimate a spatially distributed surface energy balance and found they could reliably use the ratio of G/R_n derived from remotely sensed data for these large-scale models. Values for G are relatively small compared to the other components, but not accounting for these values will lead to errors in the energy balance, and these methods produce reliable estimation methods.

Estimation of *G* provides an initial step in the estimation of soil temperature profiles within the soil. Soil temperature is relatively well behaved as a process and has a defined sinusoidal pattern throughout the day. Soil temperature has been extensively studied, and there have been several reports written on the methods to estimate soil temperature from a combination of *G*, soil properties, and water content. These were summarized by Novak (2005), who showed how soil temperatures could be estimated from different approaches. Many of these methods are incorporated into simulation models where soil temperatures are required as part of the overall simulation scheme.

Overall Radiation Balance

As described above, the balance of radiation to form net radiation (Eq. [1]) at the surface of the earth can be estimated by several methods. These have been developed as a function of observable meteorological parameters. Air temperature has been used most often because of the extensive nature of these data. Most of the incoming solar radiation models that use some sort of air temperature relationship (Table 9–1) have been shown to be fairly robust for use across a wide range of conditions. Similarly, there are temperature-based methods for incoming long-wave radiation (Table 9–2), and one of the most robust ones is also one of the original relationships developed (Brunt, 1932). The outgoing terms can be estimated using empirical relationships that describe the surface. Of these, albedo shows the largest temporal change for crop canopies because of the growth and senescence of the crop and is where efforts should be placed in improving these relationships.

The alternative to estimating these parameters is to obtain direct measurements as inputs. Both methods have errors associated with them. The errors in different measurement methods have been discussed in a series of chapters in a monograph edited by Hatfield and Baker (2005). Empirical equations extracted from the literature don't often report the associated uncertainty with the different models, and this information would be invaluable to help guide the users of these relationships. On the other hand, there is very little uncertainty assessment provided on observed data, and the user has to assume that every precaution was taken to ensure the highest quality of data possible.

Refinement of Crop Model Parameters

Crop growth models and even energy balance models often generate canopy characteristics that represent different aspects of crop growth to estimate

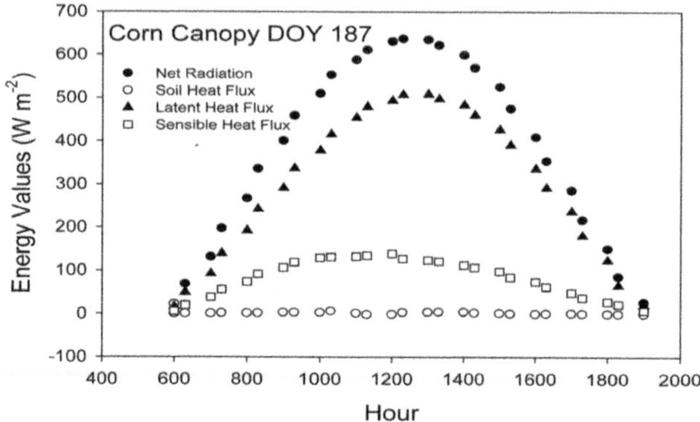

Fig. 9–2. Partitioning of the energy balance throughout the day for a corn canopy in central Iowa.

evapotranspiration or sensible heat exchange. Crop growth parameters change throughout the growing season, and these changes affect the overall energy balance. For example, as the crop develops and as ground cover increases there is a shift in the energy balance toward less soil heat flux and increased latent heat flux. The partitioning of these components throughout the day is shown in Fig. 9–2. On this day, with a leaf area index of 5 there was a very small amount of energy used for soil heat flux, and the canopy had an adequate soil water supply so that latent heat flux was the largest component, followed by sensible heat flux. With limited soil water for evapotranspiration, LE would be reduced and *H* would increase. The seasonal changes in the energy balance are affected by parameters associated with crop growth, and crop simulation models estimate these values. However, the values for these parameters are not always available for many different fields and at a fine temporal resolution expect in very intense field experiments. It is desirable to have estimates of these values to compare model performance and their capability of predicting crop growth.

Characteristics such as leaf area, biomass, ground cover, or even crop residue amounts can be derived primarily from remote sensing methods to provide a feedback to models for crop growth or a direct inputs into some models. Maas (1988) was one of the first to show that is was possible to use this type of approach to improve the performance of crop yield prediction. In his approach there was an estimate of leaf area index (LAI) from remote sensing which was compared to the modeled result and provided feedback to the model as a sort of mid-course correction. This improved the performance of the crop simulation models (Maas, 1993).

Leaf Area Index

Leaf area index can be estimated through many different remote sensing approaches. These have been summarized in various reports (Hatfield et al., 2008), and a brief summary of the literature developed during the past 20 yr still remains useful and critical today as methods to estimate LAI. Zheng and Moskal (2009) reviewed the approaches that can be used at the landscape and regional scales and concluded that incorporation of short-wave infrared (SWIR) wavebands provided an enhancement to the visible and near-infrared wavebands typically used in calculating most vegetative indices.

Leaf area index has been used to assess the ability of a plant to intercept light, and LAI is used as a critical calculation in crop growth models. The NIR/R ratio was found to be highly correlated with green LAI. However, there was a different form of the relationship of LAI = $a + b$ NIR/R for the growth and senescence portions of the growth cycle. Leaf area estimates for different crops across locations have been summarized by Wiegand and Hatfield (1988) and Wiegand et al. (1990). The LAI for wheat (*Triticum aestivum* L.) was best estimated by the NIR/R ratio or the TSAVI index using the linear relationships LAI = $a + b$ NIR/R and LAI = $a + b$ TSVAI. Thus, estimates of LAI are possible with simple linear models using NIR/R reflectance. However, there is improvement when the same parameters are used but in a different form for different growth stages, with greatly improved sensitivity. These relationships are valid across a number of crops and agronomic practices within and among locations. To be useful for standard agronomic practices these relationships need to be calibrated for a specific crop. The multi-site comparisons for corn, wheat, and grain sorghum [*Sorghum bicolor* (L.) Moench] provide a degree of confidence that remote sensing measures can be adequately used to estimate LAI.

There is a also relationship between LAI and light interception in plant canopies. One approach that has been proposed for estimating LAI is based on the relationships between fractional cover, f_C and LAI using a relatively simple exponential relationship (Choudhury, 1987):

$$f_C = 1 - \exp(-\beta \mathrm{LAI}) \quad [7]$$

where β is a function of the leaf angle distribution. He estimated β as 0.67 from an average of 18 broadleaf and grass crops. This method, although robust, has not been applied as often as NIR/R ratios because the first step in this method is to obtain an estimate of ground cover or fractional cover and then incorporated into Eq. [7]. The multiple steps for this approach have contributed to the more widespread use of the simple regression models between vegetation indexes and LAI.

Crop Biomass

Crop biomass represents the total aboveground accumulation of plant material and is a measure of net primary productivity of crop canopies. Biomass directly affects the energy balance because of the associated relationship between increasing biomass and ground cover. Remote sensing has been used to estimate dry matter accumulation or biomass estimation through a combination of NIR and red wavebands. These empirical fits have plant specific relationships because of the difference in NIR reflectance among species; therefore, this approach requires calibration for each crop and soil combination. There is a stronger relationship to green biomass with the NIR/R combinations than to total biomass, which includes stems, branches, and other non-green material.

Other approaches to estimating crop biomass have been to use a conversion factor of intercepted solar radiation to crop biomass using the following form of the relationship:

$$\text{Biomass} = \int \text{PAR fIPAR RUE } \Delta t \quad [8]$$

where PAR is the incident photosynthetic active radiation, fIPAR the fraction of intercepted PAR by the canopy, RUE the radiation use efficiency for conversion of PAR to dry biomass, and Δt the time interval. The estimation of the intercepted values of PAR has taken on many different forms for this approach.

Intercepted Solar Radiation

Estimation of crop biomass is often based on intercepted light by crop canopies and is a critical component in plant growth models. Estimation of light interception by canopies from remotely sensed data would greatly aid in comparing management systems and also in the evaluation of crop growth models. Intercepted light by a crop canopy can be related to the accumulation of biomass and harvestable grain yield. Daily estimates of intercepted light can be obtained from extrapolation of observations of the normalized difference vegetation index (NDVI) combined with daily total PAR obtained from a nearby meteorological station. It is possible to directly compare the performance of different cropping systems using this type of approach with confidence in the results. Observations collected over a wide range of crops and growing conditions suggest that LAI is best obtained from NIR/R ratios, while incident photosynthetic active radiation is best determined from NDVI obtained in remote sensing. These relationships have been developed by several different researchers (Table 9–3) and were recently summarized by Hatfield et al. (2008). Hatfield and Prueger (2010) evaluated the different vegetative indices for corn, soybean [*Glycine max* (L.) Merr.], wheat, and canola (*Brassica napus* L. var. *napus*)

Table 9–3. Summary of selected vegetation indices, wavebands, applications, and citations.†

Index	Relationship	Application	Reference
Difference	$R_{800} - R_{680}$	Biomass	Jordan (1969)
Difference	$R_{800} - R_{550}$	Biomass	Buschmann and Nagel (1993)
Ratio	R_{NIR}/R_{red}	Biomass, LAI	Birth and McVey (1968)
Perpendicular vegetative index	$(R_{NIR} - R_{red} - b)/(1 + a^2)^{1/2}$	LAI	Richardson and Wiegand (1977)
Normalized difference vegetative index	$(R_{NIR} - R_{red})/(R_{NIR} + R_{red})$	Intercepted PAR	Deering (1978)
Soil adjusted vegetative index	$(R_{NIR} - R_{red})(1 + L)/(R_{NIR} + R_{red} + L)$	LAI	Huete (1988)
Transformed soil adjusted vegetative index	$a(R_{NIR} - aR_{red} - b)/(R_{red} + aR_{NIR} - ab)$	LAI, biomass	Baret et al. (1989)

† R_{550}, R_{680}, and R_{800}, reflectance at the 550 (green), 680 (red), and 800 (near-infrared) nm wavelengths, respectively; R_{NIR}, reflectance in near-infrared wavebands; R_{red}, reflectance in red wavebands; *a, b,* and *L*, coefficients based on empirical fits; LAI, leaf area index; PAR, photosynthetic active radiation.

to evaluate the errors in using these vegetative indices and found that properly calibrated models estimated leaf area or biomass with the same error as direct plant sampling. The use of NDVI to obtain intercepted light was shown to have an uncertainty of ±10%, which is comparable to variation in values obtained from radiometers placed above and below the canopy.

One of the applications of intercepted light is determination of the rate of senescence of crop canopies. The rate of change in the NDVI can be utilized as a tool to examine how quickly plants are losing their physiological functions at the end of the growing season. This approach offers potential to determine if the rates of changes are different than expected and may indicate if there are some factors causing premature loss of green leaves in the canopy. This type of approach is often used for visual determination of premature changes in the canopy and could be easily determined from remote sensing platforms.

One component of biomass accumulation is the gross primary productivity (GPP). Gitelson et al. (2006) found that GPP relates closely to total chlorophyll content in maize and soybean. The relationship algorithm for GPP estimation provided accurate estimates of midday GPP in both crops under rainfed and irrigated conditions. This approach has not been rigorously evaluated but offers potential to improve biomass estimates.

Crop Ground Cover

One of the components often evaluated for agricultural applications is the amount of ground covered by the crop canopy, expressed as the fraction of ground area covered by the projection of standing leaf and stem area onto the ground surface. Changes in ground cover are often indicative of the health of the crop. Determination of ground cover provides a linkage between the growth of the crop and water use patterns of the crop since many evapotranspiration (ET) models use crop cover to relate Potential ET to actual ET.

Maas (1988) proposed a method of estimating canopy ground cover in cotton that combined the overall reflectance of the scene and the individual reflectance values from the soil and the crop. He developed the following model for ground cover:

$$GC = (R_{scene} - R_{soil})/(R_{canopy} - R_{soil}) \quad [9]$$

where GC is the fraction of ground cover, R_{scene} is the scene reflectance, R_{soil} is the soil reflectance, and R_{canopy} is the canopy reflectance. By rearranging Eq. [9], the scene reflectance is given as

$$R_{scene} = R_{canopy}GC + R_{soil}(1 - GC) \quad [10]$$

He used reflectance values from either red (0.6–0.7 μm) or NIR (0.8–0.9 μm) for these relationships and found that either waveband could be used. This method of estimating ground cover was independent of location and year. This method was not dependent on empirical fits of the vegetation indexes with plant parameters. Estimation of ground cover via remote sensing has proven to be fairly simple and not subject to problems associated with LAI or incident photosynthetic active radiation. The error in estimates of ground cover using these approaches has been on the order of ±5%. This level of error is acceptable for agricultural applications that require ground cover estimates. In a recent study by Rajan and Maas (2009), they used the photosynthetic vegetation index (PVI) for the canopy compared to a full canopy area ($PVI_{FC)}$ for each pixel being evaluated to derive the following relationship for ground cover (GC)

$$GC = PVI/PVI_{FC} \quad [11]$$

They found they could estimate GC with an accuracy of 3% of the true values, which would be acceptable for almost all applications. Ground cover has been used in the estimation of crop growth (Boissard et al., 1992; Asrar et al., 1992). In a recent analysis, Ritchie et al. (2010) used green/red ratios obtained from digital cameras to estimate ground cover with a correlation of ($r^2 = 0.86$). They found the effective range of ground cover estimates using this method was between 0.2 to

0.8 and was less sensitive at the lower values of ground cover. This would present a problem in the use of this index because the impact of ground cover is most significant at the lower values.

The use of remote sensing methods to derive the above crop growth characteristics either as direct input into large area models or as comparison values for feedback to evaluate and refine model parameters provides a method of comparing across scales. Doraiswamy et al. (2003) provided one of the earlier examples of blending remotely sensed data with crop yield models to estimate crop yields at the state level. There continues to be refinement of these types of approaches for crop yield estimation, and most of these use some combination of the methods described in the previous sections.

Measurement Methods: Comparing Models with Measured Variables

It is beyond the scope of this chapter to cover all of the specific details about the measurement of the different parameters required by models or generated by models that need to be compared in terms of actual versus modeled output. The collection and assembly of high quality data over long periods of time is a valuable resource for the modeling community, and the development of catalogs of available data would be a great asset. Modelers are urged to evaluate the uncertainty associated with specific variables required in the different models, and although sensitivity analyses on model inputs are often conducted, these are not always linked with the measurement errors in a particular measurement. Meek and Hatfield (1994) developed a systematic approach for data quality checking for meteorological station data to help improve the reliability of different data collected from these stations. One of the problems that exists within meteorological datasets is the evaluation of the data quality. This has become an increased emphasis over the past few years, which makes it difficult at times to fully utilize data without a thorough screening before it is used. Another type of comparison that has become more common is to conduct intercomparisons of multiple instruments as a method of being able to assess the differences that could be detected among instruments when deployed in the field. This method was described by Meek et al. (2005) for a series of eddy covariance equipment. They found they could compare multiple instruments, and detectable differences were then used to evaluate differences among locations within a series of watershed scale studies. These types of rigorous intercomparisons may be more valuable to help link observations with simulation models because of the ability to provide an analysis of the instrumentation variance and systematic biases.

Measurement of different variables required as input into simulation models also includes the instrumentation accuracy. There is not a uniform summary of the accuracy for different classes of instrumentation because of the deployment of the instrument, the efforts taken to ensure a high quality data set, the screening of the data for any potential problems, and differences among instruments. The reader is cautioned to be aware of these issues and to work closely with individuals collecting data to understand the complexities in the observational data. A summary and comparison of methods for measuring the parameters discussed in this chapter are included in the monograph by Hatfield and Baker (2005). Measurement methods need to be robust and evaluated for their accuracy and reliability across a number of crops. One thing that would help the modeling community would be to provide more assessment of the variation associated with the values reported in research papers. Then it would be easier to determine if the model predictions are within the variance of the measured results.

Challenges and Emerging Approaches

Parameterizing models requires inputs that may not be readily available from routine sources or may include derived parameters that are difficult to measure directly. The most common approaches to derive the inputs into the energy balance models begin with solar and long-wave radiation, and these have been done with various relationships with temperature. Some of the approaches are locally calibrated and developed and have a limited range of application. Others, however, tend to have a wider range of application. The challenge for the modeling community is to evaluate thoroughly the algorithm against measured data for a particular site. There is no universal method of deriving some of these parameters and perhaps even less attention is given to evaluating the performance of many of these algorithms across larger areas or different applications. The paper by Liu et al. (2009) is one of the more recent comparisons of a number of the solar radiation models. There should be a greater effort to provide these comparisons across a wider range of environments as a service to the modeling community. The approaches currently available help to provide the modeling community with more rigorous analyses of many of these estimation methods. Applying this approach to all of the different modeling components could enhance the reliability and spatial extent of many of the modeling approaches.

There are some emerging challenges that provide some opportunities. One example is the recent effort by Perez et al. (2009) to classify CO_2 concentrations using meteorological classification approaches. With the increasing interest in C dynamics of the atmosphere and the exchange of CO_2 between the surface and the atmosphere this type of approach may provide some insights into the overall model performance or dynamics of the system. The use of Doppler radar to provide estimates of the spatial variation of rainfall provides a distinct advantage over current rain-gauge networks. These efforts will continue to improve as the capabilities and calibration of the Doppler systems advance with technological innovations. Remote sensing of surface soil moisture with microwave methods will increase as satellite systems that carry these instruments become capable of providing more continuous coverage, similarly to the way in which we obtain cloud cover estimates from weather satellites today. All of these offer the potential to improve crop simulation modeling.

The major challenge will be the continued dialog between the modeling community and the measurement community to help improve the parameterization algorithms. Enhancement of models will come with improved inputs and more reliable data sources. Understanding these needs will help to advance modeling efforts and to acquire even more reliable input data. One component of that effort could be the development of a website or location where parameterization routines are posted along with their validation and calibration data sources. With time as these are used the performance evaluations could be made available as well. In the long term this would benefit all of science.

References

Abraha, M.G., and M.J. Savage. 2008. Comparison of estimates of daily solar radiation from air temperature range for application in crop simulations. Agric. For. Meteorol. 148:401–416Annandale, J.G., N.Z. Jovanic, N. Benade, and R.G. Allen. 2002. Software for missing data error analysis of Penman-Monteith reference evapotranspiration. Irrig. Sci. 21:57–67.

Asrar, G., R.B. Myneni, and B.J. Choudhury. 1992. Spatial heterogeneity in vegetation canopies and remote sensing of absorbed photosynthetically active radiation: A modeling study. Remote Sens. Environ. 41:85–103.

Baret, F., G. Guyot, and D.J. Major. 1989. TSAVI: A vegetation index which minimizes soil brightness effects on LAI and APAR estimation. Proc. IGARRS 12th Canadian Symposium on Remote Sensing, Vancouver, Canada. 3:1355–1358.

Bastiaanssen, W.G.M., M. Menenti, R.A. Feddes, and A.A.M. Holtslag. 1998. A remote sensing surface energy balance algorithm for land (SEBAL). 1. Formulation. J. Hydrol. 212–213:198–212.

Birth, G.S., and G. McVey. 1968. Measuring the color of growing turf with a reflectance spectrophotometer. Agron. J. 60:640–643.

Boissard, P., J.-G. Pointel, and J. Tranchefort. 1992. Estimation of the ground cover ratio of wheat using radiometry. Int. J. Remote Sens. 13:1681–1692.

Bristow, K.L., and G.S. Campbell. 1984. On the relationship between incoming solar radiation and daily maximum and minimum temperature. Agric. For. Meteorol. 31:159–166.

Brunt, D. 1932. Notes on radiation in the atmosphere. Q. J. R. Meteorol. Soc. 58:389–418.

Brutsaert, W. 1975. On a derivable formula for long-wave radiation from clear skies. Water Resour. Res. 11:742–744.

Buschmann, C., and E. Nagel. 1993. In vivo spectroscopy and internal optics of leaves as basis for remote sensing of vegetation. Int. J. Remote Sens. 14:711–722.

Carrer, D., B. Geiger, J.-L. Roujean, O. Hautecoeur, J. Cedilnik, J.-F. Mahfouf, C. Meurey, and O. Franchisteguy. 2008. Land surface albedo from MSG/SEVIRI: Retrieval method, validation, and application for weather forecast. Trans. Geosci. Remote Sens. 47:3841–3856.

Chen, R.S., K. Ersi, J.P. Yang, S.H. Lu, and W.Z. Zhao. 2004. Validation of five global radiation models with measured daily data in China. Energy Conserv. Manage. 45:1759–1769.

Choudhury, B.J. 1987. Relationships between vegetation indices, radiation absorption, and net photosynthesis evaluated by a sensitivity analysis. Remote Sens. Environ. 22:209–233.

Clothier, B.E., K.L. Clawson, P.J. Pinter, Jr., M.S. Moran, R.J. Reginato, and R.D. Jackson. 1986. Estimation of soil heat flux from net radiation during the growth of alfalfa. Agric. For. Meteorol. 37:319–329.

Deering, D.W. 1978. Rangeland reflectance characteristics measured by aircraft and spacecraft sensors. Ph.D. diss. Texas A&M University, College Station, TX.

DeJong, R., and D.W. Stewart. 1993. Estimating global solar radiation from common meteorological observations in western Canada. Can. J. Plant Sci. 73:509–518.

Donatelli, M., and G.S. Campbell. 1998. A simple model to estimate global solar radiation. p. 133–134. *In* Proc. ESA Congr. 5th Nitra Slovak Republic, 28 June– 2 July 1998. The Slovak Agric. Univ. Nitra, Slovak Republic.

Doraiswamy, P.C., S. Moulin, P.W. Cook, and A. Stern. 2003. Crop yield assessment from remote sensing. Photogramm. Eng. Remote Sens. 69:665–674.

Flerchinger, G.N., R.M. Aiken, K.W. Rojas, L.R. Ahuja, K.E. Johnson, and C.V. Alonso. 2000. Soil heat transport, soil freezing and snowpack conditions. p. 281–314. *In* L.J. Ahjua et al. (ed.) Root Zone Water Quality Model: Modeling management effects on water quality and crop production. Water Resources Publ., Highlands Ranch, CO.

Gitelson, A.A., A. Viña, S.B. Verma, D.C. Rundquist, T.J. Arkebauer, G. Keydan, B. leavitt, V. Ciganda, G.G. Burba, and A.E. Suyker. 2006. Relationship between gross primary production and chlorophyll content in crops: Implications for the synoptic monitoring of vegetation productivity. J. Geophys. Res. 111:D08S11, doi:10.1029/2005JD006017.

Ham, J.M. 2005. Useful equations and tables in micrometeorology. p. 533–560. *In* L.J. Hatfield and J.M. Baker (ed.) Micrometeorology in agricultural systems. Agron. Monogr. 47. ASA, CSSA, and SSSA, Madison, WI.

Hargreaves, G.H. 1981. Responding to tropical climates. p. 29–32. *In* The 1980–81 food and climate review. The Food and Climate Forum, Aspen Institute for Humanistic Studies, Boulder, CO.

Hatfield, J.L., and J.M. Baker. 2005. Micrometeorology in agricultural systems. Agron. Monogr. 47. ASA, CSSA, and SSSA, Madison, WI.

Hatfield, J.L., A.A. Gitelson, J.S. Schepers, and C.L. Walthall. 2008. Application of spectral remote sensing for agronomic decisions. Agron. J. 100:S-117–S-131.

Hatfield, J.L., and J.H. Prueger. 2010. Value of using different vegetative indices to quantify agricultural crop characteristics at different growth stages under varying management practices. Remote Sens. 2:562–578, doi:10.3390/rs2020562.

Hatfield, J.L., R.J. Reginato, and S.B. Idso. 1983. Comparison of long-wave radiation calculation methods over the United States. Water Resour. Res. 19:285–288.

Huete, A.R. 1988. A soil-adjusted vegetative index (SAVI). Remote Sens. Environ. 16(25):295–309.

Hunt, L.A., L. Kucharb, and C.J. Swanton. 1998. Estimation of solar radiation for use in crop modeling. Agric. For. Meteorol. 91:293–300.

Idso, S.B. 1981. A set of equations for full spectrum and 8–14 μm and 10.5–12.5 μm thermal radiation from cloudless skies. Water Resour. Res. 17:295–304.

Idso, S.B., and R.D. Jackson. 1969. Thermal radiation from the atmosphere. J. Geophys. Res. 74:3397–3403.

Jordan, C.F. 1969. Derivation of leaf area index from quality of light on the forest floor. Ecology 50:663–666.

Kustas, W.P., and C.S.T. Daughtry. 1990. Estimation of the soil heat flux/.net radiation ratio from spectral data. Agric. For. Meteorol. 49:205–223.

Liu, X., X. Mei, Y. Li, Q. Wang, J.R. Jensen, Y. Zhang, and J.R. Porter. 2009. Evaluation of temperature-based global solar radiation models in China. Agric. For. Meteorol. 149:1433–1446.

Maas, S.J. 1988. Using satellite data to improve model estimates of crop yield. Agron. J. 80:655–662.

Maas, S.J. 1993. Parameterized model of gramineous crop growth: II. Within-season simulation calibration. Agron. J. 85:354–358.

Meek, D.W., and J.L. Hatfield. 1994. Data quality checking for single station micrometeorological databases. Agric. For. Meteorol. 69:85–109.

Meek, D.W., J.H. Prueger, W.P. Kustas, and J.L. Hatfield. 2005. Determining meaningful differences for SMEX eddy covariance measurements. J. Hydrometeorol. 6:805–811.

Meza, F., and E. Varas. 2000. Estimation of mean monthly solar global radiation as a function of temperature. Agric. For. Meteorol. 100:231–241.

Novak, M.D. 2005. Soil temperature. p. 105–129. *In* J.L. Hatfield and J.M. Baker (ed.) Micrometeorology in agricultural systems. Agron. Monogr. 47. ASA, CSSA, and SSSA, Madison, WI.

Perez, I.A., M.L. Sanchez, M.A. Garcia, and B. de Torre. 2009. A classification of CO_2 concentrations based on a binary meteorological six variable system. Agric. For. Meteorol. 149:1686–1692.

Rajan, N., and S.J. Maas. 2009. Mapping crop ground cover using airborne multispectral digital imagery. Precis. Agric. 10:304–318.

Richardson, A.J., and C.L. Wiegand. 1977. Distinguishing vegetation from soil background information. Photogramm. Eng. Remote Sens. 43:1541–1552.

Ritchie, G.L., D.G. Sullivan, W.K. Vencill, C.W. Bednarz, and J.E. Hook. 2010. Sensitivities of normalized difference vegetation index and a green/red ratio index to cotton ground cover fraction. Crop Sci. 50:1000–1010.

Swinbank, W.C. 1963. Long-wave radiation from clear skies. Q. J. R. Meteorol. Soc. 89:339–348.

Weiss, A., C.J. Hays, Q. Hu, and W.E. Easterling. 2001. Incorporating bias error in calculating solar irradiance: Implications for crop yield simulations. Agron. J. 93:1321–1326.

Wiegand, C.L., A.H. Gerbermann, K.P. Gallo, B.L. Blad, and D. Dusek. 1990. Multisite analyses of spectral-biophysical data for corn. Remote Sens. Environ. 33:1–16.

Wiegand, C.L., and J.L. Hatfield. 1988. The spectral-agronomic multisite-multicrop analyses (SAMMA) project. Int. Arch. Photogramm. Remote Sens. 27(B7):696–706.

Wright, J.L. 1982. New evapotranspiration crop coefficients. J. Irrig. and Drain. Div. 108:57–74.

Zheng, G., and L.M. Moskal. 2009. Retrieving leaf area index (LAI) using remote sensing: Theories, methods and sensors. Sensors 9:2719–2745.

Chapter 10

Quality Assurance of Weather Data for Agricultural System Model Input: A Case Study using the Walnut Creek Watershed in Central Iowa

R.W. Malone
D.W. Meek
L. Ma
D.B. Jaynes
B.T. Nolan
D.L. Karlen

Abstract

Use of agricultural system models in field research requires quality weather data for input. This chapter provides methods to assess the integrity of rainfall, radiation, and humidity input for the Root Zone Water Quality Model (RZWQM). These methods were applied to a recent RZWQM application at Walnut Creek in central Iowa. Data bias of 10% in humidity, rainfall, and solar radiation results in long-term errors of more than 40% for nitrate loss in tile drains from these RZWQM simulations. Actual weather records show measurement errors in rain, solar radiation, and humidity of 10% or more. The quality assurance procedures described here will help model users detect erroneous weather measurements, which will help improve their model simulations through more accurate simulations and higher confidence in the weather input.

R.W. Malone (rob.malone@ars.usda.gov), D.W. Meek (dave.meek@ars.usda.gov), D.B. Jaynes (dan.jaynes@ars.usda.gov), D.L. Karlen (doug.karlen@ars.usda.gov), USDA-ARS, National Laboratory for Agriculture and the Environment, 2110 University Boulevard, Ames, IA 50011-0001; L. Ma, USDA-ARS, Agricultural Systems Research Unit, 2150 Centre Ave. Bldg. D, Fort Collins, CO 80526 (liwang.ma@ars.usda.gov); B.T. Nolan, USGS, 413 National Center, Reston, VA 20192 (btnolan@usgs.gov).

doi:10.2134/advagricsystmodel2.c10

Methods of introducing system models into agricultural research. L.R. Ahuja and L. Ma (ed.)
Advances in Agricultural Systems Modeling Series 2.

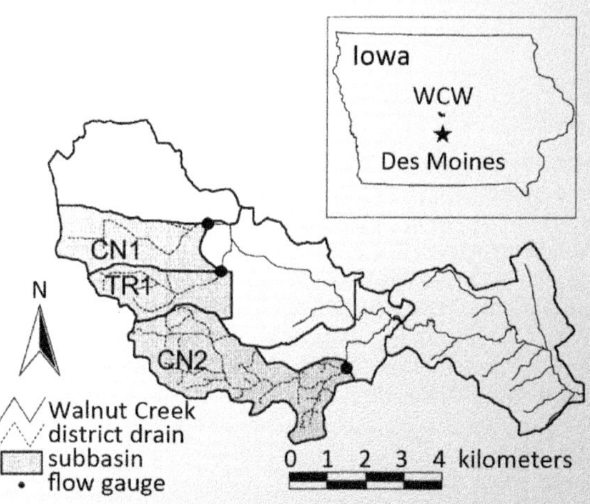
Iowa
WCW
Des Moines
N
CN1
TR1
CN2
Walnut Creek
district drain
subbasin
flow gauge
0 1 2 3 4 kilometers

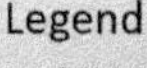
Legend
Weather Stations
Tipping Buckets

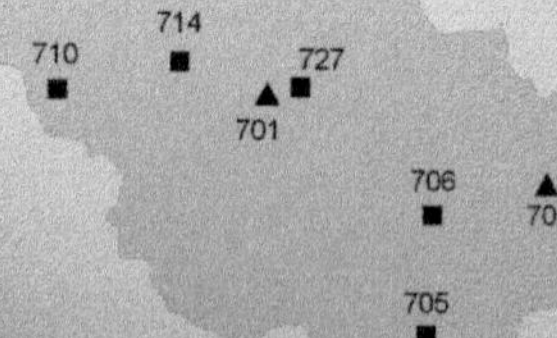
710
714
727
701
706
702
705
0 750 1,500 30,000 meters

It is well known that crop production and hydrologic variation on watersheds is weather related. This information can be found throughout the research literature. Recently, year-to-year crop yield variability was observed to be associated with weather variability such as radiation and rainfall (e.g., Malone et al., 2009). Crop- and hydrological model–simulated yield and river discharge variability are sensitive to synthetic weather generator error and variability (e.g., Dubrovský et al., 2000; Schuol and Abbaspour, 2007). In fact, Fodor and Kovacs (2005) reported that certain combinations of weather data errors resulted in more than 20% deviation in model-simulated crop yield. Also, weather input can be a more significant factor than soil input for simulated crop yields (e.g., Mathe-Gaspar et al., 2005).

Although the effect of weather variability on crop and hydrologic systems is well known, the quality of raw meteorological observations is suspect (e.g., Shafer et al., 2000; Allen, 1996; Carlson et al., 1994; Meek and Hatfield, 1994). Recently, this has led to more reported quality assurance (QA) procedures for meteorological data in evapotranspiration model research (e.g., Jabloun and Sahli, 2008; Kjaersgaard et al., 2007; Farahani et al., 2007). Rarely, however, are meteorological data quality checks reported for agricultural systems model applications (e.g., APSIM [Agricultural Production Systems Simulator], RZWQM, SWAT [Soil and Water Assessment Tool], CERES-MAIZE [Crop Environment Resource Synthesis–Maize], DSSAT [Decision Support System for Agrotechnology Transfer]).

Our objectives are to present QA procedures for agricultural system model weather data input as applied to a recent model application. More comprehensive and general analyses of QA procedures for meteorological data have recently been reported elsewhere (e.g., Fiebrich et al., 2010). Few reports, however, have discussed weather data QA procedures as applied to agricultural system model applications. We use the modeling study by Malone et al. (2010) as an example. They tested RZWQM simulations to small subbasin-scale data from the Walnut Creek watershed in central Iowa first described by Jaynes et al. (2004) . In this experiment the late spring soil nitrate test (LSNT) was used to determine fertilizer application rates in one subbasin of the watershed, and primarily fall applied N fertilizer treatments were used in adjacent subbasins.

Weather Measurement Problems

Some common problems associated with weather sensor measurements are listed in Table 10–1. These include electronic failure, physical interference (debris, spider webs, insect nests), poor calibration and sensor drift, poor location and setup, instrument malfunction (e.g., cracked bearings on tipping bucket rain gauges), high wind resulting in underpredicted rainfall, and low wind resulting in overpredicted temperature. According to field technicians at the National Laboratory for Agriculture and the Environment (NLAE), among the most common weather sensor problems at the NLAE weather stations are spider, insect, and bird activity and poor station location (Kevin Cole and Kelly Barnett, NLAE, Ames, IA, personal communication, 2010). They go on to say that the following contribute to poor data quality: rodent damage (especially to the wiring), improper maintenance (including infrequent sensor calibration), and oxidation. These problems are similar to those found in larger weather networks (e.g., Fiebrich et al., 2006). The procedures presented below can help one assess the quality of measured data (e.g., Allen, 1996; Fiebrich et al., 2010). Other benefits of QA procedures include detecting unusual weather phenomena and extreme events.

Sensitivity of Model Output to Errors in Weather Data

Malone et al. (2010) applied RZWQM to the Walnut Creek Watershed in central Iowa. Early in this analysis incorrect meteorological inputs were identified as a factor that reduced the quality of simulated crop yield, drainage, and N loss. Therefore, QA procedures were implemented. These procedures were in addition to those of Meek and Hatfield (1994) that are routinely applied to NLAE weather data. Figure 10–1 show the Walnut Creek Watershed located in central Iowa (41°55′ to 42°00′ N; 93°32′ to 93°45′ W). Weather and cropping patterns have been monitored within the 5130-ha watershed since 1991. Details of the location, geology, soils, climate, land use, and farming practices are found in Hatfield et al. (1999) and associated references. Details of the LSNT-treatment and control subbasins within the watershed are found in Jaynes et al. (2004).

The LSNT treatment was implemented for 4 yr (1997–2000) on 16 fields (300 ha) of a 366-ha subbasin within the larger watershed and was designated TR1 to retain the terminology of Jaynes et al. (2004) (see Fig. 10–1). The adjacent subbasins CN1 and CN2 were selected as controls for a paired watershed research design.

The level of tolerable weather measurement error depends on the objectives and the specific conditions of the agricultural system model application. To simply illustrate the effects of biased meteorological input with the study at Walnut Creek, Iowa (Malone et al., 2010) we adjusted relative humidity (RH), solar radia-

Table 10–1. Examples of problems associated with weather sensor measurements.

Cause	Citations†
Rainfall	
Electronic failure	Howell et al. (1984)
Tipping mechanism can ice over	Meek and Hatfield (1994)
Tipping mechanism can get clogged with debris	Shafer et al. (2000)
Tipping mechanism can become entangled w/spider webs	Fiebrich et al. (2006)
High wind can cause underestimation	Yang et al. (1998)
High rainfall rate can cause too few tips per time resulting in underestimation	Meek and Hatfield (1994) Shafer et al. (2000)
Tipping bucket design and physical problems such as cracked bearings	
Humidity	
Corrosion	Howell et al. (1984)
Insect nests and spider webs interfere with the sensor	Fiebrich et al. (2006)
Sensor drift	Fiebrich et al. (2006)
Temperature	
Insect nests and spider webs interfere with the sensor	Fiebrich et al. (2006)
Radiation errors due to low wind and poor shielding and ventilation	Tanner (2000)
Naturally ventilated shields over snow can have large errors	Tanner (2000)
Solar radiation	
Poor location (e.g., too high or low due to reflective or shadow surface)	Meek et al. (2000)
Bird droppings and debris obscure the lens	Fiebrich et al. (2006)
Sensor drift	Fiebrich et al. (2006)
High bias in historical strip chart records	Meek et al. (2000)

† Citations are examples; other sources could be used.

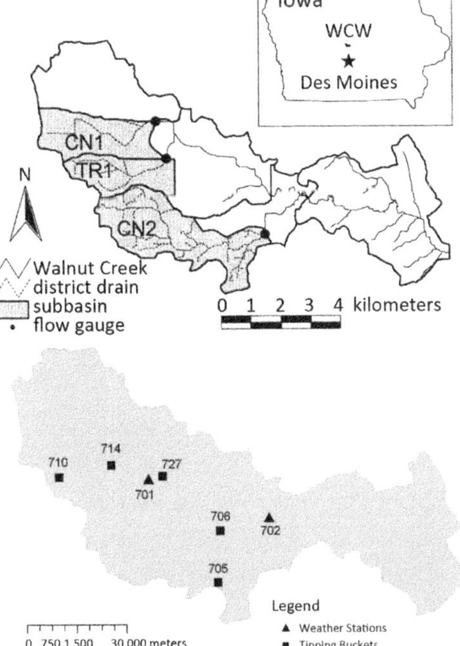

Fig. 10–1. Walnut Creek Watershed. Location of Walnut Creek Watershed (WCW) within the state of Iowa (inset), and the location of the stream, district drains, discharge gauging stations, tipping bucket rain gauges, weather stations, control subbasins (CN1 and CN2), and treatment subbasin (TR1) within WCW (Jaynes et al., 2004).

tion (SR), and rainfall +10%, 0, and −10% from 1970–2004 and used RZWQM to simulate nitrate loss in tile drainage and corn (*Zea mays* L.) yield over this period. This resulted in changes of up to +41% in N loss to drains, −70% in tile flow, +37% in N concentration, and −15% in crop yield (Table 10–2). Within the range of variable change and the conditions of the Walnut Creek watershed, single variable adjustment suggests that N loss was most sensitive to change in rainfall (−21%) and least sensitive to radiation (−5%). Corn yield was most sensitive to radiation (+8%) and least sensitive to rainfall (−1.2%).

Quality Assurance of Weather Data

Methods to assure the quality of meteorological data include double-mass analysis of rainfall, comparison of measured radiation to theoretical clear sky radiation, comparison of maximum hourly RH to 100%, and comparison of multiple humidity sensors in the area (Fiebrich et al., 2010; Allen, 1996). We applied these methods to data collected from the Walnut Creek Watershed in central Iowa.

Rainfall

Quality assurance of rainfall can be evaluated by double-mass analysis, which compares accumulated rainfall for individual gauges to that of nearby sites (e.g., Fiebrich et al., 2010; Gupta, 1989; Allen et al., 1998). We compared five rain gauges within 6 km of each other for 3 yr within the Walnut Creek Watershed (Fig. 10–1). We designated gauges 705 and 710 as the *x* axis and 706, 714, and 727 as the *y* axis. Visual inspection of these figures indicates that successive points did not follow an obvious unique straight line at five periods (Fig. 10–2), which are listed in Table 10–3. In the case of these data, all five gauges recorded rainfall during these periods while the problem gauge did not record rainfall. Field notes revealed problems such as spider webs preventing the tipping bucket from performing properly, clogging, and wiring problems (Table 10–3). The problems caused the rain gauges to miss from 25 to 182 mm of rainfall compared to gauges without noted problems on those specific dates (Table 10–3). This resulted in nearly a 20% annual bias in 2007.

To reduce rainfall input quality problems and to reduce spatial variation effects caused by selecting one gauge for rainfall input, Malone et al. (2010) used the median weekly rainfall as the most representative gauge for rainfall input to RZWQM from several gauges throughout the watershed. Therefore, the "representative" rain gauge was used for each week's RZWQM breakpoint rainfall input. Note that the subbasin was treated as a point in Malone et al. (2010).

Table 10–2. Change in RZWQM simulated output with 10% changes in weather data. The data is sorted by the smallest to largest change in N loss.†

Weather variable change			RZWQM output change			
Precipitation	Radiation	Humidity	N loss	Drainage	N concentration	Corn yield
			%			
−10	10	−10	−35.8	−53.3	37.4	−0.9
−10	0	−10	−31.9	−48.8	33.1	−7.1
−10	10	0	−25.9	−36.6	16.9	6.4
−10	−10	−10	−23.7	−44.2	36.8	−15.4
−10	0	0	−21.3	−30.8	13.7	−1.2
0	10	−10	−16.4	−27.6	15.5	3.7
−10	−10	0	−12.2	−24.8	16.7	−10.3
−10	10	10	−11.9	−14.2	2.7	10.2
0	0	−10	−11.7	−21.9	13.0	−3.6
−10	0	10	−6.2	−6.7	0.6	2.1
0	10	0	−4.8	−7.0	2.4	8.0
0	−10	−10	−4.1	−16.0	14.2	−12.6
0	0	0	0.0	0.0	0.0	0.0
10	10	−10	2.7	0.9	1.8	5.4
−10	−10	10	3.1	0.7	2.4	−7.3
10	0	−10	6.3	6.9	−0.6	−2.3
0	−10	0	8.3	7.1	1.1	−9.3
0	10	10	8.5	18.5	−8.4	11.2
10	−10	−10	13.7	13.2	0.4	−11.5
10	10	0	13.9	24.3	−8.4	8.7
0	0	10	14.8	27.1	−9.7	2.7
10	0	0	18.3	31.7	−10.2	0.6
0	−10	10	22.7	36.0	−9.7	−6.2
10	−10	0	25.6	38.9	−9.5	−8.6
10	10	10	28.3	52.8	−16.0	11.3
10	0	10	33.1	61.8	−17.7	3.1
10	−10	10	40.7	70.5	−17.5	−5.7

† The weather and RZWQM output variables are (left to right): precipitation (cm), radiation ($MJ\ m^{-2}\ d^{-1}$), humidity (%), nitrate loss in tile flow ($kg\ N\ ha^{-1}$), drainage amount (cm), flow-weighted N concentration in drainage ($mg\ N\ L^{-1}$), corn yield ($Mg\ ha^{-1}$).

Solar Radiation

Pyranometer accuracy can be evaluated by comparing against clear sky SR (R_{so}), which can be estimated from Allen (1996)

$$R_{so} = (0.75 + z0.00002)R_a \quad [1]$$

where z is station elevation (290 m) and R_a is extraterrestrial radiation computed as described in Allen (1996). R_{so} is less than R_a because of atmospheric clarity (e.g.,

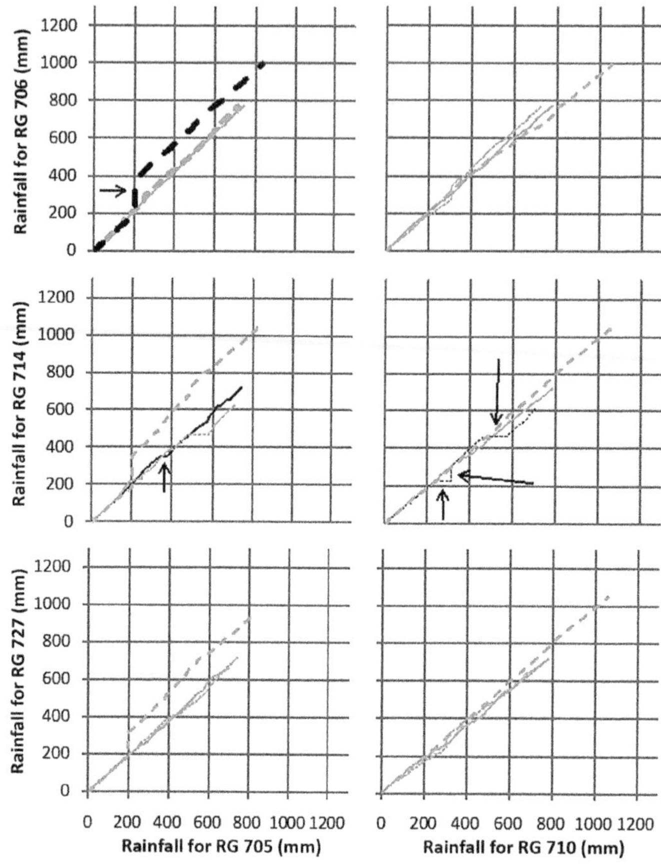

Fig. 10–2. Double-mass analysis applied to 18 series of cumulative annual precipitation for 2005 (solid line), 2006 (thin dotted line), and 2007 (thick dashed line). Therefore, six dual rain gauge (RG) comparisons for 3 yr are presented. The locations of the five stations are shown in Fig. 10–1. The arrows point to the time period for the problem RG (Table 10–3). To simplify the figure, only one dual RG comparison per year is in black; the other year–gauge comparisons are in light gray.

Table 10–3. Problems noted from double mass analysis of rainfall gauges in Walnut Creek Watershed (see Fig. 10–1 and 10–2). The rain gauge locations (e.g., 714) are given in Fig. 10–1.

Year	Problem gauge	Gauge level	Range of other four gauges increase (range percent difference)	Noted problem
		mm	mm	
2005	714	353	25–38 (34%)	Spider webs
2006	714	229	36–55 (35%)	Clogged top
2006	710	308	54–73 (26%)	Wiring
2006	714	468	87–112 (22%)	Spider webs
2007	705	201	130–182 (29%)	Wiring

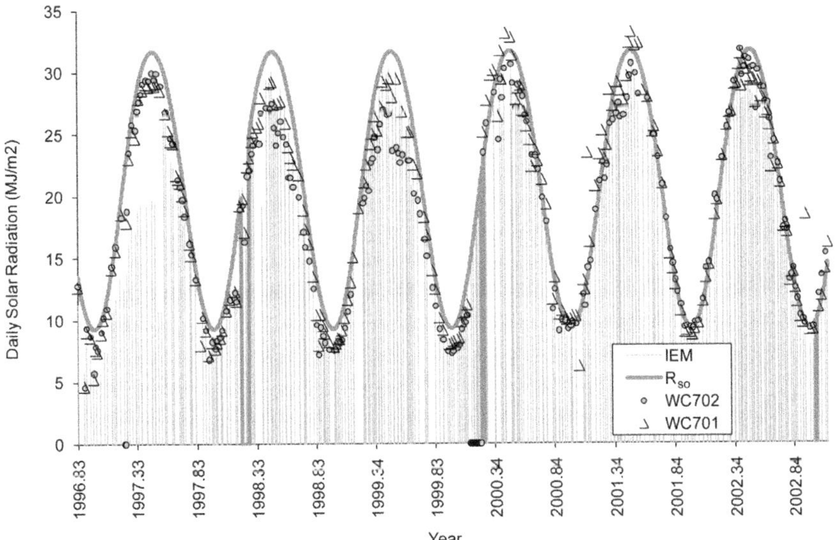

Fig. 10–3. Ten-day daily maximum solar radiation for three sensors in central Iowa compared against clear sky condition solar radiation (R_{so}). WC702 and WC701 are the eastern- and western-most weather stations located in the Walnut Creek Watershed (Fig. 10–1); IEM is the Iowa Environmental Mesonet (IEM, 2009).

dust, pollution, and humidity). More comprehensive approaches are also available (e.g., Meek, 1997).

Figure 10–3 shows Eq. [1] estimated clear sky and 10-d maximum measured daily SR from three different radiometers near the Walnut Creek Watershed. Poor radiometer performance on a routine basis is clear: IEM (Iowa Environmental Mesonet, 2009) before 1998 (low); WC702 from 1998–1999 (low); and WC701 for 2000–2001 (high). WC702 and WC701 are the eastern- and western-most weather stations located in the Walnut Creek Watershed (Fig. 10–1) To illustrate the effect of a poor performing radiometer on monthly radiation, the July 1999 average daily SR was 19.4 for WC702 and 23.4 for WC701 (17% difference of 23.4).

After eliminating these three data groups of poor performing radiometers (Fig. 10–3), we selected daily RZWQM input based on the radiometer (WC701, WC702, or IEM) that was closest to the estimated monthly SR (R_s; Fig. 10–4; Allen, 1996)

$$R_s = 0.16R_a(T_{max} - T_{min})^{0.5} \qquad [2]$$

where T_{max} and T_{min} are average monthly maximum and minimum daily temperature.

Humidity

Daily RH data were quality checked by comparing WC701, WC702, Des Moines (DSM), and IEM (Ames, IA). The different datasets report either hourly RH, actual

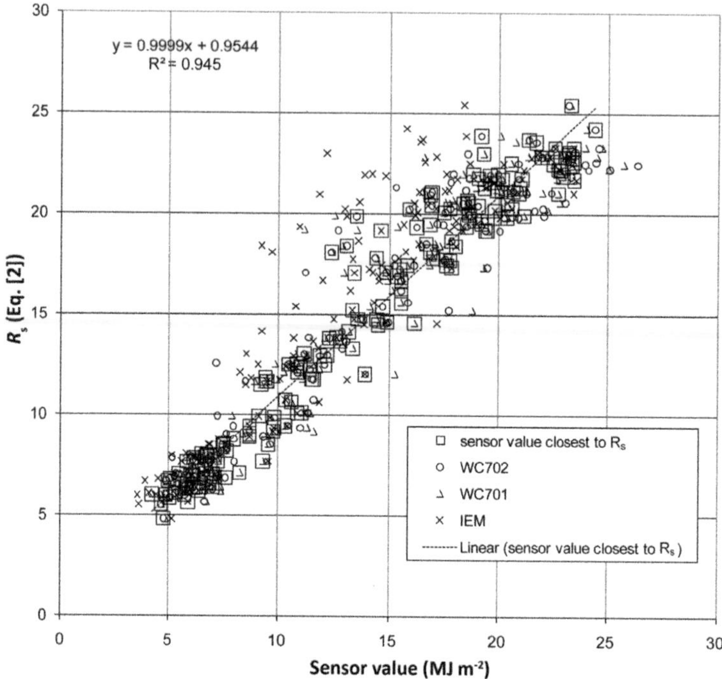

Fig. 10–4. Average monthly solar radiation from three sensors in central Iowa compared with estimated monthly solar radiation (R_s) from Eq. [2]. WC702 and WC701 are the eastern- and western-most weather stations located in the Walnut Creek Watershed (Fig. 10–1); IEM is the Iowa Environmental Mesonet (IEM, 2009).

vapor pressure (E_a), or dew point temperature. Hourly measured dew point and actual air temperature were used to calculate actual and saturation vapor pressure (E_a and E_s) from a form of the Magnus–Tetens formula:

$$E_a \text{ and } E_s = 0.611 \times 10^{[7.5T/(T+237.7)]} \quad [3]$$

where T is dew point temperature for calculation of actual vapor pressure (e.g., T_d) or T is actual air temperature for calculation of saturation vapor pressure (T_a).

When hourly humidity and temperature were reported, it was converted to hourly E_a and E_s (RH = $E_a 100/E_s$). For all datasets, daily RH was calculated as average daily E_a divided by average daily E_s.

The Des Moines airport daily RH values were consistently lower than the Walnut Creek values (data not shown). For example, the average RH in August 1996 was 77% for DSM and 86% for WC702 (10% difference). These results are consistent with other research reporting lower RH in commercial environments than

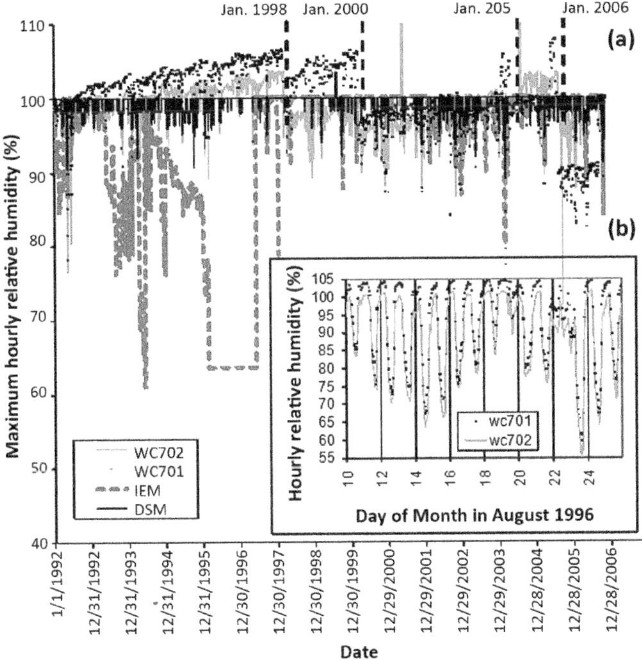

Fig. 10–5. Hourly relative humidity. (a) The 10-d running maximum hourly relative humidity for four sensors. (b) The hourly relative humidity for two weather stations within Walnut Creek Watershed (Fig. 10–1) during August 1996. DSM is Des Moines airport and IEM is Iowa Environmental Mesonet (IEM, 2009). WC702 and WC701 are the eastern- and western-most weather stations located in the Walnut Creek Watershed.

the nearby countryside (e.g., Akinbode et al., 2008). Higher countryside RH measurements complicates use of Des Moines airport measurements for this research even with the short distance from the Walnut Creek watershed to DSM (distance is approximately 48 km [30 miles]).

Relative humidity from IEM was also not used for input because before 1998 the IEM maximum daily humidity was well below 100 for extended periods of time (Fig. 10–5), raising concerns about the quality of the data (Allen, 1996). We used WC702 rather than WC701 for most RH input because the maximum hourly RH remained closer to 100% for much of the study period (Fig. 10–5). For example, the maximum hourly RH reached 105% for 701 at several periods in August 1996 (Fig. 10–5b), and the average monthly RH for August 1996 was 86 for 702 and 90% for 701. Maximum hourly WC702 RH was low in November 1998 through March 1999, when evaporation was low and had minimal effect on RZWQM simulations. Daily WC701 RH was used for May 2001 and August 2004 through April 2005 because WC702 maximum hourly RH was high for these periods (Fig. 10–5). An alternative method of QA for humidity measurements may be to compare dew points using paired observations that can be modified from Meek et al. (2009).

Summary and Conclusions

Clearly high quality weather data are required for high quality agricultural system model simulations. Problems associated with weather sensors include electronic failure; rodent damage; activity of insects, spiders, and birds; poor location, calibration, and setup; and oxidation. Weather data bias of 10% in humidity, rainfall, and solar radiation resulted in long-term errors of more than 40% for nitrate loss in tile drains from central Iowa RZWQM simulations in the Walnut Creek Watershed. Actual weather records showed measurement errors in rain of 20% (annual, 2007), solar radiation of 17% (July 1999), and humidity of 10% (August 1996). Methods used to ensure the quality of meteorological data included double-mass analysis of rainfall, comparison of measured radiation with theoretical clear sky radiation, and comparison of maximum hourly relative humidity to 100%, as well as comparison of multiple humidity sensors in the area. For our model simulations, data were chosen that appeared to have the least error from several area rainfall, radiation, and humidity records. The results suggest the methods used were capable of detecting erroneous meteorological measurements.

Acknowledgments

We appreciate the efforts and diligence of a number of individuals who helped make the complete Walnut Creek Project possible. These include Dana Dinnes, Kelly Barnett, Jeff Cook, Kevin Cole, Diane Farris, Forrest Goodman, Richard Hartwig, Kent Heikens, Amy Morrow, Jeff Nichols, Jody Ohmacht, Larry Pellack, Donna Schmitz, Margaret Smith, Heartland Co-Op of Slater, IA, and the eight cooperating farmers and their families.

References

Akinbode, O.M., A.O. Eludoyin, and O.A. Fashae. 2008. Temperature and relative humidity distributions in a medium-size administrative town in southwest Nigeria. J. Environ. Manage. 87:95–105.

Allen, R.G. 1996. Assessing integrity of weather data for reference evapotranspiration estimation. J. Irrig. Drain. Eng. 122:97–106.

Allen, R.G., L.S. Pereira, D. Raes, and M. Smith. 1998. Crop evapotranspiration: Guidelines for computing crop water requirements. Irrigation and Drainage Paper 56. FAO, Rome.

Carlson, R.E., J.W. Enz, and D.G. Baker. 1994. Quality and variability of long-term climate data relative to agriculture. Agric. For. Meteorol. 69:61–74.

Dubrovský, M., Z. Zalud, and M. Stastna. 2000. Sensitivity of CERES-maize yields to statistical structure of daily weather series. Clim. Change 46:447–472.

Farahani, H.J., T.A. Howell, W.J. Shuttleworth, and W.C. Bausch. 2007. Evapotranspiration: Progress in measurement and modeling in agriculture. Trans. ASABE 50:1627–1638.

Fiebrich, C.A., D.L. Grimsley, R.A. McPherson, and K.A. Kesler. 2006. The value of routine site visits in managing and maintaining quality data from the Oklahoma Mesoset. J. Atmos. Ocean. Technol. 23:406–416.

Fiebrich, C.A., C.R. Morgan, A.G. McCombs, P.K. Hall, and R.A. McPherson. 2010. Quality assurance procedures for mesoscale meteorological data. J. Atmos. Ocean. Technol. 27:1565–1582.

Fodor, N., and G.J. Kovacs. 2005. Sensitivity of crop models to the inaccuracy of meteorological observations. Phys. Chem. Earth 30:53–57.

Gupta, R.S. 1989. Hydrology and hydraulic systems. Prentice Hall, Englewood Cliffs, NJHatfield, J.L., D.B. Jaynes, M.R. Burkart, C.A. Cambardella, T.B. Moorman, J.H. Prueger, and M.A. Smith. 1999. Water quality in Walnut Creek watershed: Setting and farming practices. J. Environ. Qual. 28:11–24.

Howell, T.A., D.W. Meek, C.J. Phene, K.R. Davis, and R.L. McCormick. 1984. Automated weather data collection for research on irrigation scheduling. Trans. ASAE 27:386–391, 396.

Iowa Environmental Mesonet. 2009. Iowa Environmental Mesonet. Available at http://mesonet.agron.iastate.edu/ (verified 14 Apr. 2011). Iowa State Univ., Ames.

Jabloun, M., and A. Sahli. 2008. Evaluation of FAO-56 methodology for estimating reference evapotranspiration using limited climatic data application to Tunisia. Agric. Water Manage. 95:707–715.

Jaynes, D.B., D.L. Dinnes, D.W. Meek, D.L. Karlen, C.A. Cambardella, and T.S. Colvin. 2004. Using the late spring nitrate test to reduce nitrate loss within a watershed. J. Environ. Qual. 33:669–677.

Kjaersgaard, J.H., F.L. Plauborg, and S. Hansen. 2007. Comparison of models for calculating daytime long-wave irradiance using long term data set. Agric. For. Meteorol. 143:49–63.

Malone, R.W., D.B. Jaynes, L. Ma, B.T. Nolan, D.W. Meek, and D.L. Karlen. 2010. Soil-test N recommendations augmented with PEST optimized RZWQM simulations. J. Environ. Qual. 39:1711–1723.

Malone, R.W., D.W. Meek, J.L. Hatfield, M.E. Mann, R.J. Jaquis, and L. Ma. 2009. Quasi-biennial corn yield cycles in Iowa. Agric. For. Meteorol. 149:1087–1094.

Mathe-Gaspar, G., N. Fodor, K. Pokovai, and G.J. Kovacs. 2005. Crop modelling as a tool to separate the influence of the soil and weather on crop yields. Phys. Chem. Earth 30:165–169.

Meek, D.W. 1997. Estimation of maximum possible daily global solar radiation. Agric. For. Meteorol. 87:223–241.

Meek, D.W., and J.L. Hatfield. 1994. Data quality checking for single station meteorological databases. Agric. For. Meteorol. 69:85–109.

Meek, D.W., J.L. Hatfield, K.J. Cole, R.J. Jaquis, and J.H. Prueger. 2000. Long term experience with quality control procedures applied to meteorological data. Paper P1.9. p. 28–29. *In* R.H. Shaw (ed.) Proc. 24 Conf. Agric. Forest Meteorol., Davis, CA. 14–18 Aug. 2000. Am. Meteorol. Soc., Boston, MA.

Meek, D.W., T.A. Howell, and C.J. Phene. 2009. Concordance correlation for model performance assessment: An example with reference evapotranspiration observations. Agron. J. 101:1012–1018.

Schuol, J., and K.C. Abbaspour. 2007. Using monthly weather statistics to generate daily data in a SWAT model application to West Africa. Ecol. Model. 201:301–311.

Shafer, M.A., C.A. Fiebrich, D.S. Arndt, S.E. Fredrickson, and T.W. Hughes. 2000. Quality assurance procedures in the Oklahoma Mesonetwork. J. Atmos. Ocean. Technol. 17:474–494.

Tanner, B.D. 2000. Evolution of automated weather station technology through the 1980s and 1990s. p. 3–20. *In* K.G. Hubbard and M.V.K. Sivakumar (ed.) Automated weather stations for applications in agriculture and water resources management: Current use and future perspectives. Proc. International Workshop. 6–10 Mar. 2000. High Plains Climate Center, Lincoln, NE and World Meteorological Organization, Geneva, Switzerland.

Yang, D., B.E. Goodison, J.R. Metcafe, V.S. Golubev, R. Bates, T. Panburn, and C.L. Hanson. 1998. Accuracy of 8" standard nonrecording precipitation gauge: Results and application of WMO intercomparison. J. Atmos. Ocean. Technol. 15:54–68.

Chapter 11

Modeling Winter Wheat Production across Europe with WOFOST—The Effect of Two New Zonations and Two Newly Calibrated Model Parameter Sets

Joost Wolf
Rudi Hessel
Hendrik Boogaard
Allard de Wit
Wies Akkermans
Kees van Diepen

Abstract

The Crop Growth Monitoring System (CGMS) is applied for regional monitoring of growing conditions, for simulating growth and yields of the main arable crops, and for producing regional yield forecasts over Europe. An automated Calibration Platform (CALPLAT) was developed and applied for calculating model parameters. The basic ideas of the calibration approach in CALPLAT and its main structure were discussed. With CALPLAT, the model parameters TSUM1 and TSUM2 were calculated for winter wheat (*Triticum aestivum* L.) across Europe. These parameters determine the phenological development of crops simulated with the WOFOST simulation model in CGMS. The calibration analyses were done at the European scale for a recently compiled set of observed crop data and for two new zonations (i.e., NUTS-2 administrative and agro-climatic zonation) for Europe. The resulting model parameters were used for crop growth simulations across Europe. The results of the crop growth simulations for winter wheat over Europe with CGMS-WOFOST for the two newly produced crop parameter sets were compared. The successfulness of the CALPLAT approach for calibrating crop parameters in CGMS can be evaluated on the basis of these CGMS results.

J. Wolf, Wageningen Univ., group Plant Production systems, P.O. Box 430, 6700 AK Wageningen, The Netherlands (joost.wolf@wur.nl); R. Hessel, H. Boogaard, A. de Wit, and K. van Diepen, Alterra Research Institute, Droevendaalsesteeg 3, 6708 PB Wageningen, The Netherlands; W. Akkermans, Plant Research International, Bornsesteeg 47, 6700 AA Wageningen, The Netherlands.

doi:10.2134/advagricsystmodel2.11

Methods of introducing system models into agricultural research. L.R. Ahuja and L. Ma (ed.)
Advances in Agricultural Systems Modeling Series 2.

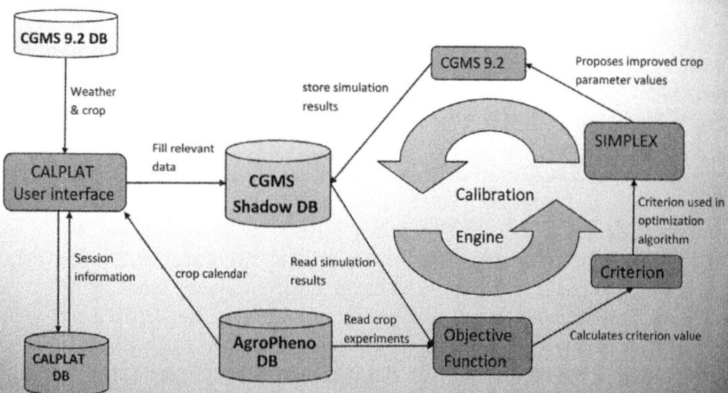
CGMS 9.2 DB
Weather
& crop
CALPLAT
User interface
Session
information
CALPLAT
DB
Fill relevant
data
crop calendar
CGMS
Shadow DB
AgroPheno
DB
store simulation
results
Read simulation
results
Read crop
experiments
CGMS 9.2
Proposes improved crop
parameter values
SIMPLEX
Calibration
Engine
Criterion used in
optimization
algorithm
Criterion
Objective
Function
Calculates criterion value

The MARS Crop Yield Forecasting System (MCYFS; Boogaard et al., 2002; Lazar and Genovese, 2004) provides the European Commission with timely and quantitative yield forecasts for the main crops over Europe. The database of crop parameters within MCYFS is to be updated regularly to guarantee the quality of the output of the system. This is especially of importance for the Crop Growth Monitoring System (CGMS) within MCYFS (Lazar and Genovese, 2004). CGMS is used for regional monitoring of the growing conditions for the main crops over Europe, for issuing alarm warnings in the case of abnormal conditions, for simulating crop growth and yields with simulation models, and for producing regional yield forecasts. Such information can be important at both the policy level (e.g., early information about wheat production in EU for the European Commission; food security in countries with strong rainfall variation) and at the farm level (e.g., adaptation of cropping strategy in relatively dry or wet years [Sadras et al., 2003; Potgieter et al., 2005]). An overview of such impacts of weather conditions and climate variability on crop growth and production and of the possibilities for applying simulation models to predict yields and to support crop and farm management decision making was given by Hoogenboom (2000).

To allow an efficient and consistent calibration of the parameters of the crop growth simulation models in CGMS for a large number of crop types and the range of conditions over Europe, an automated Calibration Platform (CALPLAT) was developed. This is roughly similar (but more focused on large-scale applications) to the automated approach for estimating genetic coefficients for the CROPGRO-Soybean model (Welch et al., 2002). CALPLAT can be applied for calibrating crop parameters within CGMS at both local and regional or European scale. For each of the main crop types, the distribution of their main varieties over Europe can be first established, and next, crop parameter sets for the simulation model WOFOST (Van Diepen et al., 1989) in CGMS are compiled for the whole range of agroclimatic conditions over Europe. WOFOST has been used in many regional land evaluation, yield potential, risk analysis, and yield forecasting studies (Supit, 1997; Van Ittersum et al., 2003), mainly in European countries and in the EU as a whole, but recently also for quantifying production potentials in China (Wu et al., 2006).

The successfulness of the CALPLAT approach for calibrating crop parameters in CGMS and for improving the yield forecasting is to be evaluated on the basis of the CGMS results. These calibration analyses were done for winter wheat for two different (i.e., agroclimatic and European administrative NUTS-2) zonations across Europe. Two new parameter sets for this crop were produced with CALPLAT for a recently compiled set of observed crop data and the two different zonations for Europe. Subsequently, wheat growth simulations over Europe with CGMS-WOFOST for the two newly produced parameter sets were performed and its simulation results were compared.

Methodology

WOFOST Model

WOFOST (WOrld FOod STudies model) is a member of the family of Wageningen crop models (Van Ittersum et al., 2003; Bouman et al., 1996). Related models are for example the model SUCROS for potential and water-limited production simulations (Van Laar et al., 1997) and the ORYZA2000 (Bouman et al., 2001) model for rice (*Oryza sativa* L.). WOFOST has originally been developed in particular for assessment of the yield potential of various annual crops in tropical countries (Van Keulen and Wolf, 1986; Van Diepen et al., 1989). In the 1990s, WOFOST was applied to simulate the production of the main annual crops across Europe under present (De Koning and Van Diepen, 1992) and future conditions (Wolf, 1993; Wolf and Van Diepen, 1995).

WOFOST is applied in CGMS for crop monitoring and yield forecasting at regional and national scales across Europe (Boogaard et al., 2002; Supit, 1997; see for the role of WOFOST in CGMS: http://mars.jrc.ec.europa.eu/mars/About-us/AGRI4CAST/Crop-yield-forecast/The-Crop-Growth-Monitoring-System-CGMS [verified 29 Apr. 2011]). The required inputs per spatial unit are daily weather data, soil characteristics, crop parameters and management practices (i.e., sowing density, planting date). WOFOST computes crop growth on a daily time step. After subtracting the respiration losses for maintenance of the crop, assimilates are partitioned over the different plant organs (i.e., roots, stems, leaves, and grains) as a function of the crop's development stage. This stage is calculated by integrating the daily development rate over time, which rate is a function of temperature. Assimilates are finally converted into structural plant material taking into account the respiration losses for this conversion. These processes that determine biomass accumulation and its distribution over the crop organs are simulated from sowing to maturity on a hectare basis.

Growth simulations for arable crops are done with the WOFOST model (Boogaard et al., 1998) for two production situations: the potential and the water-limited. The potential situation is only determined by temperature, daylength, solar radiation and crop parameters (e.g., leaf area dynamics, assimilation characteristics, dry matter partitioning). The daily crop growth is calculated as daily gross CO_2 assimilation rate minus the respiration rate. The daily gross assimilation rate is calculated from the absorbed radiation and the photosynthesis-light response curve of individual leaves by integration over the leaf layers of the canopy and over the day (Goudriaan, 1986; Spitters, 1986). For the potential situation the effect of soil moisture content on crop growth is not considered, and a continuously moist soil is assumed. In the water-limited situation soil moisture content determines whether the crop growth is limited by drought stress or not. Therefore, soil water dynamics (as dependent on rainfall and water losses by surface runoff, soil evaporation, crop transpiration, and downward percolation) are simulated over time applying the tipping bucket approach, which gives the soil moisture content in the rooted zone. Both soil evaporation and crop transpiration are calculated on the basis of the potential evapotranspiration (from the Penman approach according to Frère and Popov, 1979), taking into account both moisture content in the root zone and the degree of light interception by the crop canopy. In both, the potential and water limited situations optimal supply of nutrients is assumed, and damage caused by pests, diseases, weed, and/or extreme weather events is not considered.

Crop growth and soil calculations are done with a time step of 1 d over the growth period. For detailed information on model structure, required input data and modeled processes in WOFOST, see Boogaard et al. (1998) and Supit et al. (1994; available at http://www.treemail.nl/download/treebook7/index.htm [verified 29 Apr. 2011] from Supit and Van der Goot, and WOFOST website (http://www.wofost.wur.nl./UK/). WOFOST is a one-dimensional model, without reference to geographical scale. For regional applications, WOFOST can be linked to geographical information systems (as, e.g., in CGMS), in which unit the crop type, soil type, weather, and management (e.g., sowing date) should be specified per grid. In principle, WOFOST is a generic model that can simulate the growth of any annual crop type growing at any location. Hence, WOFOST has a modeling and data structure that allows easy application to all possible crop–soil–weather combinations, and it is particularly suitable for calculating biomass production and yield at regional and national scales.

Applied Calibration Approach in CALPLAT

To allow an efficient and consistent calibration of the parameters in WOFOST for a large number of crop types and the range of conditions over Europe, an automated Calibration platform (CALPLAT) was developed and applied for this study. For a discussion and more information about the calibration approaches in

CALPLAT, see the Appendix and the CALPLAT manual (de Wit et al., 2007). Two aspects of the CALPLAT approach are of main importance. First, the optimization of crop parameters is performed by applying a criterion (indicating sufficiency of convergence), which is basically the maximally allowed sum of squared differences between simulated and observed values for crop variables (e.g., date of maturity). Second, regional calibration of crop parameters first defines the main agroclimatic zones and next establishes the set of crop parameter values for each crop variety being representative for one agroclimatic zone, on the basis of the observed values for crop variables in that zone. An overview of the calibration levels in CALPLAT—(i) phenology, (ii) potential yield level simple, (iii) potential yield level complex, (iv) water limited yield level—are given in Table 11–1 for both regional calibration and point calibration.

Structure of CALPLAT

CALPLAT consists of three major components: the CALPLAT user interface, the Calibration Engine including the optimization algorithm Simplex (Nelder and Mead, 1965), and a CGMS 9.2 (Fig. 11–1, gray rectangles). Furthermore, various database components are necessary, such as the operational CGMS database (CGMS_DB) for providing weather data and initial values of crop parameters, the CALPLAT session database (CALPLAT_DB), and the database with crop observations (AgroPheno_DB) (Fig. 11–1, cylinders).

The Simplex algorithm is applied because it may escape out of local minima but at the price of having a slow convergence rate. It is an iterative procedure: for a certain set of parameter values it compares the CGMS model output (e.g., total biomass) with the actually observed data (biomass) and then adjusts the set of proposed parameter values in a direction that will reduce the difference. When the difference between simulated and observed data is smaller than some user-defined convergence criterion, the iterative process is stopped and the current parameter values are output as estimates.

The CALPLAT user interface takes care of the user interaction and the communication with the different database components. In addition, the user interface prepares a "shadow" CGMS database (CGMS Shadow DB), so that CGMS can be applied for simulating a specific simulation setup without interfering with the operational CGMS database. This database is filled by selecting the appropriate data from the CGMS_DB and from the AgroPheno_DB. The latter database is necessary because the starting conditions for the crop are taken from the AgroPheno_DB. The tables containing the crop calendar, weather, crop parameters, and soil parameters (in the case of water-limited calibration) are populated with the right data for the specific simulation setup selected by the

Table 11–1. Levels of calibration for regional and (multi-)point calibration in the calibration platform CALPLAT.†

Level	Regional calibration	Point calibration
Phenology	Parameters TSUM1, TSUM2 Necessary observations Regional crop calendars or observations of emergence, flowering and maturity.	Parameters TSUM1 detailed, TSUM2, TSUMEM, TBASEM, DLO, DLC, TEFFMX Necessary observations Detailed observations of emergence, flowering, maturity for different years and sowing dates at point locations.
Potential yield level–simple	Parameters AMAXTB Necessary observations Observations of total biomass under optimal conditions (with LAI-MAX above 3) at point or NUTS level.	Parameters AMAXTB, SLATB, SPAN Necessary observations Point observations of LAI-MAX and total biomass under optimal conditions.
Potential yield level– complex	Parameters SLATB, SPAN, FOTB Necessary observations Observations of maximum LAI and total biomass and/or yield under optimal conditions at point or NUTS level.	Parameters AMAXTB detailed, SLATB detailed, SPAN detailed, RGRLAI, LAIEM, TDWI, partitioning factors (FLTB, FRTB, FOTB detailed), TMPFTB, RDRRTB Necessary observations Point observations of LAI, total biomass, weights of leaves, stems, storage organs and roots over time during the growth period under optimal conditions.
Water-limited yield level	Parameters (based on complex potential calibration) CFET, RDMCR. Necessary observations Observations of total biomass and/or yield level under rain fed conditions at point or NUTS level.	Parameters (based on complex potential calibration) CFET detailed, RDMCR detailed, PERDL, DEPNR Necessary observations Point observations of LAI, total biomass, weights of leaves, stems and storage organs and possibly crop transpiration, evaporation and rooting depth over time during the growth period under water-limited conditions.

† Description of the model parameters is given in Table 11–3. For model parameters that cannot be calibrated, see Table 11–3 with the complete list of model parameters in CGMS–WOFOST.

user. The crop observations in the AgroPheno_DB used for the calibrations of winter wheat are based on a compilation of data from the following four datasets for arable crops: CGMS set (original CGMS dataset, from Boons-Prins et al., 1993),

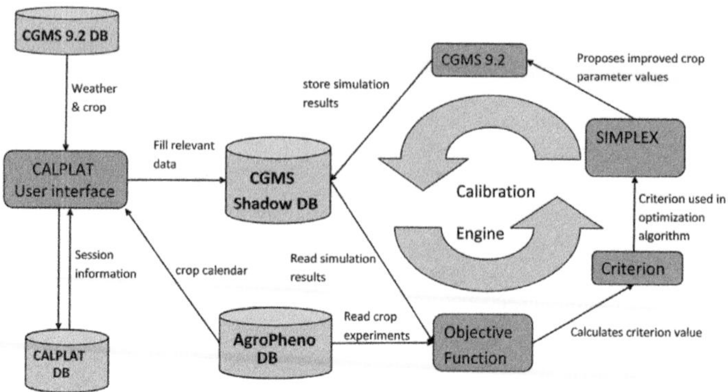

Fig. 11–1. Schematic overview of CALPLAT.

KUL set (Willekens et al., 1998), MOCA set (GISAT, 2003), and the agrophenological dataset as collected by GeoSys.

The Calibration Engine is doing an iterative processing loop that combines the different components. First, the user interface calls the Calibration Engine. Within each iteration, the CGMS executable is begun with a certain value for the crop parameters to be optimized. CGMS is run with these values, and it writes simulated output corresponding to these crop parameter values into the CGMS Shadow DB. The Calibration Engine then reads these simulation results from the CGMS Shadow DB. Next, the Objective Function compares the simulated values with the observed values in the AgroPheno_DB. It calculates a criterion value, which basically is a sum of squared differences between simulated values and observed values. Based on the sequence of criterion values, the Simplex algorithm then proposes a new crop parameter value with which CGMS again is run, and so on (Fig. 11–2). Upon convergence, control is given back to the user interface, and the user can continue to view the results in the Result Manager. More detailed information on the calibration procedure is given in the Annex and the CALPLAT manual (de Wit et al., 2007).

The choice between regional and (multi-)point calibrations directly influences the calibration strategy in CALPLAT in a number of aspects:

- For regional calibrations, a spatial zonation needs to be created to divide the area into zones with (assumed) homogeneous characteristics per crop type; calibration is done for each zone on the basis of the observed values for crop variables in that zone.
- For regional calibrations, the number of model parameters per crop type that can be calibrated is a subset of that for (multi-)point calibration (Table 11–1)

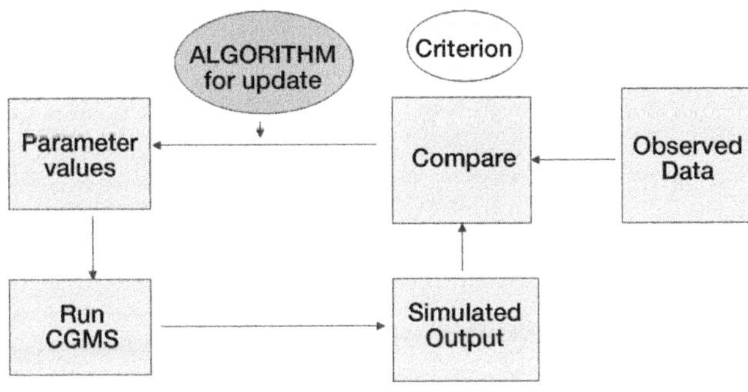

Fig. 11–2. Schematic overview of the Calibration Engine applying the Simplex algorithm.

- (Multi-)point calibrations are assumed to be done only on the basis of detailed field experiments containing times series, whereas regional calibrations can use generally available (but more uncertain) datasets (Table 11–2).

Regional Calibration for Winter Wheat over Europe

Winter wheat is calibrated with CALPLAT across the whole of Europe, from Scandinavia up to and including the northern part of the Maghreb area. For this calibration for winter wheat, two types of zonations were applied. The NUTS-2 zonation is a European administrative zonation. The agroclimatic zonation consists of a temperature sum-based zonation (10 zones) for Europe in combination with a continentality index-based zonation (5 zones; following the approach from Conrad, 1946). This zonation appeared to indicate well the differences between the main agroclimatic zones of Europe and the resulting differences in growing conditions, affecting, for example, the start date of crop growth after winter and the initial growth rate. The calibration of crop parameters was done for each separate zone, on the basis of the available observed values for crop variables in that zone.

Table 11–2. Data types for calibration which can either be used for regional calibration or (multi-) point calibration.

Data types for Calibration	Regional calibration	(Multi-)point calibration
Field experiments	No†	Yes
Regional observation	Yes	No
Expert estimate	Yes	No

† Within CALPLAT field experiments cannot be used in regional calibrations because observed time series cannot be mixed with regional data like crop calendars. However, field experiments can still be valuable for regional calibration by using, for example, crop calendar, final biomass, and maximum leaf area index.

The regional calibration in CALPLAT consists of a procedure for calculating optimal values for crop parameters, such as in this study (see upper left cell in Table 11–1) the parameters TSUM1 (temperature sum required for phenological development from crop emergence to flowering) and TSUM2 (temperature sum requirement from flowering to maturity) by optimizing the relationships between the observed crop calendars in a zone and the calculated crop calendars (with WOFOST). This optimization leads to well-calibrated TSUM1 and TSUM2 values for each zone, but is of course completely dependent on the reliability of the observed crop calendars. Note that for winter wheat (and other winter crops), TSUM1 is calculated in WOFOST from January 1. This means that the simulated crop growth and phenological development begins from the moment in winter or spring that temperatures are above 0 to 5°C. Hence, the initial biomass weight in spring (as applied in WOFOST) is the estimated weight after the end of winter. Another important point is that the regional calibration with CALPLAT was done for only the model parameters TSUM1 and TSUM2. The reason that the regional calibration over Europe with CALPLAT has not been done for the other three regional calibration levels (Table 11–1) is not related to CALPLAT but is due to the lack of observations over Europe for total biomass, yield, and maximum leaf area index (LAI) for winter wheat (and also for other crop types) under both optimal and rainfed growth conditions (i.e., with optimal nutrient supply, crop protection, and management). Hence, the check of the WOFOST outcomes for winter wheat could only be done as follows: after the temperature sum calibration the model runs over Europe were done for the standard set of model parameters (Boons-Prins et al., 1993), and if the results with respect to total biomass, harvest index, and maximum LAI were out of the expected range per crop type (being based on expert knowledge), we might change few model parameters for certain zones (which for winter wheat was not needed).

Wheat Growth Modeling

Simulation runs for winter wheat over Europe were performed for a time period of 15 yr (1991–2005). Three main crop characteristics were calculated in the simulations for winter wheat growing under potential growing conditions: (i) total aboveground biomass at harvest (BIOM, in kg dry matter ha^{-1}), (ii) harvest index (HI, equal to grain yield/BIOM), and (iii) maximum value for LAI (LAI-max, m^2 m^{-2}) during the growth period. A low value for LAI-max indicates low values for LAI during the growth period, which causes low light interception by the crop canopy and thus low biomass production. The quality of the two new parameter sets for winter wheat were evaluated by comparing the results from the simulation runs of the CGMS newly calibrated for agroclimatic zonation over Europe with those from CGMS newly calibrated for NUTS-2 zonation over Europe.

Note that the simulated values for biomass and yield production are used as inputs for the Yield Forecasting module of CGMS. These simulation results represent the interannual yield variation that results from weather variability. The yield forecasting module uses the regional yield statistics from the European Statistical Office (EUROSTAT, 2005) as main input and combines the mean regional yield, a linear (or curve-linear) time trend of regional yields, and a linear (or curve-linear) regression function explaining the interannual yield variation (as based on the simulated productions). The time trend represents the influence of long-term economic and technological dynamics, such as increased fertilizer application, improved crop management, and improvement varieties, on the regional yields and production. For more information about the Yield Forecasting module, see Genovese and Bettio (2004).

Results from the Model Calibration for Two Zonations

Results for Agroclimatic Zonation

Flowering

The calibration procedure results in estimated flowering dates for each of the agroclimatic zones of Europe (not shown). This shows that the flowering date is estimated to be generally around Day of Year 170 in northern Europe (e.g., southern Sweden and Scotland), around Day 150 in central Europe (e.g., northern France and southern Poland) and around Day 120 in southern Europe (e.g., southern Spain, southern Italy, and southern Turkey). The general patterns of estimated flowering dates over Europe correspond quite well with the pattern of observed flowering dates.

The error in the estimated flowering date (i.e., difference between observed and estimated date) is moderately large (mainly from –20 to +20 d) on average over Europe (Plate 11–1; see color insert section). Larger errors are observed in southern Turkey, the northern part of the Maghreb, Hungary, Ireland, parts of Spain, and England. The estimated flowering dates were plotted against the observed flowering dates (Fig. 11–3). The regression coefficient is moderately high (R^2 = 0.596). The main part of the estimated flowering dates are maximally 20 d earlier or later than the observed flowering dates, which is a moderately good result.

Maturity

The estimated maturity dates are generally around Day of Year 230 in northern Europe (e.g., southern Sweden and Scotland), around Day 210 in central Europe (e.g., northern France and Poland), and around Day 180 in southern Europe (e.g.,

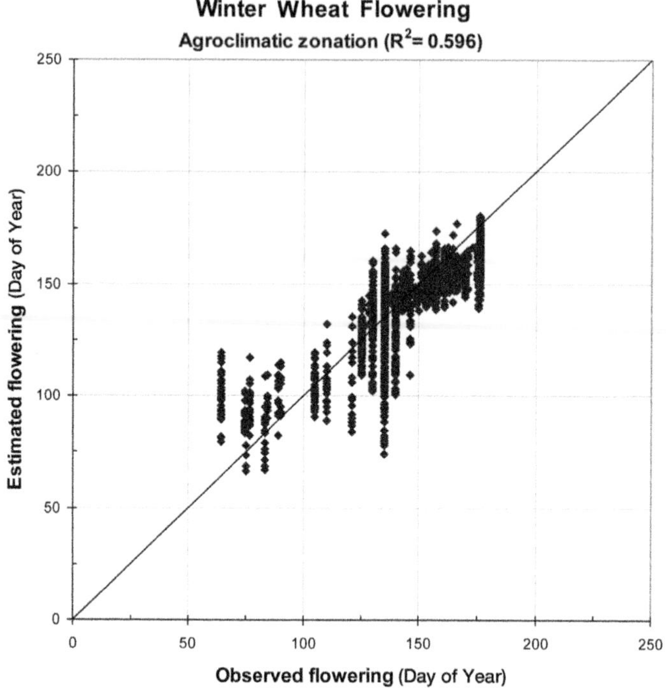

Fig. 11–3. Observed dates of flowering for winter wheat across Europe versus estimated dates as based on calibration procedure with agroclimatic zonation.

southern Spain, southern Italy, and southern Turkey). The general pattern of estimated maturity dates over Europe corresponds moderately well with the observed maturity dates.

The error in the estimated maturity date is moderately large (mainly from –20 to +20 d) on average over Europe (Plate 11–2). Larger errors do occur in central Spain, the western part of the Netherlands, northern Poland, northern Italy, part of the Ukraine, and Bulgaria. This can be explained, for example, by the late observed date of maturity for winter wheat in Spain (not shown). The estimated maturity dates are shown plotted against the observed maturity dates in Fig. 11–4. The regression coefficient is high ($R^2 = 0.750$). The main part of the estimated maturity dates are maximally 15 d earlier or later than the observed maturity dates, which is a good result.

Results for NUTS-2 Zonation

If observed dates for flowering and/or maturity for some zones are missing or are insufficient, the model parameters TSUM1 and TSUM2 cannot be calculated, and

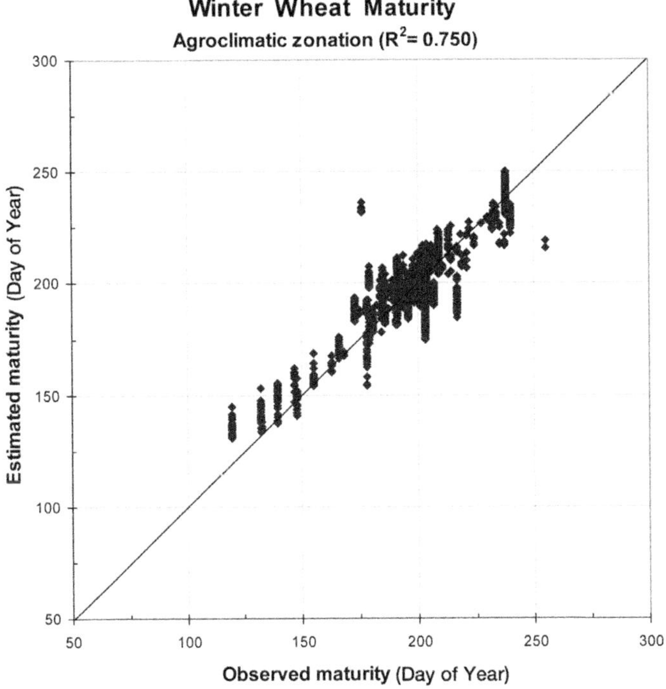

Fig. 11–4. Observed dates of maturity for winter wheat across Europe versus estimated dates as based on calibration procedure with agroclimatic zonation.

the estimation of flowering and maturity dates is not possible. This is more likely to occur for NUTS-2 than for agroclimatic zonation, as NUTS-2 zones are smaller.

Flowering

The estimated flowering dates are generally around Day of Year 175 in northern Europe (e.g., Sweden and Scotland), around Day 145 in central Europe (e.g., northern France and Poland), and around Day 120 in southern Europe (e.g., southern Spain and southern Italy). The general pattern of estimated flowering dates across Europe (not shown) corresponds well with the pattern of observed flowering dates.

Estimated flowering dates across Europe were plotted against observed flowering dates (Fig. 11–5), resulting in a very high regression coefficient ($R^2 = 0.882$). The main part of the estimated flowering dates are maximally 10 to 15 d earlier or later than the observed flowering dates, which is a good result. Larger errors are observed in Turkey, Spain, southern Italy, and Croatia, which can be explained from the large differences in altitude (giving strong temperature differences) in these countries.

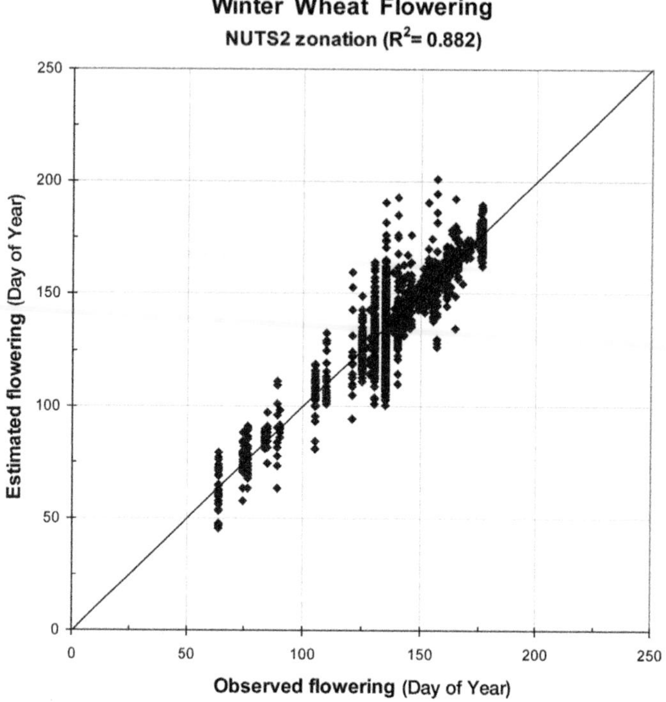

Fig. 11–5. Observed dates of flowering for winter wheat across Europe versus estimated dates as based on calibration procedure with NUTS-2 zonation.

Maturity

The estimated maturity dates are generally around Day of Year 240 in northern Europe (e.g., southern Sweden and Scotland), around Day 210 in central Europe (e.g., northern France and southern Poland), and around Day 175 in southern Europe (e.g., southern Spain and southern Italy). The general pattern of estimated maturity dates over Europe (not shown) corresponds well with the observed maturity dates.

The estimated maturity dates across Europe were plotted against the observed maturity dates (Fig. 11–6), resulting in a very high regression coefficient (R^2 = 0.973). The estimated maturity dates are maximally 5 to 10 d earlier or later than the observed maturity dates.

Results from Wheat Growth Modeling

Results from the WOFOST model in CGMS are given in the following and all presented results are mean values for the period 1991 to 2005. Two crop charac-

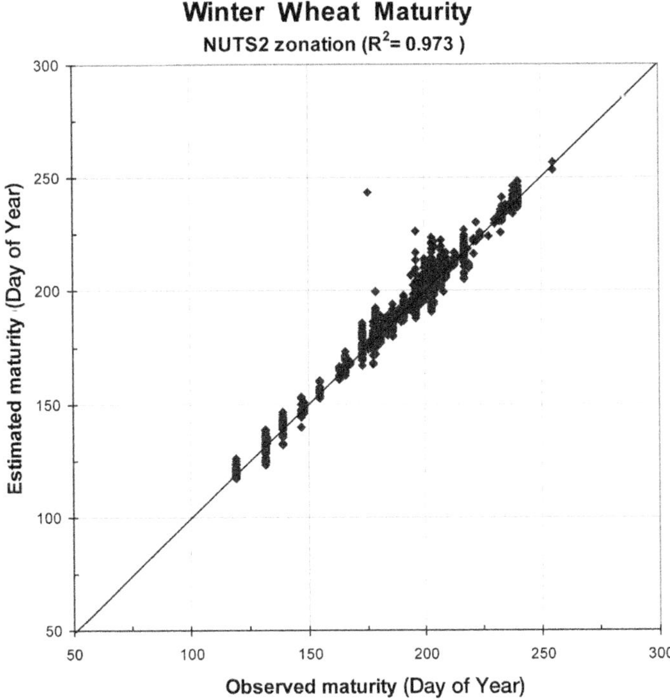

Fig. 11–6. Observed dates of maturity for winter wheat across Europe versus estimated dates as based on calibration procedure with NUTS-2 zonation.

teristics are presented as calculated for winter wheat growing under potential growing conditions over Europe: (i) BIOM (kg dry matter ha^{-1}) and (ii) HI. A third crop characteristic, LAI-max, was also established for the same wheat simulation runs over Europe, but the results are not shown. First, the results from CGMS newly calibrated for agroclimatic zonation over Europe are given and, second, the results from CGMS newly calibrated for NUTS-2 zonation. Note that for areas in which insufficient observations were available for the NUTS-2 calibration (e.g., roughly 50% of the areas of France and Germany and all of Portugal), the parameter values as based on agroclimatic zonation were used for the CGMS simulations.

Resulting Biomass Production

Agroclimatic Zonation

BIOM is higher than 19.000 kg ha^{-1} in western and southern Europe and is somewhat lower (12,000–19,000 kg ha^{-1}) in central, northern, and eastern Europe. BIOM is lowest (<10.000 kg ha^{-1}) in eastern Turkey and northern Russia.

NUTS-2 Zonation

BIOM is higher than 19.000 kg ha^{-1} in western, northern, and southern Europe and is somewhat lower (14.000–19.000 kg ha^{-1}) in central and eastern Europe. BIOM is lowest (<10.000 kg ha^{-1}) in central and eastern Turkey, northern Russia, and parts of southeastern Europe and the Maghreb area.

Resulting Harvest Index

Agroclimatic Zonation

HI is between 0.4 and 0.55 in the main parts of Europe. HI is very high (HI > 0.60) in eastern Turkey and northern Russia and rather low in southwestern Spain, southern Italy, and Morocco (HI = about 0.35).

NUTS-2 Zonation

HI is between 0.4 and 0.55 in the main parts of Europe. HI is very high (HI > 0.60) in northern Russia and parts of the Maghreb area and rather low in southern Spain, Sicily, southern Turkey, and western Greece (HI $\leq$ 0.35).

Discussion

The successfulness of the automated CALPLAT approach for calibrating crop parameters in WOFOST was evaluated. This was done by comparing estimated dates with observed dates over Europe for the flowering and maturity of winter wheat. The calibration approach was applied to two different types of zonations over Europe, an agroclimatic zonation and an administrative (NUTS-2) zonation, and the results are discussed in the following.

Different Zonations

With agroclimatic zonation, the main reasons for the errors (i.e., difference between observed and estimated flowering and maturity dates) are that first, a region can be at the border of the agroclimatic zone to which it is assigned, and second, the "observed" crop calendar for a region does not correspond well with the "observed" crop calendars for the other regions within the same agroclimatic zone. It is clear that the first type of error ("border effect") becomes larger if the number of agroclimatic zones is reduced. The second type of error is clearly demonstrated for Hungary. The observed flowering date is here around Day 115, but its agroclimatic zonation is not much different from southern Poland or Slovenia with a much later observed flowering date (about Day 150). This early observed flowering date in Hungary may be the result of (i) different and very early variety in Hungary compared to neighboring countries or (ii) estimated crop calendar from the national/regional crop expert. Note that this "national expert effect" may lead to strange model parameters (e.g., a low TSUM1 value for Hungary and regions in the same agroclimatic zone) when based on such crop "observations."

The flowering dates estimated with the CALPLAT approach correspond moderately well with the observed dates. The main part of these errors can be explained by first the "border effect" and second the "national expert effect" (see above). Errors of the first type can be reduced by increasing the number of agroclimatic zones; however, the number of agroclimatic zones for winter wheat over Europe appeared to be sufficient. Errors of the second type cannot be prevented by improving the calibration approach but only by improving the used crop "observation" data.

With NUTS-2 zonation, the flowering and maturity dates estimated with the CALPLAT approach are well related to the observed dates. This is a good result, considering that the main part of errors are caused by the quality of the "observed" mean crop calendar and thus flowering (or maturity) date in each NUTS-2 zone on the one hand and the degree of variation in environmental conditions over NUTS-2 zones on the other hand. For example, the preciseness of the (with CGMS) modeled flowering dates is determined by the quality of the observed dates for flowering over Europe. This means that even if the plot with estimated versus observed flowering dates (e.g., in Fig. 11–5) results in R^2 equal to 1.00, the modeled flowering dates over Europe may differ strongly from the actual flowering dates. This leads to an overall conclusion that applies to all calibration results in the following: the quality of the observed crop data is the most essential factor to achieve a good model calibration.

Relatively large errors are observed in regions where large differences in altitude occur. The main reason is that a mean "observed" crop calendar is assumed to be valid for a whole NUTS-2 region. Of course, the actual flowering dates will differ strongly between high and low altitude sites within one zone, and this also applies to the estimated flowering dates. For example, the estimated flowering dates in mountainous areas in Turkey are later than those along the coast. This means that the larger errors in the mountainous areas do not indicate a problem in the calibration procedure but show the limited value of the mean "observed" crop calendar for a strongly varying NUTS-2 region. This is logical as the NUTS-2 zones are administrative zones and not agroclimatically uniform zones.

Derived model parameters TSUM1 and TSUM2 (not shown) strongly differ between countries and within countries (e.g., in Spain); these differences can be explained by the "national expert" effect (as described above for the agroclimatic zonation). These differences are much stronger for the NUTS-2 zonation than those for the agroclimatic zonation, which is due to the smaller NUTS-2 zones (with less observed data per zone and less averaging out of extreme observations) for which the calibration was done.

The crop model parameters are of primary importance for the quality of the WOFOST output because the interaction between variety-specific model parameters and site-specific climate and soil conditions may considerably affect the simulated yields and the produced crop indicators. For example, a simulated variety with a long growth period may show strongly negative effects of unfavorable conditions (e.g., high temperatures and/or drought in summer or frost in autumn) on the final yield, which in practice do not occur by cultivating site-adapted crop varieties (e.g., with a short growth period).

Wheat Growth Modeling

For the analysis of the modeling results we should define an acceptable range of values. For example, the expected values for the simulated crop characteristics of winter wheat under potential growing and suitable (e.g., not too cold or too warm) agroclimatic conditions in Europe are roughly as follows: BIOM between 15.000 and 22.000 kg dry matter ha^{-1} and lower in cold and hot areas, HI between 0.40 and 0.55 with lower values under unfavorable growing conditions (e.g., too cold), and LAI-max between 3 and 7 m^2 m^{-2}.

For the Newly Calibrated CGMS with Agroclimatic Zonation

BIOM is highest in western and southern Europe and is somewhat lower in central, northern, and eastern Europe. Highest BIOM is to be expected in the temperate climate in western and south-western Europe, which has an early start of the growing season in spring and the longest growth period. BIOM is lowest in eastern Turkey and northern Russia, which can be explained from both the continental climate (i.e., late start in spring and high temperatures in summer and thus a short growth period) and the low values for LAI (not given). HI is between 0.4 and 0.55 in the main parts of Europe and is according to the expectations.

For the Newly Calibrated CGMS with NUTS-2 Zonation

BIOM is highest in western, northern, and southern Europe and is somewhat lower in central and eastern Europe. BIOM is lowest in central and eastern Turkey, northern Russia, and parts of southeastern Europe and the Maghreb area, which can be explained from the low values for LAI-max (<2.0, not given). The low LAI-max indicates low values for LAI during the growth period, which cause low light interception by the crop canopy and thus low biomass production. The pattern of BIOM over Europe shows a stronger spatial variation than the pattern of BIOM simulated with CGMS calibrated for the agroclimatic zonation. This stronger spatial variation can also be seen in the distribution of LAI-max over Europe

(LAI-max is not given). HI is between 0.4 and 0.55 in the main parts of Europe and matches expectations.

Thus, analysis of the results from CGMS modeling for winter wheat over Europe led to a number of main discussion points with respect to (i) areas with low biomass production and low LAI and their consequences for yield forecasting, (ii) consequences of strange observed crop calendars, and (iii) the best calibration approach for CGMS.

Areas with Low Biomass Production and Low Leaf Area Index

A maximum value for leaf area index (LAI-max) that is low (<2.0) indicates low values for LAI during the growth period of winter wheat, a low light interception by the crop canopy, and thus a low biomass production (i.e., BIOM) and yield. Such areas with low LAI-max, and thus a low biomass production, appear to have a moderate and slight extent over Europe in the modeling results from the CGMS calibrated for NUTS-2 zonation and the CGMS calibrated for agroclimatic zonation, respectively. The interannual variation in biomass production due to such low LAI values is used in the regression analysis (of simulated versus statistical yield data) for the yield forecasting, but this interannual variation may be a model artifact due to the underestimated LAI.

Consequences of Strange Observed Crop Calendars

The model parameters TSUM1 (determining the modeled period until flowering) and TSUM2 (determining the period from flowering to maturity) over Europe were derived on the basis of the observed crop calendars per zone. TSUM1 and TSUM2 might have values that give strange modeling results. For example, a too high TSUM1 and a too low TSUM2 (as based on a late observed flowering date used for the calibration) often results in a too high LAI-max and a too low HI, and the reverse is true—a too low TSUM1 and a too high TSUM2 often give a too low LAI-max and a too high HI, with the low LAI-max resulting in a too low biomass production.

The Best Calibration Approach for CGMS

To determine which of the two calibration approaches for CGMS (i.e., CGMS calibrated for agroclimatic and NUTS-2 zonation) gives the best simulation results, the extent of the area over Europe with crop indicators (i.e., BIOM and HI and possibly LAI-max) being within the expected range, is considered the best indicator. Hence, we prefer the CGMS calibrated for agroclimatic zonation above the other CGMS version.

Conclusions

The calibration procedure for winter wheat with agroclimatic zonation results in:

- general patterns of estimated flowering and maturity dates over Europe that are satisfactory
- errors in the estimated flowering and maturity dates that are moderately large (mainly from −20 to + 20 d) on average over Europe
- estimated flowering and maturity dates that correspond moderately well (R^2 = 0.60) and well (R^2 = 0.75), respectively, with the observed dates.

The CGMS calibration for NUTS-2 zonation often results in a large variation in the crop model parameters across Europe, as it is often being based on small sets of observed data; this large parameter variation results in a large variation in the simulated crop characteristics (e.g., biomass production and harvest index).

We prefer the CGMS calibrated for agroclimatic zonation above the other CGMS version because the CGMS calibrated for NUTS-2 zonation gives a stronger spatial variation in simulated LAI-max, BIOM, and HI over Europe.

Appendix

Discussion on Calibration Approaches in CALPLAT

Despite the aim of "automatic calibration" in CALPLAT, it is a not possible to use a large number of crop data from field experiments and do the calibration of many model parameters simultaneously. Hence, the model calibration should always be done in an organized and stepwise fashion, beginning with the main crop characteristics (e.g., growth duration), gradually considering other crop characteristics (e.g., yield level), and finally including other, more complex effects on crop growth (e.g., drought effects). If such a stepwise procedure were not followed, the derived crop parameters would often be wrong. For example, the assimilation parameters in the model should be based on the growth rate and yield level under optimal conditions. Otherwise, these model parameters would be underestimated if based on the low yield level under drought-stress conditions. This approach prevents the mixing-up of different types of crop characteristics and crop growth conditions.

A second characteristic of CALPLAT is that CALPLAT must be able to carry out calibrations for regions instead of dedicated point locations with experimental results. Traditionally, crop calibrations were performed at specific point locations with well-known crop growth conditions (weather, soil, nutrient application, drought stress, management practices) to calibrate a large number of parameters for a crop model. For regional calibrations this approach is generally not possible because regional calibrations using results from a number of experiment, may easily become very complex. This has to do with the fact that the relationship between the measured results from the experimental site and the used input data for CGMS calibration (e.g., weather and soil) is often poor. This problem is directly related to the fact that the representativeness of experimental results for large areas is generally poor due to strong variations in growing conditions. The above-mentioned problems indicate the difficulties for a regional calibration, and they also set a limit on the number of parameters that can be calibrated in a regional calibration.

A third point with regard to the CALPLAT approach is related to the definition of a "region." In practice the characteristics of crop varieties gradually change from one climate zone to another (e.g., increasing in temperature) because local farmers use crop varieties that are optimally adapted to the local climate. Ideally, the crop parameters should therefore be adapted for the specific location and conditions of each CGMS grid cell, leading to a specific crop variety for each grid cell. However, it is not possible to derive crop characteristics separately for each grid unit based on the limited number of field experiments that are usually available. In general, two approaches can be conceptualized:

- Define crop parameters locally (for single grid cells) based on available crop observations, and interpolate the parameters to the surrounding grid cells using some covariable (e.g., yearly temperature sum, yearly precipitation)
- Define zones a priori over, for example, Europe using some sort of clustering on climatic data, and derive "representative" crop parameter values that will be applied to each of the zones based on calibration using all useful crop observations within each zone.

Given the problem of representativeness of experimental results, the second approach is chosen for the regional calibrations in CALPLAT. Hence, regional calibration should first define the main agroclimatic (or other) zones. Next determine the main changes in crop characteristics (e.g., temperature sum requirements for phenological development) from one zone to the other on the basis of representative field experiments, and finally establish a number of crop varieties, each being representative for one agroclimatic zone. For example, for winter wheat over Europe, 10 agroclimatic zones are defined, and hence, 10 crop varieties with increasing temperature sum requirements (from e.g., cooler to warmer zones) are to be derived on the basis of growth duration and anthesis dates in field experiments in each zone.

The issues raised above indicate the complexity of a regional calibration and lead to a number of important recommendations for the working procedure, as incorporated in CALPLAT:

1. Any calibration should be performed in a stepwise fashion with increasing complexity.
2. Rigorous calibration of nearly all crop parameters (e.g., for a totally new crop) cannot be performed regionally and should be done on a (multi-)point scale.
3. Regional calibration should be performed for a limited number of crop parameters that are known to vary with climate.
4. Zones should be characterized a priori using, for example, clustering of climatic data.
5. Rigorous checks should be performed on experimental datasets to determine their representativeness and plausibility.

This leads to a structured approach in CALPLAT with four levels of regional calibration and four levels of (multi-)point calibration, as presented in Table 11–1.

A list of all model parameters in the WOFOST model that can (or sometimes cannot) be calibrated in CALPLAT, is given in Table 11–3. For each model parameter, this table gives the description, unit, regional or (multi)point calibration to be applied, single value or table, scaling factor to be applied for table calibration or not, required field observations, and some comments on the procedure. The description and units of these field observations are given in Table 11–4.

When performing a calibration with CALPLAT, an iterative optimization technique is used (the Simplex algorithm; Nelder and Mead, 1965) to achieve optimal fit between simulated and observed data. Optimization stops when either the user-specified maximal number of iterations was performed or the tolerance criterion was reached. The tolerance determines the accuracy of the desired solution; it must be set by the user. A larger tolerance means that a less precise result is deemed acceptable. The user has a choice of three different optimization criteria. Minimizing the root of the Mean Squared Error is applicable when the variables have more or less the same scale. Here the sum of the contributions $[y - f(\mathbf{t})]2$ is minimized as a function of the parameter vector **t**, where y is the value of an observed variable, and $f(\mathbf{t})$ is the corresponding CGMS simulated output generated using parameter vector **t**, but for a situation in which two parameters are calibrated, whose corresponding variables have values on a totally different scale, a second criterion is available, the root of the relative mean squared error. Here

Table 11–3. Model parameters, their description, type of calibration, required field observations, and comments on procedure, for WOFOST. Note that indicated regional calibration also allows (multi-)point calibration but not the opposite.

Model parameter	Description	Unit, type of calibration†	Type of parameter‡	Field observation§	Comments
TSUM1	TSum from emergence to anthesis	°C d, R	S,–	DOE, DOA	Determined from DOA–DOE. Daylength effect is not included. Winter crops start on 1 January.
TSUM1 additional	TSum from emergence to anthesis	°C d, R	S,–	DOP, DOA	Idem the calibration for TSUM1. Determined from DOA–DOE with estimated period between DOP and DOE.¶
TSUM2	TSum from anthesis to maturity	°C d, R	S,–	DOM, DOA	Determined from DOM– DOA.
TSUM2 additional	TSum from anthesis to maturity	°C d, R	S,–	DOH, DOA	Determined from DOM–DOA with estimated period between DOM and DOH.#
AMAXTB	Max. leaf assimilation rate	kg ha^{-1} h^{-1}, R	T, SCF	TAGBI	For potential growth conditions. Only if LAI-max above 3; otherwise, calibrate SLATB first.
SLATB	Specific leaf area	ha kg^{-1}, R	T, SCF	LAI-max	For potential growth conditions. If SLATB remains too low, calibrate SPAN first.
SPAN	Life span of leaves at 35°C	d, R	S,–	LAI-max	For potential growth conditions. Only if SLATB calibration is not satisfactory.
FOTB	Fraction of above-ground DM to storage organs	–, R	T, SCF	WSOI	For potential growth conditions. Requires the AMAXTB calibration first.
RDMCR	Max. crop dependent rooting depth	cm, R	S,–	TAGBR or WSOR (in this order)	For water-limited growth conditions. Requires the AMAXTB calibration first, and possibly the FOTB calibration.
CFET	Correction factor transpiration rate	–, R	S,–	TAGBR or WSOR (in this order)	Idem to the RDMCR calibration, but only if the RDMCR is fixed input.

Continued.

Table 11–3. Continued.

Model parameter	Description	Unit, type of calibration†	Type of parameter‡	Field observation§	Comments
TSUM1 detailed	TSum from emergence to anthesis	°C d, P	S,–	DOA	Daylength effect may be included (IDSL = 0, 1, or 2). Calibration for potential growth conditions of both DLO and TSUM1 or DLC, DLO, and TSUM1 for multipoints.
AMAXTB detailed	Max. leaf assimilation rate	kg ha^{-1} h^{-1}, P	T, SCF	TAGBITB	For potential growth conditions. Only if LAI-max above 3; otherwise, calibrate SLATB first.
SLATB detailed	Specific leaf area	ha kg^{-1}, P	T, SCF	LAIITB	For potential growth conditions. If SLATB remains too low, calibrate SPAN first.
SPAN detailed	Life span of leaves at 35°C	d, P	S,–	LAIITB	For potential growth conditions. Only if SLATB (detailed) calibration is not satisfactory.
FOTB detailed	Fraction of aboveground DM to storage organs	–, P	T, SCF	WSOITB	For potential growth conditions. Requires the AMAXTB and SLATB calibration first.
RDMCR detailed	Max. crop dependent rooting depth	cm, P	S,–	TAGBRTB or WSORTB (in this order)	For water-limited growth conditions. Requires first the AMAXTB and SLATB (detailed) calibrations and possibly the FOTB (detailed) calibration.
CFET detailed	Correction factor transpiration rate	–, P	S,–	TRARTB, TAGBRTB or WSORTB (in this order)	Idem to the RDMCR calibration, but only if the RDMCR is fixed input.
DLO	Optimal day length for development	h, P	S,–	DOA	Determined for multipoints with potential growth conditions and with calibration of both DLO and TSUM1.
DLC	Critical day length for development	h, P	S,–	DOA	Idem the calibration for DLO but with calibration of DLC, DLO, and TSUM1.
TSUMEM	Temperature sum from sowing to emergence	°C d, P	S,–	DOE	TEFFMAX and TBASEM are fixed inputs.

Continued.

Table 11–3. Continued.

Model parameter	Description	Unit, type of calibration†	Type of parameter‡	Field observation§	Comments
TEFFMX	Max. Effective temperature for emergence	°C, P	S,–	DOE	Idem the calibration for TSUMEM and with calibration of both TEFFMX and TSUMEM.
TBASEM	Lower threshold for emergence	°C, P	S,–	DOE	Idem the calibration for TSUMEM but with calibration of TBASEM, TEFFMX, and TSUMEM.
DTSMTB	Daily increase in temperature sum	°C, N			Important but calibration not possible; use crop-specific value.
TDWI	Initial crop dry weight	kg ha^{-1}, P	S,–	TAGBITB	Determined from initial part of TAGBITB curve.
RGRLAI	Maximum relative increase in LAI	d^{-1}, P	S,–	LAIITB	Determined from initial part of LAITB curve for potential growth conditions.
LAIEM	Leaf area index at emergence	–, P	S,–	LAIITB	Idem the calibration for RGRLAI and with calibration of both LAIEM and RGRLAI.
SPA	Specific pod area	ha kg^{-1}, N			Not important; use crop-specific value.
SSA	Specific stem area	ha kg^{-1}, N			Not important; use crop-specific value.
TBASE	Lower threshold temperature for ageing of leaves	°C, N			Not important; use crop-specific value.
KDIF	Extinction coefficient for diffuse visible light	–, N			Important but calibration not possible; use crop-specific value.
EFF	Light use efficiency of single leaf	Kg ha^{-1} h^{-1}/(J m^{-2} s^{-1}), N			Important but calibration not possible; use crop-specific value.
TMPFTB	Reduction factor of leaf assimilation rate dependent on mean temperature	–, P	T, SCF	TAGBITB	For potential growth conditions. Requires the AMAXTB and SLATB calibrations first.
TMNFTB	Reduction factor of gross assimilation rate as function of low Tmin	–, N			Not important; use crop-specific value.

Continued.

Table 11–3. Continued.

Model parameter	Description	Unit, type of calibration†	Type of parameter‡	Field observation§	Comments
CVL	Conversion efficiency to leaves	–, N			Important but calibration not possible; use crop-specific value.
CVO	Conversion efficiency to storage organs	–, N			Important but calibration not possible; use crop-specific value.
CVR	Conversion efficiency to roots	–, N			Important but calibration not possible; use crop-specific value.
CVS	Conversion efficiency to stems	–, N			Important but calibration not possible; use crop-specific value.
Q10	Relative increase in respiration rate per 10°C increase	–, N			Important but calibration not possible; use crop-specific value.
RML	Relative maintenance respiration of leaves	–, N			Important but calibration not possible; use crop-specific value.
RMO	Relative maintenance respiration of storage organs	–, N			Important but calibration not possible; use crop-specific value specific value.
RMR	Relative maintenance respiration of roots	–, N			Important but calibration not possible; use crop-specific value.
RMS	Relative maintenance respiration of stems	–, N			Important but calibration not possible; use crop-specific value.
RFSETB	Reduction factor of senescence in dependence of development stage	–, N			Not important; use crop specific values.
FSTB	Fraction of assimilates to stems	–, P			Calculated through FSOTB and FLTB.
FRTB	Fraction of assimilates to roots	–, P	T, SCF	WRTITB	For potential growth conditions. Requires the AMAXTB and SLATB calibrations first.

Continued.

Table 11–3. Continued.

Model parameter	Description	Unit, type of calibration†	Type of parameter‡	Field observation§	Comments
FLTB	Fraction of assimilates to leaves	–, P	T, SCF	WLVITB	For potential growth conditions. Requires the AMAXTB and SLATB calibrations first.
PERDL	Relative death rate of leaves due to water stress	–, P	S,-	WLVDRTB or WLVRTB or LAIRTB (in this order)	For water-limited growth conditions. Requires the AMAXTB, the SLATB, and the RDMCR calibrations first.
RDRRTB	Relative death rate of roots	–, P	T, SCF	WRTITB	For potential growth conditions. Idem the calibration for FRTB but with calibration for both RDRRTB and FRTB.
RDRSTB	Relative death rate of stems	–, N			Not important; use crop specific values.
DEPNR	Crop group number for soil water depletion	–, P	S,-	TRARTB or TAGBRTB or WSORTB (in this order)	For water-limited growth conditions. Requires first the AMAXTB and SLATB calibrations to derive TRAITB or TAGBITB for comparison.
RDI	Initial rooting depth	cm, N			Not important; use crop specific value.
RRI	Maximum daily increase in rooting depth	cm d^{-1}, N			Important but calibration not possible; use crop-specific value.
IDSL	Indicates whether preanthesis development dep. on temp., day length or both	–, N			Not important; use crop specific values.
IAIRDU	Presence air ducts in roots	–, N			Not important; use crop specific values.
DVSEND	Development stage at maturity	–, N			Not important; use crop specific values.

† R, regional; P, (multi-)point calibration; N, not possible.

‡ S, single value; T, table; SCF, scaling factor for table or not.

§ Description of field observations is given in Table 11–4.

¶ DOE is estimated from DOP using TSUMEM.

DOM is estimated from DOH using simple rules. For root crops (sugarbeet [*Beta vulgaris* L.] and potato [*Solanum tuberosum* L.]) the difference is assumed to be 7 d, for other crops 3 d.

Table 11–4. Field observations, their description, unit and type, and model parameters in the WOFOST model that can be calibrated on the basis of the observation.

Observation code	Description	Unit	Type†	Model parameters‡
DOA	Julian day of anthesis or start of tuber or beet filling for root/ tuber crops	–	S	TSUM1, DLO,DLC
DOE	Julian day of emergence	–	S	TSUM1, TSUMEM, TBASEM, TEFFMX
DOH	Julian day of harvest	–	S	TSUM2
DOM	Julian day of crop maturity	–	S	TSUM2
DOP	Julian day of planting or sowing	–	S	TSUM1
LAI-MAX	Maximum LAI during growth period of irrigated crop	$m^2\ m^{-2}$	S	SLATB, SPAN
LAIITB	Leaf area index over time of irrigated crop	$m^2\ m^{-2}$	T	SLATB, SPAN, RGRLAI, LAIEM
LAIRTB	Leaf area index over time of rainfed crop	$m^2\ m^{-2}$	T	PERDL
TAGBI	Total above-ground biomass at final harvest for irrigated crop	kg dry matter ha^{-1}	S	AMAXTB
TAGBITB	Total above-ground biomass over time for irrigated crop	kg dry matter ha^{-1}	T	AMAXTB, LAIEM, TDWI, TMPFTB
TAGBR	Total above-ground biomass at final harvest for rainfed crop	kg dry matter ha^{-1}	S	RDMCR, CFET
TAGBRTB	Total above-ground biomass over time for rainfed crop	kg dry matter ha^{-1}	T	RDMCR, CFET, DEPNR
TRARTB	Cumulative crop transpiration over time for rainfed crop	cm	T	DEPNR, CFET
WLVDRTB	Weight of dead leaves over time for rainfed crop	kg dry matter ha^{-1}	T	PERDL
WLVITB	Weight of green leaves over time for irrigated crop	kg dry matter ha^{-1}	T	FLTB
WLVRTB	Weight of green leaves over time for rainfed crop	kg dry matter ha^{-1}	T	PERDL
WRTITB	Weight of roots for irrigated crop	kg dry matter ha^{-1}	T	FRTB, RDRRTB
WSOI	Weight of storage organs for irrigated crop at final harvest	kg dry matter ha^{-1}	S	FOTB
WSOITB	Weight of storage organs over time for irrigated crop	kg dry matter ha^{-1}	T	FOTB
WSOR	Weight of storage organs for rainfed crop at final harvest	kg dry matter ha^{-1}	S	RDMCR, CFET
WSORTB	Weight of storage organs over time for rainfed crop	kg dry matter ha^{-1}	T	RDMCR, CFET, DEPNR

† S, single value; T, table vs. day number.

‡ Order of use for calibration from left to right. Description of model parameters is given in Table 11–3.

the contributions to the criterion are divided by the simulated value, so that the criterion is a sum of terms $\{[y - f(\tau)]/[f(\tau)]\}2$. Finally, a third criterion is available, which has been termed the *Shape criterion*. The criterion is applicable whenever it is considered more important to reproduce a trend or shape in the data than to reproduce the actual values of the data. The Shape criterion results in fitted values whose distance to the true values is more or less equal, so that the variance of $(y - f)$ is very small.

Acknowledgments

This study was commissioned in 2004 by the Joint Research Centre of the European Commission at Ispra, Italy (http://ipsc.jrc.ec.europa.eu/ [verified 2 May 2011]). It was carried out as part of the ASEMARS project and supports the operation and maintenance of the CGMS system. For more information about recent developments of the MCYFS, see http://mars.jrc.ec.europa.eu/mars (verified 2 May 2011).

References

Boogaard, H.L., H. Eerens, I. Supit, C.A. Van Diepen, I. Piccard, and P. Kempeneers. 2002. Description of the MARS crop yield forecasting system (MCYFS). Study commissioned by Joint Research Centre of the European Commission. Alterra, Wageningen, the Netherlands.

Boogaard, H.L., C.A. Van Diepen, R.P. Rötter, J.M.C.A. Cabrera, and H.H. Van Laar. 1998. User's guide for the WOFOST 7.1 crop growth simulation model and WOFOST control center 1.5. Technical Doc. 52. Winand Staring Centre, Wageningen, the Netherlands.

Boons-Prins, E.R., G.H.J. De Koning, C.A. Van Diepen, and F.W.T. Penning de Vries. 1993. Crop specific simulation parameters for yield forecasting across the European Community. Simulation Rep. CABO-TT no. 32. CABO-DLO, Wageningen, the Netherlands.

Bouman, B.A.M., M.J. Kropff, T.P. Tuong, M.C.S. Wopereis, H.F.M. Ten Berge, and H.H. Van Laar. 2001. ORYZA2000: Modelling lowland rice. IRRI, Los Baños, the Philippines.

Bouman, B.A.M., H. Van Keulen, H.H. Van Laar, and R. Rabbinge. 1996. The 'School of de Wit' crop growth simulation models: Pedigree and historical overview. Agric. Syst. 52:171–198.

Conrad, V. 1946. Usual formulas of continentality and their limits of validity. Trans. Am. Geophys. Union 27:663–664.

De Koning, G.H.J., and C.A. Van Diepen. 1992. Crop production potential of rural areas within the European Communities. IV. Potential, water-limited and actual crop production. Working Doc. W68. Netherlands Scientific Council for Government Policy, the Hague, the Netherlands.

de Wit, A., J. Wolf, H.L. Boogaard, R. Hessel, J. te Roller, and W. Akkermans. 2007. CALPLAT: Platform for calibrating CGMS. ASEMARS Proj. Rep. 5.2.1. Alterra, WUR, Wageningen, the Netherlands and Joint Research Centre of the European Commission, Ispra, Italy.

EUROSTAT. 2005. European regional and urban statistics—Reference guide. Office for Official Publications of the European Communities, Luxembourg.

Frère, M., and G.F. Popov. 1979. Agrometeorological crop monitoring and forecasting. FAO Plant Production and Protection Paper 17. FAO, Rome.

Genovese, G., and M. Bettio (ed.) 2004. Methodology of the MARS Crop Yield Forecasting System. Vol. 4. Statistical data collection, processing and analysis. Office for Official Publications of the European Communities, Luxembourg.

GISAT. 2003. Studies in support to the MARS project. Crop monographs on candidate countries: MOCA study. Final report. GISAT, Praha, Czech Republic.

Goudriaan, J. 1986. A simple and fast numerical method for the computation of daily crop photosynthesis. Agric. For. Meteorol. 38:249–254.

Hoogenboom, G. 2000. Contribution of agrometeorology to the simulation of crop production and its applications. Agric. For. Meteorol. 103:137–157.

Lazar, C., and G. Genovese (ed.) 2004. Methodology of the MARS Crop Yield Forecasting System. Volume 2. Agrometeorological data collection, processing and analysis. Office for Official Publications of the European Communities, Luxembourg.

Nelder, J.A., and R. Mead. 1965. A simplex method for function minimization. Comput. J. 7:308–313.

Potgieter, A.B., G.L. Hammer, H. Meinke, R.C. Stone, and L. Goddard. 2005. Three putative types of El Nino revealed by spatial variability in impact on Australian wheat yield. J. Clim. 18:1566–1574.

Sadras, V., D. Roget, and M. Krause. 2003. Dynamic cropping strategies for risk management in dry-land farming systems. Agric. Syst. 76:929–948.

Spitters, C.J.T. 1986. Separating the diffuse and direct component of global radiation and its implications for modelling canopy photosynthesis. Part II: Calculation of canopy photosynthesis. Agric. For. Meteorol. 38:231–242.

Supit, I. 1997. Predicting national wheat yields using a crop simulation and trend models. Agric. For. Meteorol. 88:199–214.

Supit, I., A.A. Hooijer, and C.A. Van Diepen (ed.) 1994. System description of the WOFOST 6.0 crop simulation model implemented in CGMS. Publ. EUR15956EN. Joint Research Centre, European Commission, Luxembourg.

Van Diepen, C.A., J. Wolf, H. Van Keulen, and C. Rappoldt. 1989. WOFOST: A simulation model of crop production. Soil Use Manage. 5:16–24.

Van Ittersum, M.K., P.A. Leffelaar, H. Van Keulen, M.J. Kropff, L. Bastiaans, and J. Goudriaan. 2003. On approaches and applications of the Wageningen crop models. Eur. J. Agron. 18:201–234.

Van Keulen, H., and J. Wolf. 1986. Modelling of agricultural production: Weather, soils and crops. Simulation Monographs, Pudoc, Wageningen, the Netherlands

Van Laar, H.H., J. Goudriaan, and H. Van Keulen. 1997. SUCROS97: Simulation of crop growth for potential and water-limited production situations. Quantitative Approaches in Systems Analysis 14. Wageningen University, Wageningen, the Netherlands.

Welch, S.M., G. Wilkerson, K. Whiting, N. Sun, T. Vagts, G. Buol, and T. Mavromatis. 2002. Estimating soybean model genetic coefficients from private-sector variety performance trial data. Trans. ASAE 45:1163–1175.

Willekens, A., J. van Orshoven, and J. Feyen. 1998. Estimation of the phenological calendar, Kc-curve and temperature sums for cereals, sugar beet, potato, sunflower and rapeseed across Pan Europe, Turkey and the Maghreb countries by means of transfer procedures. Vol. I. Project report. Ground for GIS & Institute for Land and Water Management, K.U. Leuven, Leuven, Belgium.

Wolf, J. 1993. Effects of climate change on wheat production potential in the European Community. Eur. J. Agron. 2:281–292.

Wolf, J., and C.A. Van Diepen. 1995. Effects of climate change on grain maize yield potential in the European Community. Clim. Change 29:299–331.

Wu, D., Q. Yu, C. Lu, and H. Hengsdijk. 2006. Quantifying production potentials of winter wheat in the North China Plain. Eur. J. Agron. 24:226–235.

Chapter 12

Inverse Modeling with RZWQM2 to Predict Water Quality

Bernard T. Nolan
Robert W. Malone
Liwang Ma
Christopher T. Green
Michael N. Fienen
Dan B. Jaynes

Abstract

This chapter presents guidelines for autocalibration of the Root Zone Water Quality Model (RZWQM2) by inverse modeling using PEST parameter estimation software (Doherty, 2010). Two sites with diverse climate and management were considered for simulation of N losses by leaching and in drain flow: an almond [*Prunus dulcis* (Mill.) D.A. Webb] orchard in the San Joaquin Valley, California and the Walnut Creek watershed in central Iowa, which is predominantly in corn (*Zea mays* L.)–soybean [*Glycine max* (L.) Merr.] rotation. Inverse modeling provides an objective statistical basis for calibration that involves simultaneous adjustment of model parameters and yields parameter confidence intervals and sensitivities. We describe operation of PEST in both parameter estimation and predictive analysis modes. The goal of parameter estimation is to identify a unique set of parameters that minimize a weighted least squares objective function, and the goal of predictive analysis is to construct a nonlinear confidence interval for a prediction of interest by finding a set of parameters that maximizes or minimizes the prediction while maintaining the model in a calibrated state. We also describe PEST utilities (PAR2PAR, TSPROC) for maintaining ordered relations among model parameters (e.g., soil root growth factor) and for post-processing of RZWQM2 outputs representing different cropping practices at the Iowa site. Inverse modeling provided reasonable fits to observed water and N fluxes and directly benefitted the modeling through: (i) simultaneous adjustment of multiple parameters versus one-at-a-time adjustment in manual approaches; (ii) clear indication by convergence criteria of when calibration is complete; (iii) straightforward detection of nonunique and insensitive parameters, which can affect the stability of PEST and RZWQM2; and (iv) generation of confidence intervals for uncertainty analysis of parameters and model predictions. Composite scaled sensitivities, which reflect the total information provided by the observations for a parameter, indicated that most of the RZWQM2 parameters at the California study site (CA) and Iowa study site (IA) could be reliably esti-

B.T. Nolan, USGS, 413 National Center, Reston, VA 20192 (btnolan@usgs.gov); R.W. Malone (rob.malone@ars.usda.gov) and D.B. Jaynes (dan.jaynes@ars.usda.gov), USDA-ARS, National Laboratory for Agriculture and the Environment, 2110 University Boulevard, Ames, IA 50011-0001; L. Ma, USDA-ARS, Agriculture Systems Research, 2150 Centre Ave. Bldg. D, Fort Collins, CO 80526 (liwang.ma@ars.usda.gov); C.T. Green, USGS, Bldg 15, McKelvey Building, 345 Middlefield Road, Menlo Park, CA 94025-3561 (ctgreen@usgs.gov); Michael N. Fienen, USGS, Wisconsin Water Science Center, 8505 Research Way, Middleton, WI 53562 (mnfienen@usgs.gov).

doi:10.2134/advagricsystmodel2.c12

Methods of introducing system models into agricultural research. L.R. Ahuja and L. Ma (ed.)
Advances in Agricultural Systems Modeling Series 2.

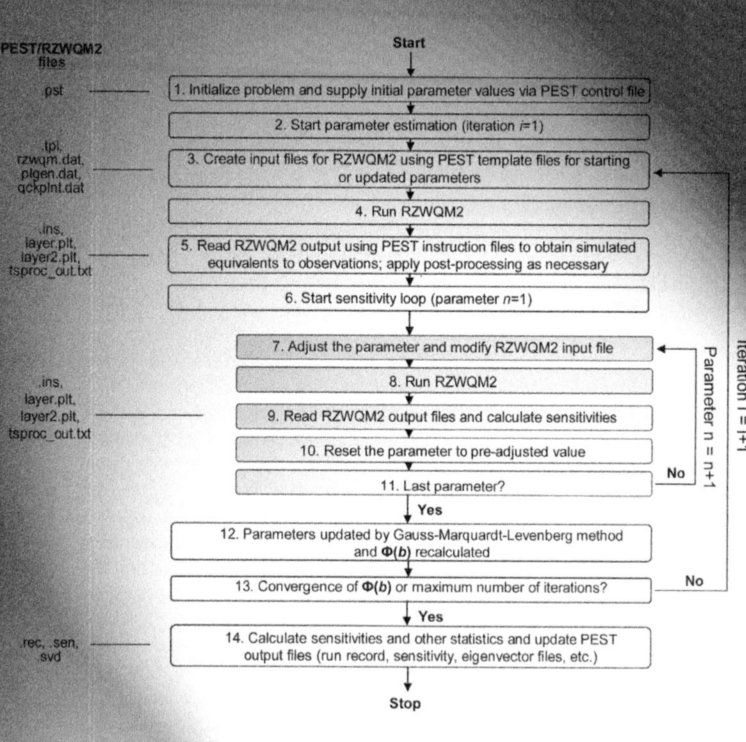
PEST/RZWQM2 files
Start
.pst
1. Initialize problem and supply initial parameter values via PEST control file
2. Start parameter estimation (iteration i=1)
.tpl, rzwqm.dat, plgen.dat, qckplnt.dat
3. Create input files for RZWQM2 using PEST template files for starting or updated parameters
4. Run RZWQM2
.ins, layer.plt, layer2.plt, tsproc_out.txt
5. Read RZWQM2 output using PEST instruction files to obtain simulated equivalents to observations; apply post-processing as necessary
6. Start sensitivity loop (parameter n=1)
7. Adjust the parameter and modify RZWQM2 input file
8. Run RZWQM2
.ins, layer.plt, layer2.plt, tsproc_out.txt
9. Read RZWQM2 output files and calculate sensitivities
10. Reset the parameter to pre-adjusted value
11. Last parameter?
No
Parameter n = n+1
Yes
12. Parameters updated by Gauss-Marquardt-Levenberg method and Φ(b) recalculated
13. Convergence of Φ(b) or maximum number of iterations?
No
Iteration i = i+1
Yes
.rec, .sen, .svd
14. Calculate sensitivities and other statistics and update PEST output files (run record, sensitivity, eigenvector files, etc.)
Stop

mated by regression. Correlations obtained in the CA case indicated that all model parameters could be uniquely estimated by inverse modeling. Although water content at field capacity was highly correlated with bulk density (−0.94), the correlation is less than the threshold for nonuniqueness (0.95, absolute value basis). Additionally, we used truncated singular value decomposition (SVD) at CA to mitigate potential problems with highly correlated and insensitive parameters. Singular value decomposition estimates linear combinations (eigenvectors) of the original process-model parameters. Parameter confidence intervals (CIs) at CA indicated that parameters were reliably estimated with the possible exception of an organic pool transfer coefficient (R45), which had a comparatively wide CI. However, the 95% confidence interval for R45 (0.03–0.35) is mostly within the range of values reported for this parameter. Predictive analysis at CA generated confidence intervals that were compared with independently measured annual water flux (groundwater recharge) and median nitrate concentration in a collocated monitoring well as part of model evaluation. Both the observed recharge (42.3 cm yr^{-1}) and nitrate concentration (24.3 mg L^{-1}) were within their respective 90% confidence intervals, indicating that overall model error was within acceptable limits.

The Root Zone Water Quality Model (RZWQM2) is a systems model that simulates N cycling processes, the fate and transport of agricultural chemicals, and crop growth (see Chapter 1, Ma et al., 2011, this volume). The model provides a detailed accounting of N losses, additions, and transformations (Ahuja et al., 2000) and has been used to simulate crop yield and N mass balances in the U.S. Midwest (Ma et al., 2007; Malone et al., 2007). These and other studies featured manual calibration of earlier versions of the model. A typical approach involves adjusting parameters one at a time in iterative fashion to try to match observed values of soil water, N, and plant growth (Hanson et al., 1999). However, systems models are complex and have numerous parameters that are unknown and difficult to estimate. Manual calibration of such models is difficult and subjective, with the result that "robust measures of uncertainty in model parameters and predictions cannot be quantified" (Schwarz et al., 2006). RZWQM2 has parameters for multiple processes representing plant growth, microbial populations, water flux, and contaminant fate and transport. Emphasizing one component of the model (e.g., crop yield), can cause other components (e.g., N leaching) to fit less well and vice versa, and without statistical controls on the calibration, it is difficult to know when effective calibration has been achieved.

Because of these difficulties, inverse methods have been identified as a future need for unsaturated zone models applied at various spatial scales (Corwin et al., 2006) and are gaining more widespread acceptance. Inverse modeling has been applied to GLEAMS, LEACHM, HYDRUS, and HYDRUS-2D to estimate field-based mobility and degradation parameters for pesticides (Close et al., 2005, 2008). An inverse algorithm also has been used to estimate soil hydraulic and N cycling parameters in LEACHM for an agricultural field, and this performed well with a few exceptions (Schmied et al., 2000). Discrepancies in model fit were attributed to processes not simulated by the model, such as macropore flow. However, inverse modeling had not been previously attempted with RZWQM2.

Chapter 1 (Ma et al., 2011, this volume) discussed several autocalibration methods, including global parameter estimation methods such as Latin Hypercube Sampling (Fang et al., 2010), and software specific to crop growth models (e.g., GENCALC). In this chapter we focus on inverse modeling by PEST parameter estimation software (Doherty, 2010). Although the method can be applied to any model, we focus on the water quality aspects of RZWQM2. Inverse modeling routines such as PEST and UCODE_2005 (Poeter et al., 2005) use an iterative, nonlinear regression approach that involves adjustment of multiple model parameters and that yields parameter correlations, sensitivities, and confidence intervals. Extreme parameter correlation and insensitivity are difficult to identify with nonautomated methods of calibration (Poeter and Hill, 1997). Sensitivities based on the observations indicate the changes in their simulated equivalents due to changes in parameter values and whether there is sufficient information in the observation data set to reliably estimate the parameters (Hill and Tiedeman, 2007). (Here, we use "simulated equivalents" or "values" to denote model simulated quantities that are compared with observations during calibration, and we use "predictions" to denote simulated quantities that are not associated with observations but are used in model evaluation.) Advanced methods, such as truncated singular value decomposition, reduce the extent of the calibration problem by compositing parameters into linear combinations and estimating only the most sensitive parameters to avoid overfitting of the model. These procedures are part of the estimation mode of PEST, which also has a predictive mode (i.e., "predictive analysis") for deriving nonlinear confidence intervals on predictions of interest. Predictive analysis is based on the optimization results and is useful in characterizing the overall uncertainty of predictions, which can arise from uncertainties in model structure, parameters, initial conditions, and measurements (Lin and Radcliffe, 2006).

In this chapter we briefly discuss inverse modeling theory and then present examples of using PEST with RZWQM2 in both estimation and predictive modes. The examples comprise two field sites having very different climatic and management regimes that are described in more detail below: (1) an almond orchard in the lower Merced River study basin in California (CA), and (2) corn–soybean rotations in the Walnut Creek watershed in Iowa (IA). The wide range of conditions provides a good test of the inverse modeling approach.

Site Descriptions

San Joaquin, CA

The CA site is an almond orchard in the lower Merced River basin, which is one of five small stream basins undergoing intensive evaluation by the USGS National Water Quality Assessment Program (NAWQA). The studies were designed and implemented in holistic fashion to better understand the transport and fate of contaminants and potential interactions in the unsaturated zones, streams, and groundwaters of small agricultural basins (Capel et al., 2008). Subsequently, three of the five sites were selected for evaluation of unsaturated zone N mass balances using PEST-RZWQM2 (Nolan et al., 2010). The CA site is semiarid, soils are predominantly sand except for a silt layer at about 3 m, and N is transported below the root zone primarily in matrix flow. Average depth to water was 6.5 m during the model simulation period (1 Jan. 2002–31 Dec. 2004), and mean annual precipitation was 31 cm. This depth to water provides an opportunity to test the extended depth capabilities of RZWQM2 (among other improvements, the simulation depth has been extended from 3 to 30 m). The site typically receives 120 cm yr^{-1} of irrigation water and 140 kg N ha^{-1} yr^{-1} of fertilizer by chemigation.

Walnut Creek, IA

In contrast to the CA site, IA is humid, has clayey soils, and N moves primarily in drain flow. Tile drains are prevalent in the area due to the presence of fine-textured soils with clay fractions of 0.20 or greater (Cambardella et al., 1999) and a high water table (depth to water = 2.7 m). The IA site consisted of three subbasins totaling 1500 ha in the Walnut Creek watershed of central Iowa. Data were collected as part of earlier experiments that described implementation of the late spring nitrate test (LSNT) to a subbasin of the watershed (Jaynes et al., 2004; Karlen et al., 2005). In this chapter we focus on an adjacent subbasin used for auto-calibration of RZWQM2 to water and N flux data (CN2). RZWQM2 simulations were conducted at the subbasin scale, which involved post-processing to aggregate simulated values from the two main cropping practices (corn and soybean) within CN2. The experimental setup is described in detail in a prior publication (Malone et al., 2010). We specified typical fertilizer application rates for the area, which were 165 kg N ha^{-1} yr^{-1} of fertilizer in the fall before the 1999 corn planting seasons (1991–1998) and 185 kg N ha^{-1} yr^{-1} thereafter (Jaynes et al., 2004). When a second consecutive year of corn was planted we assumed 200 kg N ha^{-1} yr^{-1} fertilizer application. LSNT management involved application of 56 kg N ha^{-1} yr^{-1} shortly before planting, and sidedress applications of 168, 118, 174, and 96 kg N $ha^{-1}yr^{-1}$ during 1997, 1998, 1999, and 2000, respectively.

Methods

Inverse Modeling Theory

Inverse modeling can use any of several methods to estimate model parameters, including global search methods, nonlinear regression, and combinations of these methods. Here we focus on nonlinear regression as implemented in PEST. Nonlinear regression involves estimation of model parameters in iterative fashion, and model fit is evaluated using an objective function. The process terminates when the objective function converges on a minimum value (Doherty, 2010). The objective function is minimized by a weighted least squares estimator and is expressed in general form as (modified from Doherty, 2010)

$$\Phi(\mathbf{b}) = [\mathbf{y} - \mathbf{y}'(\mathbf{b})]^T \mathbf{Q}[\mathbf{y} - \mathbf{y}'(\mathbf{b})] \quad [1]$$

where $\mathbf{Q}$ is a weight matrix containing the squares of the observation weights, $\mathbf{y}$ is a vector of observations, $\mathbf{y}'(\mathbf{b})$ is a vector of simulated values produced by the mechanistic model (RZWQM2 in this case) based on parameter vector $\mathbf{b}$, and T indicates the matrix transpose. The two vectors have m elements, where m is the number of observations plus prior information expressions in the model. Prior information reflects knowledge of the value of a parameter gained from outside the parameter estimation process and can mitigate problems with parameter nonuniqueness and numerical instability (Doherty, 2010). A penalty is imposed on the objective function when the estimated value of a parameter strays too far from an assigned prior value. The modeling examples discussed in this book chapter did not include prior information.

Parameter values that minimize Eq. [1] are obtained from normal equations derived for a nonlinear regression model that takes the form (modified from Doherty 2010)

$$\mathbf{y}' = \mathbf{M}(\mathbf{b}) \quad [2]$$

where $\mathbf{M}$ represents the mechanistic model. To simplify parameter estimation in the nonlinear case, Taylor series expansion is used at the beginning of each optimization iteration to obtain a linear approximation of Eq. [2] with the current, best parameter set (modified from Doherty 2010):

$$\mathbf{y}' \approx \mathbf{y}_0' + \mathbf{J}(\mathbf{b} - \mathbf{b}_0) \quad [3]$$

where $\mathbf{y}_0'$is the initial set of model simulated values, $\mathbf{b}_0$ is a vector of starting parameters, $\mathbf{b}$ is a vector of parameters that is slightly different from $\mathbf{b}_0$, and $\mathbf{J}$ is a "Jacobian" or "sensitivity" matrix with *m* rows and *n* columns, where each element is the derivative of a simulated equivalent of an observation with respect to one of the *n* parameters. Thus, linearization involves calculating the derivatives of all simulated equivalents with respect to all parameters. Each element of the Jacobian matrix is given by (Hill and Tiedeman, 2007)

$$J_{ij} = \frac{\partial y_i'}{\partial b_j} \quad [4]$$

which is the derivative of the simulated value for the *i*th observation with respect to the *j*th parameter in $\mathbf{b}$.

To derive the normal equations for the nonlinear least squares estimator, we take the derivative of the objective function and set it equal to zero. The resulting normal equation for the linearized form of the regression model is expressed as (modified from Doherty 2010)

$$\mathbf{u} = (\mathbf{J}^T\mathbf{Q}\mathbf{J})^{-1}\mathbf{J}^T\mathbf{Q}(\mathbf{y}' - \mathbf{y}_0') \quad [5]$$

where $\mathbf{u}$ is the parameter upgrade vector $\mathbf{b} - \mathbf{b}_0$. Because the normal equation is nonlinear in $\mathbf{b}$, the parameters cannot be solved for directly as in ordinary least squares estimation (linear case). PEST uses the Gauss–Marquardt–Levenberg method to solve the nonlinear problem (Doherty, 2010). As estimation proceeds, the linearized equation is solved for a new parameter set, which is evaluated by running the model again.

The composite sensitivity of each parameter based on all *m* observations is calculated by PEST as

$$s_j = \frac{\left(\mathbf{J}^T\mathbf{Q}\mathbf{J}\right)_{jj}^{1/2}}{m} \quad [6]$$

where s_j is the composite sensitivity of the *j*th parameter (Doherty, 2010). Parameter sensitivities indicate the amount of change in simulated values for a change in the parameter values (Poeter and Hill, 1997). The sensitivities and model residuals are used to determine the magnitude and direction of adjustments in parameters to decrease $\Phi(\mathbf{b})$. We also calculated the composite-scaled sensitivity (CSS) as described by Hill and Tiedeman (2007):

$$\text{CSS}_j = \left|s_j b_j \sqrt{m}\right| \quad [7]$$

where CSS_j is the composite-scaled sensitivity of the *j*th parameter. In the case of a log-transformed parameter, b_j is expressed as the log (base 10) of the parameter value for consistency with how PEST implements log transforms. The CSS measures the composite changes in model simulated values based on a fractional change of a parameter value (Doherty, 2010) and indicates the total information content of the data for estimation of a single parameter (Hill and Tiedeman, 2007).

Limitations to inverse modeling include the potential for local objective function minima and parameter nonuniqueness. The ideal is to locate a global minimum that is associated with a unique set of parameters, but in practice this may be difficult to achieve. Objective function surfaces may contain local minima, which can "trap" the optimization algorithm such that the true minimum is not easily found. It is advisable to try different starting values to see if the optimizations converge on the same $\mathbf{\Phi(b)}$ value and, if not, to see if lower values can be obtained.

Nonuniqueness stems from highly correlated parameters and can be mitigated either by setting one of the parameters to a fixed value, by gathering additional data that would enable uniquely estimating all of the parameters, or through alternative methods such as singular value decomposition (discussed below). A major advantage of inverse modeling is that highly correlated and insensitive parameters can be readily identified. Such problems are difficult to identify when calibration of complex models is attempted by manual methods (Poeter and Hill, 1997).

When only two parameters are adjusted, objective function contour maps can help identify parameter nonuniqueness. Valleys in such maps indicate potential problems. For example, Fig. 12–1 indicates nonuniqueness in the form of a valley along the *x* axis for $\mathbf{\Phi(b)} < 25$; when parameter 2 $\approx$ 125, parameter 1 values of $\approx$25–50 yield similar values of $\mathbf{\Phi(b)}$, such that in this range there is no unique optimal combination of the two parameters.

Extreme parameter correlation can cause numerical instability in PEST. In such cases performance can be improved by invoking singular value decomposition (SVD), which reduces the parameter space by forming linear combinations of the original process-model parameters based on elements of SVD eigenvectors. Thus, the original process-model parameters are transformed into SVD parameters for estimation purposes. Numerical stability is enhanced through truncation, whereby only sensitive combinations of parameters (those with singular values significantly greater than zero) are estimated. Parameters with singular values that are too small are excluded from estimation. A potential difficulty is determining whether a singular value is "small" (Hill and Osterby, 2003), which has implications for truncation. One approach is to set the PEST

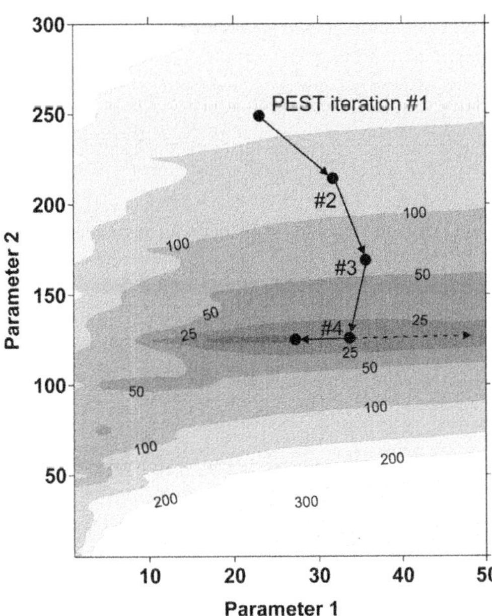

Fig. 12–1. Objective function, Φ(b), contours as a function of two parameters, and progression of PEST iterations.

variable EIGTHRESH at 10^{-6} to 10^{-7}, where EIGTHRESH is the ratio of the lowest and highest singular values at which truncation occurs. Other advanced methods such as regularization and SVD-assist are not discussed here, but interested readers can consult Doherty (2010).

We report the relative root mean square error (RRMSE) and index of agreement (*d*) as additional indicators of model fit generated outside of PEST. The RRMSE is the root mean square error normalized to the mean of the observed values:

$$\mathrm{RRMSE} = \frac{\sqrt{\frac{1}{m}\sum_{i=1}^{m}\left(y_i - y_i'\right)^2}}{\left|\overline{y}\right|} \quad [8]$$

where $\overline{y}$ is the mean of the observations. The RRMSE is interpreted similarly to the coefficient of variation, except that RMSE is substituted for the standard deviation. The index of agreement is defined as

$$d = 1 - \frac{\sum_{i=1}^{m}\left(y_i - y_i'\right)^2}{\sum_{i=1}^{m}\left(\left|y_i' - \overline{y}\right| + \left|y_i - \overline{y}\right|\right)^2} \quad [9]$$

The index of agreement is more sensitive than traditional correlation measures to differences between observed and predicted means and variances; d varies between 0 and 1, with higher values indicating better fit (Legates and McCabe, 1999).

Model Setup

California

We used version 1.80.2009 of RZWQM2 to predict N losses and transformations for the entire unsaturated zone thickness at CA (6.5 m). The RZWQM2 setup at CA consisted of compiling weather data, establishing boundary conditions, setting up the crop model, and specifying initial values of parameters for adjustment by PEST as described by Nolan et al. (2010). Weather data were measured in 2004 at an on-site weather station and supplemented with off-site data to extend the period of record to 2002–2003 for modeling purposes (Domagalski et al., 2008). We specified a unit gradient boundary condition for the lower boundary to allow free drainage and to keep the water table below the simulated profile. RZWQM2's minimum input option was specified for soil hydraulic parameters, which requires estimates of saturated hydraulic conductivity (K_s), water content at field capacity (WFC), and bulk density (BD). WFC and BD are used in a subroutine that applies scaling factors to built-in water-retention functions, $\theta(h)$ representing different soil types. Knowledge of K_s and $\theta(h)$ allow estimation of the unsaturated hydraulic conductivity function $K(h)$ (Ahuja et al., 2000). Initial values of these parameters are described below. Because the simulations were comparatively short (3 yr), we used RZWQM2's organic matter initialization wizard to help establish crop residues, organic matter pools, and microbial populations; a minimum of 10 yr was specified. Two simulations were performed consecutively to stabilize the hydraulic state of the model, which also helped stabilize the C and N pools after each perturbation of model parameters. We used Quicktree, a simple crop growth model in RZWQM2, to simulate the effects of the almond crop on water and N fluxes (Nolan et al., 2010). Because Quicktree does not predict crop yield, inverse modeling calibration was limited to soil hydraulic and chemical data.

Iowa

The RZWQM2 setup at IA is described in Malone et al. (2010), and we briefly summarize it here. At IA, median weekly rainfall from several gauges was used as input to RZWQM2 to represent the scale of the Walnut Creek watershed, as described in earlier work (Hatfield et al., 1999). Solar radiation and humidity data were checked using procedures described in an earlier study (Allen, 1996). The IA model simulated lateral flow to a drain, and therefore a high, fluctuating water table and constant flux boundary condition were specified. We used the full Brooks–Corey function to describe soil hydraulics, which requires estimates of the pore size distribution index (A2), an exponent for the hydraulic conductivity curve (N2), and several other parameters (Malone et al., 2010). RZWQM2's detailed crop growth model (DSSAT) was used to predict corn and soybean yields. The DSSAT model has been previously used to simulate long-term corn yield and N dynamics in the region (Thorp et al., 2007).

PEST Setup

We used PEST v. 11.8 (available at http://www.pesthomepage.org/, verified 13 May 2011) at both CA and IA, where inverse modeling proceeded according to the flow chart in Fig. 12–2. To help initialize the problem and to supply reasonable starting values of parameters to PEST (Step 1 in the flow chart), we anticipate that users will have performed some manual calibration as described in Chapter 2 to "ballpark" the model. We present initial parameter values for CA and IA below, along with discussion of key files and variables for PEST usage in estimation mode. Following calibration, we used predictive analysis at CA to evaluate the model based on comparison with independently measured quantities not used in calibration, and to characterize the uncertainty of the predictions. The objective of the evaluation phase was to see if the measured quantities were within the uncertainty estimates of the predictions.

California

Inverse modeling at CA was based on 1670 observations comprising the following groups: nitrate and bromide concentrations from lysimeters, soil nitrate and organic matter content, and soil moisture content and water tension collected during 2003–2004 (Table 12–1). Therefore the weighted least-squares objective function took the following form:

$$\begin{aligned}\Phi(\mathbf{b}) = & \sum_{i=1}^{m_1} w^2{}_{\mathrm{sm},i}\left(y_{\mathrm{sm},i} - y'_{sm,i}\right)^2 + \sum_{i=1}^{m_2} w^2{}_{\mathrm{t},i}\left(y_{\mathrm{t},i} - y'_{\mathrm{t},i}\right)^2 \\ & + \sum_{i=1}^{m_3} w^2{}_{\mathrm{an},i}\left(y_{\mathrm{an},i} - y'_{\mathrm{an},i}\right)^2 + \sum_{i=1}^{m_4} w^2{}_{\mathrm{ab},i}\left(y_{\mathrm{ab},i} - y'_{\mathrm{ab},i}\right)^2 \\ & + \sum_{i=1}^{m_5} w^2{}_{\mathrm{sn},i}\left(y_{\mathrm{sn},i} - y'_{\mathrm{sn},i}\right)^2 + \sum_{i=1}^{m_6} w^2{}_{\mathrm{om},i}\left(y_{\mathrm{om},i} - y'_{\mathrm{om},i}\right)^2\end{aligned} \quad [10]$$

where y is the observed value, y' is the predicted value, w is the observation weight, and the subscript i denotes ith observation, sm is soil moisture, t is soil

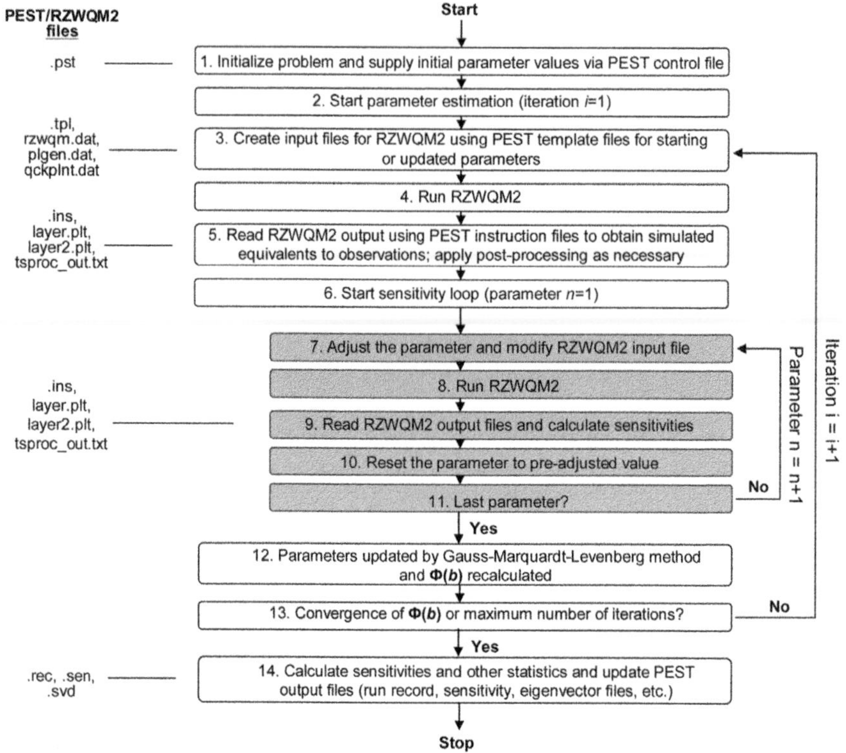

Fig. 12–2. Major steps in inverse modeling applied to RZWQM2 (modified from Poeter et al., 2005).

tension, an is aqueous nitrate, ab is aqueous bromide, sn is soil nitrate, om is organic matter, and $m_{1\text{-}6}$ are the numbers of observations associated with each of the preceding observation groups. Observation weights are discussed in more detail below.

Parameters selected for adjustment by inverse modeling were anticipated to affect water and N fluxes and N transformations, and initially included soil hydraulic parameters, all of the organic pool transformation constants, all of the Arrhenius coefficients, and N uptake rates (Table 12–2). To avoid numerical problems associated with wide ranges of parameter sensitivities, selected parameters identified in preliminary PEST runs were excluded from the final optimizations. The numerical problems stem from difficulties in inverting the matrix in the parameter upgrade vector (Eq. [2.23] in Doherty, 2010). Several of the parameters in Table 12–2 had sensitivities below the threshold recommended by Hill and Tiedeman (2007), wherein CSS values <1% of the maximum CSS obtained for a set of parameters indicates insensitivity. At CA, use of the following parameters in

Table 12–1. Observed data used in inverse calibration of RZWQM2 at California and Iowa. Sampling and analytical procedures at CA were described by Capel et al. (2008), and those at IA were described by Jaynes et al. (2004) and Karlen et al. (2005).

Observation group	N	Data source
California		
Soil moisture	603	Water content reflectometry probes
Soil water tension	1005	Heat dissipation probes
Aqueous bromide conc.	29	Ceramic porous-cup suction lysimeters followed by analysis at USGS NWQL†
Aqueous nitrate conc.	27	Ceramic porous-cup suction lysimeters followed by analysis at USGS NWQL†
Soil nitrate	5	Extraction of sediment cores followed by analysis at USGS soil physics laboratory
Soil organic matter	1	Estimated from organic carbon (combustion of surficial soil sample followed by gas chromatography at RSIL‡)
Iowa		
Drain flow	103	Flowtote meter where drain empties into Walnut Creek
N loss	103	Weekly samples at flow gauge followed by colorimetric analysis
Annual flow-weighted nitrate concentration	103	Sum of weekly product of drain flow and nitrate concentration
Corn yield	8	Yield monitors on farmers' combines

† National Water Quality Laboratory.

‡ Reston Stable Isotope Laboratory.

combination with SVD promoted model stability: K_s, WFC, and BD for each of the five simulation layers and the constant that controls transformation of organic matter from the medium soil humus pool to the slow pool (R45) (Table 12–3). Initial estimates of K_s and WFC and upper and lower bounds were taken from probability distributions of soil hydraulic properties (Meyer et al., 1997) for soil types indicated by sediment cores obtained at the site. Initial estimates of bulk density came from Natural Resource Conservation Service classes corresponding to measured soil texture at the site (USDA-NRCS, 2009). We used initial values of interpool transformation coefficients (e.g., R45 = 0.4) based on prior experience with RZWQM in the U.S. Midwest.

Porosity is a function of BD; therefore, we used the PAR2PAR utility in PEST to calculate porosity each time BD was adjusted, before it was written to the model input file. PAR2PAR commonly is used to enforce relationships among parameter values. For example, it can force a parameter to always have a value greater than another through definition of parameter ratios. Use of PAR2PAR at

Table 12–2. All model parameters initially adjusted by PEST at California (CA) and/or Iowa (IA).

Name	Parameter	Units	RZWQM2 file	Site
Ks	Saturated hydraulic conductivity	cm h^{-1}	rzwqm.dat	CA, IA
WFC	Water content at field capacity	cm^3 cm^{-3}	rzwqm.dat	CA
BD	Soil bulk density	g cm^{-3}	rzwqm.dat	CA, IA
ADEC	Arrhenius coefficient for OM† decay	s d^{-1}	rzwqm.dat	CA
ANIT	Arrhenius coefficient for nitrification	s d^{-1}	rzwqm.dat	CA
ADEN	Arrhenius coefficient for denitrification	s d^{-1}	rzwqm.dat	CA, IA
AHYD	Arrhenius coefficient for hydrolysis of urea	s d^{-1}	rzwqm.dat	CA
CNU1	Maximum N uptake by the plant	g $plant^{-1}$ d^{-1}	rzwqm.dat	CA
TSUP	Total seasonal N uptake	kg ha^{-1}	qckplnt.dat	CA
R14	Transformation coefficient, slow residue to intermediate OM	fraction	rzwqm.dat	CA, IA
R23	Transformation coefficient, fast residue to fast OM	fraction	rzwqm.dat	CA, IA
R34	Transformation coefficient, fast OM to intermediate OM	fraction	rzwqm.dat	CA, IA
R43	Transformation coefficient, intermediate OM to fast OM	fraction	rzwqm.dat	CA
R45	Transformation coefficient, intermediate OM to slow OM	fraction	rzwqm.dat	CA, IA
R53	Transformation coefficient, slow OM to fast OM	fraction	rzwqm.dat	CA
LHG	Lateral hydraulic gradient	cm cm^{-1}	rzwqm.dat	IA
RD	Soil root growth factor	fraction	mzdssat.rzx	IA
A2	Pore size distribution index	dimensionless	rzwqm.dat	IA
N2	Exponent of the Brooks–Corey unsaturated hydraulic conductivity curve	dimensionless	rzwqm.dat	IA
S1	Bubbling pressure for unsaturated hydraulic conductivity curve	cm	rzwqm.dat	IA
WS	Saturated water content	cm^3 cm^{-3}	rzwqm.dat	IA
SNO	Infiltrating snowmelt	fraction	[case].sno	IA
G2	Maximum number of kernels per plant	kernels $plant^{-1}$	mzcer040.cul	IA
G3	Kernel filling rate during the linear grain filling stage	mg d^{-1}	mzcer040.cul	IA
PHINT	Phylochron interval between successive leaf tip appearances	degree days	mzcer040.cul	IA

† OM, organic matter.

IA is discussed in more detail below. Quicktree does not predict crop yield for comparison with typical yields in the area; thus, we did not adjust crop growth parameters during inverse modeling at CA.

Table 12–3. Inverse modeling results at California and Iowa.†

Name	Simulation depth	Initial parameter value	Lower bound	Upper bound	Parameter estimate	Composite-scaled sensitivity
	cm					
California						
Ks1	0–10	9.65	5	50	24.07	3.40
Ks2	10–200	13.66	5	50	6.37	2.11
Ks3	200–330	9.86	5	50	16.08	1.52
Ks4	330–400	7.16	5	20	20.00	0.90
Ks5	400–650	64	5	65	21.69	1.32
WFC1	Same as above	0.10	0.06	0.11	0.11	19.51
WFC2		0.11	0.06	0.11	0.11	12.20
WFC3		0.08	0.06	0.09	0.08	4.95
WFC4		0.24	0.15	0.25	0.24	2.20
WFC5		0.04	0.04	0.06	0.04	2.24
BD1	Same as above	1.59	1.40	1.60	1.45	3.10
BD2		1.66	1.40	1.67	1.67	5.95
BD3		1.59	1.20	1.60	1.47	1.37
BD4		1.10	1.10	1.60	1.10	0.14
BD5		1.69	1.40	1.69	1.69	1.23
R45	NA	0.10	0.1	0.9	0.10	0.50
Iowa						
RD1	15–30	0.3	0.1	0.9	0.81	0.58
RD2	30–45	NA	NA	NA	0.23	NA
RD3	45–60	NA	NA	NA	0.10	NA
RD4	60–90	NA	NA	NA	0.01	NA
A21	0–2	NA	NA	NA	0.19	NA
A22	2–20	NA	NA	NA	0.20	NA
A23	20–130	0.1	0.07	0.15	0.15	9.50
A24	130–268	NA	NA	NA	0.08	NA
N21	Same as above	NA	NA	NA	2.19	NA
N22		NA	NA	NA	2.20	NA
N23		NA	NA	NA	2.15	NA
N24		NA	NA	NA	2.08	NA
R14	NA	0.3	0.1	0.9	0.86	0.60
R23	NA	0.6	0.4	0.9	0.40	1.77
R34	NA	0.6	0.4	0.9	0.43	3.89
R45	NA	0.7	0.4	0.9	0.74	2.23
LHG	NA	1.0×10^{-5}	2×10^{-6}	2×10^{-5}	3.62×10^{-6}	10.89

† Parameter definitions are shown in Table 12–2. 1 = first soil layer, 2 = second layer, and so on; NA, not applicable. All parameters were $\log_{10}$ transformed (values are supplied to PEST in the original untransformed units, and PEST returns the optimized parameters in the original units). IA parameters with composite-scaled sensitivity shown as NA were calculated from independently adjusted parameters using the PAR2PAR utility.

Iowa

Because DSSAT was used to predict crop yields at IA, parameters representing three major components of RZWQM2 (crop growth, hydrology, N dynamics) could be optimized simultaneously to harmonize their interaction. For example, N uptake by corn is often the largest component of the annual nitrogen budget, and N uptake is sensitive to both N mineralization and root growth factors. Observed data available for inverse modeling calibration included monthly N loss in drain flow, monthly drain flow amount, monthly flow-weighted nitrate concentration (FWNC) in drain flow, and annual corn data collected during 1997–2000 (Table 12–1). The objective function at Iowa took the following form:

$$\Phi(\mathbf{b}) = \sum_{i=1}^{m_1} w^2_{\mathrm{nl},i}\left(y_{\mathrm{nl},i} - y'_{\mathrm{nl},i}\right)^2 + \sum_{i=1}^{m_2} w^2_{\mathrm{df},i}\left(y_{\mathrm{df},i} - y'_{\mathrm{df},i}\right)^2 + \sum_{i=1}^{m_3} w^2_{\mathrm{nc},i}\left(y_{\mathrm{nc},i} - y'_{\mathrm{nc},i}\right)^2 + \sum_{i=1}^{m_4} w^2_{\mathrm{cy},i}\left(y_{\mathrm{cy},i} - y'_{\mathrm{cy},i}\right)^2 \quad [11]$$

where subscript i is the monthly or annual ith observation, nl is N loss, df is drain flow, nc is flow-weighted nitrate concentration, and cy is corn yield.

Model parameters initially adjusted at IA comprised soil hydraulic parameters, including those of the Brooks–Corey water retention and unsaturated hydraulic conductivity functions, N and organic matter transformation coefficients, and maize parameters in DSSAT (Table 12–2). After some initial experimentation in PEST, several of these were fixed because of parameter interdependence and/or insensitivity. SVD was not used at IA. Independently adjusted parameters in later PEST runs comprised the soil root growth factor (RD) in layer 1 for corn, four interpool transformation coefficients for the organic matter pools (R14, R23, R34, and R45), the lateral hydraulic gradient (LHG), and A2 for layer 3 (Table 12–3). Starting values of these parameters came from previous studies (Ma et al., 2008; Thorp et al., 2007). The remaining parameters in Table 12–3 were adjusted using the PAR2PAR utility based on the following constraints: RD values for deeper soil layers were less than shallower layers, A2 was within a factor of 0.7 to 1.4 for adjacent soil layers, N2 was calculated from A2, and porosity was calculated from bulk density. PAR2PAR manipulations at IA are described in more detail below.

PEST Files and Variables

PEST operations and related model simulations are conducted in DOS mode, so the user must be familiar with running RZWQM2 outside of the Windows interface. PEST operations require a control (.pst) file, one or more template

Table 12–4. PEST files and model executables used for inverse modeling at California.

PEST file	Target file	Function
rzsanj.pst	NA	Main PEST control file
rzwqm.bat	NA	Model execution file for RZWQM2 and post-processor sed2water.exe
par2par_in.tpl	par2par.in	PEST template file that writes input to the PAR2PAR utility
par2par.in	NA	PEST utility to calculate porosity from PEST adjusted bulk density values
rzsanj.tpl	rzwqm.dat	PEST template file that reads the main RZWQM2 input file
bulkdens.tpl	bulkdens.txt	PEST template file to write adjusted bulk densities to input file used by sed2water.exe
rzsanjm.ins	layer.plt	PEST instruction file that reads predicted soil moisture
rzsanjb.ins	layer.plt	PEST instruction file that reads predicted aqueous boron conc.
rzsanjp.ins	layer.plt	PEST instruction file that reads predicted soil water tension
rzsanjn.ins	layer2.plt	PEST instruction file that reads predicted aqueous nitrate conc.
rzsanjom.ins	layer.plt	PEST instruction file that reads predicted soil organic matter
rzsanjsn.ins	layer.plt	PEST instruction file that reads predicted soil nitrate conc.
mblwat.ins	mblwat.out	PEST instruction file that reads predicted deep seepage
sed2water.exe	bulkdens.txt	Post-processor that computes aqueous nitrate concentration and writes file layer2.plt (not a PEST executable)

† NA, not applicable.

files (.tpl), and one or more instruction files (.ins), as indicated in Fig. 12–2 and Table 12–4. The PEST control file (Step 1 in the flow chart) controls the overall operation of PEST and lists parameter groups, parameter data, observation groups, observation data, the model command line, and model input and output. An excerpt of the file is shown in the Appendix Table 12–A1 for the CA model (only three of the 1670 observed values are shown). We highlight the main features of the control file here; the reader is advised to consult the PEST manual (Doherty, 2010) for detailed descriptions of key PEST files and typical values of PEST variables.

Major components of the PEST control file include the parameter data section (lines 22–48), data observation groups (lines 50–55), data observations and weights (lines 57–63), and input and output files manipulated by PEST (lines 67–74) (Table 12–A1). These latter files include template files to supply parameter values to RZWQM2 and instruction files to read simulated values corresponding to the data observation groups (Table 12–4).

The example control file includes a section for implementing SVD (lines 12–14) (Table 12–A1). SVD is optional, but if not invoked the user will have to pay particular attention to parameter correlations and sensitivities and likely take alternative measures to ensure numerical stability. We set the PEST variable EIGTHRESH moderately high (10^{-5}) to mitigate problems encountered with insensitive or correlated parameters.

Lower and upper bounds in the parameter data section (Table 12–A1) should be chosen to avoid unrealistic combinations of model parameters. Certain combinations of parameters, as well as parameter values of adjacent soil layers that are very different, can cause Richards' equation convergence problems in RZWQM2. The Richards equation is highly nonlinear and relations among moisture content, soil water tension, and hydraulic conductivity are complex. Additionally, field soils are heterogeneous. Therefore solution of Richards' equation is not possible for certain reasonable sets of constitutive relations, parameter values, and environmental conditions (Miller et al., 1998). The extreme divergence of parameter values in adjacent layers can be controlled to some extent with prior information. Alternatively, for unsaturated zone models, transitional soil layers can be inserted in the simulation profile to minimize differences between adjacent layers.

The observation weights control the relative influence of each group in estimating model parameters. Here the weights were manually adjusted so that the observation groups in Eq. [10] and [11] either did not dominate or were not dominated by the other groups, following the strategy of prior researchers (Lin and Radcliffe, 2006). The weights in general were inversely proportional to the variance of the observations. Soil tension measurements received small weights (e.g., 0.0025) because they had high variance and ranged over thousands of centimeters. In contrast, soil moisture measurements received large weights (e.g., 110) because they varied between 0 and 1. Soil nitrate and organic matter had far fewer observations (1–5) and received weights such that they had nominal contributions to $\mathbf{\Phi}(\mathbf{b})$. The end result was that soil moisture, soil tension, aqueous nitrate, and aqueous bromide had about the same sum of squares contribution to $\mathbf{\Phi}(\mathbf{b})$, with smaller contributions by soil nitrate and organic matter. Alternatively, weights may be assigned in more rigorous fashion by setting them equal to the inverse of the standard deviation of the data measurement error in PEST (Doherty, 2010) or to the inverse of the variance of the measurement error as described by Hill and Tiedeman (2007). However, the true measurement variance or standard deviation is difficult to know with certainty and is confounded by structural noise (Doherty and Welter, 2010; Hill and Tiedeman, 2007). Therefore, a certain level of subjectivity is inevitable when assigning weights.

Certain of PEST's utilities ensure that the overall suite of files has been set up properly. TEMPCHEK checks the integrity of the template file, INSCHEK verifies that the instruction files are reading the model output files properly, and PESTCHEK conducts an overall check of the inverse modeling setup by alerting the user of filename syntax errors, inconsistencies between values of PEST variables and information supplied in various sections of the control file, and other problems. At a minimum, PESTCHEK should be run before starting any PEST project to avoid unnecessary crashes.

PEST terminates when $\Phi(\mathbf{b})$ converges or when the user-specified maximum number of iterations has been reached, at which point optimization is complete and final parameter values, sensitivities, eigenvectors, and other statistics are written to output files (Steps 13 and 14 of the flow chart in Fig. 12–2). In the CA example, convergence occurs based on either of two criteria specified by the user in the control file: (i) when the last three $\Phi(\mathbf{b})$ values are within 1%; or when the change in the parameters is within 1% for three iterations (Table 12–A1, line 9). PEST then produces a run record file (.rec) containing interim results, final parameter estimates, and other information (Step 14 in the flow chart). Normally this file includes 95% confidence limits on parameters and parameter covariances and correlations. However, this information is omitted when SVD is used because the eigenvectors are uncorrelated and because linear confidence intervals (CIs) on the parameter estimates may be inaccurate. Such CIs require assumptions of normality of the parameter estimates and linearity of simulated values with respect to the parameters, which are frequently violated in natural systems (Hill and Tiedeman, 2007). Therefore, linear CIs generated by PEST (when SVD is not invoked) should be interpreted with caution. To obtain these CIs after parameter estimation by SVD, the optimal parameters can be copied from the parameter value file to a PEST control file using the PEST utility PARREP, and PEST is run again without SVD but with PEST variable NOPTMAX set to −1. This halts PEST operation after calculating the Jacobian matrix one time and writes parameter CIs, correlations, and other statistics to the run record file (Doherty, 2010).

If RZWQM2 convergence problems are encountered, it may be necessary to modify the parameter bounds or supply prior information. If problems with numerical instability in PEST are encountered (caused by a singular or nearly singular matrix), then SVD can be invoked, or highly insensitive and/or correlated parameters can be dealt with individually by setting them to specified values. In the CA example, best results were obtained using SVD.

Use of PAR2PAR and Post-Processing Utilities at Iowa

PEST-RZWQM setup at IA used PAR2PAR manipulations extensively because several of the Brooks–Corey parameters are related. Additionally, PAR2PAR was used to prevent A2 values in adjacent soil layers from diverging too greatly and to maintain ordered relations among RD values. We introduced ratios as new parameters in the IA control file (Table 12–A2) to establish relations among the model parameters being adjusted by PEST. For example, the ratio "par11" on line 33 was defined as a parameter and assigned lower and upper bounds of 0.1 and 1 to establish depth-ordered relations among root-growth factors (RD1 $\geq$ RD2 $\geq$ RD3 $\geq$ RD4). In the PAR2PAR input file, RD2 was set equal to RD1 $\times$ par11 such that RD2 $\leq$ RD1. In other words, the estimate of RD2 was constrained within a factor of 0.1 to 1$\times$ RD1. RD3 was constrained to be within 0.1 to 1$\times$ RD2, and so on.

Extensive post-processing of RZWQM2 simulated values at IA was required for two reasons. First, daily values had to be converted to monthly values to match monthly observations of drain flow and N flux at subbasin CN2. Second, because water quality data at the outlet of the subbasin reflected two different crop rotations, subbasin-scale fluxes had to reflect the proportion of each crop. These operations involved use of TSPROC, a PEST post-processing utility that performs operations on model-generated time series (available at http://www.pesthomepage.org/Home.php, verified 13 May 2011). TSPROC manipulates time series and performs mathematical operations between the terms of any number of series to generate a new time series.

We used TSPROC to process the daily simulated runoff and tile drainage from the CN2 corn–soybean (CN2_CS) and soybean–corn rotations (CN2_SC) into proportional monthly composite N loss in drain flow and drain flow amount (DFX) using the formula

$$\mathrm{DFX} = \sum_{i=1}^{n} \left(T_i + \mathrm{RO}_i \right) f_i \qquad [12]$$

where T is daily tile flow amount (cm) or N loss in tile flow (kg N ha^{-1}), RO is daily runoff amount (cm) or N loss in runoff (kg N ha^{-1}), i represents the crop rotation (CN2_CS or CN2_SC), n is the number of crop rotations (2), and f is the fraction of subbasin area in a given rotation (0.526 for CN2_SC and 0.474 for CN2_CS). The fractions in Eq. [12] were determined as CN2 area (ha) in SC or CS divided by sum of area in SC and CS; for example, f = 258/(258 + 233) for SC. Drain flow is the sum of subsurface (tile) drainage and runoff because the watershed samples represented both, which was partly due to the many surface inlets into the subsurface drainage system. Monthly values for DFX were then computed, followed by monthly FWNC in drain flow (mg L^{-1}), where FWNC = (N loss in drain flow $\times$ 10)/(drain flow amount). Interested readers may consult the PEST Surface Water Utilities manual for more information on TSPROC (Doherty, 2008).

Predictive Analysis at California

So far we have discussed use of PEST in parameter estimation mode, which seeks a unique set of parameters associated with an objective function minimum. In predictive analysis mode, PEST evaluates alternative combinations of parameters to quantify the uncertainty of a prediction of interest. We used predictive analysis at CA to develop nonlinear calibration-constrained confidence intervals as described in prior research (Vecchia and Cooley, 1987) to characterize the overall uncertainty of the prediction. The objective function obtained from parameter estimation (Φ_{min}) is allowed to increase by a small amount (δ) as model parameters are varied, so as to define a $\Phi_{min} + \delta$ contour on the objective function surface. The increment δ is small and provides "headroom" to allow PEST to try different combinations of parameters while maintaining the model in a nearly calibrated state. The maximum model prediction on the $\Phi_{min} + \delta$ contour is referred to as the "critical point" and is the maximum that can be achieved while the model is still considered calibrated (Doherty, 2010). This critical point is the upper confidence interval for the prediction of interest at a specified significance level. Alternatively, the minimum achievable prediction on the $\Phi_{min} + \delta$ contour is the lower confidence interval. The increment δ is calculated as

$$\delta = n\sigma^2 F_{\alpha,n,m-n} \qquad [13]$$

where n is the number of adjustable parameters (16 in this PEST run), m is the number of observations (1670), σ^2 is the calibration variance = $\boldsymbol{\Phi}(\mathbf{b})/m - n$, and F is the statistic of the F distribution for significance level α (0.10) and n, $m - n$ degrees of freedom (1.47). For the final optimization where several parameters were fixed, $\boldsymbol{\Phi}(\mathbf{b})$ = 19,724, and therefore δ = 281. The original PEST control file (Table 12–A1) was altered for predictive analysis and the resulting control file is shown in Table 12–A3.

Results

Parameter Estimates, Sensitivities, Correlations, and Confidence Intervals

California

Several PEST optimizations were performed to ensure that the final parameter estimates were not associated with a local minimum in the objective function surface. The final PEST run required four optimization iterations and 124 model calls to stabilize $\boldsymbol{\Phi}(\mathbf{b})$ (Fig. 12–3). The three most sensitive soil hydraulic parameters (KS1, WFC1, and BD2) oscillated somewhat initially, and then were mostly stable during the last three iterations. At this point, convergence of $\boldsymbol{\Phi}(\mathbf{b})$ was achieved, and final statistics were calculated according to Steps 13 and 14 of the

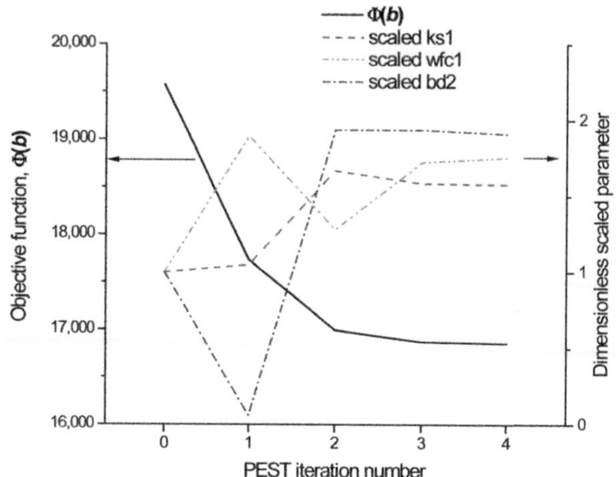

Fig. 12–3. Objective function response during PEST optimization at California and scaled values of the most sensitive soil hydraulic parameters. Parameter definitions are shown in Table 12–2.

flow chart (Fig. 12–2). In Fig. 12–3, we scaled the parameters by dividing by the initial parameter value and applying a power transform to compensate for the different scales of measurement; the actual units of these parameters are shown in Table 12–2.

Estimates of independently adjusted parameters obtained by inverse modeling for CA are shown along with their CSS values in Table 12–3. In Fig. 12–4, the CSS values were normalized by expressing them as a percentage of the maximum CSS to facilitate comparison between CA and IA. Water content at field capacity in the first two soil layers had the highest CSS values at CA, generally followed by BD. Adjustments to these parameters tended to cause greater changes in the simulated values. Both WFC and BD are used by RZWQM2 to estimate water retention functions for calculation of unsaturated-zone water fluxes. The CA model generally was less sensitive to N cycling parameters than to soil hydraulic parameters. The comparatively low normalized CSS value of R45 (2.6%) suggested smaller precision; a parameter is considered imprecisely estimated when CSS is less than 1% of the maximum value (Hill and Tiedeman, 2007). In such cases, more data may be required to reliably estimate the parameter by nonlinear regression. The CA study emphasized collection of soil hydraulic and water quality data, and fewer measurements of soil properties (e.g., organic matter) were available for use in calibration.

Although SVD was invoked at CA to mitigate parameter correlation problems, it is useful to examine correlations between the untransformed process-model parameters (Mary C. Hill, USGS, personal communication). Parameter correlations (Table 12–5) and CIs were obtained for the process-model parameters by running PEST again without SVD as described above and setting NOPTMAX

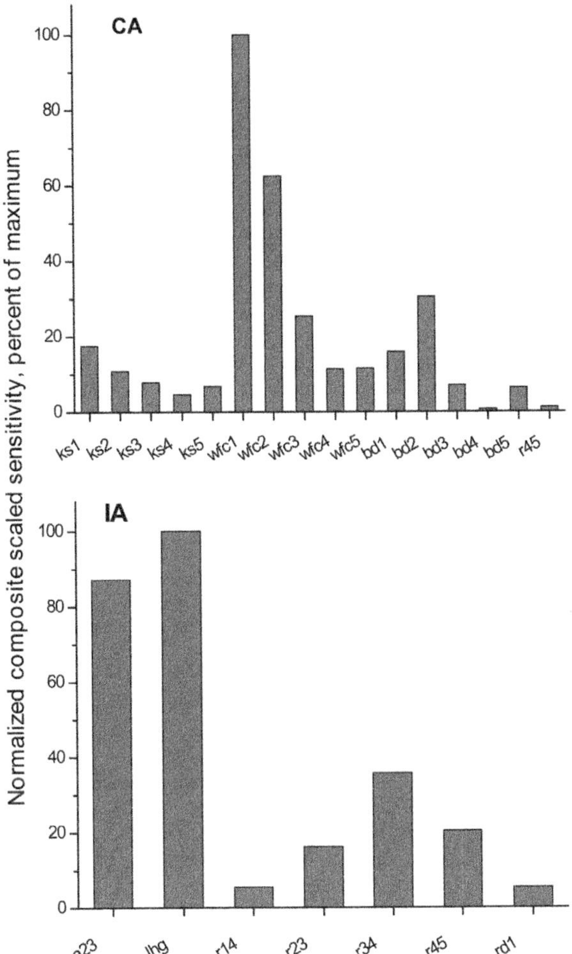

Fig. 12–4. Normalized composite-scaled sensitivities of independently adjusted model parameters at California and Iowa. Parameter definitions are shown in Table 12–2.

= −1. The highest correlations on an absolute value basis were between BD1 and WFC1 (−0.94) and KS2 and BD2 (−0.84). The higher value approaches a nonuniqueness threshold. However, parameter correlations <0.95 (absolute value basis) typically indicate that the parameters can be uniquely estimated (Poeter and Hill, 1997). The parameter correlations are consistent with the identifiability statistic (Fig. 12–5), which is calculated from the squared elements of SVD eigenvectors (Doherty and Hunt, 2009). In Fig. 12–5, the height of the bar corresponds to parameter identifiability, and the shades correspond to the contribution from each eigenvector. WFC1 and BD1 have similar identifiability values (0.51 and 0.47, respectively), which is consistent with the high correlation between these two parameters. Additionally, the most dominant or identifiable parameters

Table 12–5. Parameter correlations from inverse modeling at California. Parameter definitions are shown in Table 12–2.

	Ks1	Ks2	Ks3	Ks4	Ks5	WFC1	WFC2	WFC3	WFC4	WFC5	BD1	BD2	BD3	BD4	BD5	R45
Ks1	1.00	−0.12	−0.21	0.33	0.13	0.18	−0.32	−0.18	−0.16	−0.33	−0.43	0.20	0.02	0.23	−0.10	−0.18
Ks2		1.00	−0.12	−0.08	−0.21	−0.05	0.23	−0.06	−0.27	−0.11	0.16	−0.84	0.23	−0.17	0.20	−0.35
Ks3			1.00	0.25	−0.21	−0.12	0.01	−0.15	−0.09	−0.03	0.14	0.09	−0.17	−0.26	0.03	0.23
Ks4				1.00	0.24	−0.07	−0.01	0.03	0.01	0.07	0.00	0.09	−0.16	0.12	−0.13	0.09
Ks5					1.00	0.14	−0.14	0.14	0.11	0.17	−0.23	0.34	−0.26	0.14	−0.35	−0.21
WFC1						1.00	−0.11	0.01	0.02	−0.04	−0.94	0.32	−0.07	−0.11	−0.03	−0.11
WFC2							1.00	0.38	0.13	0.39	0.30	−0.20	0.33	−0.27	−0.06	0.08
WFC3								1.00	0.51	0.53	−0.03	0.03	−0.59	0.10	0.01	−0.22
WFC4									1.00	0.33	0.00	0.29	−0.46	0.26	0.28	−0.04
WFC5										1.00	0.07	0.07	−0.32	−0.07	−0.37	−0.21
BD1											1.00	−0.40	0.21	0.01	0.13	0.23
BD2												1.00	−0.29	0.17	−0.19	0.24
BD3													1.00	−0.29	−0.12	0.24
BD4														1.00	−0.01	0.15
BD5															1.00	0.21
R45																1.00

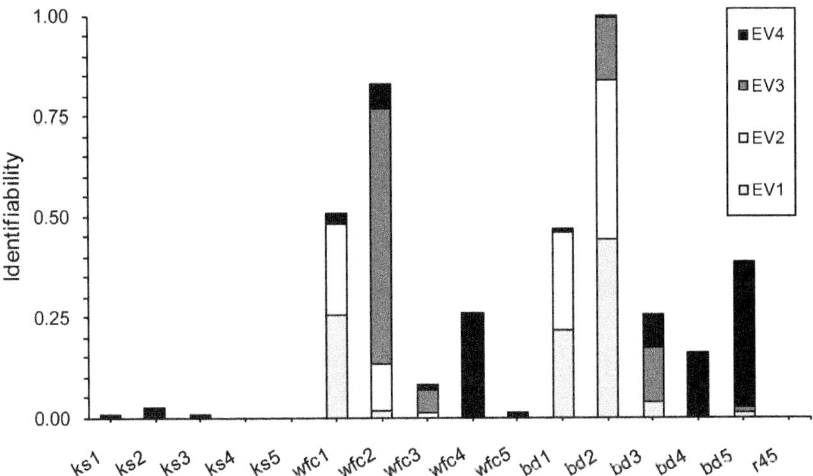

Fig. 12–5. Identifiability of model parameters at California based on singular value decomposition. Parameter definitions are shown in Table 12–2.

are BD2, WFC2, and WFC1, which corroborates the CSS values obtained for the untransformed process-model parameters at CA (Fig. 12–4). SVD integrates the effects of parameter sensitivity and interdependence, while CSS and parameter correlations evaluate these effects individually (Mary C. Hill, USGS, personal communication). Identifiability values in Fig. 12–5 are based on truncation at four out of a possible 16 maximum eigenvectors.

The width of parameter CIs in relation to the magnitude of the process-model parameters (indicated by the height of the bar in Fig. 12–6) provides information on the uncertainty of the estimate. CIs for K_s, WFC, and BD were comparatively narrow, suggesting that these parameters were reasonably estimated (Fig. 12–6). In contrast, R45 had a wide CI relative to its magnitude, suggesting greater uncertainty. However, the CI for R45 (0.03–0.35) overlaps with the range of values used in previous RZWQM studies (0.1–0.7) (Cameira et al., 2007; Ma et al., 2007). Therefore, it is unlikely that there was significant model error with respect to this parameter. Confidence intervals that fail to overlap reasonable ranges of parameters suggest such errors (Hill and Tiedeman, 2007).

Linear CIs are approximate and should be interpreted with caution because required assumptions may not be met with environmental data (Hill and Tiedeman, 2007). Nonlinear CIs on predictions of interest are discussed below in the Predictive Analysis at California section.

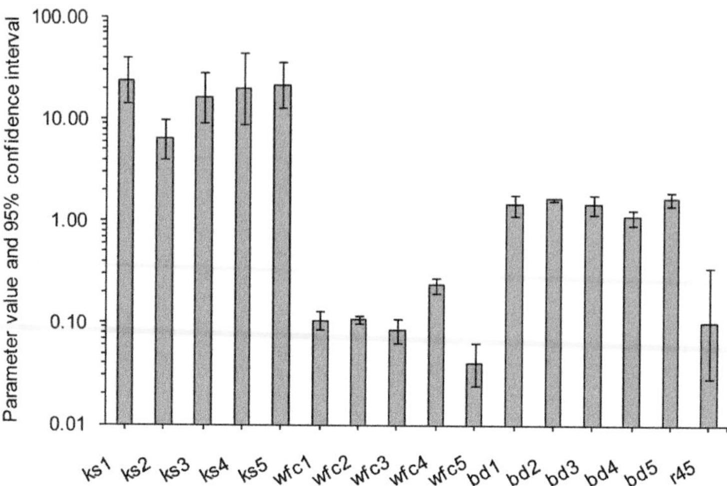

Fig. 12–6. Linear confidence intervals for independently adjusted model parameters at California. Parameter definitions are shown in Table 12–2.

Iowa

Parameter estimates obtained by inverse modeling for the IA model are shown in Table 12–3. As at CA, water flux parameters (LHG, A23) had higher normalized CSS values than N transformation parameters (R14, etc.), and root growth factor (RD1) was not particularly sensitive (Fig. 12–4). All of the parameters shown had normalized CSS values >5% of maximum CSS, which is above Hill and Tiedeman's guidance of a 1% threshold for a precisely estimated parameter. Thus, the measured data appeared sufficient to estimate all of the independently adjusted parameters by regression at IA.

RZWQM2 Fit to Observation Groups

California

Inverse modeling resulted in reasonable model fits to soil moisture, soil tension, and aqueous bromide and nitrate at CA (Nolan et al., 2010). For example, predicted moisture contents at 0.30 m closely matched the observed values, which increased sharply in response to irrigation (Fig. 12–7). Considering all measurement depths (0.30–0.91 m), RRMSE and d corresponding to measured and predicted soil moisture contents were 0.17 and 0.74, respectively (Table 12–6). Soil tension was reasonably well simulated by the calibrated model except for the highly negative tensions (to −2620 cm) measured at 0.91 m in early October 2004. Using the full Brooks–Corey water retention function might have improved the fit during this period. However, model fit to soil tension data at

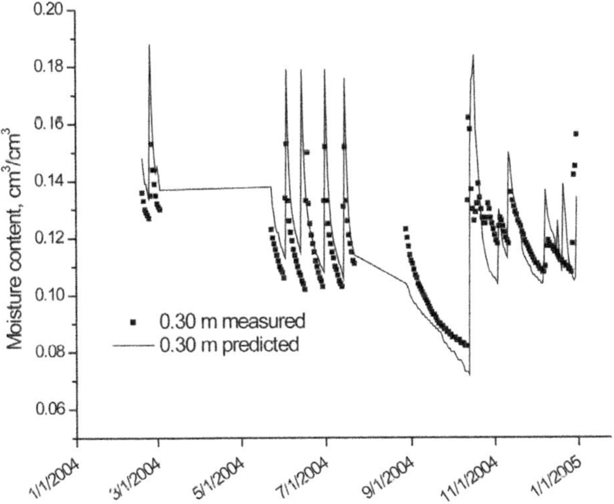

Fig. 12–7. Observed and predicted soil moisture contents at California.

6.10 m was very good, which is promising because this is the first application of RZWQM2 at depths >3 m (Nolan et al., 2010). Considering all five depths (0.30–6.10 m), RRMSE = 2.68 and *d* = 0.40 for soil tension data (Table 12–6). Bromide concentration in lysimeters was reasonably well simulated overall (*d* = 0.71). The measured data indicated rapid transport in irrigation water applied to highly permeable soils. The predicted breakthrough curve had more dispersion, however, compared with the measured values, which inflated the RRMSE (2.57) at this site (Nolan et al., 2010). Nitrate concentration generally was well simulated but was underpredicted at 0.91 m in June 2004. A single large measured value at this time (50.0 mg L^{-1}) may have been caused by preferential flow, which was not invoked in the model because soils at the site are predominantly sand (Nolan et al., 2010). Considering all lysimeter sampling depths, RRMSE = 0.40 and *d* = 0.60 for nitrate (Table 12–6).

Iowa

RZWQM2 reasonably simulated drain flows, N losses, and FWNC at IA. For example, drain flows predicted by the model closely matched observed monthly losses for a 9-yr period (Fig. 12–8). RRMSE was about 0.6 or less and *d* was 0.78 or greater for all three observation groups (Table 12–6). Although the observation groups at CA and IA were different and therefore not directly comparable, *d* values were higher at IA (0.78–0.97) than at CA (0.40–0.74), indicating closer agreement with the observed data. Model fit at IA may have benefitted from post processing of

Table 12–6. Model fit statistics after calibration by inverse modeling at CA and IA (RMSE, root mean square error; FWNC, flow-weighted nitrate concentration).

Observation group	RMSE	Relative RMSE	Index of agreement
California			
Soil moisture (cm^3 cm^{-3})	0.02	0.17	0.74
Soil tension (cm)	714.60	2.68	0.40
Nitrate (mg L^{-1})	8.86	0.40	0.60
Bromide (mg L^{-1})	20.35	2.57	0.71
Iowa subbasin CN2			
Drain flow (cm)	1.04	0.59	0.97
FWNC (mg L^{-1})	5.06	0.38	0.78
N loss (kg N ha^{-1})	1.45	0.62	0.95

daily simulated values, which were averaged according to the areas of CN2 subbasins in corn or soybean in a particular year and then aggregated to monthly and annual values. In contrast, all comparisons of predicted and observed values at CA were on a daily basis at a single field site.

Corn yield was simulated within 500 kg ha^{-1} (±5%) of observed values for 11 of 12 treatment scenarios. The corn yield calibration scenarios included LSNT (96–174 kg N ha^{-1} as sidedress), minimum (56 kg N ha^{-1}), and maximum (220 kg N ha^{-1}) application rates during the years of LSNT treatment (1997– 2000). The least accurate simulation was for the maximum rate in 1998, which saw underprediction of corn yield by about 1200 kg ha^{-1}. More than 30 cm of precipitation occurred in June 1998; therefore, the low predicted yield might have been caused by simulation of excessive water logging.

Predictive Analysis at California

Predictive analysis involved obtaining nonlinear confidence intervals for quantities predicted by RZWQM2 and independently measured at the site but not used in the calibrations: annual water flux at the bottom of the simulated profile and average nitrate concentration in deep seepage from the profile. Predicted annual water flux (cm) was obtained by calculating average daily predicted deep seepage of water for the length of the simulation and multiplying by 365. This quantity was compared with estimates of annual groundwater recharge obtained by the water table fluctuation method (WTF). Average predicted nitrate concentration (mg L^{-1}) at the bottom of the profile was obtained by dividing deep seepage nitrate flux (kg N ha^{-1}) by cumulative water flux (cm) for the simulation period, for comparison with median nitrate concentration in the on-site monitoring well.

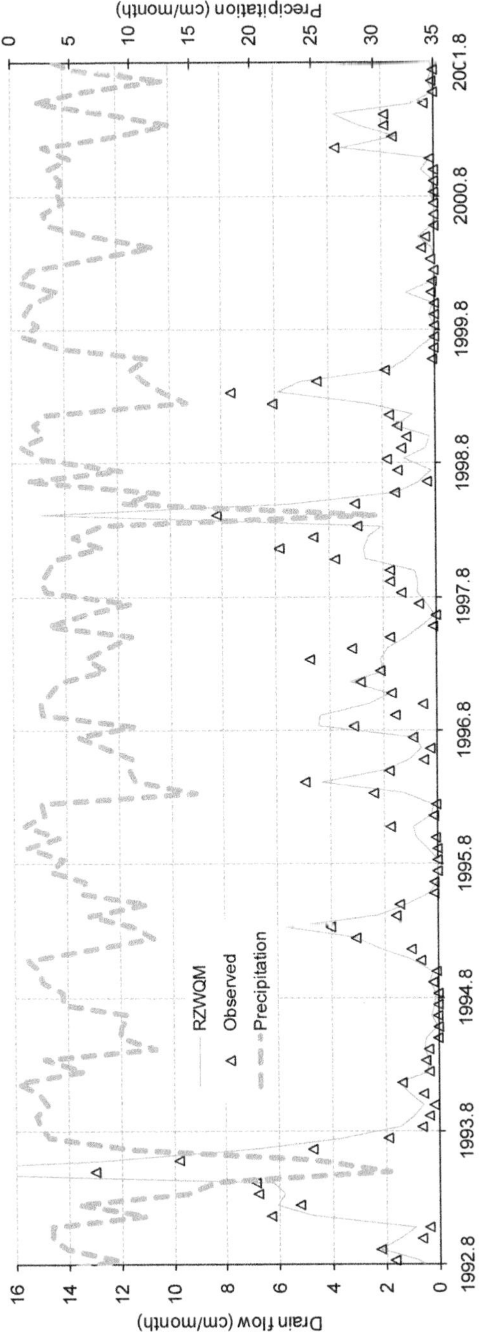

Fig. 12–8. Observed and predicted monthly drain flows at Iowa.

Table 12–7. Evaluation of PEST-RZWQM2 at California by independently measured water fluxes and groundwater nitrate concentrations. The 90% confidence interval from predictive analysis is shown in parentheses next to the predicted value obtained with the optimal parameters.

Annual water flux	
cm yr^{-1}	
Observed	42.3
Predicted	38.1 (33.1, 48.5)
Average groundwater nitrate	
mg L^{-1}	
Observed	24.3
Predicted	18.7 (16.1, 37.1)

Predicted water flux (38.1 cm yr^{-1}) was slightly less than the WTF estimate of recharge (42.3 cm yr^{-1}) (Fisher and Healy, 2008) (Table 12–7), but the latter was within the 90% confidence interval obtained by predictive analysis (33.1–48.5 cm yr^{-1}). The WTF method is based on changes in groundwater elevation and thus is a more integrative measure of recharge than a point estimate at a specific location. Because WTF is more likely to capture preferential flow paths, the somewhat higher value by that method is reasonable. Predicted nitrate concentration at the water table at CA (18.7 mg L^{-1}) was somewhat less than the value observed at the monitoring well (24.3 mg L^{-1}). This difference might reflect limitations in the simulation of preferential flow by RZWQM2 for sandy soil (Nolan et al., 2010). However, the latter value is within the 90% confidence interval (16.1–37.1 mg L^{-1}), indicating that the prediction error is acceptable (Table 12–7).

Conclusions

Inverse modeling with PEST resulted in effective calibration of RZWQM2 for two sites with markedly different soils, climate, management practices, and spatial extent. Modeling was conducted at the field scale for an almond orchard near Merced, CA, and at the watershed scale for corn–soybean rotations in central Iowa. The examples illustrate the usefulness of the approach, namely: (i) simultaneous adjustment of multiple parameters, versus one-at-a-time manual approaches; (ii) clear indication when calibration is complete, according to convergence criteria; (iii) straightforward detection of nonunique and insensitive parameters, which can affect operational stability; and (iv) generation of confidence intervals for uncertainty analysis of both parameters and model predictions.

At CA, PEST was applied in both estimation and predictive modes (predictive analysis). The estimation mode yielded an optimal parameter set, as well as sensitivities that yielded insight into which parameters could be most reliably estimated by regression. Predictive analysis at CA effectively identified critical points on an objective function surface for estimating calibration-constrained, nonlinear confidence intervals. Independent measurements of annual groundwater recharge and average nitrate concentration fell within the 90% CIs, indicating that overall model error was acceptable.

Common pitfalls associated with inverse modeling (local minima in the objective function response and extreme parameter correlation and insensitivity) were largely avoided by using different starting parameter values for different optimization runs and by invoking truncated singular value decomposition in PEST's estimation mode. The latter procedure reduced the parameter space through linear combinations of parameters (eigenvectors). Truncated SVD stabilized PEST matrix operations and also may have minimized RZWQM2 convergence problems. SVD identifiability statistics were consistent with correlations and composite-scaled sensitivities computed for the untransformed process-model parameters.

At IA, PEST utilities (PAR2PAR, TSPROC) provided an effective means of maintaining ordered relations among parameters (e.g., root growth factors at shallow depth were held greater than those at deeper depths during adjustment by PEST), converting daily simulated values at IA into monthly time series, and averaging predictions for individual crop practices at the subbasin scale.

The calibrated models fit the observed water and N fluxes at CA and IA (and observed crop yields at IA) reasonably well, according to model fit criteria. The index of agreement between predictions and observations was 0.60 or greater for three of four observation groups at CA (soil moisture and nitrate and bromide concentrations) and was 0.78 or greater for all three observation groups at IA (drain flow, flow-weighted nitrate concentration, and N loss). Model fit at IA may have benefitted from compositing daily predictions into monthly values and averaging RZWQM2 outputs at the subbasin scale. The CA results are encouraging because this is the first application of RZWQM2 to the deeper unsaturated zone (6.5 m). Previous versions of the model were limited to a simulation depth of 3 m.

Appendix

Table 12–A1. PEST control file rzsanj.pst for parameter estimation at California.

Line no.	PEST control file rzsanj.pst
1	pcf
2	* control data
3	restart estimation
4	27 1670 5 0 6
5	2 6 single point 1 0 0
6	0 2.0 0.3 0.03 1
7	3.0 3.0 0.001
8	0.1
9	30 0.01 3 3 0.01 3
10	1 1 1
11	* singular value decomposition
12	1
13	17 1.0E-5
14	1
15	* parameter groups
16	gt1 relative 0.01 0.001 switch 2 parabolic
17	lt1 relative 0.01 0.0001 switch 2 parabolic
18	omtrans relative 0.01 0.001 switch 2 parabolic
19	Acoeff relative 0.01 1E-08 switch 2 parabolic
20	cnuptak relative 0.01 0.001 switch 2 parabolic
21	* parameter data
22	ks1 log factor 9.64735 5 50 gt1 1 0 1
23	ks2 log factor 13.6569 5 50 gt1 1 0 1
24	ks3 log factor 9.85991 5 50 gt1 1 0 1
25	ks4 log factor 7.16074 5 20 gt1 1 0 1
26	ks5 log factor 64 5 65 gt1 1 0 1
27	wfc1 log factor 0.100185 0.06 0.11 lt1 1 0 1
28	wfc2 log factor 0.11 0.06 0.11 lt1 1 0 1
29	wfc3 log factor 8.16E-02 0.06 0.09 lt1 1 0 1
30	wfc4 log factor 0.24 0.15 0.25 lt1 1 0 1
31	wfc5 log factor 4.09E-02 0.04 0.06 lt1 1 0 1
32	bd1 log factor 1.59 1.40 1.60 gt1 1 0 1
33	bd2 log factor 1.659 1.40 1.67 gt1 1 0 1
34	bd3 log factor 1.595 1.20 1.60 gt1 1 0 1
35	bd4 log factor 1.10000 1.10 1.60 gt1 1 0 1
36	bd5 log factor 1.68680 1.40 1.69 gt1 1 0 1
37	r14 fixed relative 0.3 0.1 0.9 omtrans 1 0 1

```
r23     fixed   relative   0.6          0.1          0.9          omtrans   1 0 1
r34     fixed   relative   0.7          0.1          0.9          omtrans   1 0 1
r43     fixed   relative   0.0          0.0          0.9          omtrans   1 0 1
r45     log     relative   0.10         0.1          0.9          omtrans   1 0 1
r53     fixed   relative   0.0          0.0          0.9          omtrans   1 0 1
adec    fixed   relative   8.187E-009   8.187E-011   8.187E-007   acoeff    1 0 1
anit    log     factor     1.000E-009   1.000E-011   1.000E-007   acoeff    1 0 1
aden    fixed   factor     1.000E-013   1.000E-015   1.000E-011   acoeff    1 0 1
ahyd    fixed   relative   0.00025      0.000025     0.0025       acoeff    1 0 1
cnu1    fixed   relative   0.5          0.01         20           cnuptak   1 0 1
tsup    fixed   relative   250          200          400          cnuptak   1 0 1
* observation groups
 moisture
 bromide
 tension
 nitrate
 orgmat
 soilno3
* observation data
.
.
.
 M1     0.136   100    moisture
 M2     0.133   100    moisture
 M3     0.130   100    moisture
.
.
.
 sne1   0.510   10.0   soilno3
* model command line
 RZWQM2.BAT
* model input/output
 par2par_in.tpl   par2par.in
 bulkdens.tpl  bulkdens.txt
 rzsanjm.ins   layer.plt
 rzsanjb.ins   layer.plt
 rzsanjp.ins   layer.plt
 rzsanjn.ins   layer2.plt
 rzsanjom.ins  layer.plt
 rzsanjsn.ins  layer.plt
* prior information
```

Table 12–A2. PEST control file comb.pst for parameter estimation at Iowa.

PEST control file comb.pst

```
pcf
* control data
restart  estimation
34     317     5       0      4
2      7       single  point  1      0      0
5.0    2.0     0.3     0.03   10
3.0    3.0     0.001
0.1
30     0.01    3       3      0.01  3
1      1       1
* parameter groups
lg     relative   0.01    0.001     switch   2   parabolic
llg    relative   0.1     0.01      switch   2   parabolic
lllg   relative   1.0     0.1       switch   2   parabolic
ssm    relative   0.01    0.000001  switch   2   parabolic
sssm   relative   1e-13   1e-15     switch   2   parabolic
* parameter data
.
.
.
parD    log   factor   .91       .7      1.4     lg        1  0  1
parC    log   factor   1.4       .85     1.4     lg        1  0  1
a23     log   factor   0.146     0.07    0.15    lg        1  0  1
.
lhg     log   factor   3.62E-6   2E-6    2E-5    ssm   1   0     1
.
R14     log   factor   0.86      .1      0.9     lg        1  0  1
R23     log   factor   0.4       .4      0.9     lg        1  0  1
R34     log   factor   0.43      .4      0.9     lg        1  0  1
R45     log   factor   0.74      .4      0.9     lg        1  0  1
.
RD1     log   factor   0.81      .1      0.9     lg        1  0  1
par11   log   factor   0.28      0.1     1       lg        1  0  1
par12   log   factor   .46       0.1     1       lg        1  0  1
par13   log   factor   .13       0.1     1       lg        1  0  1
par14   log   factor   .13       .05     1       lg        1  0  1
.
.
.
* observation groups
nitrogen
tile_flow
fw_nconc
Yield
* observation data
n1 1.030022925 2    nitrogen
n2 0.465729653 2    nitrogen
n3 4.347677521 2    nitrogen
.
.
.
y8 8542     0.015  yield
* model command line
C:\PEST\wc_tom_y_n_comb\rzwqmb.bat
* model input/output
C:\PEST\wc_tom_y_n_comb\par2par_rzdat.tpl C:\PEST\wc_tom_y_n_comb\par2par_rzdat.in
C:\PEST\wc_tom_y_n_comb\par2par_rd.tpl  C:\PEST\wc_tom_y_n_comb\par2par_rd.in
C:\PEST\wc_tom_n\csctrl.ins   "C:\PEST\wc_tom_n\tsproc_out.txt"
C:\PEST\wc_tom_n\csctrl2.ins   "C:\PEST\wc_tom_n\tsproc2_out.txt"
C:\PEST\wc_tom_n\csctrl3.ins   "C:\PEST\wc_tom_n\tsproc3_out.txt"
C:\PEST\wc_tom_yield\cs_min.ins   "C:\RZWQM2\wc_tom_n\C_S_min\MANAGE.OUT"
C:\PEST\wc_tom_yield\sc_min.ins   "C:\RZWQM2\wc_tom_n\S_C_min\MANAGE.OUT"
C:\PEST\wc_tom_yield\cs_nlim.ins   C:\RZWQM2\wc_tom_n\C_S_nlim\MANAGE.OUT"
C:\PEST\wc_tom_yield\sc_nlim.ins   C:\RZWQM2\wc_tom_n\S_C_nlim\MANAGE.OUT"
* prior information
```

Table 12–A3. PEST control file rzsanj_pa.pst for predictive analysis at California.

Line no.	PEST control file rzsanj_pa.pst
1	pcf
2	* control data
3	restart prediction
4	26 1671 5 0 7
5	4 7 single point 1 0 0
6	5 2.0 0.3 0.03 10 999
7	3.0 3.0 0.001
8	0.1
9	30 0.01 3 3 0.01 3
10	1 1 1
11	* parameter groups
12	.
13	.
14	.
15	* observation groups
16	moisture
17	bromide
18	tension
19	nitrate
20	orgmat
21	soilno3
22	predict
23	* observation data
24	.
25	.
26	.
27	wf 0.000 0.0 predict
28	* model command line
29	RZWQM2_PA.BAT
30	* model input/output
31	par2par_in.tpl par2par.in
32	bulkdens.tpl bulkdens.txt
33	rzsanjm.ins layer.plt
34	rzsanjb.ins layer.plt
35	rzsanjp.ins layer.plt
36	rzsanjn.ins layer2.plt
37	rzsanjom.ins layer.plt
38	rzsanjsn.ins layer.plt
39	mblwat.ins mblwat.out
40	* prior information
41	* predictive analysis
42	1
43	20000 21000 30000
44	0.0 0.005 1 1.5 1
45	0.0 0.05
46	3 0.0 0.02 3

Acknowledgments

The authors gratefully acknowledge Claire R. Tiedeman and three anonymous reviewers for insightful comments on the draft manuscript. We also sincerely thank John Doherty, Mary C. Hill, Randall J. Hunt, and Eileen P. Poeter for invaluable insight and assistance during the course of this research and in preparation of the manuscript.

References

Ahuja, L.R., K.W. Rojas, J.D. Hanson, M.J. Shaffer, and L. Ma. 2000. Root Zone Water Quality Model—Modelling management effects on water quality and crop production. Water Resources Publ., Highlands Ranch, CO.

Allen, R. 1996. Assessing integrity of weather data for reference evapotranspiration estimation. J. Irrig. Drain. Eng. 122:97–106.

Cambardella, C.A., T.B. Moorman, D.B. Jaynes, J.L. Hatfield, T.B. Parkin, W.W. Simpkins, and D.L. Karlen. 1999. Water quality in Walnut Creek watershed: Nitrate-nitrogen in soils, subsurface drainage water, and shallow groundwater. J. Environ. Qual. 28:25–34.

Cameira, M.R., R.M. Fernando, L.R. Ahuja, and L. Ma. 2007. Using RZWQM to simulate the fate of nitrogen in field soil–crop environment in the Mediterranean region. Agric. Water Manage. 90:121–136.

Capel, P.D., K.A. McCarthy, and J.E. Barbash. 2008. National, holistic, watershed-scale approach to understand the sources, transport, and fate of agricultural chemicals. J. Environ. Qual. 37:983–993.

Close, M.E., R. Lee, G.N. Magesan, M.K. Stewart, G. Skuse, and G. Bekesi. 2005. Field study of pesticide leaching in a Himatangi sand (Manawatu) and in a Kiripaka bouldery clay loam (Northland). 1. Results. Aust. J. Soil Res. 43:457–469.

Close, M.E., R. Lee, A.K. Sarmah, L. Pang, R. Dann, G.N. Magesan, J.P.C. Watt, and K.W. Vincent. 2008. Pesticide sorption and degradation characteristics in New Zealand soils—A synthesis from seven field trials. N.Z. J. Crop Hortic. Sci. 36:9–30.

Corwin, D.L., J. Hopmans, and G.H. De Rooij. 2006. From field- to landscape-scale vadose zone processes: Scale issues, modeling, and monitoring. Vadose Zone J. 5:129–139.

Doherty, J. 2008. PEST surface water utilities. Watermark Numerical Computing, Brisbane, Australia and University of Idaho, Idaho Falls.

Doherty, J. 2010. PEST—Model independent parameter estimation. Watermark Numerical Computing, Brisbane, Australia.

Doherty, J., and R.J. Hunt. 2009. Two statistics for evaluating parameter identifiability and error reduction. J. Hydrol. 366:119–127.

Doherty, J., and D. Welter. 2010. A short explanation of structural noise. Water Resour. Res. 46:W05525, doi:10.1029/2009WR008377.

Domagalski, J.L., S.P. Phillips, E.R. Bayless, C. Zamora, C. Kendall, R.A. Wildman, and J.G. Hering. 2008. Influences of the unsaturated, saturated, and riparian zones on the transport of nitrate near the Merced River, California, USA. Hydrogeol. J. 16:675–690.

Fang, Q., T.R. Green, L. Ma, R.H. Erskine, R.W. Malone, and L.R. Ahuja. 2010. Optimizing soil hydraulic parameters in RZWQM2 under fallow conditions. Soil Sci. Soc. Am. J. 74:1897–1913.

Fisher, L.H., and R.W. Healy. 2008. Water movement within the unsaturated zone in four agricultural areas of the United States. J. Environ. Qual. 37:1051–1063.

Hanson, J., K. Rojas, and M. Shaffer. 1999. Calibrating the Root Zone Water Quality Model. Agron. J. 91:171–177.

Hatfield, J.L., J.H. Prueger, and D.W. Meek. 1999. Spatial variation of rainfall over a large watershed in central Iowa. Theor. Appl. Climatol. 64:49–60.

Hill, M.C., and O. Osterby. 2003. Determining extreme parameter correlation in ground water models. Ground Water 41:420–430.

Hill, M.C., and C.R. Tiedeman. 2007. Effective groundwater model calibration with analysis of data, sensitivities, predictions, and uncertainty. John Wiley and Sons, Hoboken, NJ.

Jaynes, D.B., D.L. Dinnes, D.W. Meek, D.L. Karlen, C.A. Cambardella, and T.S. Colvin. 2004. Using the late spring nitrate test to reduce nitrate loss within a watershed. J. Environ. Qual. 33:669–677.

Karlen, D.L., D.L. Dinnes, D.B. Jaynes, C.R. Hurburgh, C.A. Cambardella, T.S. Colvin, and G.R. Rippke. 2005. Corn response to late-spring nitrogen management in the Walnut Creek watershed. Agron. J. 97:1054–1061.

Legates, D.R., and G.J. McCabe. 1999. Evaluating the use of "goodness-of-fit" measures in hydrologic and hydroclimatic model evaluation. Water Resour. Res. 35:233–241.

Lin, Z., and D.E. Radcliffe. 2006. Automatic calibration and predictive uncertainty analysis of a semi-distributed watershed model. Vadose Zone J. 5:248–260.

Ma, L., L.R. Ahuja, S.A. Saseendran, R.W. Malone, T.R. Green, B.T. Nolan, P.N.S. Bartling, G.N. Flerchinger, K.J. Boote, and G. Hoogenboom. 2011. A protocol for parameterization and calibration of RZWQM2 in field research. p. 1–64. *In* L.R. Ahuja and L. Ma (ed.) Methods of introducing system models into agricultural research. Advances in Agricultural Systems Modeling 2. ASA, CSSA, and SSSA, Madison, WI.

Ma, L., R.W. Malone, P. Heilman, D.L. Karlen, R.S. Kanwar, C.A. Cambardella, S.A. Saseendran, and L.R. Ahuja. 2007. RZWQM simulation of long-term crop production, water and nitrogen balances in Northeast Iowa. Geoderma 140:247–259.

Ma, L., R.W. Malone, D.B. Jaynes, K.R. Thorp, and L.R. Ahuja. 2008. Simulated effects of nitrogen management and soil microbes on soil nitrogen balance and crop production. Soil Sci. Soc. Am. J. 72:1594–1603.

Malone, R.W., D.B. Jaynes, L. Ma, B.T. Nolan, D.W. Meek, and D.L. Karlen. 2010. Soil-test N recommendations augmented with PEST-optimized RZWQM simulations. J. Environ. Qual. 39:1711–1723.

Malone, R.W., L. Ma, P. Heilman, D.L. Karlen, R.S. Kanwar, and J.L. Hatfield. 2007. Simulated N management effects on corn yield and tile-drainage nitrate loss. Geoderma 140:272–283.

Meyer, P.D., M.L. Rockhold, and G.W. Gee. 1997. Uncertainty analyses of infiltration and subsurface flow and transport for SDMP Sites. Report NUREG/CR-6565 PNNL-11705. U.S. Nuclear Regulatory Commission, Washington, DC.

Miller, C.T., G.A. Williams, C.T. Kelley, and M.D. Tocci. 1998. Robust solution of Richards' equation for nonuniform porous media. Water Resour. Res. 34:2599–2610.

Nolan, B.T., L.J. Puckett, L. Ma, C.T. Green, E.R. Bayless, and R.W. Malone. 2010. Predicting unsaturated zone nitrogen mass balances in agricultural settings of the United States. J. Environ. Qual. 39:1051–1065.

Poeter, E.P., and M.C. Hill. 1997. Inverse models: A necessary next step in ground-water modeling. Ground Water 35:250–260.

Poeter, E.P., M.C. Hill, E.R. Banta, S. Mehl, and S. Christensen. 2005. UCODE_2005 and six other computer codes for universal sensitivity analysis, calibration, and uncertainty evaluation. USGS Techniques and Methods 6-A11. USGS, Reston, VA

Schmied, B., K. Abbaspour, and R. Schulin. 2000. Inverse estimation of parameters in a nitrogen model using field data. Soil Sci. Soc. Am. J. 64:533–542.

Schwarz, G.E., A.B. Hoos, R.B. Alexander, and R.A. Smith. 2006. The SPARROW surface water-quality model: Theory, application, and user documentation. USGS Techniques and Methods Report, Book 6, Chapter B3. USGS, Reston, VA.

Thorp, K.R., R.W. Malone, and D.B. Jaynes. 2007. Simulating long-term effects of nitrogen fertilizer application rates on corn yield and nitrogen dynamics. Trans. ASABE 50:1287–1303.

USDA-NRCS. 2009. NSSH Part 618—Soil properties and qualities. Subpart B—Exhibits

Available at http://soils.usda.gov/technical/handbook/contents/part618ex.html (accessed November 2009, verified 12 May 2011).

Vecchia, A.V., and R.L. Cooley. 1987. Simultaneous confidence and prediction intervals for nonlinear regression models with application to a groundwater flow model. Water Resour. Res. 23:1237–1250.

Chapter 13

Estimating DSSAT Cropping System Cultivar-Specific Parameters Using Bayesian Techniques

James W. Jones
Jianqiang He
Kenneth J. Boote
Paul Wilkens
C.H. Porter
Z. Hu

Abstract

Crop models are highly useful for simulating crop and soil processes in response to variations in climate and management. However, if one wishes to simulate a crop's performance in a specific soil and climate for a particular set of management inputs, cultivar-specific parameters (CSPs) are needed because of the genetic variations among cultivars of any crop. In this chapter, we summarized methods that have been used to estimate CSPs for the CERES and CROPGRO-based models in the Decision Support System for Agrotechnology Transfer (DSSAT) cropping system model. We primarily described a Bayesian parameter estimation procedure (the Generalized Likelihood Uncertainty Estimation, or GLUE) for use in estimating CSPs in DSSAT. The procedure is simple to use, requiring only that users select a crop, a cultivar, and the data for use in the estimation procedure from a list of data available for that cultivar in the DSSAT system. Results are displayed for users to view and copy to the standard cultivar file in DSSAT for the crop involved. The procedure does require a large number of model runs; we recommend 6000 but users can optionally change this number. Two cultivars, 'Prisma' maize (*Zea mays* L.) and 'Williams' soybean [*Glycine max* (L.) Merr.] were selected to demonstrate the performance of DSSAT GLUE program. For Prisma maize, two experiments conducted in Zaragosa, Spain in 1995 and 1996 were selected; for Williams soybean, three experiments individually conducted in Wooster, OH and Gainesville, FL, were selected for the demonstration. Results showed that the GLUE method performed better than the arbitrary default CSPs and slightly better than the hand-calibrated CSPs in simulating these maize and soybean cultivars when using one time measurements, such as phenology dates, final dry matter yield, maximum leaf area index, and grain yield. For example, in the Prisma maize experiments in Zaragosa, Spain in 1995 and 1996, the average relative absolute error (RAE) values between the simulated and measured output variables were only 3 and 8%, respectively, while they were between 4 and 10% for hand-calibrated CSPs and above 16% for the default CSPs.

J.W. Jones (jimj@ufl.edu), C.H. Porter, and Z. Hu, Dep. of Agricultural and Biological Engineering; K.J. Boote, Agronomy Dep., Univ. of Florida, Gainesville, FL 32611. Jianqiang He, INRA, UMR1095, Génétique, Diversité et Ecophysiologie des Céréales, F-63100 Clermont-Ferrand, France; P. Wilkens, International Fertilizer Development Center, Muscle Shoals, AL 35662.

doi:10.2134/advagricsystmodel2.c13

Methods of introducing system models into agricultural research. L.R. Ahuja and L. Ma (ed.)
Advances in Agricultural Systems Modeling Series 2.

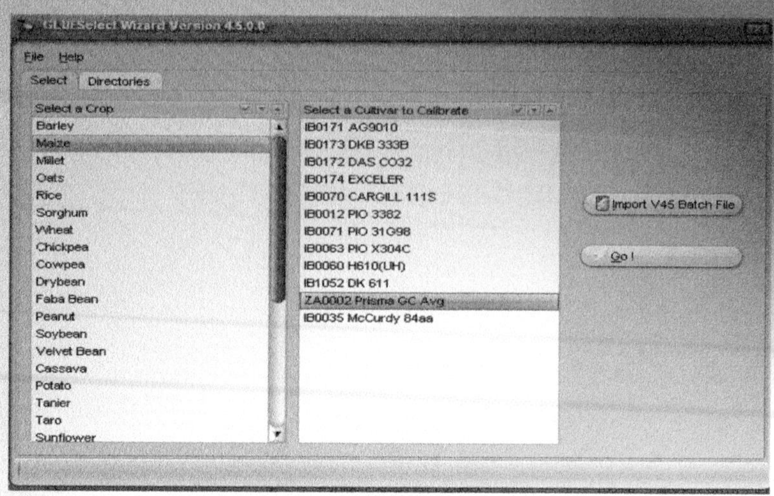
GLUESelect Wizard Version 4.5.0.0
File Help
Select Directories
Select a Crop
Barley
Maize
Millet
Oats
Rice
Sorghum
Wheat
Chickpea
Cowpea
Drybean
Faba Bean
Peanut
Soybean
Velvet Bean
Cassava
Potato
Tanier
Taro
Sunflower
Select a Cultivar to Calibrate
IB0171 AG9010
IB0173 DKB 333B
IB0172 DAS CO32
IB0174 EXCELER
IB0070 CARGILL 111S
IB0012 PIO 3382
IB0071 PIO 31G98
IB0063 PIO X304C
IB0060 H610(UH)
IB1052 DK 611
ZA0002 Prisma GC Avg
IB0035 McCurdy 84aa
Import V45 Batch File
Go !

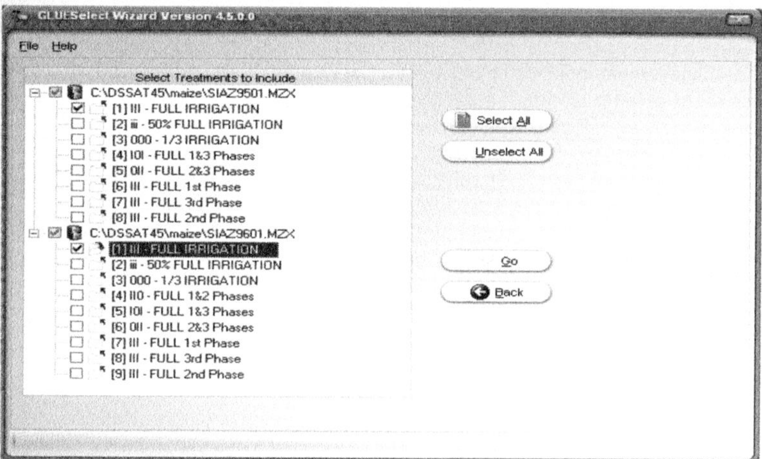
GLUESelect Wizard Version 4.5.0.0
File Help
Select Treatments to Include
C:\DSSAT45\maize\SIAZ9501.MZX
[1] III - FULL IRRIGATION
[2] iii - 50% FULL IRRIGATION
[3] 000 - 1/3 IRRIGATION
[4] I0I - FULL 1&3 Phases
[5] 0II - FULL 2&3 Phases
[6] III - FULL 1st Phase
[7] III - FULL 3rd Phase
[8] III - FULL 2nd Phase
C:\DSSAT45\maize\SIAZ9601.MZX
[1] III - FULL IRRIGATION
[2] iii - 50% FULL IRRIGATION
[3] 000 - 1/3 IRRIGATION
[4] II0 - FULL 1&2 Phases
[5] I0I - FULL 1&3 Phases
[6] 0II - FULL 2&3 Phases
[7] III - FULL 1st Phase
[8] III - FULL 3rd Phase
[9] III - FULL 2nd Phase
Select All
Unselect All
Go
Back

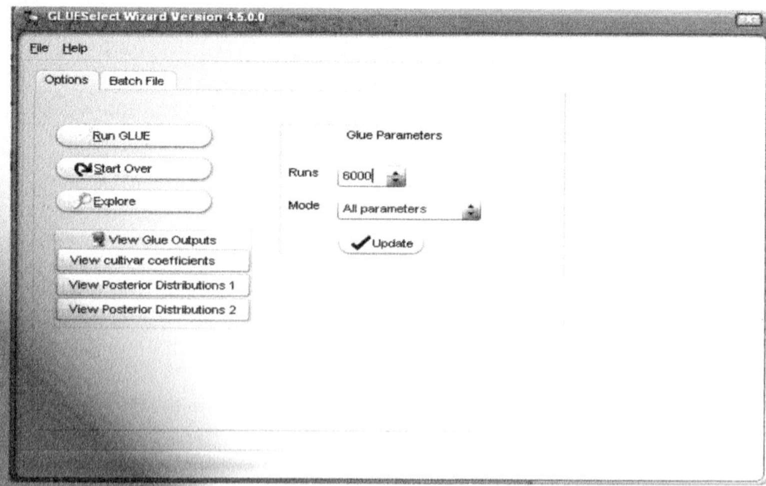
GLUESelect Wizard Version 4.5.0.0
File Help
Options Batch File
Run GLUE
Start Over
Explore
View Glue Outputs
View cultivar coefficients
View Posterior Distributions 1
View Posterior Distributions 2
Glue Parameters
Runs
6000
Mode
All parameters
Update

Crop models are highly useful for simulating crop and soil processes in response to variations in climate and management. The basic concept is that simulating crop growth and yield using dynamic crop models will produce results that represent how a real crop growing under specific environmental and management conditions would perform. Furthermore, if simulated results are sufficiently accurate, one can perform experiments using the models to evaluate likely responses to climate, soil properties, and crop and soil management. However, there are practical limitations that must be considered before making use of this approach in any study. One main limitation is that crop models do not contain all of the factors in the field that may influence crop yield. For example, crop diseases, weeds, and spatial variability of soils and management implementation can cause large differences in yield, and these factors are seldom included in crop simulation analyses. Another limitation is that inputs must be accurate or else simulated outputs are unlikely to match observations from the field. Attempts to evaluate the predictability of a crop model thus require that weather, management, and soil inputs are measured in the field where the evaluation experiments are conducted. Furthermore, model evaluation experiments would ideally be designed to eliminate yield-reducing factors that are not included in the model. Finally, parameters that are used to model the dynamics of soil and crop processes need to be accurate for comparison with observed field data. For example, if one uses a crop model to simulate crop yield responses to water or N management using incorrect soil water parameters, results will show that the model fails to mimic results from field experiments or, more problematically, will provide results that may mislead researchers or other model users.

If, however, one wishes to simulate crop growth and yield for large areas in which soils and climate vary, the input soil, weather, and management conditions should represent the spatial variability that exists over the area to provide reliable estimates at aggregate scales. In this case, many possible fields may be simulated without an attempt to mimic any particular real field. The objectives of such studies may be to compare aggregate effects of crops, varieties, and management systems under the range of soil and climate conditions that existing in an area. Options can be evaluated for changing crop management in an area to achieve a goal, such as to study management options for adapting to potential changes in climate. For these broad studies, model inputs need to represent the spatial variability that exists over the landscape, and crop and variety parameters should accurately represent those of a particular field from which observations were collected for comparisons with the model.

In this chapter, we focus on methods for estimating parameters for the DSSAT cropping system model (Tsuji et al., 1994; Jones et al., 2003; Hoogenboom et al., 2003; Boote et al., 2003) using data collected from field studies on real cropping systems. When field observations are used to estimate model parameters, however, the resulting parameters may not be transferrable to other conditions. This model transferability depends on its robustness, the parameters used in model equations, and the quality of observations used as model inputs and response variables. Parameters in models are often highly related to their testing conditions and are less universal than expected. Therefore, all crop models should be calibrated and validated for the environment of interest if results are to be credible (Timsina and Humphreys, 2006).

Emphasis is given on estimating the so-called genetic coefficients (Ritchie et al., 1986; Hunt et al., 1993; Boote et al., 2003), or what are more accurately referred to as cultivar-specific parameters (CSPs). The implicit assumption in the models is that there are parameters for a given crop species that remain the same for all cultivars and that there are parameters that vary among cultivars and allow simulating differences in yield or other traits when different cultivars are grown in the same environments and management conditions (Bertin et al., 2009; Boote et al., 2001). This is important because there are many existing cultivars, and new ones are released each year. Because the CSPs are not known for most cultivars, model users need to estimate them using field data. The emphasis in this chapter on CSPs should not be interpreted as a statement that they are more important than soil parameters. As stated above, when one attempts to evaluate simulated results with field observations, soil physical and chemical properties and initial conditions are very important. In some cases, crop yield responses are more strongly affected by uncertainties in soil parameters than by CSPs (Jones et al., 2010; Mavromatis et al., 2001).

One should not expect a crop model to simulate soil water and nutrient dynamics or growth and yield of a crop accurately unless soil physical and chemical properties and CSPs are accurate. The methods presented in this chapter can also be used to estimate soil parameters if appropriate soil measurements are taken in the field where an experiment is conducted (He et al., 2009), but the procedure implemented in DSSAT only estimates CSPs. Furthermore, a model may not be sensitive to some of its parameters, and in such cases, one may need to put less effort into estimating those parameters than others. A sensitivity analysis is generally recommended before estimating parameters for models (e.g., see Jawitz et al., 2008; Monod et al., 2006; Makowski et al., 2006; Muñoz-Carpena et al., 2007, 2010; Saltelli et al., 2004). However, we used prior experience to select the CSPs to estimate in the DSSAT implementation, so it is not necessary for users to do this before using the DSSAT GLUE program.

The objectives of this chapter are to summarize methods that have been used to estimate CSPs for the CERES and CROPGRO-based models in DSSAT and to present a Bayesian parameter estimation procedure (the Generalized Likelihood Uncertainty Estimation, or GLUE) that is in the latest release of DSSAT (version 4.5, Hoogenboom et al., 2010). Some results are presented to highlight the capabilities and limitations of this approach.

General Parameter Estimation Approaches

Model parameter estimation, sometimes referred to as calibration, is the process of estimating parameters to obtain a match between observed and simulated system behavior (Oliva, 2003). This requires a set of observed data from the real system, the model that simulates the system's behavior, a criterion for determining the best parameters, and a method to determine the best parameter set. Generally, the criterion is to minimize the error between observed and simulated variables. Many methods have been developed and used, such as conjugate gradient-descent search (Bhalla and Bower, 1993), stochastic-search (Foster et al., 1993), genetic algorithms (Baldi et al., 1998; Eichler-West and Wilcox, 1997; Vanier and Bower, 1996), and simulated annealing (Baldi et al., 1998; and Vanier and Bower, 1996). Perhaps the most common approach is trial and error (Wallach et al., 2001). Various parameter values are tested until a set of values is found that gives an acceptable fit to the data. For example, when Müller et al. (2003) calibrated the DAISY model to simulate decomposition of plant residues in soil, they used a stepwise trial and error process. The criterion of each iteration step was either to minimize the root mean square error (RMSE) or to maximize modeling efficiency (EF). The trial and error process has been criticized because it is unreliable and difficult to replicate (Lyneis and Pugh, 1996). Manual calibration may also be very tedious and time-consuming, depending on the number of model parameters and the degree of parameter interaction. Thus, a great deal of research has been directed to development of more effective and efficient automatic calibration procedures (Madsen et al., 2002).

There are two general approaches used to estimate parameters, frequentist and Bayesian (Makowski et al., 2006). The frequentist approach uses estimation methods to approximate assumedly true and fixed parameter values by using a sample of data. Prior information on parameter values is not taken into account. The use of a frequentist method gives a single, deterministic estimate of each parameter. In contrast, Bayesian methods estimate parameters from two types of information, a sample of data (like the frequentist method) and prior information about parameter values. The results of Bayesian methods are probability distributions of parameter values. All Bayesian methods proceed in two steps. The first step is to define a "prior" parameter probability distribution based on literature or expert knowledge. The second step involves calculating a new parameter probability distribution from both the prior distribution and the available data set. This new distribution, termed the *posterior distribution*, is computed by using Bayes theorem (Makowski et al., 2006).

Bayesian methods are becoming increasingly popular for estimating parameters for complex mathematical models (e.g., Campbell et al., 1999) because these methods provide a coherent framework for dealing with uncertainty. This is also due to the increase in speed of computers and the development of new algorithms (Malakoff, 1999). One commonly used Bayesian method is the GLUE method (Beven and Binley, 1992; Franks et al., 1998; Shulz et al., 1999). The GLUE method assumes that, in the case of large models with many parameters, there is no exact inverse solution. Hence, the estimation of a unique set of parameters, which optimizes a goodness-of-fit criterion given the observations, is not possible (Romanowicz and Beven, 2006). The main principle of this method is to discretize the parameter space by generating a large number of parameter values from the prior distribution. Likelihood values are then calculated for each parameter set using field observations. Probabilities, an empirical posterior distribution of the parameters, are calculated using Bayes' equation.

Since it was introduced in 1992, the GLUE framework has found widespread application in environmental modeling (Blasone et al., 2008). The popularity of GLUE is largely due to its conceptual simplicity, relative ease of implementation and use, and its ability to handle different error structures and models without major modifications to the method itself (Blasone et al., 2008). Example applications include those for rainfall–runoff (Beven and Binley, 1992; Freer et al., 1996; Lamb et al., 1998), soil erosion (Brazier et al., 2001), tracer dispersion in a river reach (Hankin et al., 2001), groundwater and well capture zone delineation (Feyen et al., 2001; Jensen, 2003), unsaturated zone (Mertens et al., 2004), flood inundation (Romanowicz et al., 1996; Aronica et al., 2002), land–surface–atmosphere interactions (Franks et al., 1997), soil freezing and thawing (Hansson and Lundin, 2006), crop yields and soil organic carbon (Wang et al., 2005), ground radar rainfall estimation (Tadesse and Anagnostou, 2005), and distributed hydrology (McMichael et al., 2006; Muleta and Nicklow, 2005). He et al. (2009, 2008) used the GLUE method to estimate soil parameters and CSPs in the DSSAT CERES-Maize model.

There are a number of other Bayesian methods for estimating parameters, including the Metropolis–Hastings method (Makowski et al., 2002). We implemented the Metropolis–Hastings method in one study and found that it is more difficult to implement, but more efficient than GLUE (Hu and Jones, unpublished data, 2010). The major difference between these methods is that the Metropolis–Hastings method uses a Markov Chain–Monte Carlo (MCMC) parameter search method, whereas the GLUE method uses a Monte Carlo random search method. In this chapter, we present the GLUE method that is implemented in the latest release of the DSSAT software (version 4.5, Hoogenboom et al., 2010).

Materials and Methods

Overview of DSSAT

The DSSAT is a software system that combines crop models with observed data from field trials and tools that help users enter data from experiments, evaluate the models, estimate CSPs, conduct sensitivity analyses, analyze economic risk and uncertainty of alternative management options, and graphically present simulated and observed results (Uehara, 1998; Jones et al., 1998; Hoogenboom et al., 2003). One of the unique features of this system is that it has databases that connect to the crop models and are used for evaluating model performance and estimating CSPs. At the heart of DSSAT is the cropping system model (DSSAT-CSM), which incorporates all crops as modules using a single soil model (Jones et al., 2003). The DSSAT version 4.5 can simulate more than 20 crops, including maize, wheat (*Triticum aestivum* L.), rice (*Oryza sativa* L.), sorghum [*Sorghum bicolor* (L.) Moench], soybean, and peanut (*Arachis hypogaea* L.). The CSM simulates growth, development, and yield of a crop growing on a uniform area of land under specified management. The dynamics of soil water, carbon, nitrogen and phosphorus that take place in the cropping system through time are also simulated. The model is structured using the modular approach described by Jones et al. (2001) and Porter et al. (2000) and consists of a main driver program, a land unit module, and primary modules for weather, management, soil, plant, and soil–plant–atmosphere interface components.

A number of methods have been used in the past to estimate CSPs for the different crops in the DSSAT-CSM. In many cases, trial and error methods are used to select a set of parameters that visually fits the observed data and/or produces an acceptably low RMSE between simulated and observed variables, often using a combination of time-series and end-of-season data concurrently (e.g., Boote,

1999; Boote et al., 2003). Other methods include the Simplex method (Grimm et al., 1993), Simulated Annealing (Mavromatis et al., 2001, 2002), and the K-Nearest Neighbor approach (Bannayan and Hoogenboom et al., 2008). A CSP estimation tool was developed by Hunt et al. (GENCALC, Hunt et al., 1993) and integrated into DSSAT v3.0, and v3.5. This tool automated a systematic search of parameters that minimized RMSE between simulated and observed variables; the criteria for selecting parameter values depended on the parameter being estimated in a sequential search process. A new updated version of this tool is also available in DSSAT v4.5.

Implementation of GLUE in DSSAT

The GLUE CSP estimation method was integrated into DSSAT using the R language (R Development Core Team, 2009; http://www.R-project.org, verified 13 May 2011), a free software environment for statistical computing and graphics. The program is simple to use in that users only have to select a crop, a cultivar (from a list of cultivars included in the DSSAT database for that crop), and the treatments from the various experiments in which that cultivar was grown (Fig. 13–1). The ranges of CSPs are stored in a file so that users do not have to specify these values. Optionally, users can choose to estimate only phenology CSPs, only growth CSPs, or both of these simultaneously, and set the number of runs to make. The interface also allows users to view the final estimated parameters and their distributions. The GLUE calculations are made using the GLUE R program, which calls the DSSAT-CSM, connects the data to the model, generates all parameter sets, and performs all of the necessary calculations to select the CSPs.

Theoretical Basis of the GLUE Method in DSSAT v4.5

Although users do not need to know all of the calculations that are done to arrive at the estimated CSPs when they use the GLUE program, we summarize the main theoretical basis for the method and the steps used internally in the program. In the GLUE procedure, a likelihood function is used as the criterion for estimating CSPs similar to that used by He et al. (2009). A likelihood value is computed for each observation, given each particular set of randomly generated CSP sets. The likelihood function is the product of these individual likelihood values. In DSSAT, the Gaussian likelihood function (Eq. [1]) (Beven and Binley, 1992; Romanowicz et al., 1994, 1996; Makowski et al., 2002; He et al. (2009, 2008, 2010) was used:

$$L[\theta_i \mid O] = \prod_{j=1}^{M} \frac{1}{\sqrt{2\pi\sigma_o^2}} \exp\left\{ -\frac{\left[O_j - Y(\theta_i)\right]^2}{2\sigma_o^2} \right\}, \qquad (i = 1,2,3\ldots N) \qquad [1]$$

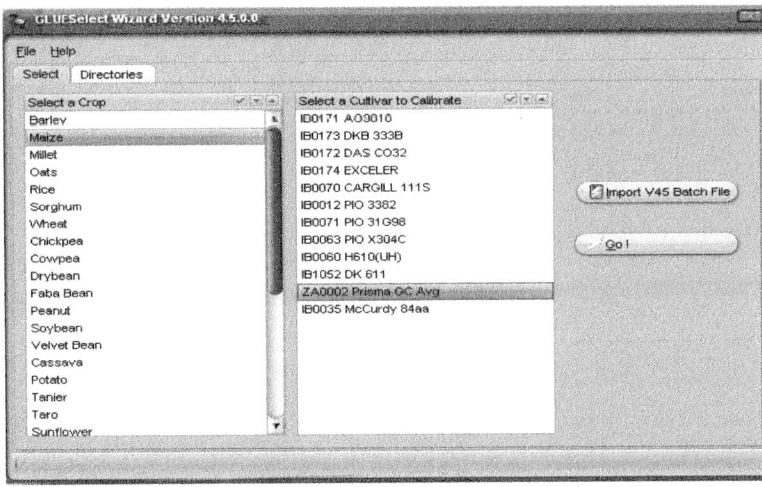

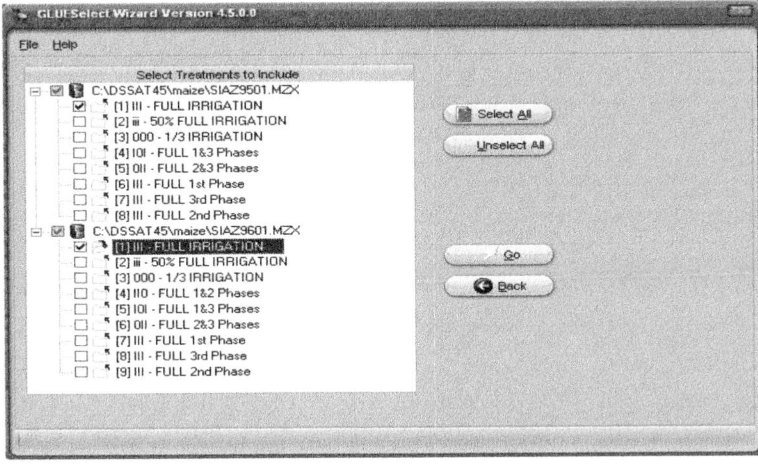

Fig. 13–1. Screen shots of the DSSAT GLUE program user interface. (Top) Users first select a crop, then a cultivar, (middle) then select which treatments they wish to use for estimating the cultivar-specific parameters (CSPs), and (bottom) then run the procedure. After the program finishes, users view the coefficients and can copy and paste the estimated CSPs into the crop cultivar file. Users may also view the standard deviations of CSP estimates from this last screen.

where θ_i is the ith parameter set, N is the total number of parameter sets generated by the program, $Y(\theta_i)$ is the model output using parameter set θ_i, O is the observation; O_j is the jth observation of O; σ_o^2 is the variance of model error, and M is the number of observations. The probability p_i of each parameter set is computed with the equation (He et al., 2009):

$$p(\theta_i) = \frac{L(\theta_i \mid Y)}{\sum_{i=1}^{N} L(\theta_i \mid Y)} \qquad [2]$$

where $p(\theta_i)$ is the probability or likelihood weight of the ith parameter set θ_i, and $L(\theta_i|y)$ is the likelihood value of parameter set θ_i, given observations Y.

The GLUE methodology (Beven and Binley, 1992) thus develops an approximate discrete posterior probability distribution, designated by (θ_i, p_i), $i = 1,...,N$, $\sum_{i=1}^{N} p_i = 1$, where p_i is the probability associated with the parameter vector $\boldsymbol{\theta}_i$, and N is the total number of generated parameter vectors.

Equations [1] and [2] are used to construct the posterior distributions of the CSPs. The implementation in DSSAT uses two iterations of GLUE, one to estimate phenological development parameters and the others to estimate growth parameters. This was done for practical reasons, mainly due to the time required when estimating all parameters simultaneously. This can be done in the DSSAT models because development is largely independent of growth but not vice versa. Thus, the development parameters are first estimated followed by those that affect biomass growth and yield in a two-step sequence. Thus, there are two posterior distributions, one for each step.

The main steps of the GLUE procedure in DSSAT are based on Beven and Binley (1992) and are summarized as follows:

1. *Develop prior parameter distributions.* In this study, the CSPs of two crops (maize and soybean) in the DSSAT database (Hoogenboom et al., 2003) were analyzed to determine the prior range of each parameter. Because we do not have additional information, the prior distributions are assumed to be independent and uniformly distributed between the minimum and maximum values across all cultivars previously calibrated for each crop.
2. *Generate random parameter sets from the prior parameter distributions.* A large number (e.g., 6000) of parameter sets are created by randomly generating each CSP in each of 6000 CSP vectors independently, according to the prior uniform distribution of each CSP. The number of runs can be modified by the user, but at least 3000 runs are recommended (He, 2008) to ensure that CSPs are each estimated accurately and the calculated posterior distributions are reliable.

3. *Run the model with the randomly generated parameter sets.* The model is run with the parameter sets generated above. The standard genetic input file is changed to simulate every random parameter set in sequence. Model outputs (e.g., anthesis date, maturity date, dry matter yield, leaf area index, and leaf number, which could be selected by model users) for each parameter set are tabulated for use in the GLUE likelihood calculations.

4. *Calculate the likelihood values.* The observations (Y) from the selected data provided by model users are used along with the corresponding simulated outputs to compute the likelihood value, $L(\theta_i|Y)$, for each generated parameter vector $\boldsymbol{\theta}_i$.

5. *Construct posterior distribution and statistics.* The pairs of parameter sets and probabilities, (θ_i, p_i), $i = 1,\ldots,N$, are computed and used to construct the posterior distribution and to compute the mean and variance of the selected parameters using following equations:

$$\hat{\mu}_{\text{post}}(\theta) = \sum_{i=1}^{N} p(\theta_i)\cdot\theta_i \quad [3]$$

$$\hat{\sigma}^2_{\text{post}}(\theta) = \sum_{i=1}^{N} p(\theta_i)\cdot\left(\theta_i - \hat{\mu}_{\text{post}}\right)^2 \quad [4]$$

where $\hat{\mu}_{\text{post}}(\theta)$ and $\hat{\sigma}^2_{\text{post}}(\theta)$ are the estimated mean and variance of the posterior distribution of parameters θ, $p(\theta_i)$ is the probability of the ith parameter set θ_i calculated by Eq. [1], and N is the number of random parameter sets.

Application of GLUE to Estimate Cultivar-Specific Parameters of Maize and Soybean

Cultivar-specific parameters were estimated for maize and soybean to demonstrate the method's performance and characteristics of results that were obtained. In particular, we show observed and simulated values derived from the CSPs selected using the GLUE method. We also present uncertainties in the estimated CSPs and in simulated outputs.

Table 13–1 lists the CSPs of the CERES-Maize model in DSSAT (Jones et al., 2003) that are estimated in the GLUE procedure. The CSPs P1, P2, and P5 determine the timing of phenological events, such as anthesis date and maturity date of maize. Coefficients G2 and G3 control the yield-related outputs, such as grain dry matter yield, grain size, and canopy weight. Our two-step procedure requires that all CSPs estimated in the first step be completely determined by phenological development data. The CSPs estimated in the first step plus all other CSPs in the second step affect growth responses. Phenological development is assumed to be independent from growth in the model, but growth is affected by phenologi-

Table 13–1. Cultivar-specific parameters in the DSSAT CERES-Maize model (Jones et al., 2003) that are estimated in the DSSAT GLUE procedure.

Coefficient	Minimum	Maximum	GLUE flag†	Definition
P1	140	365	1	Degree days (base 8°C) from emergence to end of juvenile phase
P2	0.0	1.0	1	Photoperiod sensitivity coefficient (0–1.0)
P5	600	990	1	Degree days (base 8°C) from silking to physiological maturity
G2	500	908	2	Potential kernel number
G3	5	15	2	Potential kernel growth rate mg/(kernel d)

† GLUE flag is an indicator to show in which round of the procedure the parameter will be estimated.

cal development. This assumption is reasonable unless major stresses occur. The CERES-Maize CSP PHINT is not included in the procedure because its value is similar across many cultivars and because it influences both phenological development and yield. PHINT was assumed to be 48.0 for all cultivars.

Table 13–2 shows the CSPs for the CROPGRO-Soybean model in the DSSAT-CSM. There are 18 total CSPs in soybean (Boote et al., 2003), but the 11 coefficients in the table below are those that vary most among cultivars and determine differences in soybean cultivars' responses to their environments. The model is highly sensitive to these CSPs and less sensitive to the other ones when they are near their nominal values in the cultivar file available in DSSAT.

In this chapter, two cultivars (Prisma and Williams shown in Table 13–3) of two crops (maize and soybean, respectively) were selected as examples to evaluate the performance of DSSAT GLUE program. For Prisma maize, the experiments conducted in Zaragosa, Spain in 1995 and 1996 were used. There are a total of eight and nine treatments based on different irrigation levels in 1995 and 1996, respectively, but only the fully irrigated treatments were selected so that development and growth were influenced only by weather, not by water or nutrient stresses. Similarly, three experiments, conducted individually in Wooster, OH and Gainesville, FL, were selected for Williams soybean. Only irrigated treatments were used to estimate the CSPs. A summary of experimenal treatment characteristics, including planting date, N application, irrigation, and available observations is presented in Table 13–3.

The monthly average weather data were summarized for growing season months for these five different experiments (Table 13–4). Generally, solar radiation was higher and rainfall (mm d^{-1}) was lower for this maize experiment site than the other sites. The Gainesville, FL, 1979, experiment had relatively higher minimum and maximum temperatures and lower rainfall, which is due the subtropical climate of Florida. In the Ohio soybean experiments, solar radia-

Table 13–2. Cultivar-specific parameters for the soybean crop in the DSSAT-CSM (from Boote et al., 2003) that are estimated in the GLUE procedure.

Coefficient	Minimum	Maximum	GLUE flag†	Definition
CSDL	11.78	14.6	1	Critical Short Day Length below which reproductive development progresses with no day length effect (for short day plants) (h)
PPSEN	0.129	0.349	1	Slope of the relative response of development to photoperiod with time (positive for short day plants) (1/h)
EM-FL	15.5	23.5	1	Time between plant emergence and flower appearance (R1) (PTD‡)
FL-SD	12	16	1	Time between first flower and first seed (R5) (PTD)
SD-PM	29.5	37.5	1	Time between first seed (R5) and physiological maturity (R7) (PTD)
LFMAX	1	1.4	2	Maximum leaf photosynthesis rate (at 30C, 350 vpm CO_2, high light, mg CO_2 m^2 s^{-1})
SLAVR	350	425	2	Specific leaf area of cultivar under standard growth conditions (cm^2 g^{-1})
SIZELF	140	230	2	Maximum size of full leaf (three leaflets) (cm^2)
WTPSD	0.155	0.195	2	Maximum weight per seed (g)
SFDUR	17	25.5	2	Seed filling duration for pod cohort at standard growth conditions (PTD)
SDPDV	1.7	2.44	2	Average seed per pod under standard growing conditions (no. pod^{-1})

† GLUE flag is an indicator to show in which round of the procedure the parameter will be estimated.

‡ PTD, photothermal days, comparable to calendar days if at optimum temperature and CSDL.

tion was higher and rainfall less than in Florida. Because irrigation was used in all of the experiments, water stress likely did not reduce yields relative to the potential yields.

The CSPs estimated using the GLUE procedure were compared with those from two other sources—an arbitrary default set of CSPs used to initialize the GLUE procedure and the hand-calibrated CSPs that were in the DSSAT v4.5 cultivar database. Field observations from the selected experiments (Table 13–3) in DSSAT were compared with the model-simulated output variables derived from the three sources of CSPs. The relative absolute error (RAE percentage; Eq. [5]) was used as a measure to evaluate the differences between simulated and observed variables for each set of CSPs:

$$\mathrm{RAE} = \frac{\left|\hat{Y} - Y\right|}{Y} 100\% \qquad [5]$$

where $\hat{Y}$ and Y are simulated and measured variables, respectively.

Table 13–3. Summary of experiment details for the example maize and soybean cultivars and experiments in DSSAT v4.5 used for estimating Cultivar-Specific Parameters in this study (Hoogenboom et al. 2010). Also shown are the field observations used in each experiment.

Crop	Cultivar	Experiment	Experimental items	Details†
Maize	Prisma		Treatment	Fully irrigated
			Location	Zaragosa, Spain
		Zaragosa, Spain	Planting date	17 May 1995
		1995	N fertilizer	100 and 200 kg urea-N ha^{-1} 16 May 1995 and 16 June 1996
			Irrigation	A sum of 568 mm water in 9 events
			Available observations	ADAP, MDAP, HWAM, HWUM, H#UM, CWAM, LAIX, HIAM
			Observations	CWAM, LAIX
			Treatment	Fully irrigated
			Location	Zaragosa, Spain
		Zaragosa, Spain	Planting date	16 May 1996
		1996	N fertilizer	100 and 200 kg urea-N ha^{-1} 15 May 1996 and 15 May 1996
			Irrigation	A sum of 505 mm water in 8 events
			Available observations	ADAP, MDAP, HWAM, HWUM, H#UM, CWAM, LAIX, HIAM
			Observations	CWAM, LAIX
Soybean	Williams	Wooster, OH 1988	Planting date	1 May 1988
			N fertilizer	No N applied
			Irrigation	A sum of 595 mm water in 18 events
			Available observations	ADAP, PD1P, MDAP, HWAM, HWUM, CWAM, LAIX
		Wooster, OH 1990	Planting date	30 Apr. 1990
			N fertilizer	No N applied
			Irrigation	Not irrigated (not necessary)
			Available	ADAP, PD1P, MDAP, HWAM, PWAM, PWAM, PWAM, PWAM, H#AM, HWUM, H#UM, CWAM, BWAM, LAIX, HIAM, THAMS
			Observations	HWUM, CWAM, LAIX
		Gainesville, FL 1979	Planting date	15 Mar. 1979
			N fertilizer	No N applied
			Irrigation	A sum of 144 mm water in 14 events
			Available observations	ADAP, PD1P, MDAP, HWAM, HWUM, CWAM, LAIX

† These abbreviations represent the model output variables in DSSAT: ADAP, anthesis date (days after planting); MDAP, maturity date (days after planting); HWAM, yield at harvest maturity (kg [dm]/ha); HWUM, unit weight at maturity (g [dm]/unit); CWAM, tops weight at maturity (kg [dm]/ha); LAIX, maximum leaf area index; PD1P, first pod day (days after planting); PWAM, pod/ear/panicle weight at maturity (kg [dm]/ha).

Table 13–4. Monthly average weather data of the five experiments of maize and soybean (Hoogenboom et al. 2010).†

	Month	Apr.	May	June	July	Aug.	Sept.	Oct.
Zaragosa, 1995	Tmin	5.1	10.9	14.0	17.2	16.7	11.7	9.4
	Tmax	21.1	24.8	27.4	32.8	30.0	24.0	24.2
	Rain	2.0	1.2	0.3	0.1	0.5	0.3	0.1
	SRAD	22.1	21.5	25.1	25.6	22.9	17.8	12.5
Zaragosa, 1996	Tmin	6.4	10.0	14.3	15.9	16.0	11.8	8.1
	Tmax	19.3	23.2	28.4	30.5	28.7	24.2	20.4
	Rain	1.2	1.6	0.6	1.0	1.1	0.3	0.2
	SRAD	20.4	23.0	25.9	25.7	21.4	17.3	12.5
Wooster, OH 1988	Tmin	3.8	7.6	10.7	15.9	16.3	10.6	–
	Tmax	16.6	23.1	28.3	32.2	29.0	23.4	–
	Rain	1.9	1.1	0.4	5.2	2.8	2.5	–
	SRAD	15.0	20.3	23.9	20.3	18.0	13.6	–
Wooster, OH 1990	Tmin	7.3	13.2	15.5	14.6	11.2	5.6	–
	Tmax	19.6	25.7	27.4	26.8	22.9	17.7	–
	Rain	4.2	2.0	5.5	3.7	3.7	3.9	–
	SRAD	17.4	20.7	20.1	16.7	13.1	9.5	–
Gainesville, FL 1979	Tmin	15.1	17.0	20.1	22.6	–	–	–
	Tmax	28.2	30.0	32.6	34.0	–	–	–
	Rain	6.9	2.8	3.9	3.6	–	–	–
	SRAD	19.1	21.3	20.4	21.1	–	–	–

† Tmin, minimum temperature (°C); Tmax, maximum temperature (°C); Rain, average daily rainfall (mm); and SRAD, solar radiation (MJ/m^2/d).

Results and Discussion

Maize Results

Comparison of Simulated and Observed Crop Variables

A comparison of results between field observed and model simulated output variables of Prisma maize from three different sources of CSPs is summarized in Table 13–5. The GLUE-estimated CSPs did a better job than the arbitrary generic maize cultivar default CSPs and performed as well as the hand-calibrated CSPs in simulating this maize cultivar. The average RAE values were only 3 and 8% in 1995 and 1996, respectively, while they were 4 and 10% for the hand-calibrated CSPs and above 15% for the default CSPs. Using both the GLUE-estimated and hand-calibrated CSPs, the maize model predicted phenology dates very well. In 1995, there were 0 and 2 d differences between the GLUE simulated and observed anthesis and maturity dates respectively, while these values were 1 and 2 d in 1996. Results for the hand-calibrated CSPs were similar. However, the errors in phenology dates simulations with the default CSPs were high. For example, the

Table 13–5. Comparison between observations and model output variables of 'Prisma' maize grown at Zaragosa, Spain

Source of outputs	Phenology				Aboveground						
	Anthesis		Maturity		Grain yield		Biomass		Max LAI		Avg.
	Value	RAE†	Value	RAE	Value	RAE	Value	RAE	Value	RAE	RAE
	——— d ———				——— kg/ha ———						
Zaragosa, Spain in 1995											
Field observations	78	–	141	–	10,960	–	23,970	–	6.15	–	–
Default CSPs	62	21%	113	20%	12,679	16%	26,428	10%	5.19	16%	17%
Hand-calibrated CSPs	78	0%	141	0%	10,255	6%	25,648	7%	5.61	9%	4%
GLUE-calibrated CSPs	78	0%	139	1%	10,866	1%	26,437	10%	5.65	8%	3%
Zaragosa, Spain in 1996											
Field observations	78	–	147	–	12,340	–	22,730	–	4.75	–	
Default CSPs	67	14%	129	12%	13,626	10%	28,973	27%	5.69	20%	16%
Hand-calibrated CSPs	79	1%	151	3%	10,202	17%	25,327	11%	5.55	17%	10%
GLUE-calibrated CSPs	79	1%	149	1%	10,791	13%	26,079	15%	5.56	17%	8%

† The RAE values are the relative absolute error between field observed and model simulated variables from three different sources of CSPs.

anthesis and maturity dates simulated with default CSPs were 11 and 18 d sooner than the field observations in 1996. Although the hand-calibrated CSPs for this cultivar performed about as well as the GLUE-estimated CSPs, it is dangerous for users to automatically use CSPs for a particular cultivar without evaluating them for their own conditions, even if the cultivar's parameter set is contained in the DSSAT database. Data from regions or locations where the model is to be applied should be used to evaluate simulated results relative to observations or used to estimate a new set of CSPs.

The average absolute error (RAE) between simulated and observed data was lower in 1995 for the GLUE-estimated CSPs (3%) than for the default or hand-calibrated CSPs (17 and 4%, respectively) (Table 13–5). In addition, the RAE values were lower for all but one variable (biomass) when GLUE-estimated parameters were used in 1995 and 1996. In 1996, the GLUE-estimated parameters performed better on average as well; the RAE was 8% in comparison with 16 and 10% for default and hand-calibrated CSPs, respectively.

Comparison of CSPs from GLUE with Hand-Calibrated and Default CSPs

The parameter values of Prisma maize obtained from different sources were compared (Table 13–6). There were large differences among the values of P1 and P2, which determine the thermal time of the vegetative and reproductive stages of maize. Thus, it is not surprising to see large differences among the

Table 13–6. Comparison between cultivar-specific parameters (CSPs) of 'Prisma' maize.

Source of CSPs	P1	P2	P5	G2	G3
Default	200.0	0.300	800.0	700.0	8.50
Hand-calibrated	280.0	0.300	789.0	650.0	6.03
GLUE-calibrated values (standard deviations)	268.2 (18.45)	0.758 (0.291)	770.8 (35.72)	675.1 (59.47)	6.439 (0.234)

predicted anthesis and maturity dates with these different sources of CSPs. The standard deviations were mostly less than 10% of the estimated CSPs (Table 13–6). An exception was for the parameter P2, the photoperiod sensitivity coefficient, which was nearly 40% of the estimated value. This is one indication that the model was not sensitive to this CSP for these two experiments, which was apparently due to the fact that the experiments were in the same location both years, and there was little difference in daylengths that the two crops experienced.

Comparing In-Season Simulated Results

Although time series data were not used to estimate CSPs in the GLUE procedure, we show time-series model outputs of in-season predictions for the three sources of CSPs for one season of 1995 (Fig. 13–2). In the maize experiments, no in-season data were collected, so it is not possible to draw conclusions about how well each set of CSPs performed in this example. However, it is interesting to show how different the in-season results are for two reasons. First, it is clear that similar or the same end of season simulations can be obtained with different CSPs, but that considerable differences may occur among in-season results. This can be seen in the GLUE simulated end point results for biomass and grain yield in Fig. 13–2. Second, in-season measurements may be very useful to refine the CSPs so that they simulate the time courses of growth and yield during a season. The differences among the simulated values of leaf area index (LAI) and grain weight (kg ha^{-1}) can be seen clearly.

Soybean Results

Comparison of Simulated and Observed Crop Variables

Similarly, comparisons of the results for Williams soybean are tabulated in Table 13–7 for observed and simulated output variables for the three experiments in two locations. The GLUE and hand-calibrated RAE results averaged across all measurements were nearly equal. In about one-half of the RAE values for individual measurement variables, the GLUE method performed better, and in the other half, the hand-calibrated CSPs performed better. For soybean, Dr. K.J. Boote

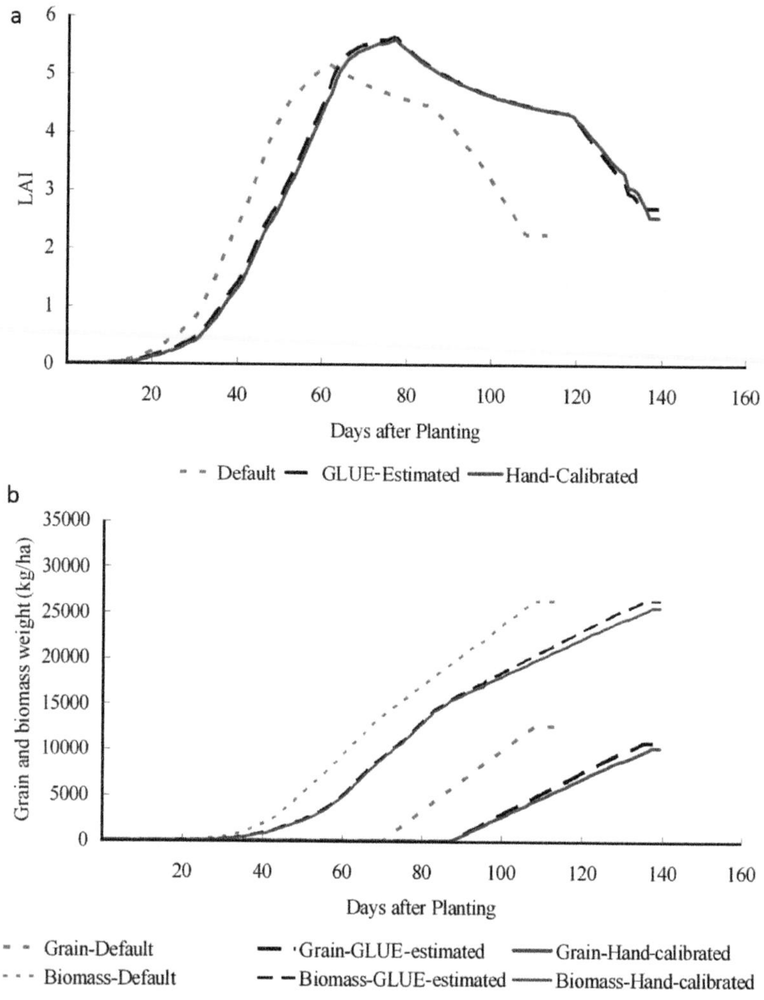

Fig. 13–2. Graphs of simulated results using the three sources of cultivar-specific parameters (CSPs) for (a) leaf area index and (b) total biomass and grain yield for the experiment in Zaragosa, Spain in 1995. "Default" represents the outputs from default genetic coefficients, "GLUE-estimated" represents the outputs from GLUE-estimated genetic coefficients, and "Hand-calibrated" represents the outputs from hand-calibrated genetic coefficients.

of the Agronomy Department of the University of Florida (personal communication, 2010) estimated the CSPs using both end of season and phenology data with considerable emphasis on in-season measurements. Thus, the GLUE method did a good job in estimation of CSPs, since the average RAE values were equal to the hand-calibrated CSPs and clearly less than most of the default CSPs. One exception was the maximum LAI (MaxLAI), for which the RAE obtained from the GLUE-estimated CSPs was 31%. The GLUE method also improved the accuracy

Table 13–7. Comparison between observations and model output variables of 'Williams' soybean. Default CSPs are those for "generic" MG 4 cultivar.

Source of outputs	Phenology				Aboveground												
	Anthesis		Maturity		Grain yield		Biomass		Max LAI		First pod		Pod wt.		Unit wt. grain		Avg.
	Value	RAE†	Value	RAE	Value	RAE	Value	RAE	Value	RAE	Value	RAE	Value	RAE	Value	RAE	RAE
	d				kg/ha						d		kg/ha		g		
Wooster, OH, 1988																	
Observed	71	–	145	–	3976	–	8090	–	7.65	–	91	–	5194	0%	0.151	–	–
Default CSPs	87	23%	–	–	3138	21%	8356	3%	7.67	0%	101	11%	4490	14%	0.16	6%	11%
Hand-calibrated CSPs	73	3%	141	3%	3781	5%	7892	2%	6.89	10%	90	1%	5145	1%	0.157	4%	4%
GLUE-calibrated CSPs	72	1%	145	0%	3893	2%	7664	5%	6.74	12%	89	2%	5163	1%	0.161	7%	4%
Wooster, OH, 1990																	
Observed	–	–	150	–	3149	–	7113	–	7.21	–	96	–	4915	0%	0.164	–	–
Default CSPs	88	–	162	8%	3718	18%	8387	18%	7.47	4%	104	8%	5127	4%	0.168	2%	9%
Hand-calibrated CSPs	73	–	143	5%	3552	13%	7450	5%	6.82	5%	91	5%	4883	1%	0.146	11%	6%
GLUE-calibrated CSPs	71	–	148	1%	3713	18%	7274	2%	6.61	8%	90	6%	4957	1%	0.156	5%	6%
Gainesville, FL, 1979																	
Observed	38	–	93	–	2474	–	3893	–	3.15	–	48	–	–	–	0.175	–	0%
Default CSPs	40	5%	114	23%	3570	44%	5700	46%	3.38	7%	50	4%	4936	–	0.183	5%	19%
Hand-calibrated CSPs	37	3%	96	3%	2831	14%	4303	11%	2.55	19%	48	0%	3893	–	0.156	11%	9%
GLUE-calibrated CSPs	37	3%	99	6%	2588	5%	3855	1%	2.16	31%	47	2%	3416	–	0.169	3%	7%

† The RAE values are the relative absolute error between field observed and model simulated variables from three different sources of CSPs.

of phenology date predictions relative to the other sources of CSPs. The differences between observed and GLUE simulated anthesis and maturity dates were all less than 2 d, except for the maturity date of the experiment in Gainesville, FL, in 1979, which was about 6 d.

Comparison of CSPs from GLUE with Hand-Calibrated and Default CSPs

The parameter values of Williams soybean obtained from different sources are compared in Table 13–8. The default CSPs were assumed to be those of a generic maturity 4 soybean variety from the DSSAT database. Except for parameters SLAVR and SIZELF, the differences between these three types of CSPs were very small. The largest relative differences between hand-calibrated and GLUE-estimated CSPs were for PPSEN (0.285 vs. 0.342, respectively). The values of CSDL and PPSEN for Williams in hand-calibrated CSP were not hand-calibrated, in fact, but were estimated using a least squares simplex method from a much larger (n > 100) data set (Grimm et al., 1993). Thus, the GLUE-estimated photoperiod sensitivity coefficient (PPSEN) was obtained from a much smaller set of environments and is less robust across regions than that obtained by Grimm et al. (1993) for the hand-calibrated CSPs. This highlights a fundamental issue that transcends methods for calibration—that CSPs estimated from field data may not be robust if the size of the dataset used is small. Other relatively large differences occurred for SLAVR and SIZELF. Nevertheless, the standard deviations of the GLUE-estimated CSPs were relatively low, indicating that the parameters worked well across the locations and years for this soybean cultivar.

Comparison of In-Season Simulated and Observed Soybean Results

The time-series model outputs were also compared for Williams soybean cultivar (Fig. 13–3, 13–4, 13–5). All three experiments in this study provided field observations for some time-series output variables, such as LAI, biomass, and grain weight. These observations can help us to evaluate the reliability of the predictions from different CSPs sources. Most model outputs simulated using the GLUE-estimated and hand-calibrated CSPs were similar for all variables that were measured during the growing season. This result provides strong evidence that the GLUE method provided reliable CSP estimates, similar in performance to those that were estimated by an expert. However, this result also demonstrates the concept of "equifinality" (Beven and Freer, 2001; Hansson and Lundin, 2006; Shulz et al., 1999), which means that different combinations of CSPs will result in the same quality of simulation. The values of the GLUE CSPs were different from those estimated by K.J. Boote, yet simu-

Table 13–8. Comparison among cultivar-specific parameters (CSPs) of 'Williams' soybean. Default CSPs are those for "generic" MG 4 cultivar.

Source of CSPs	CSDL	PPSEN	EM-FL	FL-SD	SD-PM	LFMAX	SLAVR	SIZELF	WTPSD	SFDUR	SDPDV
Default	13.09	0.294	19.4	15.0	34.00	1.030	375.0	180.0	0.190	23.0	2.20
Hand-calibrated	13.40	0.285	19.0	13.8	32.20	1.000	385.0	180.0	0.180	26.0	2.40
GLUE-calibrated	13.67	0.342	18.7	15.4	36.23	1.004	425.0	143.1	0.174	23.4	2.23
values (SD)	(0.384)	(0.059)	(1.205)	(1.132)	(2.209)	(0.025)	(15.60)	(13.98)	(0.008)	(1.445)	(0.21)

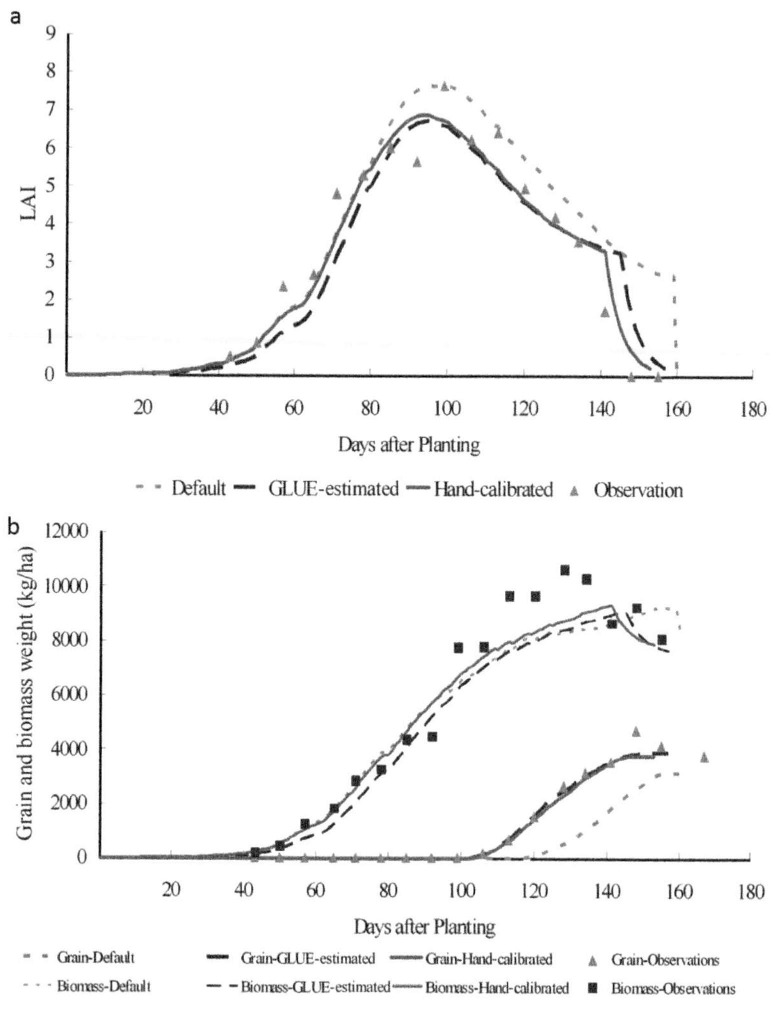

Figure 13–3. Graphs of observed and simulated results using the three sources of cultivar-specific parameters (CSPs) for (a) leaf area index and (b) total biomass and grain yield for the experiment in Wooster, OH in 1988. "Default" represents the outputs from default genetic coefficients, "GLUE-estimated" represents the outputs from GLUE-estimated genetic coefficients, and "Hand-calibrated" represents the outputs from hand-calibrated genetic coefficients by K.J. Boote.

lated results and errors in prediction were nearly equal showing that different "best" CSP sets might exist. However, the GLUE-estimated CSPs did not correctly predict LAI for the Gainesville, FL 1979 experiment. It is not surprising to see that the GLUE method did not improve time-series prediction because there were no time-series observations involved in estimating the CSPs in the GLUE approach used in this study.

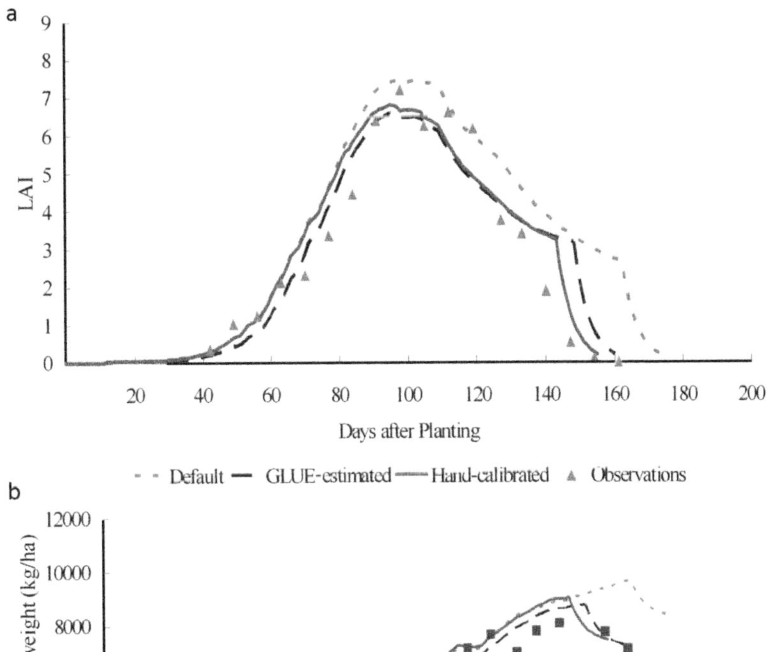

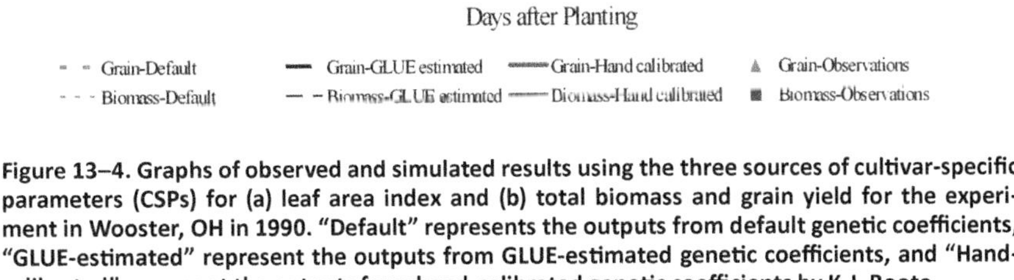

Figure 13–4. Graphs of observed and simulated results using the three sources of cultivar-specific parameters (CSPs) for (a) leaf area index and (b) total biomass and grain yield for the experiment in Wooster, OH in 1990. "Default" represents the outputs from default genetic coefficients, "GLUE-estimated" represent the outputs from GLUE-estimated genetic coefficients, and "Hand-calibrated" represent the outputs from hand-calibrated genetic coefficients by K.J. Boote.

Conclusions

The use of field observations to estimate CSPs of cropping system models is necessary for practical uses in predicting crop performance under different soil, climate, and management scenarios. Various methods can be used to estimate CSPs, each with advantages and disadvantages. We presented the GLUE method in this chapter that has been implemented for estimating CSPs for all crops in

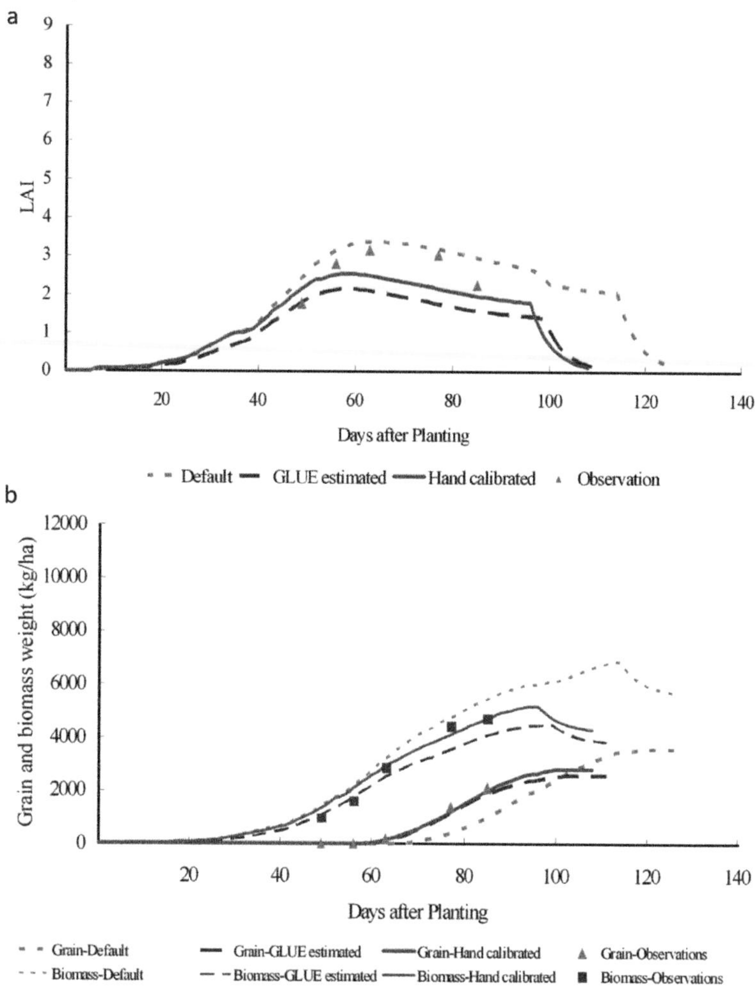

Figure 13–5. Graphs of observed and simulated results using the three sources of cultivar-specific parameters (CSPs) for (a) leaf area index and (b) total biomass and grain yield for the experiment in Gainesville, FL in 1979. "Default" represents the outputs from default genetic coefficients, "GLUE-estimated" represent the outputs from GLUE-estimated genetic coefficients, and "Hand-calibrated" represent the outputs from hand-calibrated genetic coefficients by K.J. Boote.

DSSAT v4.5, using field observations made once during a season. We showed that this method works well for two of the crops in DSSAT. However, there are important theoretical considerations that users should be aware of when using any method to estimate CSPs using field data. First, any calibration process may result in CSPs that are not generally applicable, particularly if the range of environments in which data were collected is narrow or if the crops experience drought, nutrient, or other stresses that are not adequately simulated by the model. In the first situation, having a narrow range of environments may result in parameters that work well only for those environments. For example, if datasets with a narrow range of daylengths are used, the CSPs that determine photoperiod response in soybean (CSDL and PPSEN) are not likely to be robust for use across locations with daylengths outside that range. In the second situation, it is likely that the estimation process will attempt to set CSP values to compensate for water, nutrient, or pest stresses if they occur and are not adequately simulated in the model. This problem can happen when the model does not include mechanisms for simulating those stresses, or when the soil parameters or initial conditions are not accurate. This is a main reason for recommending the use of irrigated and well-fertilized treatments for estimating CSPs. This is true for the GLUE and GENCALC methods available in DSSAT as well as any other procedure where the two described situations occur. Another principle that is important regardless of method used is that the model itself may not adequately represent the crop development and growth processes adequately. In this case, one may obtain estimates from a particular set of field data that work well for that set of data but may not be robust. Thus, it is important for users to evaluate any existing CSPs for their own conditions if they were estimated in other environments.

In this chapter, two cultivars were selected, one for each of two crops (maize and soybean), to evaluate the performance of GLUE program. It was shown that the GLUE-estimated CSPs for maize were different from the default values and from the manually estimated parameters. Similarly, the soybean CSPs estimated by the GLUE procedure were different from the arbitrary default set. However, soybean CSPs were similar to those manually estimated by an expert model user. Simulated outputs using the GLUE-estimated CSPs and hand-calibrated CSPs were similar to observed maize values. Both compared more favorably to observed maize values than those obtained from the default CSPs. For soybean, the simulated outputs were superior to the default set and equivalent to those obtained using the hand-calibrated CSPs. The average RAE values were all smaller than those for the default set and generally equal to those for the hand-calibrated CSPs in this study. There are limitations of the GLUE program as implemented in DSSAT. First, it does not include time-series measurements. As shown in the comparisons with time-series observations, the GLUE method did not always improve the accuracy of in-season model predictions, but it also did not deteriorate accuracy. This limitation can be overcome in future implementations of methods to estimate CSPs. Second, the GLUE program requires a large number of model runs, which may require more than 1 h on modern microcomputers. Nevertheless, implementation of the GLUE procedure in DSSAT v4.5 provides users an easy to use option for estimating CSPs from field observations for cultivars grown in their region.

Acknowledgments

This contribution was partially supported by the Office of Natural Resources Management and Office of Agriculture in the Economic Growth, Agriculture, and Trade Bureau of the U.S. Agency for International Development, under terms of Grant No. LAG-G-00-97-00002-00. Special appreciation is expressed to Dr. Michael D. Dukes of the Department of Agricultural and Biological Engineering, University of Florida for his help on this project.

References

Aronica, G., P.D. Bates, and M.S. Horritt. 2002. Assessing the uncertainty in distributed model predictions using observed binary pattern information within GLUE. Hydrol. Processes 16:2001–2016.

Baldi, P.F., M.C. Vanier, and J.M. Bower. 1998. On the use of Bayesian methods for evaluating compartmental neural models. J. Comput. Neurosci. 5:285–314.

Bannayan, M., and G. Hoogenboom. 2008. Weather analog: A tool for real-time prediction of daily weather data realizations based on a modified K-nearest neighbor approach. Environ. Model. Softw. 23:703–713.

Bertin, N., P. Martre, M. Génard, B. Quilot, and C. Salon. 2009. Under what circumstances can process-based simulation models link genotype to phenotype for complex traits? Case study of fruit and grain quality traits. J. Exp. Bot. 61:955–967, doi:10.1093/jxb/erp377.

Beven, K., and A. Binley. 1992. The future of distributed models: Model calibration and uncertainty prediction. Hydrol. Processes 6:279–298.

Beven, K., and J. Freer. 2001. Equifinality, data assimilation, and uncertainty estimation in mechanistic modeling of complex environmental systems using the GLUE methodology. J. Hydrol. 249:11–29.

Bhalla, U.S., and J.M. Bower. 1993. Exploring parameter space in detailed single neuron models: Simulations of the mitral and granule cells of the olfactory-bulb. J. Neurophysiol. 69:1948–1965.

Blasone, R.S., J.A. Vrugt, H. Madsen, D. Rosbjerg, B.A. Robinson, and G.A. Zyvoloski. 2008. Generalized likelihood uncertainty estimation (GLUE) using adaptive Markov Chain Monte Carlo sampling. Adv. Water Resour. 31:630–648.

Boote, K.J. 1999. Concepts for calibrating crop growth models. p. 179–200. *In* G. Hoogenboom, P.W. Wilkens, and G.Y. Tsuji (ed.) DSSAT Version 3. A Decision Support system for Agrotechnology Transfer. Vol. 4. University of Hawaii, Honolulu.

Boote, K.J., J.W. Jones, W.D. Batchelor, E.D. Nafziger, and O. Myers. 2003. Genetic coefficients in the CROPGRO–Soybean model: Links to field performance and genomics. Agron. J. 95:32–51.

Boote, K.J., M.J. Kropff, and P.S. Bindraban. 2001. Physiology and modeling of traits in crop plants: Implications for genetic improvement. Agric. Syst. 70:395–420.

Brazier, R.E., K.J. Beven, S.G. Anthony, and J.S. Rowan. 2001. Implications of model uncertainty for the mapping of hillslope-scale soil erosion predictions. Earth Surf. Process. Landf. 26:1333–1352.

Campbell, E.P., D.R. Fox, and B.C. Bates. 1999. A Bayesian approach to parameter estimation and pooling in nonlinear flood event models. Water Resour. Res. 35:211–220.

Eichler-West, R., and G. Wilcox. 1997. Robust parameter selection for compartmental models of neurons using evolutionary algorithms. *In* J.M. Bower (ed.) Computational neuroscience: Trends in research 1997. Plenum Publishing, New York.

Feyen, L., K.J. Beven, F. De Smedt, and J. Freer. 2001. Stochastic capture zone delineation within the generalized likelihood uncertainty estimation methodology: Conditioning on head observations. Water Resour. Res. 37:625–638.

Foster, W.R., L.H. Ungar, and J.S. Schwaber. 1993. Significance of conductances in Hodgkin-Huxley models. J. Neurophysiol. 70:2502–2518.

Franks, S.W., K.J. Beven, P.F. Quinn, and I.R. Wright. 1997. On the sensitivity of soil–vegetation–atmosphere transfer (SVAT) schemes: Equifinality and the problem of robust calibration. Agric. For. Meteorol. 86:63–75.

Franks, S.W., P. Gineste, K.J. Beven, and P. Merot. 1998. On constraining the predictions of a distributed model: The incorporation of fuzzy estimates of saturated areas into the calibration process. Water Resour. Res. 34:787–797.

Freer, J., K.J. Beven, and B. Ambroise. 1996. Bayesian estimation of uncertainty in runoff prediction and the value of data: An application of the GLUE approach. Water Resour. Res. 32:2161–2173.

Grimm, S.S., J.W. Jones, K.J. Boote, and J.D. Hesketh. 1993. Parameter estimation for predicting flowering date of soybean cultivars. Crop Sci. 33:137–144.

Hankin, B.G., R. Hardy, H. Kettle, and K.J. Beven. 2001. Using CFD in a GLUE framework to model the flow and dispersion characteristics of a natural fluvial dead zone. Earth Surf. Process. Landf. 26:667–687.

Hansson, K., and C. Lundin. 2006. Equifinality and sensitivity in freezing and thawing simulations of laboratory and in situ data. Cold Reg. Sci. Technol. 44:20–37.

He, J. 2008. Best management practice development with the CERES-Maize model for sweet corn production in North Florida. Ph.D. diss. Agricultural and Biological Engineering Dep., University of Florida, Gainesville.

He, J., M.D. Dukes, J.W. Jones, W.D. Graham, and J. Judge. 2009. Applying GLUE for estimating CERES-Maize genetic and soil parameters for sweet corn production. Trans. ASABE 52:1907–1921.

He, J., J.W. Jones, W.D. Graham, and M.D. Dukes. 2010. Influence of likelihood function choice for estimating crop model parameters using the generalized likelihood uncertainty estimation method. Agric. Syst. 103:256–264, doi:10.1016/j.agsy.2010.01.006.

Hoogenboom, G., J.W. Jones, C.H. Porter, P.W. Wilkens, K.J. Boote, W.D. Batchelor, L.A. Hunt, and G.Y. Tsuji. 2003. Decision Support System for Agrotechnology Transfer Version 4.0 [CD]. Univ. of Hawaii, Honolulu.

Hoogenboom, G., J.W. Jones, P.W. Wilkens, C.H. Porter, K.J. Boote, L.A. Hunt, U. Singh, J.L. Lisazo, J.W. White, O. Uryasev, F.S. Royce, R. Ogoshi, A.J. Gijsman, and G.Y. Tsuji. 2010. Decision Support System for Agrotechnology Transfer Version 4.5 [CD]. Univ. of Hawaii, Honolulu.

Hunt, L.A., S. Pararajasingham, J.W. Jones, G. Hoogenboom, D.T. Imamura, and R.M. Ogoshi. 1993. Gencalc: Software to facilitate the use of crop models for analyzing field experiments. Agron. J. 85:1090–1094.

Jawitz, J.W., R. Muñoz-Carpena, S. Muller, K.A. Grace, and A.I. James. 2008. Development, Testing, and Sensitivity and Uncertainty Analyses of a Transport and Reaction Simulation Engine (TaRSE) for Spatially Distributed Modeling of Phosphorus in South Florida Peat Marsh Wetlands. Scientific Investigations Rep. 2008-5029. USGS, Reston, VA.

Jensen, J.B. 2003. Parameter and uncertainty estimation in groundwater modeling. Ph.D. diss. Series Paper 23. Dep. of Civil Engineering, Aalborg Univ., Denmark.

Jones, J.W., G. Hoogenboom, C. Porter, K.J. Boote, W.D. Batchelor, L.A. Hunt, P. Wilkens, U. Singh, A. Gijsman, and J.T. Ritchie. 2003. The DSSAT model cropping system model. Eur. J. Agron. 18:235–265.

Jones, J.W., B.A. Keating, and C.H. Porter. 2001. Approaches to modular model development. Agric. Syst. 70:421–443.

Jones, J.W., J.B. Naab, D. Fatondji, K. Dzotsi, S. Adiku, and J. He. 2010. Uncertainties in simulating crop performance in degraded soils and low input production systems. p. 218–219. *In* M. van Ittersum et al. (ed.) Integrated assessment of agricultural and sustainable development: Setting the agenda for science and policy. (AgSAP 2009), 10–12 Mar. 2009. Egmond aan Zee, The Netherlands.

Jones, J.W., G.Y. Tsuji, G. Hoogenboom, L.A. Hunt, P.K. Thornton, P.W. Wilkens, D.T. Imamura, W.T. Bowen, and U. Singh. 1998. Decision support system for agrotechnology transfer; DSSAT v3. p. 157–177. *In* G.Y. Tsuji et al. (ed.) Understanding options for agricultural production. Kluwer Academic Publ., Dordrecht, the Netherlands.

Lamb, R., K. Beven, and S. Myrab. 1998. Use of spatially distributed water table observations to constrain uncertainty in a rainfall–runoff model. Adv. Water Resour. 22:305–317.

Lyneis, J.M., and A.L. Pugh. 1996. Automated vs. hand calibration of system dynamics models: An experiment with a simple project model. p. 317–320. *In* G.P. Richardson and J.D. Sterman (ed.) Proceedings of the 1996 International System Dynamics Conference. System Dynamics Society, Cambridge, MA.

Madsen, H., G. Wilson, and H.C. Ammentorp. 2002. Comparison of different automated strategies for calibration of rainfall–runoff models. J. Hydrol. 261:48–99.

Makowski, D., J. Hillier, D. Wallach, B. Andrieu, and M.H. Jeuffroy. 2006. Parameter estimation for crop models. p. 101–150. *In* D. Wallach et al. (ed.) Working with dynamic crop models. Elsevier, Amsterdam.

Makowski, D., D. Wallach, and M. Tremblay. 2002. Using a Bayesian approach to parameter estimation: Comparison of the GLUE and MCMC methods. Agronomie 22:191–203.

Malakoff, D. 1999. Bayes offers a 'New' way to make sense of numbers. Science 286:1460–1464.

Mavromatis, T., K.J. Boote, J.W. Jones, A. Irmak, D. Shinde, and G. Hoogenboom. 2001. Developing genetic coefficients for crop simulation models with data from crop performance trials. Crop Sci. 41:40–51.

Mavromatis, T., K.J. Boote, J.W. Jones, G.G. Wilkerson, and G. Hoogenboom. 2002. Repeatability of model genetic coefficients derived from soybean performance trials across different states. Crop Sci. 42:76–89.

McMichael, C.E., A.S. Hope, and H.A. Loaiciga. 2006. Distributed hydrological modeling in California semiarid shrublands: MIKE SHE model calibration and uncertainty estimation. J. Hydrol. 317:307–324.

Mertens, J., H. Madsen, L. Feyen, D. Jacques, and J. Feyen. 2004. Including prior information in the estimation of effective soil parameters in unsaturated zone modeling. J. Hydrol. 294:251–269.

Monod, H., C. Naud, and D. Makowski. 2006. Uncertainty and sensitivity analysis for crop models. p. 55–99. *In* D. Wallach et al. (ed.) Working with dynamic crop models. Elsevier, Amsterdam.

Muleta, M.K., and J. Nicklow. 2005. Sensitivity and uncertainty analysis coupled with automatic calibration for a distributed watershed model. J. Hydrol. 306:127–145.

Müller, T., J. Magid, L.S. Jensen, and N.E. Nielsen. 2003. Decomposition of plant residues of different quality in soil—DAISY model calibration and simulation based on experimental data. Ecol. Model. 166:3–18.

Muñoz-Carpena, R., G.A. Fox, and G.J. Sabbagh. 2010. Parameter importance and uncertainty in predicting runoff pesticide reduction with filter strips. J. Environ. Qual. 39:1–12, doi:10.1016/10.2134/jeq2009.0300.

Muñoz-Carpena, R., Z. Zajac, and Y.-M. Kuo. 2007. Evaluation of water quality models through global sensitivity and uncertainty analyses techniques: Application to the vegetative filter strip model VFSMOD-W. Trans. ASABE 50:1719–1732.

Oliva, R. 2003. Model calibration as a testing strategy for system dynamics models. Eur. J. Oper. Res. 151:552–568.

Porter, C., J.W. Jones, and R. Braga. 2000. An approach for modular crop model development. International Consortium for Agricultural Systems Applications, Honolulu, HI.

R Development Core Team. 2009. R: A language and environment for statistical computing. R Foundation for Statistical Computing, Vienna, Austria.

Ritchie, J.T., J.R. Kiniry, C.J. Jones, and P.T. Dyke. 1986. Model inputs. p. 37–48. *In* C.A. Jones and J.R. Kiniry (ed.) CERES-Maize: A simulation model of maize growth and development. Texas A&M Univ. Press, College Station, TX.

Romanowicz, R., and K.J. Beven. 2006. Comments on generalized likelihood uncertainty estimation. Reliab. Eng. Syst. Saf. 91:1315–1321.

Romanowicz, R., K.J. Beven, and J. Tawn. 1994. Evaluation of predictive uncertainty in nonlinear hydrological models using a Bayesian approach. p. 297–317. *In* V. Barnett and K.F. Turkman (ed.) Statistics for the environment II. Water related issues. John Wiley and Sons, New York.

Romanowicz, R.J., K.J. Beven, and J. Tawn. 1996. Bayesian calibration of flood inundation models. p. 333–360. *In* M.G. Anderson et al. (ed.) Floodplain processes. John Wiley and Sons, New York.

Saltelli, A., S. Tarantola, F. Campolongo, and M. Ratto. 2004. Sensitivity analysis in practice: A guilde to assessing scientific models. John Wiley and Sons, Chichester, UK.

Shulz, K., K. Beven, and B. Huwe. 1999. Equifinality and the problem of robust calibration in nitrogen budget simulations. Soil Sci. Soc. Am. J. 63:1934–1941.

Tadesse, A., and E.N. Anagnostou. 2005. A statistical approach to ground radar–rainfall estimation. J. Atm. Ocean Technol. 22:1055–1071.

Timsina, J., and E. Humphreys. 2006. Performance of CERES-Rice and CERES-Wheat models in rice–wheat system: A review. Agric. Syst. 90:5–31.

Tsuji, G.Y., G. Uehara, and S. Balas. 1994. DSSAT V.3. University of Hawaii, Honolulu.

Uehara, G. 1998. Synthesis. p. 389–392. *In* G.Y. Tsuji et al. (ed.) Understanding options for agricultural production. Kluwer Academic Publ., Dordrecht, The Netherlands.

Vanier, M., and J. Bower. 1996. A comparison of automated parametersearching methods for neural models. *In* J.M. Bower (ed.) Proceedings of the 1995 Computational Neuroscience Conference (CNS*95), Monterey (CA). Academic Press, New York.

Wallach, D., B. Goffinet, J.E. Bergez, P. Debaeke, D. Leenhardt, and J.N. Aubertot. 2001. Parameter estimation for crop models: A new approach and application to a corn model. Agron. J. 93:757–766.

Wang, X., X. He, J.R. Williams, R.C. Izaurralde, and J.D. Atwood. 2005. Sensitivity and uncertainty analyses of crop yields and soil organic carbon simulated with EPIC. Trans. ASAE 48:1041–1054.

Chapter 14

The STICS Crop Model and Associated Software for Analysis, Parameterization, and Evaluation

S. Buis
D. Wallach
S. Guillaume
H. Varella
P. Lecharpentier
M. Launay
M. Guérif
J.-E. Bergez
E. Justes

Abstract

The STICS model simulates the behavior of soil–crop systems for a large range of crops. It is widely used for analyzing, evaluating, and designing cropping systems. Because of the complexity of such models, it is important to adapt methods and tools for analyzing, improving and evaluating them. This chapter presents software providing these services for the STICS model. Its main features are: (i) handling of multiple simulations, (ii) uncertainty and sensitivity analysis, (iii) parameter estimation, and (iv) comparison with observed data. Parameter estimation of crop models presents particular problems. The choice of an appropriate statistical model for model error is one of them. The software proposes options for choosing among a collection of possible assumptions concerning model error including zero or non-zero expectations, independent or correlated errors, known or unknown variances...These options are available both for the frequentist and Bayesian parameter estimation methods implemented. The software also includes special features specifically adapted to crop models, such as automatic procedure for parameter estimation in several stages and simultaneous estimation of varietal and specific parameters. Numerous indicators are provided to help the modeler analyze results. This software thus has two major objectives: (i) to help modelers using and improving the model and (ii) to allow one to compare various methods for parameter estimation on real problems. In this chapter we detail the design, features, and ways of using the software. In closing, we present examples of sensitivity analysis and parameter estimation applications with the STICS crop model.

S. Buis (samuel.buis@avignon.inra.fr), P. Lecharpentier, and M. Guérif, INRA, UMR 1114 EMMAH, Domaine Saint Paul- Site Agroparc, 84914 Avignon Cedex 9, France; D. Wallach, S. Guillaume, J.-E. Bergez, and E. Justes, INRA, UMR 1248 AGIR, BP 52627, 31326 Castanet Tolosan Cedex, France; S. Guillaume, Arvalis-Institut du végétal, 6 chemin de la côte vieille, 31450 Baziege, France; H. Varella, Météo-France, CNRM-GMAP, 42 av Gaspard Coriolis, 31057 Toulouse, France; M. Launay, INRA, Unité Agroclim, Domaine Saint Paul- Site Agroparc, 84914 Avignon Cedex 9, France.

doi:10.2134/advagricsystmodel2.c14

Methods of introducing system models into agricultural research. L.R. Ahuja and L. Ma (ed.)
Advances in Agricultural Systems Modeling Series 2.

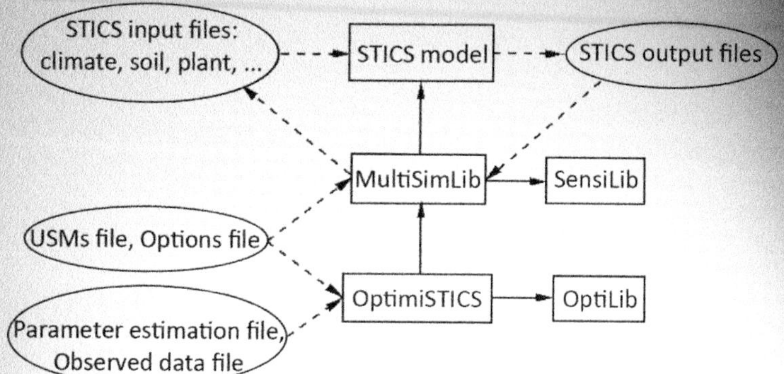

STICS input files: climate, soil, plant, ...
STICS model
STICS output files
MultiSimLib
SensiLib
USMs file, Options file
OptimiSTICS
OptiLib
Parameter estimation file, Observed data file

The STICS model (Brisson et al., 1998, 2009) simulates the behavior of the soil–crop system over one crop cycle or a sequence of several crop cycles, using a daily time step. STICS is an acronym that stands for "Simulateur mulTIdisciplinaire pour les Cultures Standard." It can simulate a wide range of crops, which is important when studying a range of cropping systems. The model takes advantage of the fact that simulating different crops involves a limited number of processes that are common to many or all crops. The basic processes simulated are described in the following section.

Because of the complexity of crop models, it is important to consider methods for analyzing (through uncertainty and sensitivity analysis), improving (through parameter estimation), and evaluating these models. Many methods for sensitivity analysis, parameterization, and evaluation exist and are available as general purpose computer packages. However, these packages often require that the models to be studied have a specific form, for example, as callable subroutines, which would require reprogramming. Furthermore, as we will see below, there are specific aspects of crop models that are often not taken into account in these packages. Finally, using a crop model can itself be quite a complicated task. It is difficult to require that users also make the effort of coupling the crop model with a battery of other programs. These considerations have led us to develop dedicated software for providing these services for the STICS model. More functionalities are still being added. The software is freely available for research purposes on request to the authors.

Parameter estimation presents particular problems. In general one makes hypotheses about the statistical properties of measured data and model error and from that derives the appropriate goodness-of-fit criterion for model calibration. For crop models, however, the statistical properties of the error are usually not clear. Our software therefore makes available a collection of different goodness-of-fit criteria, corresponding to different hypotheses.

In the following section, we describe briefly the STICS model. Then we present the different methods of sensitivity analysis, parameter estimation, and model evaluation that are furnished by the software. The software itself is described next. Finally we present examples of sensitivity analysis and parameter estimation done with the software.

The STICS Model

STICS is organized into modules and submodules. As the examples of sensitivity analysis and parameter estimation applications detailed in this chapter concern a wheat (*Triticum aestivum* L.) crop, we briefly describe in the next subsections the main modules for the case of this crop. Among the large number of parameters involved in the different equations, only the parameters we focused on in the examples are presented, which in this case only concern plant parameters (soil parameters should be addressed also, as in Guérif et al., 2006; Varella et al., 2010b). For a more detailed description, see Brisson et al. (1998, 2002, 2003, 2009).

The required input variables for STICS are permanent soil characteristics (texture, physical and chemical properties, water-holding capacity), daily climate (e.g., maximum and minimum temperature, solar radiation), initial conditions (initial soil water and soil inorganic N), and crop management (tillage depth and date, species and variety sown, sowing date and density, dates and amounts of irrigation and fertilization, harvest date). In addition, there are optional input variables. The dates at which the various development stages are realized can be specified as input variables when available or can be calculated by the model. Also, daily values of leaf area index (LAI) can be specified as inputs. In general one only has a few LAI measurements during a growing season. STICS offers an automatic nonlinear interpolation to obtain daily values based on the few measured LAI values. If LAI is not provided, it is calculated by the model.

A specific combination of input variables—climatic data, initial soil, crop input data, parameters for soil and crop functioning—is termed a *USM* (Unit of SiMulation). Each USM corresponds to one execution of the STICS model.

In the following, the parameters assumed to be varietal are identified by a subscript v, meaning that they may have different values for different varieties. Otherwise, the same value applies for all varieties of a given crop.

Development Module

Two phenology scales are used: one for vegetative growth and the other for grain filling. During vegetative growth, the model calculates normalized leaf development units, ULAI. ULAI is equal to 1 at emergence and 3 at maximum leaf growth. The increase in development units is mainly driven by air temperature, but vernalization, photoperiod, and water and nitrogen stresses can slow down or accelerate development.

Root Growth Module

Daily increase in rooting depth is a linear function of soil temperature and a varietal parameter $CROIRAC_v$. Water stress or excessive bulk density reduce the rate of root depth increase. The increase in root depth stops when soil depth or an obstacle is reached.

The daily increase in total length of roots is calculated as the sum of two components. The first is root length growth above the rooting front. It is calculated as a function of development stage using a logistic function analogous to that used for describing leaf growth dynamics (see Eq. [2] below). The asymptotic value is given by the parameter DRACLONG. Temperature, plant density and anoxia influence this growth rate. The second component is growth at the rooting front, which depends on the rate of increase of the rooting depth and the root density at the root front parameter LVFRONT.

Each day the additional root length is distributed among soil layers. Each layer can have constraints that limit the additional amount of root length assigned to that layer. Constraints include soil dryness, soil compaction (as measured by bulk density), and waterlogging.

After a species-dependent lifetime, the roots senesce and enter the mineralization process as crop residue at the end of the crop cycle.

Shoot Growth Module

Leaf Dynamics Submodule

The increase in leaf area on day *d*, DELTAI(*d*), is calculated as a product of factors that depend on development stage ($\text{DELTAI}_{\text{dev}}$), crop temperature ($\text{DELTAI}_{\text{T}}$), plant density ($\text{DELTAI}_{\text{dens}}$), and stresses ($\text{DELTAI}_{\text{stress}}$):

$$\text{DELTAI}(d) = \text{DELTAI}_{\text{dev}}(d)\ \text{DELTAI}_{\text{T}}(d)\ \text{DELTAI}_{\text{dens}}(d)\ \text{DELTAI}_{\text{stress}}(d) \qquad [1]$$

The development stage factor $\text{DELTAI}_{\text{dev}}$ is a logistic function of development units (ULAI), with asymptote DLAIMAXBRUT up to ULAI = 3. Denominator terms are PENTLAIMAX, the slope at the inflection point, and VLAIMAX, the value of ULAI at the inflection point.

$$\text{DELTAI}_{\text{dev}}(d) = \frac{\text{DLAIMAXBRUT}}{1+\exp\left\{\text{PENTLAIMAX}\left[\text{VLAIMAX-ULAI}(d)\right]\right\}} \qquad [2]$$

For ULAI greater than three, $\text{DELTAI}_{\text{dev}}$ is null.

The other factors express limiting factors, with values lying between 0 and 1. DELTAI_{T} represents the effect of crop temperature, modeled as a piecewise linear function. The plant density factor $\text{DELTAI}_{\text{dens}}$ represents the reduction of LAI increase due to plant competition after a defined LAI threshold. Water and N status affect leaf growth rate through the stress factor $\text{DELTAI}_{\text{stress}}$.

Senescence is based on the concept of lifespan. The leaf area and part of the biomass produced on a given day are lost once the lifespan has elapsed. The basic lifespan depends on the development stage—it varies as a linear function of development units between the duration of first leaves (DURVIEI) and that of last leaves ($DURVIEF_v$). Water and N stresses shorten the lifespan, while overfertilization (with INN > 1, see below) increases lifespan. Temperature affects the course of senescence, with the same minimal, optimal, and maximal temperatures as for development.

Biomass Production Submodule

The daily production of shoot biomass, DLTAMS(*d*) is given by

$$\mathrm{DLTAMS}(d)=\left[\mathrm{EBMAX}\ \mathrm{RAINT}(d)-\mathrm{COEFB_G}\ \mathrm{RAINT}(d)^2\right] \\ \mathrm{FTEMP}(d)\ \mathrm{SWFAC}(d-1)\ \mathrm{INNS}(d-1)\ \mathrm{EXOBIOM}(d-1)\ \mathrm{FCO2} \quad [3]$$

The term in brackets is a parabolic function of radiation interception in day *d*, RAINT(*d*), which is calculated using Beer's law. This implicitly assumes a homogeneous canopy with randomly distributed leaves. The calculation of DLTAMS(*d*) also takes into account various factors that influence photosynthesis and respiration, namely water (SWFAC) and N (INNS) stresses, temperature (FTEMP), anoxia (EXOBIOM), and CO_2 concentration (FCO2). EBMAX is radiation use efficiency, which can differ with development stage due to the different chemical composition of different organs. There are thus three radiation use efficiency parameters, EFCROIJUV, EFCROIVEG, and EFCROIREPRO for the juvenile, vegetative, and reproductive phases, respectively.

The mass corresponding to the chaff depends linearly on the number of grains through a coefficient ENVFRUIT. This is transformed into an equivalent leaf surface by using a specific "ear" area SEA.

Yield Formation Module

The number of grains for wheat depends linearly on the mean growth rate of the canopy during the 30-d period preceding the onset of grain filling, thus taking into account the effects of the various stresses during this period. The slope of the linear relation is given by the parameter CGRAIN. Two other parameters define the limits of variation of the number of grains.

The quantity of dry matter accumulated in grains is calculated as the product of a potential harvest index function and plant dry weight. This potential harvest index increases linearly with time from the onset of filling to maturity, with a slope VITIRCARB, and cannot exceed a maximum value. The actual grain filling dynamics are affected by suboptimal temperatures and high temperatures; there is no carbon accumulation in grains at temperatures above the threshold TMAXREMP. The weight per grain cannot exceed a varietal maximum value.

The harvest index concept is also applied to N accumulation in the seeds. The parameter VITIRAZO is analogous to VITIRCARB. It determines the daily increase in the N harvest index. As for weight increase, grain N filling is affected by nonoptimal temperatures.

Water Balance Module

Crop water requirements are calculated as a fraction of potential evapotranspiration TETP, with a crop coefficient approach. The fraction is given by a logistic function of LAI, which involves the maximum crop coefficient KMAX.

$$\mathrm{EO}(d) = \mathrm{TETP}(d)\left(1 + \frac{\mathrm{KMAX} - 1}{1 + \exp\left\{-1.5\left[\mathrm{LAI}(d) - 3\right]\right\}}\right) \qquad [4]$$

The quantity of water actually transpired by the crop depends on both the quantity of water available in the soil and the capacity of the plant to extract this water. The water stress index is calculated as the ratio between actual and potential transpiration.

Nitrogen Balance Module

Nitrogen uptake is the smaller of soil availability and crop demand. The crop demand as well as the "optimal" N content are defined by dilution curves as proposed by Lemaire and Gastal (1997) and adapted to wheat and durum wheat [*Triticum turgidum* L. subsp. *durum* (Desf.) Husn.] by Justes et al. (1994, 1997). These curves relate the N concentration in shoots with plant biomass. A critical N concentration (NC) can be defined below which the plant metabolism is affected. The corresponding critical dilution curve can be used to diagnose the N nutrition status of a crop: points below the curve express N deficiency, while points above the curve express no limitation of growth by N (Justes et al., 1994). A maximum dilution curve, corresponding to maximal crop N content NMAX, is defined and allows determining the daily N demand as the product of the crop growth rate and the derivative of NMAX.

Nitrogen uptake depends on root distribution in the soil and N demand. Uptake depends on the specific absorption capacity of the plant VABS, which increases with nitrate concentration CONCN according to double Michaelis–Menten kinetics, corresponding to two types of transport system, a high affinity transport system and a low affinity system. The specific absorption capacity involves two parameters for each system, the constant of nitrate affinity KMABS1 or KMABS2 and the maximal nitrate uptake rate VMAX1 or VMAX2.

$$\mathrm{VABS} = \frac{\mathrm{VMAX1\ CONCN}}{\mathrm{KMABS1\ CONCN}} + \frac{\mathrm{VMAX2\ CONCN}}{\mathrm{KMABS21\ CONCN}} \qquad [5]$$

A nitrogen nutrition index, INN, is calculated as the ratio of actual plant N content to critical plant N content corresponding to the same biomass (MASECABSO). An instantaneous INN named INNI is also calculated based on daily accumulation of nitrogen VABSN, nitrogen dependent biomass DELTABSO, and the ratio between daily variation of critical N content *d*NC and the corresponding biomass *d*MASECABSO.

$$\text{INNI} = \frac{\text{VABSN}(d)}{\text{DELTABSO}(d)\left[\dfrac{d\text{NC}(d)}{d\text{MASECABSO}(d)}\right]} \qquad [6]$$

Three different N stress indices are calculated. They reduce radiation use efficiency (INNS) and LAI growth (INNLAI), and accelerate senescence (INNSENES). Their values depend on the value of the variable INNI and of the parameters INNmin, INNimin, and INNSEN or INNTURGMIN.

$$\text{INNS} = \frac{\text{1-INNmin}}{\text{1-INNimin}}\text{INNI} + \frac{\text{INNimin} - \text{INNmin}}{\text{INNimin} - 1} \qquad [7]$$

$$\text{INNLAI} = \max\left[\text{INNmin}, \frac{\text{INNTURGMIN} - 1}{\text{INNmin} - 1}\text{INNS} + \left(1 - \frac{\text{INNTURGMIN} - 1}{\text{INNmin} - 1}\right)\right] \qquad [8]$$

$$\text{INNSENES} = \max\left[\text{INNmin}, \frac{\text{INNSEN} - 1}{\text{INNmin} - 1}\text{INNS} + \left(1 - \frac{\text{INNSEN} - 1}{\text{INNmin} - 1}\right)\right] \qquad [9]$$

The N demand by storage organs depends on the N status of the crop. Demand for N is increased if INNS is nearly optimal (between INNgrain1 and INNgrain2).

Methods

Uncertainty Analysis, Sensitivity Analysis

Uncertainty and sensitivity analysis methods are important tools when constructing or using a model. They can be used for various objectives, including exploration and evaluation of model behavior, identification of the parameters or input variables that can be fixed at a nominal value or on the contrary must be accurately estimated or measured, and simplification of the model.

Uncertainty analysis quantifies the variability in a model output variable that results from uncertainty in one or several input factors (a factor can be a parameter, an input variable or an option that affects model structure). It involves four steps: (i) definition of the distribution of each uncertain input factor, (ii) generation of a sample of values from these distributions, (iii) simulation of the corresponding values of the output variables, and (iv) computation of statistics (mean, variance, quantiles…) for the sample of values of the output variables.

Sensitivity analysis is the study of how the total variation in an output variable can be apportioned to the different input factors. Various methods exist for doing sensitivity analysis (Saltelli et al., 2000). The extended FAST (Fourier Amplitude Sensitivity Test) method (Saltelli et al., 1999) is implemented in our software. This method has the advantage of being global (the effect of one factor is evaluated on average for the possible values of the other factors) and model free (no assumptions are made concerning the behavior of the model). This is a variance-based method: the variance of model output is the basis for the calculated sensitivity indices. The indices represent the proportion of the output variance explained by the different factors. The method calculates, for each factor, first-order and total sensitivity indices (which include all interactions with other factors). The computation of these indices typically requires several thousand model simulations, which is acceptable for a model having low computational cost like STICS.

Parameter Estimation

It is a common practice to calibrate crop models for each new range of situations or for a new crop or cultivar. This involves fitting the model to data from those situations.

A first step is choosing the parameters to estimate. Crop models typically have a large number of parameters and it is not possible to fit them all. The choice of the most important parameters to adjust is generally based on a combination of expert knowledge of the model and sensitivity analysis. Attention should be paid to relations between parameters with respect to available observed variables since compensatory effects between parameters may lead to high variances for their estimators.

A second step concerns the statistical aspects of parameter estimation. Parameter estimation of a crop model can be treated as a regression problem; it is thus necessary to define a statistical model of error. This aspect of crop models has received little attention. Below we propose a range of hypotheses about model error, adapted to the case of crop models, and present the equations for parameter estimation that result. Both frequentist and Bayesian methods are considered here. All these methods and options are implemented in our software.

Frequentist Methods

Independent Errors, Single Response Variable

Consider first the simple case where all model errors ε_{ijk} are independent and identically distributed (iid), with a normal distribution and zero expectation and with the same variance. The statistical model is

$$Y_{ijk} = f_{jk}\left(X_i;\boldsymbol{\theta}_0 \right) + \varepsilon_{ijk}$$
$$\varepsilon_{ijk} \underset{\text{iid}}{\sim} N\left(0,\sigma^2 \right) \quad [10]$$

Y_{ijk} is the observed value of output variable j at the kth measurement date in USM i, $f_{jk}(X_i;\theta)$ is the corresponding prediction of STICS, which depends on X_i, which represents all the input variables of STICS for USM i, and $\boldsymbol{\theta}_0$, which is the vector of parameters to be estimated.

The assumption of constant variance may be reasonable in the case of a single measured value per USM, for example, yield. In the case of time-dependent data, for example, LAI and biomass, a log transformation would probably be necessary to achieve constant variance (Sokal and Rohlf, 1981). Given the assumptions of Eq. [10], the maximum likelihood estimator is the same as the least squares estimator (Seber and Wild, 1989) and is given by

$$\hat{\theta} = \underset{\theta}{\arg\min} \sum_{ijk} \left[Y_{ijk} - f_{jk}\left(X_i;\boldsymbol{\theta}_0 \right) \right]^2 \quad [11]$$

The parameter estimator is the parameter vector that minimizes the sum of squared errors.

Independent Errors, Several Response Variables, Known Variances

Suppose now that we maintain the assumption that all errors are independent, but we now assume that we want to simultaneously fit several different response variables. Each response variable can have a different variance of model error. The statistical model is:

$$Y_{ijk} = f_{jk}\left(X_i;\boldsymbol{\theta}_0 \right) + \varepsilon_{ijk}$$
$$\varepsilon_{ijk} \underset{\text{iid}}{\sim} N(0,\sigma_j^2) \quad [12]$$

Here σ_j^2 is the error variance for response variable j. The maximum likelihood estimate of $\boldsymbol{\theta}_0$ is then

$$\hat{\theta} = \underset{\theta}{\arg\min} \sum_j \left[1/\left(2\sigma_j^2 \right) \right] \sum_{i,k} \left[Y_{ijk} - f_{jk}\left(X_i;\theta \right) \right]^2 \quad [13]$$

The goodness-of-fit criterion is now a weighted sum of squares, and $1/\sigma_j^2$ is the weight for response variable j.

If one has estimated values for the error variances, they can be substituted into the above equation. For example, Van Oijen et al. (2005) took the model error variance as equal to the (known) measurement error variance.

Independent Errors, Several Response Variables, Unknown Variances

If one does not have an estimate of the σ_j^2, one can use the method of concentrated likelihood (Seber and Wild, 1989). The maximum likelihood estimator of θ is then:

$$\hat{\theta} = \arg\min_{\theta} \prod_{j} \left\{ \left(1/n_j \right) \sum_{i,k} \left[Y_{ijk} - f_{jk}\left(X_i ;\theta \right) \right]^2 \right\}^{n_j/2} \quad [14]$$

where n_j is the total number of measurements of response variable j. This expression allows one to combine information from different response variables, without having to input the weight to each. The variances are estimated as

$$\hat{\sigma}_j^2 = \left(1/n_j \right) \sum_{i,k} \left[Y_{ijk} - f_{jk}\left(X_i ;\hat{\theta} \right) \right]^2 \quad [15]$$

Independent Errors, Several Response Variables, Non-zero Expectation

Equations [13] and [14] assume that the same parameters would lead to an error with zero expectation for every measured response variable, which is a strong assumption. Suppose that there is no such value of $\boldsymbol{\theta}_0$. The statistical model is said to be misspecified (White, 1981). In this case, parameter values minimizing prediction error for the different observed response variables can be different.

This leads us to define the estimator

$$\hat{\theta} = \arg\min_{\theta} \sum_{j} w_j \sum_{i,k} \left[Y_{ijk} - f_{jk}\left(X_i ;\theta \right) \right]^2 \quad [16]$$

The goodness-of-fit criterion here is still a weighted sum of squares, but the weights w_j are chosen by the user to reflect his priorities. The greater the weight given to response j, the greater the priority given to improving prediction accuracy for that response.

Correlated Errors

The model errors for multiple measurements in the same USM are probably correlated. A general procedure in this case is generalized least squares. However, that involves the variance–covariance matrix of model errors, which is not known and would in general be very difficult to estimate in the case of crop models.

The practical question here is how to weight USMs with different numbers of measurements. If all the errors were independent, then one would use ordinary or weighted least squares. USMs with many measurements would contribute many terms to the sum of squares and thus would have more influence than USMs with few measurements. On the other hand, if there is very high correlation between errors, then additional measurements in the same USM do not add much extra information, so roughly speaking each USM should receive the same weight regardless of the number of measurements.

This led Wallach et al. (2001) to propose a simplified approach in the case of high correlations. It is still assumed that errors in different USMs or for different response variables in the same USM are independent. However, the sum of squared errors for a response *j* in USM *i* is replaced by the average squared error for response *j* in USM *i*. In Eq. [13–16] then we make the replacement

$$\sum_{i,k}\left[Y_{ijk}-f_{jk}(X_i;\theta)\right]^2 \rightarrow \sum_{i}\left\{\left(1/n_{ij}\right)\sum_{k}\left[Y_{ijk}-f_{jk}\left(X_i;\theta\right)\right]^2\right\} \qquad [17]$$

where n_{ij} is the number of measurements of output variable *j* in USM *i*. For example Eq. [11] gives

$$\hat{\theta}=\underset{\theta}{\arg\min}\sum_{j}\sum_{i}\left\{\left(1/n_{ij}\right)\sum_{k}\left[Y_{ijk}-f_{jk}\left(X_i;\theta\right)\right]^2\right\} \qquad [18]$$

Also, in Eq. [14] and [15], we replace the total number of measurements of output variable *j*, n_j, by the number of USMs where *j* is measured.

Bayesian Approach

In a Bayesian approach the parameter is treated as a random variable and one seeks to determine its probability distribution, called the posterior distribution. The uncertainty in the estimated parameter is thus central in this approach. An advantage of the Bayesian approach is that it uses prior information about the parameter values. In the case of crop models, this would be information based on studies of individual processes. Jansen and Hagenaars (2004) have discussed the advantages of a Bayesian approach for crop models, and Van Oijen et al. (2005) applied a Bayesian approach to a forest model. The GLUE (generalized likelihood uncertainty estimation) approach is similar in that it also focuses on the uncertainty in the estimated parameter values (Beven, 2006; Makowski et al., 2002; Beven et al., 2008). This method has been applied in (Guérif et al., 2006) for estimating soil parameters for the STICS crop model.

The Bayesian method implemented is Sampling Importance Resampling (Carlin and Louis, 1996). A sample of values is generated from the parameter prior distribution defined by the user. For each value of the parameter vector in the sample, the likelihood is calculated according to one of the statistical models presented in the section "Uncertainty Analysis, Sensitivity Analysis." The corresponding likelihoods are not detailed here for the sake of brevity. The likelihood value is the weight assigned to that parameter vector. The weighted sample is an approximation to the posterior distribution. Since it is more convenient to work with an unweighted sample, there is a resampling step. A new sample is drawn with replacement from the approximate posterior, of size equal to that of the original sample.

Evaluation

Many different criteria have been proposed for evaluating crop models. Each brings out a somewhat different aspect of model quality, so our software calculates a range of criteria.

A common measure of model quality is mean squared error (MSE). MSE for response variable *j* is

$$\mathrm{MSE} = \left(1/n\right) \sum_{ik} \left[Y_{ijk} - f_{jk}\left(X_i;\theta\right)\right]^2 \qquad [19]$$

where n is the total number of measurements of that variable. When MSE is calculated using the same data as those used for calibration, MSE measures the quality of adjustment of the model to the data. When MSE is calculated using independent data, it is a measure of prediction error. It is often convenient to work with the root mean squared error, $\mathrm{RMSE} = \sqrt{\mathrm{MSE}}$, since this has the same units as the output variable in question. The software also computes the decomposition of MSE into two components: a systematic and an unsystematic part, as proposed in Willmott (1981). Another decomposition of MSE used in the example presented "Parameter Estimation" below, is into three components, namely the squared bias (bias^2), the squared difference between the standard deviation of the measurements and the standard deviation of the predictions (SDSD) and a residual term that depends on the correlation between predicted and observed values (LCS) (Wallach, 2006).

Other useful criteria are proposed by the software: the regression coefficients and R^2 when simulated values are expressed as a linear function of observed values, and the modeling efficiency, defined as

$$\mathrm{eff} = 1 - \frac{\sum_{ik} \left[Y_{ijk} - f_{jk}\left(X_i;\theta\right)\right]^2}{\sum_{ik}^{n} \left[Y_{ijk} - \overline{Y}_j\right]^2} \qquad [20]$$

where $\overline{Y}_j$ is the average of the measured values for output variable *j*. A perfect model has eff = 1, while eff ≤ 0 means that the model is not useful for describing the variability between different situations. It is an equivalent (eff = 0) or worse (eff < 0) predictor than the average of observed values.

Software for Multiple Simulations, Analysis, and Parameter Estimation for The STICS Crop–Soil Model

Software Design

The software design and usage are illustrated in Fig. 14–1. The software is a library of functions implemented in Matlab (Mathworks, Natick, MA). It is divided into several packages:

- The OptiLib and SensiLib packages contain the implementation of the mathematical methods used for uncertainty analysis, sensitivity analysis, model evaluation, and parameter estimation plus some functions for generating graphics. These packages are fully independent of the model used, and thus the methods implemented can be applied to other crop models interfaced with Matlab.
- MultiSimLib contains functions for the management of the STICS simulations and their inputs and outputs plus functions for automating multiple simulations, uncertainty analysis, and sensitivity analysis for the STICS model. The STICS model must be installed in this library. For each USM to simulate, the executable version of the model is run via system calls. Inputs are furnished to the model and outputs are recovered via its input and output ASCII files.
- OptimiSTICS contains functions for automating parameter estimation for the STICS model and comparison between STICS results and measured data (Justes et al., 2009).

The software has been tested on Windows and Unix-like environment and works with Matlab V7.2 or later.

Using the Software

The software does not contain a graphical user interface. One uses it by entering the Matlab environment and typing in command line mode the main function of MultiSimLib or OptimiSTICS corresponding to the operation one wants to perform. These operations are detailed in the next section.

The information required by MultiSimLib or OptimiSTICS is organized in input files:

- STICS input files: climate, soil, crop management and plant parameters input files required by STICS for all USMs to be simulated.
- Observed data files: comma-separated values (CSV) files containing observed data (one file per USM). This is required if the user wants to perform parameter estimation and/or comparisons between STICS results and observations.
- USMs file: A CSV file describing each USM to be simulated. For each USM one must specify the corresponding available soil, weather, plant, and crop management STICS input files and, if available, the corresponding observed

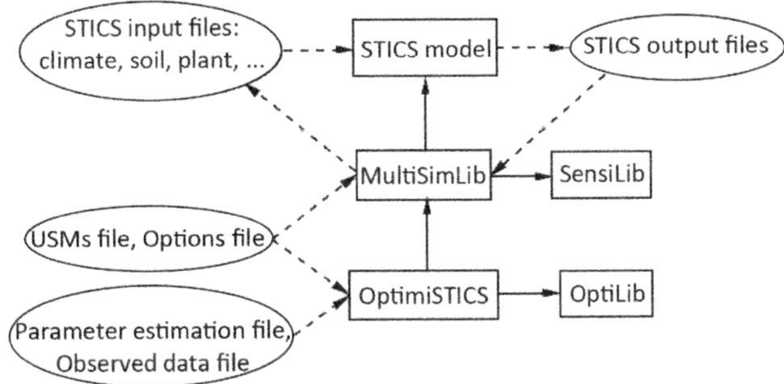

Fig. 14–1. Design and use of the software. Rectangles represent the different subpackages and ellipses the different files. Output files of the software are not detailed here for the sake of simplicity. Plain arrows represent dependences between subpackages and dashed arrows read or write operations on files.

data file, plus the initial and final dates of simulation and the initial values of the STICS state variables.

- Parameter estimation file: A CSV file that describes the different stages of the parameter estimation process (see "Parameter Estimation" sections).
- Options file: An XML file that specifies information and options for OptimiSTICS and MultiSimLib (name and path of the input and output files, options for sensitivity and parameter estimation methods...).

Execution of the main functions produces general information for the user written in the Matlab window and also saved in an ASCII file. The numerical results produced by the main functions are saved in CSV, EXCEL and/or MAT (Matlab) files, and the graphics are saved in FIG (Matlab), postscript, and/or PDF files. When errors occur during execution, a directory is generated that contains the error messages and the model input and output files of the corresponding simulations.

Main Features

A summary of the software capability is shown in Table 14–1.

Multiple Simulations

MultiSimLib allows one to automatically perform the simulations of the STICS model for a list of USMs. The user can control output by specifying the calculated variables and the dates for output in the options file.

Uncertainty and Sensitivity Analysis

For uncertainty analysis, the list of factors, the probability distribution describing the uncertainty of each factor, and the list of output variables are given in the options file. MultiSimLib supports various choices for the method for sam-

Table 14–1. Software capabilities.

Function	Main choices	
Multiple simulations	List of USMs	
	Response variables and dates	
Uncertainty and sensitivity analysis	Factors (parameters and/or input variables)	
	Distribution of each factor	
	Sampling scheme	
	Sample size	
	Response variables and dates	
Parameter estimation	Frequentist	Parameters
		Parameter bounds
		Response variables
		List of USMs
		Statistical model of error
		Log transformation
		Options for numerical minimization
		Special features
		· multiple stages
		· varietal and nonvarietal parameters
	Bayesian	Parameters
		Prior distributions
		Response variables
		List of USMs
		Statistical model of error
		Log transformation
		Sample size
		Special features
		· multiple stages
		· varietal and non varietal parameters
Evaluation	List of USMs	

pling from the distributions of the inputs: full factorial design, Latin Hypercube Sampling or Quasi-random LP_T sampling (Helton and Davis 2000). The result is a sample of values of the output variables that corresponds to the sample of values of the input factors. This distribution can then be characterized using Matlab graphical and statistical tools.

The MultiSimLib main function for sensitivity analysis generates the extended FAST experimental design, performs the corresponding model simulations, and computes main and total sensitivity indices for the input factors and output variables specified in the options file. That file must also contain the probability distribution for each input factor.

Parameter Estimation

The process of parameter estimation implemented in the software is illustrated in Fig. 14–2.

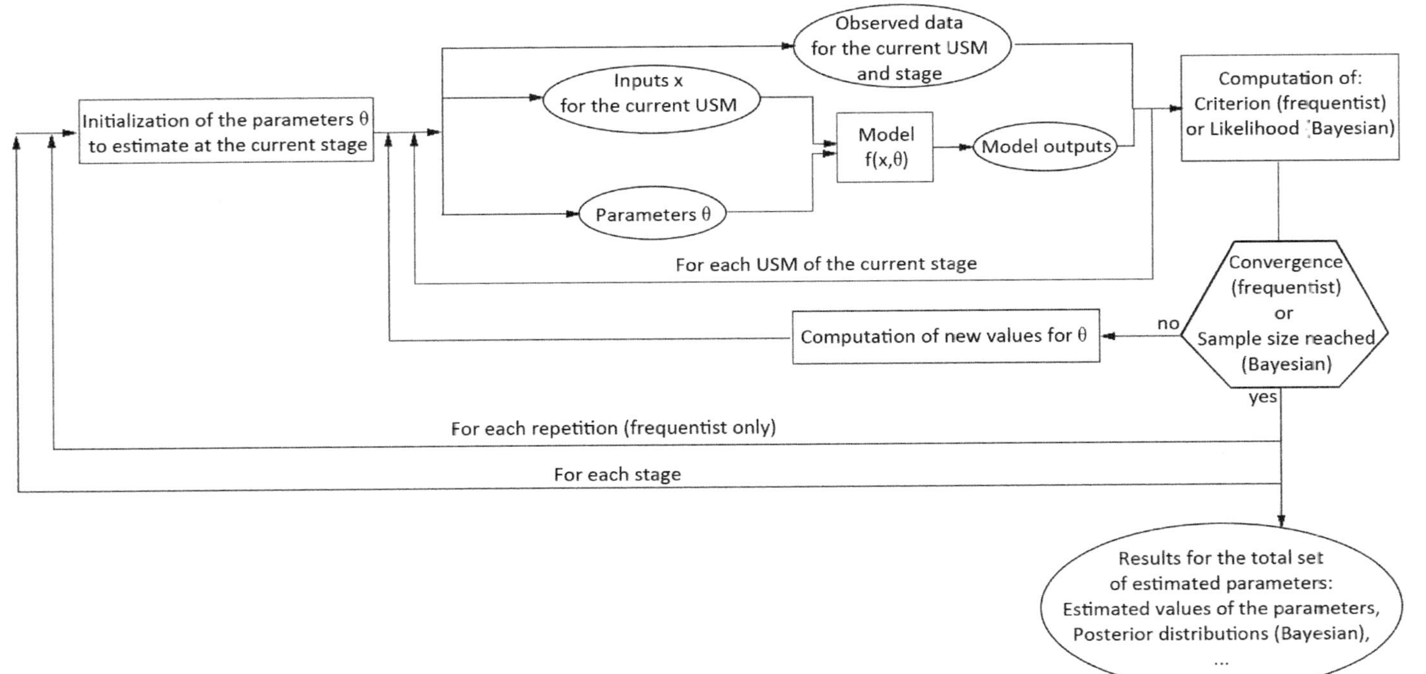

Fig. 14–2. Parameter estimation process implemented in the software.

All the frequentist estimators presented in the previous section are implemented. In each case, the frequentist estimator involves minimizing a goodness-of-fit criterion. Crop models often have discontinuities that make it difficult to use gradient-based minimization methods such as Gauss–Newton or Levenberg–Marquardt. A common approach, adopted in OptimiSTICS, is to use the Nelder–Mead simplex algorithm, which is adapted to non-smooth functions because the search of the optimum is not based on the computation of the function's gradient. Although few theoretical results are available (Lagarias et al., 1996), this algorithm is very popular because it can be used for multidimensional minimization for essentially any function. The simplex algorithm used in this study is a modified version of the function "fminsearch" provided by the Matlab software that implements a bound-constrained simplex algorithm. As the Nelder–Mead simplex is a local optimization method, OptimiSTICS proposes to repeat automatically the minimization with several different starting parameter values to minimize the risk of converging to a local minimum. The user specifies the number of different starting points as well as the starting values in the options file. If starting parameter values are not specified, they are generated at random within the bounds defined for each parameter. To help analyze the behavior of the algorithm, the software generates plots of final estimated parameter values versus initial values. It also creates plots showing the evolution of the criterion to be minimized and of the parameter values during the minimization. Examples of these plots are given in Fig. 14–3.

The Bayesian approach implemented uses the Sampling Importance Resampling algorithm described in the previous section. The user specifies the prior distributions and selects the appropriate statistical model for error. The output furnished includes the sample which approximates the posterior distribution and summary statistics (mean, mode, variance, 95% confidence interval, correlations among parameters). The software also generates histograms of the prior and posterior distributions of the parameters and of selected variables of interest. Examples of these plots are given in Fig. 14–4.

Two special features, specifically adapted to crop models, are included in the software. The first concerns simultaneous estimation of varietal and nonvarietal parameters. The varietal parameters are estimated using only USMs of corresponding variety, while all USMs are used to fit the nonvarietal parameters. To do so, the user defines in the option file the group of USMs to use for each varietal parameter.

The second special feature is that one can automatically carry out parameter estimation in several stages, which is a common practice (Abeledo et al., 2008; Anothai et al., 2008; Ducheyne et al., 1998). Instead of simultaneously estimat-

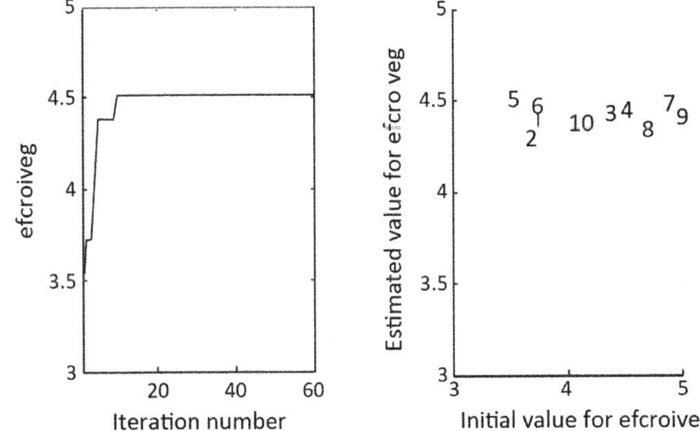

Fig. 14–3. Examples of plots for the simplex minimization. (Left) Parameter value versus iteration number; (right) final parameter value versus starting value. The name of the parameter (EFCROIVEG in this example) is given in the axes labels. Symbols in the right plot correspond to the index of the repetition of the minimization for different starting values.

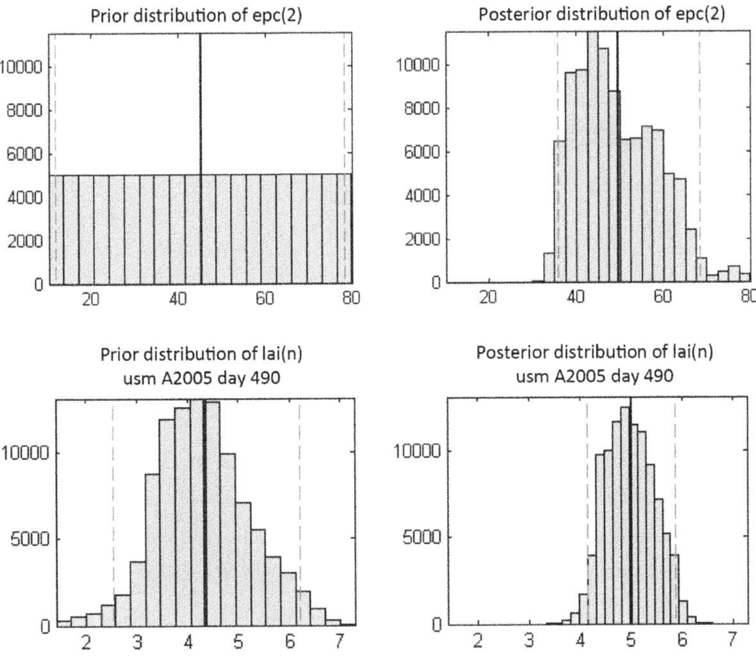

Fig. 14–4. Examples of plots for Sampling Importance Resampling method. (Top) prior (left) and posterior (right) distribution of a parameter (epc(2) in this example); (bottom) prior (left) and posterior (right) distribution of an output variable (LAI in this example) for given USM and day of simulation. Mean values and 95% intervals are indicated.

ing all the unknown parameters from the entire set of observations combined in a complex goodness-of-fit function, the user can decompose the estimation into several stages, each one being dedicated to the estimation of a subset of the unknown parameters from a subset of the observed variables. This can improve the quality of the estimates because the number of parameters simultaneously estimated is reduced and the goodness-of-fit function is simplified. The parameters to be estimated, the data (response variables names and USMs for which the corresponding observed data are available), the statistical model and options for running the STICS model in each stage are defined by the user in the parameter estimation file and define the calibration procedure (see "Parameter Estimation" and Table 14–3). The software does the estimation for each stage in turn, using the parameters estimated in the previous stages.

Evaluation

The OptimiSTICS package provides main functions for comparing simulated and measured values for a set of USMs selected by the user. The simulations corresponding to the USMs are performed automatically, and then the evaluation criteria presented in the "Evaluation" section are calculated. In addition, the software produces several different plots: (i) simulated values versus time for each USM (observed values are also shown), (ii) model residuals as a function of observed value, and (iii) simulated versus observed values of each variable. Figure 14–5 illustrates these plots.

Examples

The examples detailed below illustrate sensitivity analysis of above plant N to soil parameters and estimation of plant parameters for a durum wheat crop. This is just an illustration—the methods and software presented in this chapter can be used for any crop, parameters, and output variables included in the STICS model.

Sensitivity Analysis

Soil characteristics are important inputs to the STICS model. However, their measurement is difficult and costly, in particular in the context of precision agriculture, which requires a good spatial representation of these characteristics. It is therefore important to know which characteristics have a large effect on the outputs of interest and thus require accurate determination, from direct measurements or indirectly, from crop measurements and STICS model inversion (Varella et al., 2010a). In this example, the Extended Fast method was used to determine the sensitivity of aboveground plant N at the beginning of stem elongation and at flowering to soil characteristics. The study has been conducted for a precision

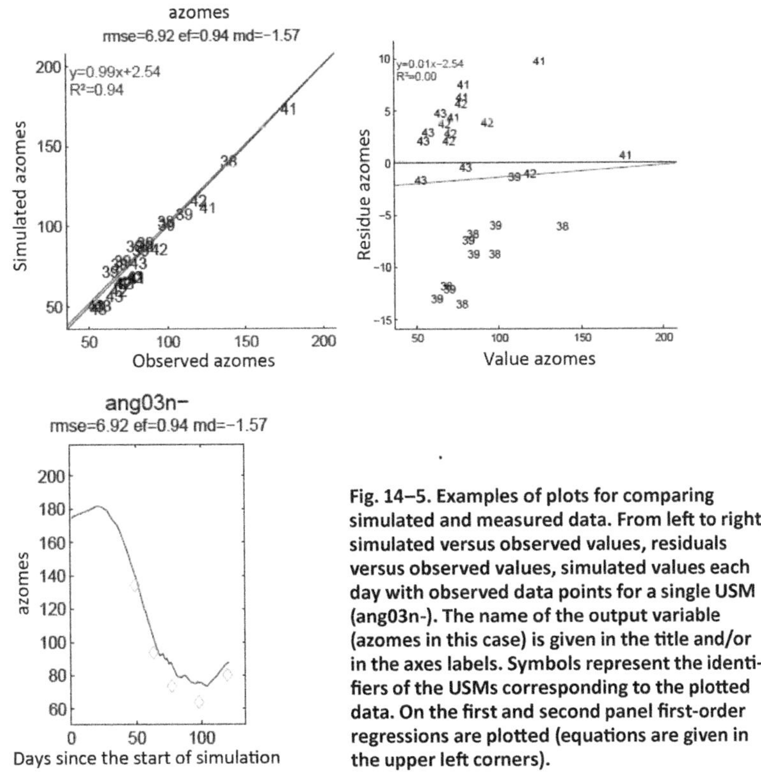

Fig. 14–5. Examples of plots for comparing simulated and measured data. From left to right: simulated versus observed values, residuals versus observed values, simulated values each day with observed data points for a single USM (ang03n-). The name of the output variable (azomes in this case) is given in the title and/or in the axes labels. Symbols represent the identifiers of the USMs corresponding to the plotted data. On the first and second panel first-order regressions are plotted (equations are given in the upper left corners).

agriculture experimental site in northern France near Laon, Picardie (Chambry 49.35°N, 3.37°E) for two different years. For both years the crop is wheat sown on October 30th after a sugarbeet (*Beta vulgaris* L. subsp. *vulgaris*) crop, with 200 kg N ha^{-1} fertilizer.

The soil is divided into two horizontal layers. The top layer has a thickness of 30 cm. On the basis of measurements made at the site, we derived relations between initial content of water, Hinit, and of mineral nitrogen, NO3init, of the two soil layers in the model. Thus, in the following, the values of these initial conditions are only considered for the first layer. This left 13 soil characteristics to study (Table 14–2). Among the permanent characteristics, clay and organic N content of the top layer are involved mainly in organic matter decomposition processes; water content at field capacity of both layers affects the water (and N) movements and storage in the soil reservoir and the thickness of the second layer defines the volume of the reservoir. The initial conditions correspond to the water and N content at the beginning of the simulation. The uncertainties in these characteristics were assumed independent and uniformly distributed (upper and

Table 14–2. Soil parameters and ranges for sensitivity analysis example.

Parameter	Definition	Range	Unit	Label
ARGI	Clay content of the first layer	14–37	%	ar
NORG	Organic N content of the first layer	0.049–0.131	%	N
CALC	Limestone content of the first layer	0–28	%	c
ALBEDO	Albedo of the bare dry soil	0.13–0.31	–	al
Q0	Threshold of daily evapotranspiration	7.5–14.5	mm	q
RUISOLNU	Fraction of rainfall submitted to runoff on a bare soil	0–0.065	–	r
EPC(2)	Thickness of the second layer	0–70	cm	e
DA(1)	Bulk density (first layer)	1.22–1.42	–	D1
DA(2)	Bulk density (second layer)	1.39–1.59	–	D2
HCC(1)	Water content at field capacity (first layer)	14–30	$g\ g^{-1}$	H1
HCC(2)	Water content at field capacity (second layer)	14–30	$g\ g^{-1}$	H2
Hinit	Initial water content (first layer)	4–29	% of weight	h
NO3init	Initial mineral N content (1rst layer)	4–21.5	kg N ha^{-1}	n

lower bounds are given in Table 14–2). These distributions correspond to the spatial variability observed for the study site (Guérif et al., 2001).

The sensitivity analysis was performed for two different climates obtained from the meteorological station of Roupy (49.48°N, 3.11°E). The first climate sequence is from a relatively dry period (1975–1976) with 343, 167, 222, and 219 mm rainfall in spring, summer, autumn, and winter, respectively. The second sequence is from a relatively wet period (1990–1991) with seasonal rainfall amounts of 361, 248, 239, and 316 mm.

Figure 14–6a shows the sensitivity indices calculated for the dry climate sequence. At the early stage of growth the effect of initial water content is dominant because there is relatively little early rain. At flowering roots explore both soil layers, so that now the major soil characteristics are epc(2), HCC(1), and HCC(2). The characteristics involved in the turnover of organic N in the soil have negligible effect because water stress is the dominant limiting factor. Furthermore, the mineralization processes are reduced under dry conditions.

The situation is quite different under wet conditions, where water stress is much less important. Here maximum LAI is equal to 3.61 on the average, compared to 2.57 in dry conditions. The roots grow more rapidly at the beginning of the season; average root depth 3 mo after sowing is 56 cm under wet conditions compared to 46 cm in dry conditions. The second soil layer is explored quite early, and so epc(2) is already important at the start of stem elongation. Moreover, in these conditions, the mineralization of the soil organic matter is increased. Cumulative mineral N arising from humus is on the average 24 kg ha^{-1} up to the date of

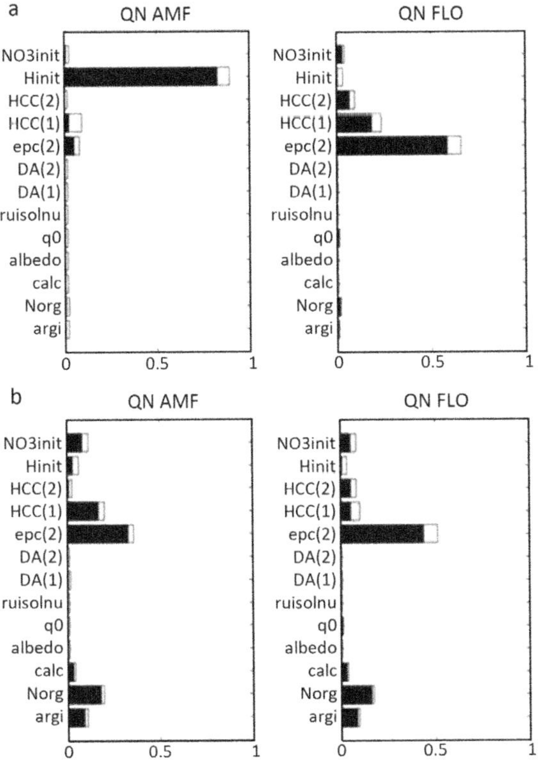

Fig. 14–6. Sensitivity indices of soil characteristics for the output variable aboveground plant N (QN) at beginning of stem elongation (AMF) and at flowering (FLO) for (a) dry climate and (b) wet climate. The first-order effects are represented in dark and interaction effects in light.

maximum leaf area index, compared to 18 kg ha^{-1} in dry conditions. As a result, the parameters involved in mineralization, namely argi, Norg, and NO3init, contribute significantly to the uncertainty in the response variable.

In this example, only a few soil characteristics have a significant effect on aboveground plant N at a given development stage. If the uncertainty in the response variable is considered too large, one can concentrate on reducing uncertainty in those soil characteristics. However it should be noted that the most important soil characteristics are different under different conditions.

Parameter Estimation

The example presented here is part of a project to use STICS to evaluate combined fertilization and irrigation strategies for durum wheat (Guillaume et al., 2009). A major objective is to predict yield and grain protein content correctly, but the prediction of variables such as LAI, biomass, or N content over time is also important because they appear in the decision rules to be tested.

An initial step in the project is model calibration using data from the region of interest (southwestern France) and evaluation of the calibrated model. The intention is to compare various different calibration procedures, using the software described in the previous section. Here we present the results of one particular procedure.

Overall, there are data from 230 USMs in the database dedicated to model calibration. Seven different varieties are represented. A second database, with 138 USMs not used for calibration, is used for evaluating prediction error. Each USM can have measurements of several different output variables at several times.

Parameter estimation was done in 11 successive stages (Table 14–3) using the frequentist method. At each stage only a single output variable, log transformed (Stages 2, 4, 8, 10, and 11) or not (Stages 1, 5, 6, 7, and 9), was used for fitting, using the estimator of Eq. [18]. Depending on the stage, LAI was input or calculated, and only data from unstressed or from all plots was used. If LAI is input then only USMs with LAI data are used. If only unstressed plots are used, this also limits the USMs used in the stage. Sixteen nonvarietal parameters and six varietal parameters were fit to the data. Since there are seven varieties, the total number of parameters is $16 + (6 \times 7) = 58$ (Table 14–3). It seemed reasonable, given the large database, to estimate a large number of parameters. Nevertheless, it is important to have a separate database for evaluating prediction accuracy. Five repetitions of the minimization were performed for each stage and, in total, several hundred thousand model runs were needed to estimate all the parameter values.

The initial and final values of MSE and modeling efficiency for each measured response variable for the calibration database are shown in Table 14–4. For every variable, calibration reduces MSE and increases efficiency. Figures 14–7 through 14–9 show calculated versus observed values for initial and final parameter values for LAI, yield, and grain N concentration.

It seems that the model captures much of the variability in the data (modeling efficiency close to 1 and SDSD only a small part of MSE) for soil water, soil N, LAI, aboveground biomass, and aboveground N. These are all time-dependent variables, so the variability includes both variability within USMs over time and between USMs. Efficiency is negative for grain N concentration and grain number, and only mildly positive for yield. For these variables, the variability is only between USMs. It seems that the model has difficulty capturing this variability. Efficiency is also negative for the index of N nutrition (INN), which is difficult to model because it is a ratio of aboveground N to aboveground biomass.

Table 14–5 shows MSE and model efficiency, calculated using the second database, for evaluating predictive accuracy. The evaluation database has no measurements of soil water or soil N. Thus, the model was not evaluated on these variables. Calibration reduced MSE somewhat for LAI and greatly reduced MSE for grain number and yield. Most of the improvement for grain number and yield comes from a reduction in bias. It seems that the reduction in MSE observed

Table 14–3. Stages for parameter estimation.

Stage	Response variable (STICS name)	Only unstressed or all USM	Number of USM	Number of data	Input LAI	Estimated parameters†	Units	Estimated value‡
1	Aboveground biomass (masec)	unstressed	9	84	yes	efcroirepro	g MJ^{-1}	3.34
						efcroiveg	g MJ^{-1}	3.95
						sea	cm^2 g^{-1}	147
						envfruit	–	0.5
2	Relative humidity of soil (resmes)	unstressed	6	61	yes	kmax	–	1.21
3	Aboveground N (QNplante)	all	28	136	yes	Kmabs1	μmol cm $root^{-1}$	33
						Vmax1	μmol cm^{-1} h^{-1}	0.0002
						Kmabs2	μmol cm $root^{-1}$	39923
						Vmax2	μmol cm^{-1} h^{-1}	0.05
4	LAI (lai)	unstressed	10	84	no	dlaimaxbrut	m^2 leaf $plant^{-1}$ degree d^{-1}	2.8 10^{-4}
5	Aboveground biomass (masec)	unstressed	161	490	no	$durvieF_v$	°C expressed in Q10	150, 204, 250, 222, 221, 180, 179
						durviel	°C	0.8
6	Aboveground biomass (masec)	all	28	197	yes	INNimin	–	0.33
						INNmin	–	0.5
7	LAI [lai(n)]	all	28	171	no	innsen	–	–2
						innturgmin	–	0.53
8	Grain number (chargefruit)	all	226	226	no	cgrain	g^{-1} m^2	0.06
9	Aboveground N (QNplante)	all	162	567	no	$inngrain1_v$	–	0.78, 0.8, 0.3, 0.79, 0.8, 0.76, 0.73
						$inngrain2_v$	–	0.82, 1.5, 0.8, 1.5, 1.5, 1.07, 1.00
10	Grain yield (mafruit)	all	230	230	no	$vitircarb_v$	g grain g $plant^{-1}$ d^{-1}	$1.09.10^{-2}$, $1.26.10^{-2}$, $1.23.10^{-2}$, $1.04.10^{-2}$, $1.97.10^{-2}$, $2.00.10^{-2}$, $1.0.10^{-2}$
						$tmaxremp_v$	°C	39, 38, 33, 40, 37, 31, 38
11	Grain N concentration (CNgrain)	all	229	229	no	$vitirazo_v$	g N g $plant^{-1}$ d^{-1}	$2.02.10^{-2}$, $1.98.10^{-2}$, $1.80.10^{-2}$, $1.80.10^{-2}$, $2.90.10^{-2}$, $2.60.10^{-2}$, $2.13.10^{-2}$

† "v" subscript refers to varietal parameter. See text for the definition of the parameters.

‡ Values for varietal parameters are, respectively, for the durum wheat cultivars: Nefer, Neodur, Biensur, Acalou, Amarillo, Lloyd, and Montsegur.

Table 14–4. Mean squared error and model efficiency based on calibration database.

Variable (STICS variable names)	Units	Initial MSE (bias², SDSD, LCS)	Initial model efficiency	Final MSE (bias², SDSD, LCS)	Final model efficiency
Soil water (resmes)	mm	824 (48, 94, 682)	0.91	710 (45, 50, 615)	0.92
Soil N (azomes)	kg N ha^{-1}	369 (6, 42, 321)	0.76	343 (77, 28, 238)	0.78
LAI [lai(n)]	cm^2 leaf cm^{-2} soil	0.50 (0.01, 0.03, 0.46)	0.75	0.32 (0.01, 0.02, 0.29)	0.84
Aboveground biomass (masec)	t ha^{-1}	2.96 (0.00, 0.00, 2.96)	0.90	2.91 (0.04, 0.01, 2.86)	0.90
Aboveground N (QNplante)	kg N ha^{-1}	918 (197, 25, 696)	0.83	601 (6, 8, 587)	0.89
INN (INN)	–	0.16 (0.08, 0.02, 0.06)	−1.93	0.11 (0.03, 0.02, 0.06)	−1
Grain number (chargefruit)	grain m^2	26996895 (19621055, 88601, 7287239)	−2.51	13990398 (977071, 1273957, 11939370)	−0.82
Grain yield (mafruit)	t ha^{-1}	4.01 (2.49, 0.03, 1.48)	−1.55	1.24 (0.12, 0.00, 1.12)	0.21
Grain N concentration (CNgrain)	%	0.19 (0.02, 0.02, 0.15)	−0.69	0.17 (0.00, 0.02, 0.15)	−0.48

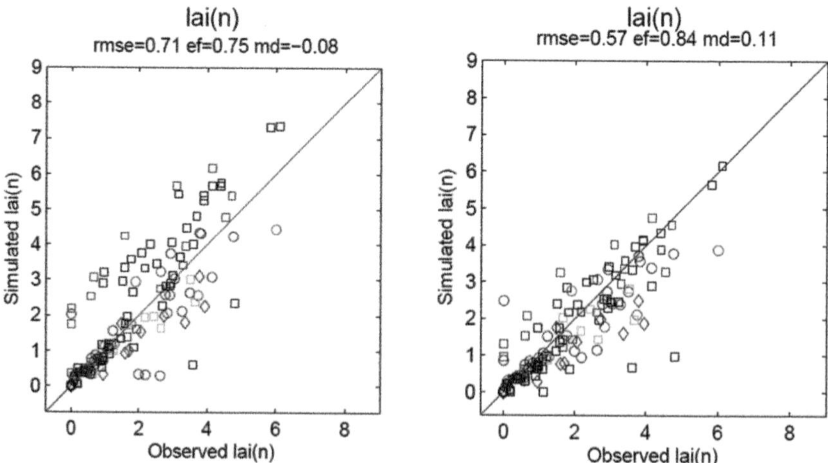

Fig. 14–7. Simulated versus observed values for LAI [lai(n)] with initial (left) and final parameter values (right). Different symbols correspond to different varieties.

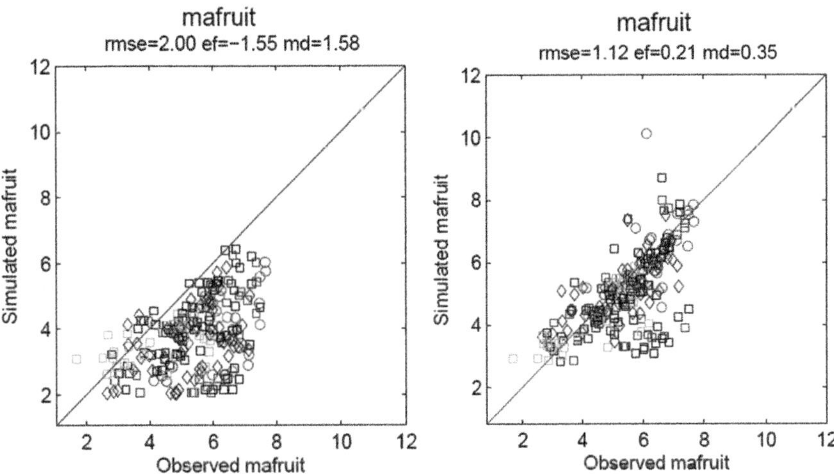

Fig. 14–8. Simulated versus observed values for grain yield (mafruit) for initial (left) and final (right) parameter values. Different symbols correspond to different varieties.

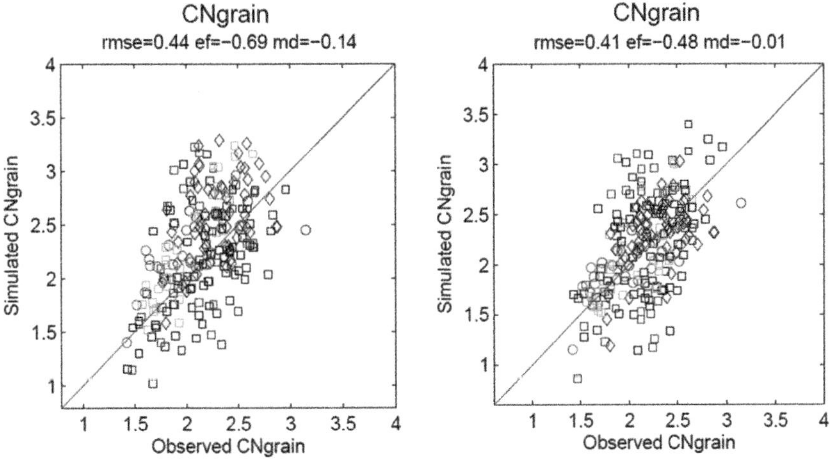

Fig. 14–9. Simulated versus observed values with initial (left) and final parameter values (right) for grain N concentration (CNgrain). Different symbols correspond to different varieties.

with the calibration data implies an improvement in predictions and not just an improved fit to past data for these variables.

However, calibration increased MSE of aboveground biomass, aboveground N, and grain N concentration for the evaluation database. Note that MSE reduction following calibration in the calibration database was already relatively small for aboveground biomass and grain N concentration. Moreover, only subsets of the available USMs were used to estimate model parameters from aboveground bio-

Table 14–5. Mean squared error and model efficiency based on database used for evaluating predictive quality.

Variable (STICS variable names)	Units	Initial MSE (bias², SDSD, LCS)	Initial model efficiency	Final MSE (bias², SDSD, LCS)	Final model efficiency
LAI [lai(n)]	cm^2 leaf cm^{-2} soil	0.35 (0.07, 0.01, 0.27)	0.53	0.26 (0.08, 0.04, 0.14)	0.66
Aboveground biomass (masec)	t ha^{-1}	2.83 (0.39, 0.04, 2.40)	0.91	3.83, (0.44, 0, 3.39)	0.87
Aboveground N (QNplante)	kg N ha^{-1}	859 (20, 8, 831)	0.86	963 (130, 115, 718)	0.84
INN (INN)	–	0.14 (0.05, 0.01, 0.08	−1.32	0.11 (0.02, 0.01, 0.08)	−0.78
Grain number (chargefruit)	grain m^{-2}	40829725 (29670844, 4403748, 6755126)	−1.51	10479543 (784, 270675, 10208084)	0.36
Grain yield (mafruit)	t ha^{-1}	5.23 (3.61, 0.52, 1.10)	−0.58	1.6 (0.31, 0.03, 1.26)	0.52
Grain N concentration (CNgrain)	%	0.13 (0.01, 0.02, 0.11)	−0.64	0.23 (0.00, 0.1, 0.13)	−1.78

mass and aboveground N data (Table 14–3), and these variables reach higher values in the evaluation database than in the calibration database (results not shown). Thus, these results may be explained by a lack of representativity of the measures used in the calibration for these variables with respect to the evaluation database.

Figures 14–10 through 14–12 show MSE, separated into three contributions (see "Methods" section) after each stage of the calibration procedure, for LAI, yield, and aboveground N. MSE is reduced after those steps where the model is fitted to the output variable in question. For example, in Steps 4 and 7 the model is fitted to LAI data, and MSE for LAI is reduced in those steps. MSE for yield is reduced in Step 10, where the model is fitted to yield (but MSE is also reduced in Step 3, where aboveground N is fitted, and in Step 8, where grain number is the target variable). MSE for grain N concentration is reduced in Step 11, where the model is fitted to grain N concentration, and also in Step 3 (where aboveground N is the target variable).

On the other hand, fitting the model to one response variable can worsen fit to another response variable. For example, MSE for yield is substantially increased in Step 6, where the model is fitted to the aboveground biomass data. MSE for grain N concentration is nearly doubled in Step 10, where the model is fitted to grain yield. Thus, one cannot automatically assume that improving the fit to one response variable necessarily improves the fit to all other response variables.

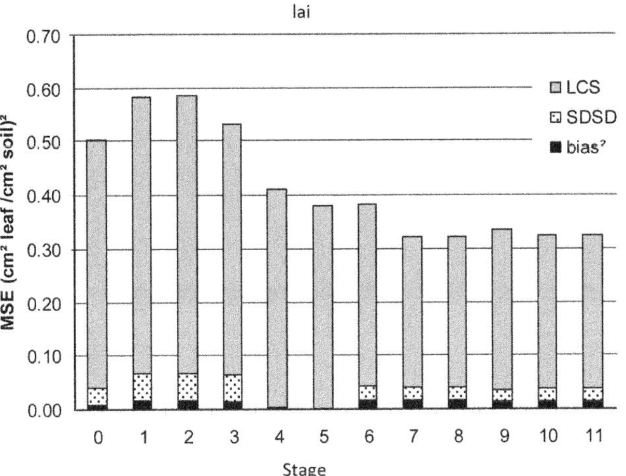

Fig. 14–10. Mean squared error and components after each stage of calibration, for LAI. (Stage 0 represents initial values). LCS, SDSD, and bias^2 are presented in the "Evaluation" section.

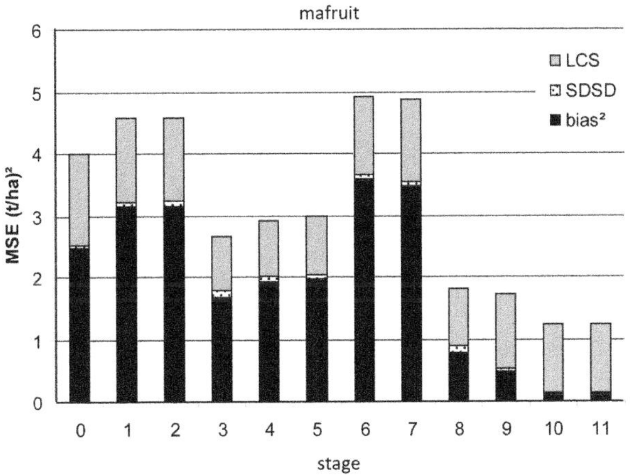

Fig. 14–11. Mean squared error and components after each stage of calibration, for grain yield. (Stage 0 represents initial values). LCS, SDSD, and bias^2 are presented in the "Evaluation" section.

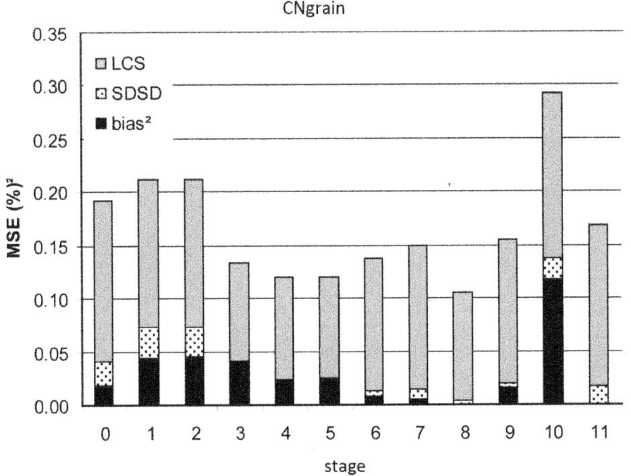

Fig. 14–12. Mean squared error and components after each stage of calibration, for grain N concentration. (Stage 0 represents initial values). LCS, SDSD, and bias^2 are presented in the "Evaluation" section.

An approach with a single calibration stage involving the different response variables and the several parameters would avoid the problem of increasing MSE in some steps. However, estimating many parameters simultaneously can also be problematic. The comparison between these approaches will be the goal of a future study. Note that the software described here can handle multiple response variables, and so is well adapted for such a comparison.

Conclusions

The STICS model is oriented toward analyzing and evaluating cropping systems. To serve this purpose, it is important to have a model like STICS that can simulate a large range of crops, taking into account differences between cultivars and taking into account a wide variety of management practices.

Crop models like STICS are necessarily complicated because they aim to describe complex systems. It is thus important to have tools for analyzing, improving, and evaluating such models. The tools must be easy to use and adapted to the specific problem of simulating cropping systems.

The software described here provides such tools for STICS. It has two major objectives. The first is to help modelers develop better and more useful models. The second objective is more methodological. Since the inception of crop modeling, the emphasis in research has been on describing the processes in cropping systems in terms of equations. It is now becoming more apparent that this is essential but not sufficient. Tools for working with models are also essential—to better understand the models, to better parameterize them, and to better appreciate how well they can predict. To make progress in this area, it is necessary to have software tools that can implement a variety of methods, so that one can evaluate the usefulness and importance of different approaches and cumulate experience while applying these different approaches to various problems. Our software is a response to that challenge.

The software already contains a quite complete set of tools to perform in a single framework the main operations when using a complex crop model such as STICS. However, several features could still be added to improve its use and functionalities. Among them, screening methods such as the Morris method (Morris, 1991; Campolongo et al., 2007) could be useful for performing sensitivity analysis on a large set of factors. Also, parameter selection could be automated using methods such as penalized likelihood or cross validation (Wallach et al., 2001; Tremblay and Wallach, 2004). Many of the current and future features need a large number of independent model simulations. Although the simulation time of STICS is short, operations such as sensitivity analysis or parameter estimation needs several hours or days of computations. Thus, automated parallelization of independent model simulations will have to be considered to exploit the full potential of new multicore computers and offer users an even more efficient tool.

References

Abeledo, L.G., R. Savin, and G.A. Slafer. 2008. Wheat productivity in the Mediterranean Ebro Valley: Analyzing the gap between attainable and potential yield with a simulation model. Eur. J. Agron. 28:541–550.

Anothai, J., A. Patanothai, S. Jogloy, K. Pannangpetch, K.J. Boote, and G. Hoogenboom. 2008. A sequential approach for determining the cultivar coefficients of peanut lines using end-of-season data of crop performance trials. Field Crops Res. 108:169–178.

Beven, K. 2006. A manifesto for the equifinality thesis. J. Hydrol. 320:18–36.

Beven, K.J., P.J. Smith, and J.E. Freer. 2008. So just why would a modeller choose to be incoherent? J. Hydrol. 354:15–32.

Brisson, N., C. Gary, E. Justes, R. Roche, B. Mary, D. Ripoche, D. Zimmer, G. Sierra, P. Bertuzzi, P. Burger, F. Bussière, Y.M. Cabidoche, P. Cellier, P. Debaeke, J.P. Gaudillère, C. Hénault, F. Maraux, B. Seguin, and H. Sinoquet. 2003. An overview of the crop model STICS. Eur. J. Agron. 18:309–322.

Brisson, N., M. Launay, B. Mary, and N. Beaudoin (ed.) 2009. Conceptual basis, formalisations and parameterization of the STICS crop model. Editions Quae, Versailles.

Brisson, N., B. Mary, D. Ripoche, M.H. Jeuffroy, F. Ruget, B. Nicoullaud, P. Gate, F. Devienne-Baret, R. Antonioletti, C. Durr, G. Richard, N. Beaudoin, S. Recous, X.P.D. Tayot, P. Cellier, J.-M. Machet, J.-M. Meynard, and R. Delecolle. 1998. STICS: A generic model for simulating crops and their water and nitrogen balances. I. Theory and parameterization applied to wheat and corn. Agronomie 18:311–346.

Brisson, N., F. Ruget, P. Gate, J. Lorgeou, B. Nicoullaud, X. Tayot, D. Plenet, M.H. Jeuffroy, A. Bouthier, D. Ripoche, B. Mary, and E. Justes. 2002. STICS: A generic model for the simulation of crops and their water and nitrogen balances. II. Evaluation according to experimental data. Agronomie 22:69–92.

Campolongo, F., J. Cariboni, and A. Saltelli. 2007. An effective screening design for sensitivity analysis of large models. Environ. Model. Softw. 22:1509–1518.

Carlin, B.P., and T.A. Louis. 1996. Bayes and empirical bayes methods for data analysis. Chapman and Hall/CRC Press, Boca Raton, FL.

Ducheyne, S., M. Vanclooster, and J. Feyen. 1998. A systematic calibration and validation procedure applied to an integrated soil–crop model. p. 461–468. *In* H.J. Heineke et al. (ed.) Land information systems: Developments for planning the sustainable use of land resources. European Soil Bureau Research Rep. 4. EUR17729EN. Office for Official Publications of the European Communities, Luxembourg.

Guérif, M., N. Beaudoin, C. Durr, J.M. Machet, B. Mary, D. Michot, D. Moulin, B. Nicoullaud, and G. Richard. 2001. Designing a field experiment for assessing soil and crop spatial variability and defining site specific management strategies. p. 677–682. *In* S. Blackmore and G. Gilbert (ed.) Proc. European Conference on Precision Agriculture, 3rd, Montpellier, France. 18–20 June 2001. AGRO Montpellier, Montpellier.

Guérif, M., V. Houlès, D. Makowski, and C. Lauvernet. 2006. Data assimilation and parameter estimation for precision agriculture using the crop model STICS. p. 395–401. *In* D. Wallach et al. (ed.) Working with dynamic crop models. Evaluation, analysis, parameterization, and applications. Elsevier, Amsterdam.

Guillaume, S., J.E. Bergez, and E. Justes. 2009. Modelling water and nitrogen interactions in durum wheat: A first step towards the conception of joint management strategies for irrigation and nitrogen fertilization. p. 212–213. *In* M.K. Van Ittersum et al. (ed.) Proc. Integrated Assessment of agriculture and sustainable development; Setting the agenda for science and policy. Egmond aan Zee, The Netherlands. 10–12 Mar. 2009. Available at http://www.conference-agsap.org/PDFs/1-Final-Proceedings-AgSAP2009.pdf (verified 24 May 2011). Wageningen University and Research Centre, Wageningen.

Helton, J.C., and F.J. Davis. 2000. Sampling-based methods. p. 101–153. *In* A. Saltelli et al. (ed.) Sensitivity analysis. John Wiley and Sons, New York.

Jansen, M.J.W., and T.J. Hagenaars. 2004. Calibration in a Bayesian modelling framework. p. 47–55. *In* M.A.J.S. van Boekel et al. (ed.) Bayesian statistics and quality modelling in the agro-food production chain. Springer-Verlag, Berlin.

Justes, E., M.H. Jeuffroy, and B. Mary. 1997. Wheat Barley and Durum wheat. p. 73–91. *In* G. Lemaire (ed.) Diagnosis of the nitrogen status in crops. Springer-Verlag, Berlin.

Justes, E., P. Lecharpentier, S. Buis, M. Launay, J. Bourges, J. Soudais, P. Clastre, J.E. Bergez, P. Debaeke, D. Wallach, and M. Guérif. 2009. OptimiSTICS, a software for STICS crop model parameter estimation and evaluation which was build up to function with other dynamical models. p. 326–327. *In* M.K. Van Ittersum et al. (ed.) Proc. Integrated Assessment of agriculture and sustainable development; Setting the agenda for science and policy. Egmond aan Zee, The Netherlands. 10–12 Mar. 2009. Available at http://www.conference-agsap.org/PDFs/1-Final-Proceedings-AgSAP2009.pdf (verified 24 May 2011). Wageningen University and Research Centre, Wageningen.

Justes, E., B. Mary, J.M. Meynard, J.M. Machet, and L. Thelier-Huché. 1994. Determination of a critical nitrogen dilution curve for winter wheat crops. Ann. Bot. (Lond.) 74:397–407.

Lagarias, J.C., J.A. Reeds, M.H. Wright, and P.E. Wright. 1996. Convergence properties of the Nelder–Mead simplex algorithm in low dimensions. SIAM J. Optim. 9:112–147.

Lemaire, G., and F. Gastal. 1997. N uptake and distribution in plant canopies. p. 3–43. *In* G. Lemaire (ed.) Diagnosis of the nitrogen status in crops. Springer-Verlag, Berlin.

Makowski, D., D. Wallach, and M. Tremblay. 2002. Using a Bayesian approach to parameter estimation; comparison of the GLUE and MCMC methods. Agronomie 22:191–203.

Morris, M.D. 1991. Factorial sampling plans for preliminary computational experiments. Technometrics 33:161–174.

Saltelli, A., K. Chan, and E.M. Scott (ed.) 2000. Sensitivity analysis. John Wiley and Sons, New York.

Saltelli, A., S. Tarantola, and K. Chan. 1999. A quantitative model-independent method for global sensitivity analysis of model output. Technometrics 41:39–56.

Seber, G.A.F., and C.J. Wild. 1989. Nonlinear regression. John Wiley and Sons, New York.

Sokal, R.R., and F.J. Rohlf. 1981. Biometry. W.H. Freeman, San Francisco.

Tremblay, M., and D. Wallach. 2004. Comparison of parameter estimation methods for crop models. Agronomie 24:351–365.

Van Oijen, M., J. Rougier, and R. Smith. 2005. Bayesian calibration of process-based forest models: Bridging the gap between models and data. Tree Physiol. 25:915–927.

Varella, H., M. Guérif, and S. Buis. 2010a. Global sensitivity analysis measures the quality of parameter estimation: The case of soil parameters and a crop model. Environ. Model. Softw. 25:310–319.

Varella, H., M. Guérif, S. Buis, and N. Beaudoin. 2010b. Soil properties estimation by inversion of a crop model and observations on crop improves the prediction of agro-environmental variables. Eur. J. Agron. 33:139–147.

Wallach, D. 2006. Evaluating crop models. p. 11–53. *In* D. Wallach et al. (ed.) Working with dynamic crop models. Evaluation, analysis, parameterization, and applications. Elsevier, Amsterdam.

Wallach, D., B. Goffinet, J.E. Bergez, P. Debaeke, D. Leenhardt, and J.-N. Aubertot. 2001. Parameter estimation for crop models: A new approach and application to a corn model. Agron. J. 93:757–766.

White, H. 1981. Consequences and detection of misspecified nonlinear regression models. J. Am. Stat. Assoc. 374:419–433.

Willmott, C.J. 1981. On the validation of models. Phys. Geogr. 2:184–194.

Chapter 15

A Synthesis of Current Parameterization Approaches and Needs for Further Improvements

L.R. Ahuja
Liwang Ma

Abstract

Parameterization of a system model includes both calibration and evaluation. Calibration is to derive a set of parameters that adequately describe a dataset from an experiment, whereas evaluation (or validation in the liberal sense) is to predict the behavior of the calibrated model under different experimental conditions (or independent datasets). However, due to the complexity of system models, there has not been a standard method to parameterize a system model. Methods reported in this book and elsewhere are often model and user dependent. Although manual calibration requires much experience on the users' side, autocalibration is not an easy task because of the high level of skills needed for using sophisticated computer software and for constructing an objective function upon which to optimize. In addition, automated calibration procedure may vary with the data available for calibration and preference on constructing the objective function. However, these chapters clearly set up the principles for model parameterization in terms of sequential calibration of a model, quality control of input data, balanced calibration among system components, and needs for datasets that cover a wide range of experimental conditions. This chapter also identifies research needs for further improvement in model parameterization, such as better collaboration between modelers and field scientists, taking into account the spatial and temporal variability of field measurements, development of science modules in a common platform for better communication among modelers, and better education for the next generation of system modelers.

L.R. Ahuja (Laj.Ahuja@ars.usda.gov) and L. Ma (Liwang.ma@ars.usda.gov), USDA-ARS, Agricultural Systems Research Unit, Fort Collins, CO 80526.

doi:10.2134/advagricsystmodel2.c15

Methods of introducing system models into agricultural research. L.R. Ahuja and L. Ma (ed.)
Advances in Agricultural Systems Modeling Series 2.

"Each model has its own strengths and weaknesses
for specific applications,
as well as parameterization requirements....
The development of a systematic
and hopefully a common protocol
is needed to help users."

Agricultural system models began to take shape with the release of CREAMS (later GLEAMS) (Knisel, 1980), EPIC (Williams et al., 1983), CERES (Jones and Kiniry, 1986), SOYGRO (Wilkerson et al., 1983), PNUTGRO (later CROPGRO) (Boote et al., 1987), WOFOST (van Diepen et al., 1989) in the 1980s and DAISY (Hansen et al., 1991), SOIL-SOILN (Bergström et al., 1991), RZWQM (Ahuja et al., 1993), ecosys (Grant, 1995), HERMES (Kersebaum, 1995), APSIM (McCown et al., 1996), STICS (Brisson et al., 1998), and DAYCENT (Parton et al., 1998) in the 1990s. Later, these field-scale models were extended to address large-scale problems by linking to geographic information systems (GIS) or adding spatial components or attributes. Each model has its own strengths and weaknesses for specific applications, as well as parameterization requirements. The process of selecting an appropriate model and parameterization methods is heavily dependent on users' experience and example applications in the literature. The development of a systematic and hopefully a common protocol is needed to help users.

Usually these system models have been developed to be used in a predictive mode, which means that the users have to provide site-specific model inputs and parameters, with guidance from databases and default parameters provided with the model. The process includes calibration of some key parameters that are not easily measurable or derivable from available information. The whole process of parameterization becomes a formidable task for model users, especially field scientists, which is further aggravated by the following:

1. There are often multiple sets of parameters that can provide acceptable simulations for a particular dataset (i.e., nonuniqueness) due to interdependencies among parameters.
2. There is seldom an agricultural field that can be represented with a unique set of measurements of soil properties, as a result of large spatial variability, even within distances of a few meters.
3. There are multiple parameters to be calibrated for a single output variable (e.g., yield, biomass, soil moisture).
4. Interactions among system components are largely empirical in the models (e.g., increased canopy growth under low sunlight and enhanced root growth under water stress) and may not be applicable quantitatively for all conditions.

5. Biological processes are generally more complex and hence empirical (e.g., water and N uptake) than physical processes (e.g., water movement).
6. Calibration of one system component (e.g., plant yield) may imbalance calibration of others (e.g., plant N uptake or soil moisture) due to interaction among parameters.
7. Many reports of model calibration in the literature are not complete because of a tendency for researchers to not list all parameters used in a model in a journal paper. Therefore, the results are usually unrepeatable unless exactly the same input files and model version are used.

Due to some of the above factors, there are very few cases when even the well-calibrated model parameters (e.g., plant parameters) are transferrable to other locations and climate conditions without further calibration for the local data. There have been studies to apply multiple models for the same datasets to evaluate which model performed the best for the datasets (Diekkrüger et al., 1995; Jamieson et al., 1998; Smith et al., 1997) and also to apply one model at multiple locations (Rosenzweig et al., 1999; Saseendran et al., 2009). Although these applications pointed to inadequacy in the process descriptions in the models, these studies were not systematically designed and have not generated the synergy to bring system models to a new plateau in terms of theoretic advancement and practical applications. With contributions from several model developers, this book is focused on parameterization and calibration of selected models with the hope of developing a common protocol for model parameterization and to help both model users and developers striving for the next generation of system models.

The USDS-ARS Root Zone Water Quality Model (RZWQM) is a field-scale model that simulates crop production and environmental quality under various agricultural management practices. It contains detailed plant growth processes, soil water movement, soil carbon and nitrogen dynamics, pesticide retention and transport, surface energy balance, and their interactions. The model is fully supported and is constantly improving to meet users' requests. Efforts have been made to link RZWQM2 to GIS for spatially distributed modeling. Chapter 1 by Ma et al. (2011, this volume) illustrates a step-by-step calibration of RZWQM coupled with DSSAT crop growth modules (RZWQM2). After defining the inputs into such an agricultural system model for weather conditions, soil properties, and management practices, the authors go through the system component by component and provide guidelines on how to calibrate the parameters (or processes) that affect the modeling results of interest. It is vital to ensure correct input information for the model before calibrating any parameters. This chapter emphasizes the iterative nature of model calibration because of the interdependency among system components. Another point in the chapter is to balance calibration among system components. It is important to check simulation of each system component (e.g., water balance, N balance, plant development and growth, and chemical transport) to ensure that all components are simulated reasonably well to avoid good results for bad reasons. The RZWQM2 provides default model parameters (plant, soil, and pesticide) and the reasonable ranges of simulated values for selective processes as guidance for model users. The authors point out that, in order for plant parameters to be transferrable to other locations or experiments, similar approaches to estimating soil hydraulic properties and soil nutrient status should be followed for all locations and experiments. Long-term simulation and multiple treatments to investigate water and N interactions are also important for model parameterization. The authors discuss the criteria for an acceptable calibration, which has general applications to all system models. This chapter does not replace the technical manual that documents the science and the user manual that provides detailed guidance on how to operate the model via a Windows user interface.

HERMES was developed originally as a nitrogen management tool and later became a whole system model. The strength of the model is in its simulation of soil N processes with simple approaches to plant growth. Calibration of the HERMES model (Chapter 2, Kersebaum, 2011, this volume) is summarized after a brief description of the model. Since the model is intended to be used by nonscientific personnel for making recommendations for N application, it has only a few parameters to calibrate. The model contains default soil parameters based on soil texture and four methods of estimating potential evapotranspiration. The model can extract soil information from a spatial database and has a Graphical User Interface (GUI). For a good calibration, the author suggests the use of detailed measurements that reflect the dynamics of the observed system in time and space. For soil hydraulic properties, users may estimate field capacity in the winter after rainfall and wilting point after a long period of drought in the summer. For plant growth and soil water balance, effective rooting depth is very important, which may be derived from measured soil water dynamics. The authors also emphasize the wholeness of reasonable simulation for all system components and the importance of comparing simulated results with literature and measured values.

LEACHM is known for its ability to address environmental quality issues with detailed soil water and chemical (N, P, pesticide) balance simulations. Like RZWQM2 and HERMES, LEACHM is a one-dimensional model. Jabro et al. (2001, Chapter 3, this volume) present calibration, validation, and sensitivity analysis for the LEACHM model and an example of its performance using field-measured water drainage and N leaching in soils under a continuous corn (*Zea mays* L.) cropping system. Other prominent features of the model (e.g., phosphorus and pesticide) are not invoked. Water flow is calibrated first, followed by parameters controlling N transformation in the model. Although the water and N measurements were collected in a corn field, the chapter does not discuss calibration of plant growth and plant water and N uptake, which affect soil water and N balance. The sensitivity analysis at the end of the chapter is valuable for identifying the important model parameters for calibration.

The SPACSYS (Chapter 4, Wu and Shepherd, 2011, this volume) is a three-dimensional model and requires more inputs and parameters than one-dimensional models. However, it can be used for one-dimensional simulation as well. It contains detailed simulation of soil N dynamic processes, including dissolved soil organic matter. The uniqueness of the model is its ability to simulate three-dimensional root growth and mechanistic root–soil interactions. Data availability is the limiting factor for calibrating this model. When limited data are available, it is recommended to reduce the model to apply it for a one-dimensional simulation. Parameterization through a user interface makes the calibration easier. Unlike the other chapters, Wu and Shepherd present each process in detail with equations and parameters to calibrate. Validation of each process is also detailed in the chapter.

DayCent is the daily version of the CENTURY model (Daily CENTURY) and is known for its soil C and N simulation at longer time scale. Soil water and plant growth are relatively simple. Del Grosso et al. (2011, Chapter 5, this volume) present an overview of the DayCent model and recommend calibration in the order of soil water, crop growth, soil organic carbon, and N losses. The authors suggest aggregation of the results on seasonal basis rather than on a daily basis for some of the more uncertain outputs (N_2, NO_x, and NH_3). Web-accessible databases and datasets are employed to facilitate model calibration and model comparison. Initialization of the soil C pools is the major step in calibrating DayCent, and the authors recommend a "spin up" simulation for 1000 yr. The authors warn against arbitrary model parameter adjustments without a solid reason and without biological and physical meanings.

The Environmental Policy Impact Climate (EPIC) model may be one of the most used models in agriculture due its simplicity and parameters for many crops. It was initially developed mainly to investigate the effect of soil erosion on soil quality and crop production. It simulates all major processes in agriculture related to crop production and soil and water quality, as well as pesticide fate in soils. Economics is a plus. The EPIC model is a field scale model and is now extended to multiple fields (APEX). Wang et al. (2011, Chapter 6, this volume) describe EPIC and its extension APEX with detailed calibration procedures. A list of most sensitive parameters based on the literature is provided that should be very helpful to users, but on-site sensitivity analysis is recommended to identify a set of sensitive parameters for a study. The authors recommend checking water balance for the whole simulated watershed before examining individual fields. They emphasize that the reasonableness should be judged based on actual information rather on expectation. A valid calibration should consider water balance, erosion, crop growth, and nutrient cycling as a whole. The chapter elaborates each process and its parameters that should be calibrated. An example illustrates APEX model application in a watershed with the help of web databases and GIS tools. With uncertainty in model inputs, model structure, and experimental observations, the model user is discouraged to overcalibrate a model for one component; rather one should strive for a balanced calibration among all system components.

GPFARM-Range is evolved from the GPFARM model with emphasis on the range component. It is recoded in Java language and modularized (Andales et al., 2005, 2006). In Chapter 7, Adiku et al. (2011, this volume) briefly describe a rangeland simulation model, its inputs and parameter requirements, and methods of obtaining the parameters. Two main methods of parameterization of the model for a new location are presented and discussed: manual adjustment of default parameters and their direct experimental determination. They give a step-by-step procedure for manual adjustment of the most sensitive default parameters, in the order of soil, plant functional groups, and animal parameters, followed by iteration. The procedures are illustrated with examples of calibration for a new location. They also give examples of how to go about determining these parameters directly from experimental data available in the literature or as a part of the future experimental research.

The next two chapters present how experimental data can be used to obtain better estimates of certain parameters. Wendroth et al. (2011, Chapter 8, this volume) investigate to what extent the DSSAT model captures the spatially variable model input of soil texture and N management in a farmer's field and compare experimental results of soil and crop state variables collected over the wheat (*Triticum aestivum* L.) growing season with model results. The results of the uncalibrated model do not reflect effects of soil variation that are obvious in measurements. However, crop sensor measurements taken in early spring as integrative state variables do reflect effects of spatial soil variation besides N management differences and are closely related to measured grain yield. The study points to the need for improving model parameterization for response to spatially variable soil and of variable management based on the experimental results, and possibly also the need for improving the effective average parameters for the field. Hatfield (2011, Chapter 9, this volume) presents the current status of methods used to estimate the required energy balance inputs for the models, such as solar radiation, albedo, and long-wave radiation, from common measurements of air temperature, altitude, precipitation, and relative humidity, at appropriate spatial and temporal scales represented in the models. He recommends wider evaluation of these methods and their further improvement. He also shows how we can use the estimates of crop residue cover, vegetation cover, biomass, and leaf area obtained by remote sensing to further improve crop growth parameters, as well as to validate the crop growth models at field to larger scales. He provides the current status of the remote sensing methods, possibilities for improving them further, and their extension to measurement of rainfall by Doppler radar and soil moisture by remote microwave sensors.

Malone et al. (2011, Chapter 10, this volume) showcase the importance of the quality of the weather data that drive the agricultural system models, with examples of model simulations in Iowa. They note the common errors in measurement of rainfall, solar radiation, and relative humidity, on the order of 20% on annual basis, as a result of electronic failures, rodent damage, activity of insects and birds, poor location, poor calibration, and oxidation. They show that weather data bias of 10% in rainfall, radiation, and humidity resulted in long-term errors of more than 40% in nitrate loss to tile drains and 15% in corn yield in Iowa simulated by a validated model. They describe the commonly recommended quality check and correction procedures for different elements of the weather data.

The last four chapters demonstrate the automated procedures for the calibration of model parameters with respect to some measured outputs. These calibration procedures are either "local" or "global." The local methods start with the best initial estimates of the parameter values and search for a minimum value of a specified objective function (say, the root mean square error between the simulated and observed data) in the vicinity by varying these values based on the direction of change of the objective function. The global methods assume that each parameter lies within a distribution of values, and they sample the entire distribution in searching for the values that minimize the objective function. The local methods are more efficient than the global methods in that they require far fewer model runs to find the optimum parameter values. However, they can get stuck in a local minimum around the starting values, which can generally be resolved by starting from several different initial values. The global methods have the advantage of finding the global minimum. Wolf et al. (2011, Chapter 11, this volume) describe the "local" calibration platform, CALPLAT, which is used to calibrate parameters, generally one at a time, in a stepwise procedure starting with a reasonable initial value and using some observed output data appropriate for each parameter and a Simplex optimization algorithm. The Simplex algorithm is basically an iterative gradient approach to find the parameter value that gives the sum of squares of deviations between the simulated results and observed data below a preset limit. The method can be applied at the point or field scale with detailed measurements of crop growth or at an agroclimatic region scale with a set of representative observed values within the region. Wolf et al. (2011, this volume) use CALPLAT to calibrate two winter wheat phenology parameters, thermal units from emergence to flowering and from flowering to maturity, over agroclimatic zones across Europe. These calibrated parameters are used with a standard set of other parameters in the WOFOST crop model to simulate winter wheat biomass, harvest index, and maximum leaf area index in each region. The results are satisfactory.

Nolan et al. (2011, Chapter 12, this volume) autocalibrate the RZWQM2 using the "local" PEST parameter estimation software, which uses an iterative nonlinear regression approach for simultaneous adjustment of multiple parameters to match the observed data. This inverse approach also yields parameter correlations and sensitivities (which makes it straightforward to identify interdependent and insensitive parameters), and confidence intervals around estimates of parameters and around the simulated output. The method uses truncated singular value decomposition to mitigate potential problems of highly correlated and insensitive parameters by estimating linear combinations (eigenvectors) of the original process-model parameters. The method is applied to data for daily water and N fluxes to groundwater in California and to monthly water and N fluxes to tile drainage in Iowa. The estimated 16 parameters in the California simulations provided a reasonable fit to the observed data. The results were better for the monthly water and N fluxes, which may have benefitted from temporal and spatial averaging of RZWQM2 inputs and outputs.

Jones et al. (2011, Chapter 13, this volume) use one of the Bayesian parameter estimation procedures, called GLUE, the Generalized Likelihood Uncertainty Estimation, to estimate cultivar-specific parameters for corn and soybean [*Glycine max* (L.) Merr.] in the DSSAT modeling package. The GLUE method is "global" and involves: (i) developing a prior distribution of each parameter from current knowledge (this may just be the range of a parameter within which the parameter is uniformly distributed), (ii) generating random parameter sets (e.g., 6000 or more) from the prior distributions, (iii) running the model for each parameter set, (iv) calculating the likelihood values based on deviation of simulated from the set of observed values, and (v) constructing a posterior distribution of each parameter from the likelihood values. Jones et al. (2011) apply GLUE first to estimate cultivar-specific phonological or developmental parameters in DSSAT (3 for corn and 5 for soybean), and then a second time to estimate growth parameters (2 for corn and 5 for soybean). Average relative absolute error values between observed and simulated output variables are between 3 to 8% for 2 yr of corn data for the GLUE method and 4 to 10% for the manual calibration method.

STICS is a one-dimensional French model that simulates a wide range of crops, crop rotations, and management practices, using a daily time step. Buis et al. (2011, Chapter 14, this volume) describe recent software associated with the French STICS multi-crop model, for sensitivity and uncertainty analyses, parameter estimation, and comparative evaluation of the observed and simulated results for various modeling scenarios. For parameter estimation, the software has options for several different assumptions about the model error (zero or non-zero, independent or correlated, known or unknown variance). These options can be implemented for frequentist (parameters assumed deterministic or single valued, i.e., the "local" method) or Bayesian (the parameters have a distribution, i.e., the "global" method) scenarios. One can fit several different output variables simultaneously or choose an automated sequential approach. The authors present an example of this latter option. The parameters are estimated in several stages by fitting the model to a single output variable at each stage. The authors note, however, that fitting at the new stage may degrade the fit to output variables at previous stages.

It should be noted that the manual and autocalibration methods go hand in hand. Generally, the model parameters should be approximated with the manual methods for use in the autocalibration methods as starting values. Supplying unreasonable starting values to the latter methods may not yield convergence to the desired optimal values.

Action and Research Needs for Further Improvements

1. It should always be borne in mind that the model simulation errors result from three sources: model errors caused by inadequate scientific description of the processes and their interaction, parameter errors due to uncertainty in the parameter values, and experimental errors in the measured or observed data that are used to calibrate the parameters and evaluate the calibrated model. The efforts to improve the model performance should, therefore, be balanced. Emphasis on just the parameterization issue may not be enough.

2. Agricultural system models describe very complex systems, so it is natural that they can have sizeable errors. It is important to try to reduce these errors, but it is also important to quantify them, to indicate the level of confidence one can have in the results. We need standard criteria and procedures for evaluating model uncertainty, and in particular the contribution of parameter uncertainty to the overall uncertainty.

3. Field researchers need to make enough measurements to characterize spatial variability of soil properties and soil inputs needed by a model, as well as of the output variables (e.g., crop leaf area index and biomass at different growth stages, soil moisture, soil residual N), in terms of mean, variance, and trends in an experimental area. This information will greatly improve estimation of parameters and simulation results. We recognize that field researchers have limited time and resources available to allow this, but doing this will greatly improve the cause–effect analyses of their results and their extension to multiple years of weather and other soils and climates with the help of modeling. It will pay off in the long run—at least, do as much as possible.
4. A dataset for parameterizing the models should also cover varying weather conditions at a given location over a number of years. This will ensure a better representation of climate–soil–plant interactions in the parameters. Additionally, or alternatively, it will be good to have data from several locations with different climate and soil conditions in a dataset.
5. We should continue to further improve all methods of parameter estimation, both manual and automated calibrations, to make them easier for field researchers to use. Standard protocols for parameterizing system models should be developed.
6. There is also a need to develop a distribution of the most important parameters and, if possible, the effective average value of these parameters in a heterogeneous field. Due to natural variability, it may be more realistic to have a distribution of values for some if not all parameters than a single value and to provide confidence intervals of simulated results.
7. To reduce the dependence on calibration in the future, further research should be focused on developing methods for obtaining soil, crop, atmospheric, nutrient, and other parameters and their variability directly from experimental measurements or from measurements of related properties or processes, as much as possible. In Chapter 9, Hatfield (2011, this volume) provides examples of obtaining energy balance parameters from remote sensed measurements. Sensitivity analyses of parameters can help target the appropriate parameters.
8. There are numerous knowledge gaps in our understanding of various processes in the system models and the interactions among them, for which the modelers make simplifying assumptions or use empirical factors. Inadequate accounting of the interactions affects the values of the so-called optimized parameters for a given set of data. Thus, the optimized crop or soil parameters may not be generic, and the crop parameters may not be transferable to other datasets even for the same cultivar. These knowledge gaps certainly increase uncertainty and errors in the simulation results, an issue not addressed in this book. Many of the existing process modules in almost all models need to be tested against good data and improved.

For this purpose, there needs to be enhanced and continuous interaction between the modelers and field researchers. This will also help improve data collection and application of the models. The modelers need to take the initiative for these purposes. This could be achieved through one-on-one collaborations, as well as by sharing experimental and model results at regular sessions organized at the national and international society meetings, conducting training courses on the use of models, and demonstrating the model application results. Field experiments and measurements should be planned with due consideration of the spatial and temporal variability of certain processes and parameters, and how a deterministic model could be applied to the data.

9. The science in the models would be more easily improved if the modelers adopted a common modular structure for processes and their interactions in a system model. This structure would make it easier to test and improve or replace a given module. This would also enable intercomparison of modules among the models and sharing of modules among modelers. Some intercomparison of big models has been done, but, because of complex interactions, it is very difficult to pinpoint where the differences among the models originate. The sharing of modules and datasets among modelers will speed up the advancement of science and application of the models. The modular structure will also permit having modules at different levels of scientific detail and parameter requirement (simple to complex), whereby one could quickly build a model tailored to the nature of the problem and the scale of application. A simpler model with fewer parameters may be adequate for a specific problem or for a large-scale application.

10. In the future, it would also be helpful if the modelers followed a standard coding convention for the modules, documented the code well, had it peer-reviewed, and shared the code and test datasets with other modelers and experimenters openly.

11. It is critical to train the next generation of scientists in the graduate schools on models and their applications to help better understanding and solve the practical problems and issues.

References

Adiku, S.G.K., L.R. Ahuja, G.H. Dunn, J.D. Derner, A.A. Andales, L. Garcia, and P.N.S. Bartling. 2011. Parameterization of the GPFARM-Range model for simulating rangeland productivity. p. 209–228. *In* L.R. Ahuja and L. Ma (ed.) Methods of introducing system models into agricultural research. Advances in Agricultural Systems Modeling 2. ASA, CSSA, and SSSA, Madison, WI.

Ahuja, L.R., D.G. DeCoursey, B.B. Barnes, and K.W. Rojas. 1993. Characteristics of macropore transport studied with the ARS root zone water quality model. Trans. ASAE 36:369–380.

Andales, A.A., J.D. Derner, P.N.S. Bartling, L.R. Ahuja, G.H. Dunn, R.H. Hart, and J.D. Hanson. 2005. Evaluation of GPFARM for simulation of forage production and cow-calf weights. Rangeland Ecol. Manage. 58:247–255.

Andales, A.A., J.D. Derner, L.R. Ahuja, and R.H. Hart. 2006. Strategic and tactical prediction of forage in northern mixed-grass prairie. Rangeland Ecol. Manage. 59:576–584.

Bergström, L., H. Johnsson, and G. Torstensson. 1991. Simulation of soil nitrogen dynamics using the SOILN model. Fert. Res. 27:181–188.

Boote, K.J., J.W. Jones, G. Hoogenboom, and G.G. Wilkerson. 1987. PNUTGRO v 1.0, Peanut crop growth and yield model. Technical documentation. Dep. of Agronomy and Agricultural Engineering, Univ. of Florida, Gainesville.

Brisson, N., B. Mary, D. Ripoche, M.H. Jeuffroy, F. Ruget, B. Nicoullaud, P. Gate, F. Devienne-Barret, R. Antonioletti, C. Durr, G. Richard, G. Beaudoin, S. Recous, X. Tayot, D. Plenet, P. Cellier, J.M. Machet, J.M. Meynard, and R. Delécolle. 1998. STICS: A generic model for the simulation of crops and their water and nitrogen balance. I. Theory and parameterization applied to wheat and corn. Agronomie 18:311–346.

Buis, S., D. Wallach, S. Guillaume, H. Varella, P. Lecharpentier, M. Launay, M. Guérif, J.-E. Bergez, and E. Justes. 2011. The STICS crop model and associated software for analysis, parameterization, and evaluation. p. 395–426. *In* L.R. Ahuja and L. Ma (ed.) Methods of introducing system models into agricultural research. Advances in Agricultural Systems Modeling 2. ASA, CSSA, and SSSA, Madison, WI.

Del Grosso, S.J., W.J. Parton, C.A. Keough, and M. Reyes-Fox. 2011. Special features of the DayCent modeling package and additional procedures for parameterization, calibration, validation, and applications. p. 155–176. *In* L.R. Ahuja and L. Ma (ed.) Methods of introducing system models into agricultural research. Advances in Agricultural Systems Modeling 2. ASA, CSSA, and SSSA, Madison, WI.

Diekkrüger, B., D. Söndgerath, K.C. Kersebaum, and C.W. McVoy. 1995. Validity of agroecosystem models a comparison of results of different models applied to the same data set. Ecol. Model. 81:3–29.

Grant, R.F. 1995. Salinity, water use and yield of maize: Testing of the mathematical model ecosys. Plant Soil 172:309–322.

Hansen, S., H.E. Jensen, N.E. Nielsen, and H. Svendsen. 1991. Simulation of nitrogen dynamics and biomass production in winter wheat using the Danish simulation model DAISY. Fert. Res. 27:245–259.

Hatfield, J.L. 2011. Parameterization of energy balance components and remote sensing in systems modeling. p. 261–282. *In* L.R. Ahuja and L. Ma (ed.) Methods of introducing system models into agricultural research. Advances in Agricultural Systems Modeling 2. ASA, CSSA, and SSSA, Madison, WI.

Jabro, J.D., J.L. Hutson, and A.D. Jabro. 2011. Parameterizing LEACHM model for simulating water drainage fluxes and nitrate leaching losses. p. 95–116. *In* L.R. Ahuja and L. Ma (ed.) Methods of introducing system models into agricultural research. Advances in Agricultural Systems Modeling 2. ASA, CSSA, and SSSA, Madison, WI.

Jamieson, P.D., J.R. Porter, J. Goudriaan, J.T. Ritchie, H. van Keulen, and W. Stol. 1998. A comparison of the models AFRCWHEAT2, CERES-Wheat, Sirius, SUCROS2, and SWHEAT with measurements from wheat grown under drought. Field Crops Res. 55:23–44.

Jones, C.A., and J.R. Kiniry. 1986. CERES-Maize: A simulation model of maize growth and development. Texas A&M Univ. Press, College Station.

Jones, J.W., J. He, K.J. Boote, P. Wilkens, C.H. Porter, and Z. Hu. 2011. Estimating DSSAT cropping system cultivar-specific parameters using Bayesian techniques. p. 365–394. *In* L.R. Ahuja and L. Ma (ed.) Methods of introducing system models into agricultural research. Advances in Agricultural Systems Modeling 2. ASA, CSSA, and SSSA, Madison, WI.

Kersebaum, K.C. 1995. Application of a simple management model to simulate water and nitrogen dynamics. Ecol. Model. 81:145–156.

Kersebaum, K.C. 2011. Special features of the HERMES model and additional procedures for parameterization, calibration, validation and applications. p. 65–94. *In* L.R. Ahuja and L. Ma (ed.) Methods of introducing system models into agricultural research. Advances in Agricultural Systems Modeling 2. ASA, CSSA, and SSSA, Madison, WI.

Knisel, W.G. 1980. CREAMS: A field-scale model for chemicals, runoff, and erosion from agricultural management systems. Conservation Research Rep. 26. USDA, Science and Education Administration, Washington, DC.

Ma, L., L.R. Ahuja, S.A. Saseendran, R.W. Malone, T.R. Green, B.T. Nolan, P.N.S. Bartling, G.N. Flerchinger, K.J. Boote, and G. Hoogenboom. 2011. A protocol for parameterization and calibration of RZWQM2 in field research. p. 1–64. *In* L.R. Ahuja and L. Ma (ed.) Methods of introducing system models into agricultural research. Advances in Agricultural Systems Modeling 2. ASA, CSSA, and SSSA, Madison, WI.

Malone, R.W., D.W. Meek, L. Ma, D.B. Jaynes, B.T. Nolan, and D.L. Karlen. 2011. Quality assurance of weather data for agricultural system model input: A case study using the Walnut Creek watershed in central Iowa. p. 283–296. *In* L.R. Ahuja and L. Ma (ed.) Methods of introducing system models into agricultural research. Advances in Agricultural Systems Modeling 2. ASA, CSSA, and SSSA, Madison, WI.

McCown, R.L., G.L. Hammer, J.N.G. Hargreaves, D.L. Holzworth, and D.M. Freebairn. 1996. APSIM: A novel software system for model development, model testing, and simulation in agricultural systems research. Agric. Syst. 50:255–271.

Nolan, B.T., R.W. Malone, L. Ma, C.T. Green, M.N. Fienen, and D.B. Jaynes. 2011 Inverse modeling with RZWQM2 to predict water quality. p. 327–364. *In* L.R. Ahuja and L. Ma (ed.) Methods of introducing system models into agricultural research. Advances in Agricultural Systems Modeling 2. ASA, CSSA, and SSSA, Madison, WI.

Parton, W.J., M. Hartman, D.S. Ojima, and D.S. Schimel. 1998. DAYCENT: Its land surface submodel: Description and testing. Global Planet. Change 19:35–48.

Rosenzweig, C., A. Iglesias, G. Fischer, Y. Liu, W. Baethgen, and J.W. Jones. 1999. Wheat yield functions for analysis of land-use change in China. Environ. Model. Assess. 4:115–132.

Saseendran, S.A., D.C. Nielsen, D. Lyon, L. Ma, D.G. Felter, D.D. Baltensperger, G. Hoogenboom, and L.R. Ahuja. 2009. Modeling responses of dryland spring triticale, proso millet, and foxtail millet to initial soil water in the High Plains. Field Crops Res. 113:48–63.

Smith, P., J.U. Smith, D.S. Powlson, W.B. McGill, J.R.M. Arah, O.G. Chertov, K. Coleman, U. Franko, S. Frolking, D.S. Jenkinson, L.S. Jensen, R.H. Kelly, H. Klein-Gunnewiek, A.S. Komarov, C. Li, J.A.E. Molina, T. Mueller, W.J. Parton, J.H.M. Thornley, and A.P. Whitmore. 1997. A comparison of the performance of nine soil organic matter models using datasets from seven long-term experiments. Geoderma 81:153–225.

Van Diepen, C.A., J. Wolf, H. Van Keulen, and C. Rappoldt. 1989. WOFOST: A simulation model of crop production. Soil Use Manage. 5:16–24.

Wang, X., A. Kemanian, and J. Williams. 2011. Special features of the EPIC and APEX modeling package and procedures for parameterization, calibration, validation, and applications. p. 177–208. *In* L. Ahuja and L.R. Ma (ed.) Methods of introducing system models into agricultural research. Advances in Agricultural Systems Modeling 2. ASA, CSSA, and SSSA, Madison, WI.

Wendroth, O., K.C. Kersebaum, G. Schwab, and L. Murdock. 2011. Spatial relationships of soil properties, crop indices, and nitrogen application pattern with wheat growth and yield in a field. p. 229–260. *In* L.R. Ahuja and L. Ma (ed.) Methods of introducing system models into agricultural research. Advances in Agricultural Systems Modeling 2. ASA, CSSA, and SSSA, Madison, WI.

Wilkerson, G.G., J.W. Jones, K.J. Boote, K.T. Ingram, and J.W. Mishoe. 1983. Modeling soybean growth for crop management. Trans. ASAE 26:63–73.

Williams, J.R., P.T. Dyke, and C.A. Jones. 1983. EPIC—A model for assessing the effects of erosion on soil productivity. p. 553–572. *In* Proceedings third international conference on state-of-the-art in ecological modeling. 24–28 May 1982. Colorado State Univ., Fort Collins.

Wolf, J., R. Hessel, H. Boogaard, A. de Wit, W. Akkermans, and K. van Diepen. 2011. Modeling winter wheat production across Europe with WOFOST—The effect of two new zonations and two newly calibrated model parameter sets. p. 297–326. *In* L.R. Ahuja and L. Ma (ed.) Methods of introducing system models into agricultural research. Advances in Agricultural Systems Modeling 2. ASA, CSSA, and SSSA, Madison, WI.

Wu, L., and A. Shepherd. 2011. Special features of the SPACSYS modeling package and procedures for parameterization and validation. p. 117–154. *In* L.R. Ahuja and L. Ma (ed.) Methods of introducing system models into agricultural research. Advances in Agricultural Systems Modeling 2. ASA, CSSA, and SSSA, Madison, WI.

Index

Advances in Agricultural Systems Modeling

A Series of Transdisciplinary Research, Synthesis, and Applications

Agriculture today is complicated by growing environmental concerns, increasingly limited water available for agriculture, market-based global competition that challenges traditional production systems, more frequent droughts and climate change, and the growing production of bio-energy crops. Whole system–based quantitative planning and decision tools are needed to guide the optimum management of resources while addressing these problems. Process-level models of agricultural systems, integrated with focused field research, are required to develop these tools and extend their application to different soils, climates, and situations.

Agricultural system modeling has made substantial progress, but there are still many critical gaps in our knowledge of various processes, and especially their interactions. Most of these gaps occur at the boundaries of disciplines, and further transdisciplinary research is needed. There is also the need for better synthesis and quantification of knowledge at the whole system level, both to improve system models and realize a more systematic and collaborative approach to building models of the future. At the same time, we need to facilitate application of these models to address and solve current real-world problems. For this purpose, we need to understand the different scales of application and the scale-dependence of parameters.

The American Society of Agronomy, Crop Science Society of America, and Soil Science Society of America are taking a leadership role in encouraging transdisciplinary and interdisciplinary research and its synthesis to solve practical problems. We believe that the future breakthroughs in science and technology lie, indeed, at the boundaries of disciplines. Therefore, the Societies have initiated the new series Advances in Agricultural Systems Modeling, A Series of Transdisciplinary Research, Synthesis, and Applications.

The purpose of the series is to:

- **Encourage and advance critical transdisciplinary research, and its synthesis and quantification, through publication of anonymously peer-reviewed papers by top researchers worldwide in a given knowledge gap area.** The writing of papers and publication will generally be preceded by a workshop of these researchers, where the papers will be presented and discussed. The papers will contain both new research and new or improved concepts for synthesis.
- **Encourage collaboration among the top researchers in new research, synthesis, and building and improvement of model components.** As often as possible and where appropriate, the authors will be asked to provide computer code of the components to be shared by researchers and modelers.

- **Encourage and advance the application of system models to solve practical problems through publishing case studies of such applications, along with illustrated instructions on how the models are used.** For example, an application may be the use of particular models to optimize the water and nutrients under limited water conditions and to evaluate the effects of climate change on agriculture.
- **Encourage better instruction in these models and their application.** The state-of-the-science syntheses given by each author will be highly useful for both undergraduate and graduate teaching. The new research in the papers and model component codes will be valuable for graduate level teaching, research, and training of student in the use of models.

L. R. Ahuja
Series Editor